MEDICAL INTELLIGENCE UNIT

Growing Bone
Second Edition

James F. Whitfield, Ph.D., FRSC
Institute for Biological Sciences
National Research Council of Canada
Ottawa, Ontario, Canada

CRC Press
Taylor & Francis Group
Boca Raton London New York

CRC Press is an imprint of the
Taylor & Francis Group, an **informa** business

GROWING BONE, SECOND EDITION

Medical Intelligence Unit

CRC Press
Taylor & Francis Group
6000 Broken Sound Parkway NW, Suite 300
Boca Raton, FL 33487-2742

First issued in hardback 2019

© 2007 by Taylor & Francis Group, LLC
CRC Press is an imprint of Taylor & Francis Group, an Informa business

No claim to original U.S. Government works

ISBN 13: 978-1-58706-156-1 (hbk)

Visit the Taylor & Francis Web site at
http://www.taylorandfrancis.com

and the CRC Press Web site at
http://www.crcpress.com

While the authors, editors and publisher believe that drug selection and dosage and the specifications and usage of equipment and devices, as set forth in this book, are in accord with current recommendations and practice at the time of publication, they make no warranty, expressed or implied, with respect to material described in this book. In view of the ongoing research, equipment development, changes in governmental regulations and the rapid accumulation of information relating to the biomedical sciences, the reader is urged to carefully review and evaluate the information provided herein.

Library of Congress Cataloging-in-Publication Data

Whitfield, James F.
   Growing bone / James F. Whitfield. -- 2nd ed.
       p. ; cm. -- (Medical intelligence unit)
   Includes bibliographical references and index.
   ISBN-13: 978-1-58706-156-1
   1. Osteoporosis--Prevention. 2. Osteoporosis--Hormone therapy. 3. Anabolic steroids--Therapeutic use. 4. Parathyroid hormone--Therapeutic use. I. Title. II. Series: Medical intelligence unit (Unnumbered : 2003)
   [DNLM: 1. Bone and Bones--metabolism. 2. Bone Regeneration--physiology. 3. Osteoporosis--prevention & control. 4. Peptides--therapeutic use. WE 200 W595g 2007]
   RC931.O73W547 2007
   616.7'16--dc22
                                        2007034205

CONTENTS

PREFACE

In the new millennium, humans will be rocketing from Earth into the radiation hazards of space and the dangers of alien worlds with antediluvian skeletons designed to respond to microgravity by self-destructing. Meanwhile on Earth growing numbers of men and many more women are suffering from crippling bone loss called osteoporosis. By 2050 50% of Americans over 50 will be at risk of, or actually have, osteoporosis (Surgeon General, 2004). During the first decade after menopause all women lose bone, which in some of them is great enough to result in the crushing of vertebrae and fracturing of various bones by ordinary body movements. This is osteoporosis, which all too often requires prolonged and expensive hospitalization treatment and causes sustained demoralization and often death. The bones' ominously accelerating microarchitectural deterioration and with it rising microfracturing by the pulling of muscles in postmenopausal women is caused by the loss of estrogen. The slower development of osteoporosis in aging men is also partly due to estrogen being made in ever smaller amounts in bone cells from the declining level of circulating testosterone which is needed for bone maintenance just as it is in women. The estrogen decline at some point sets off a vicious cycle of a runaway resorption which would have normally been balanced by the bones' microcrack-removing ⇒ refilling (or "remodeling") mechanism, which now instead of preventing microcracks from accumulating to structural failure and fracture, increases bone fragility and thus activity-driven microfracturing. This in turn, further stimulates the unbalanced repair mechanism and so on ultimately to large-scale fracturing of the increasingly stress-hypersensitive bones.

This vicious cycle is due to the increasing generation, longevity and activity of the bone-eating, microfracture-excavating osteoclasts in the growing numbers of unbalanced crack-"repairing" osteoclast-osteoblast teams in the increasingly microcracking-prone bones. The accelerating overproducion of hyperactive osteoclasts can be suppressed by estrogens and selective estrogen receptor modulators (SERMs) as well as by osteoclast-disabling and/or killing anti-catabolic bisphosphonates and calcitonin. These agents harden bone by prolonging mineralization by bone-building osteoblasts and they indirectly cause a small amount of bone growth because murdering osteoclasts enables the unaffected osteoblasts to overfill the existing holes with mineralized bone without having to compete with oversized teams of overzealous osteoclasts rapidly digging more holes. Unfortunately these many agents do not directly stimulate osteoblasts to make bone. But the rapidly swelling crowd of seniors with fragile, fracturing bones can only be met effectively with a new "anabolic" (bone-growing) drug that can actually improve their deteriorated microstructure, replace lost bone, accelerate fracture healing, and accelerate and strengthen the initial anchorage of artificial knees and hips to bone and prevent their subsequent loosening.

Here the reader will meet the newest real and possible bone builders and learn how they might work. These include novel steroids, an osteogenic growth peptide (OGP), leptin from both fat cells and osteoblasts and the many kinds of statin that are widely used to reduce blood cholesterol and seem to prevent Alzheimer's disease. But the spotlight must be directed onto the currently most promising bone growers, the 84-amino acid parathyroid hormone (PTH) and three of its 31- and 34-amino-acid fragments one of which, Lilly's Forteo™, (recombinant hPTH-(1-34)), is already being used by osteoporotics.

James F. Whitfield, Ph.D., FRSC

INTRODUCTION

A small band of astronauts may soon be touring Mars and perhaps someday more distant planets and moons (Cohen and Stewart, 2003). However, they will be doing so with skeletons that were designed millions of years ago to bear terrestrial loads but not prolonged exposure to microgravity for many Earth-years in a space ship or to the much lower gravity on smaller planets or moons. The bones' strain-sensing mechanisms will respond to the lack of load by mobilizing hordes of osteoclasts to get rid of the unloaded, and therefore seemingly unneeded, bone (Marie et al., 2000; Vico et al., 2001). Without using in-flight drugs to kill osteoclasts and protect load-bearing bones from their destructive action, this could result in breaking bones upon return to Earth, or even worse, stepping out into substantial gravity on an alien planet such as Mars with now fragile bones and limited or no facilities to fix the almost inevitable fractures. What is needed are long-term, oral in-flight drugs to prevent the ravages of osteoclasts and short term oral or injectable bone-growth-stimulators to accelerate accidental fracture healing while on a mission and rapidly rebuild untreated bones upon return to Earth.

But the more immediate and very serious problem is that back on Earth, a swelling mob of seniors is banging on the doors of the medical facilities of the so-called 'First World'. Coming with them are failing body parts that are difficult and very expensive to replace or repair. A massive and growing problem for aging men and women, but particularly for women during their first postmenopausal decade, is the severe loss of bone, especially cancellous (trabecular) bone, and the 'spontaneous' (i.e., muscle-inflicted non-traumatic) or low-trauma or impact fracturing of the remaining bone known as osteoporosis that can lead to extended hospitalization, crippling and all too often despair and death particularly because of broken hips (Baylink et al., 1999; Reginster and Burlet, 2006; Stevenson and Lindsay, 1999; Surgeon General, 2004). Indeed, the risk of mortality among hip fracture victims increases 2.8 to 4-fold during the first 3 months after the fracture (Surgeon General, 2004). As measures of this growing menace, about 30% of the current populations of postmenopausal women in the USA and the European Union have osteoporosis, more than 40% of them will have fractures, and by 2001 the costs of the hospitalization and nursing home resulting from these fractures had already reached $17 billion in the USA and 32 billion euros in the European Union (Reginster and Burlet, 2006).

To make effective, safe bone growers for rebuilding deteriorated bones in aging persons we must first find out what controls bone growth and resorption and then how to make microstructurally strong bone when needed. But this is not easy. A bone is a veritable cellular Tower of Babel, a fiendishly complex, polyglot community of intricately networking cells with dozens of factors and unbelievably talking to each other like neurons with the languages of neurotransmitters and glutamate receptors that were until very recently believed to be found only in brains. The result is a seemingly endlessly thickening Devil's brew of variously interacting growth and other factors that cannot be imitated with short-lived bone cultures or dysregulated lines of semi-neoplastic, self-driving cyto-psychotic bone cells. Indeed such crazy cells are as valid models of normal bone cell interactions and responses as convicts in solitary confinement are valid models for studying normal human social relations. Moreover, bones, like trees, do nothing in a hurry. They do things with a glacier-like speed that require patience, longevity and teams of biomechanical engineers, biomaterials scientists, cell biologists, molecular biologists physiologists and clinicians to understand

The daunting complexity of the Real Bone World has driven researchers to fish cells and factors out of the seething stew and with them spin tales of what makes bone cells proliferate, make bone and die. Indeed one is reminded of the tale of the blind wise men examining an elephant and each picturing the beast from the viewpoint of the part of it he is feeling. However, their job is far too easy—we blind osteoseers have been given the job of trying to picture a baroque boney Millipede rather than just a simple four-footed elephant!!

The neuroscientist Rudolfo Llinás has said in another context that anyone attempting a synthesis risks failure, but "…..without which there are only fields of dismembered parts" (Llinás, 2001). In what follows I will try to build a picture of bones and the things that control them with what we currently seem to know about postmenopausal osteoporosis simply because it is the most important kind of bone loss for us the aging and from which most of our knowledge about bone building and demolition is currently flowing. But I won't neglect fractures, bone implants, rheumatoid arthritis, ossified arteries and even Alzheimer's disease. However, my main mission is to tell about the newest of the bone-growing drugs such the PTHs which we call "anabolics", the trade names of some of which will soon be scribbled on physicians' prescription pads and may sooner or later arrive on drugstore shelves alongside PTHs, one of which osteoporotic patients are now injecting into themselves.

But before starting I must thank my former colleagues Cynthia Allen, Jean-René Barbier, Tony Candeliere, Hervé Jouishomme, Rob Langille, Sue MacLean, Godfrey Marchand, Paul Morley, Lillith Ohanessian-Barry, Ray Rixon, Virginia Ross, Gord Willickæand my wife Barbara and our friend Vanessa for helping me to try to understand how bones grow and to have the temerity to write about it.

James F. Whitfield, Ph.D., FRSC

CHAPTER 1

What Is Osteoporosis?

Osteoporosis is *"...a disease characterized by low bone bone mass, microarchitectural deterioration of bone tissue, and a consequent increase in fracture risk"*—Copenhagen Consensus Conference (1990).

Osteoporosis is *"...a bone density 2.5 standard deviations below the mean for young white adult women at lumbar spine, femoral neck or forearm"*—The World Health Organization (1994).

Osteoporosis is *"...a systemic skeletal disease characterized by low bone mass and microarchitectural deterioriation of bone tissue with a consequent increase in bone fragility and susceptibility of fractures"*—The Hong Kong and Amsterdam Consensus Conferences (1996).

Osteoporosis is *"...characterized by a failure to maintain bone architecture sufficiently robust to withstand the loading of everyday life without substantial risk of fracture"*—Ehrlich and Lanyon (2002).

Osteoporosis is *"...a skeletal disorder characterized by compromised bone strength predisposing to an increased risk of fracture"*—National Institutes of Health (cited by Pearson et al., 2002).

Osteoporosis is *"...a disease of decreased bone strength rather than a disease of decreased bone mass"*—Riggs and Parfitt (2005).

Osteoporotic postmenopausal women don't need to fall or hit something to break their fragile bones. Their, hips, ribs, wrists and especially vertebrae are apt to be broken or crushed by bending spines, muscle pullings and the low-impact bumps of ordinary daily activities. In fact the greatest bone breakers are the person's muscles because as R.B. Martin et al. (1998) pointed out *"....the muscle forces acting on the skeleton are quite large relative to the forces exerted on it by the outside world"*. Those women who were lucky enough to have put enough into their bones when young to keep them above the "spontaneous" fracture threshold during the postmenopausal years of accelerated bone loss may only become "osteopenic" or pseudo-osteoporotic. Their osteopenic bones are more likely to be broken by falls and other bumps and blows due to poor eyesight and balance (Boxsein, 1999). But these bones are still strong enough to resist being broken by the weakening muscles of older people during their declining activities. However, weakening muscles are not unmixed blessings for bones because there has to be enough muscle loading and strain to placate the bones' thrift-minded mechanostatic mechanism which is programmed to destroy understrained bone rather than waste hard-earned resources to maintain it (Noble and Reeve, 2000; Skerry, 1999).

Osteoporosis in women is the most devastating of the many consequences of the estrogen drop at menopause. Perhaps surprisingly the osteoporosis that develops more slowly in men (in Europe the fracture incidence in men is about half that in women between the ages of 60 and 79) is also at least partly due to an estrogen decline, in this case to the the declining amounts of estrogen being made from the gradually dropping testosterone by the bone cells' enzyme called "aromatase" (Bilezekian, 2002; Khosla, 2002; Klein, 1999; Troen, 2003). Of course, the estrogen made by the fat cells' aromatase might also moderate the debilitating effects of the lack of ovarian estrogen on the bones of those postmenopausal women whose adrenal glands still make a significant amount of androgens.

The message is loud and clear—aging populations urgently need something that can significantly improve deteriorated bone microstructure and to restore lost bone mass and strength. PTH, the parathyroid hormone, and certain of its fragments, are currently by far the most promising answers to this need. We will also meet OGP (the osteogenic peptide) and learn that there is some evidence that the cholesterol-lowering, heart attack-preventing, statins might join the exclusive "Bone Builders Club" and maybe Alzheimer's disease. Also, calcium's big brother strontium might join the "Club". But we will also learn in great detail that none of these have the now massively studied potent anabolic punch of the PTHs.

There is another kind of bone fragility or osteoporosis that will not greatly affect humankind in the near future, but could prevent a few of us from exploring our neighboring planets. But it would be hugely important in the unlikely event that *Homo "sapiens"* survives its suicidal inter-tribal wars, Snowball Earths (Walker, 2003) and other climatic and plate tectonic challenges such as volcanic eruptions and tsunamis. Then *H."sapiens"* would eventually have to flee the incinerating embrace of a bloated, dying Sun. Unfortunately for us our load-bearing bones are designed to use gravitation-imposed strain to maintain them and at the same time avoid wasting expensive metabolic resources by hanging onto unused bone. The design of the strain-sensing device is elegantly simple—cells called osteocytes, equipped with tiny bendable flow-metering antennae and moored to the walls of the bones of our lower limbs with sensitive tuggable integrin signal "wires" (i.e., matricrine signaling), need the constant ebb and flow of extracellular fluid caused by the rib-squeezing strains of breathing and the limb long-bone-sqeezing heel strikes to get the oxygen and nutrients cells need to function and to carry away their waste. If this gravitation-driven pumping and extracelluar fluid sloshing should fall below a critical level, the now starving and suffocating osteocytes drowning in their own sewage self-destruct and in their death throes emit signalers that summon osteoclasts to get rid of their unneeded bone. So unless we develop effective chemical gravity, astronaut/cosmonaut explorers/emigrants will be stepping out onto another planet after many months or years in the microgravity of space with fracture-prone bones but without the facilities to fix them.

To try to understand how these new bone-making tools might work in the complex world of bones, we must start by learning how bones are made and kept from failing from constant wear and tear.

CHAPTER 2

BMUS—The Microcrack Fixers

"...if we increase the resolution of our investigation and focus on scales of microns to millimeters, we see that bone is a highly dynamic tissue, continually adjusting to its physiologic and mechanical environment by changes in its composition and microscopic architecture. An important principle of skeletal physiology is that bones are able to sense the mechanical loads which they bear and modify their structures to suit changes in these loads."—R.B. Martin et al. (1998; p.31).

"In the adult, bone homeostasis is maintained by a tight balance between the supply, activity, and life span of osteoclasts and osteoblasts within the basic multicellular unit (BMU), a temporary anatomic structure within which bone remodeling occurs."—van Bezooijen et al. (2005b; p.33)

"In the face of inevitable wear and tear, no designed thing persists for long without renewal and replication."—D.C. Dennett (2006; p.156).

i. Strains, Microcracks, Macrocracks and Molecular Screams

At first sight bones are inert, rock-like things that store 99% of the body's calcium and consist of a hard shell, the cortex, that encloses a deceptively delicate lattice of struts and plates (Jee, 2001) (Fig. 1).The cortex is made of hard, 5%-10% porous, calcified armor plate with embedded cells called osteocytes that are fed via blood vessels and and informed by nerves running through a system of conduits. The cortex provides the attachment sites for tendons and muscles. The internal cancellous bone is 75%-95% porous with cavities filled with red (hematopoietic or blood-forming) marrow and/or yellow fatty marrow and appropriately porous blood vessels through which the newly minted mature blood cell products can get into the circulation. The trabecular lattice has a huge surface for interactions of the bone cells with various hormones and cytokines and for releasing Ca^{2+} into the blood when needed. In fact the trabeculae serve as a calcium store that can be mobilized to feed the voracious calcium needs of a developing fetus. Moreover, the struts and plates of the trabecular lattice are aligned with their bones' stress forces and thus provide maximum strain reduction with minimum mass. And, as we shall learn further on, the multipurpose trabeculae also have the niches (or nests) carpeted with bone-lining cells that provide the anchorages and restraints for the primitive hematopoietic stem cells.

But wait! Look more closely and you may be surprised, perhaps even shocked,to see that, like the more media-attractive, charismatic and mercurial networking neurons in the brain, bone cells live in networks and can talk to each other, share and store memories of past strains and modify their responsiveness to strain by recording their past experiences in their bones' structures (Robling et al., 2006; Turner et al., 2002). As I will show below they are equipped like neurons to secrete, reaccumulate and respond to neurotransmitters such as CGRP (calcitonin gene-related peptide), GABA (γ-amino butyric acid), glutamate and serotonin (Mason, 2004; Spencer et al., 2004; A.F.Taylor, 2002). Indeed bone cells respond to strain by activating ionotropic glutamate NMDA (N-methyl-D-aspartic acid) and AMPA (DL-α-amino-3-hydroxy-5-methylisoxasole-4-propionic acid) receptors in direct proportion to intermittent strain frequency with the same CaMKII (calmodulin-dependent protein ki-

---

Growing Bone, Second Edition, by James F. Whitfield. ©2007 Landes Bioscience.

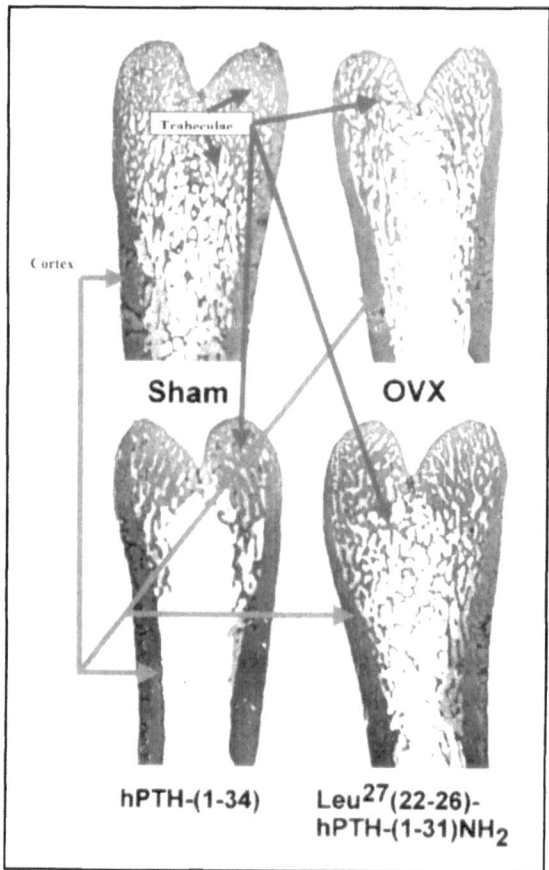

Figure 1. Typical specimens of demineralized (i.e., the calcium apatite was acid-extracted) distal femurs of OVXed (ovariectomized) rats showing that injecting 6.0 nmoles of PTH-(1-34)/1 kg of body weight once a day from the end of the 2nd week to the end of the 8th week after the operation did not stop the destruction and loss of trabeculae while injecting the same molar dose of [Leu^{27}]cyclo[Glu^{22}-Lys^{26}]hPTH-(1-31)NH$_2$ (Ostabolin C™) reduced the loss of trabeculae. However, by 6 weeks of injections both fragments had nearly doubled (1.8-2.0-fold) the thickness of the remaining trabeculae compared to the the mean thickness of the femurs in the trabeculae in the femurs of vehicle-injected OVXed control rats. The specimens were prepared at the end of the 6th week of the series of injections (i.e., 8 weeks after OVX). The lines—red for cortical bone and blue for trabecular or cancellous bone—on these photographs are meant to acquaint the reader with the universal basic anatomy of a long bone such as the femur. The reader should also consult Kerr (1999) and Netter (1997) to learn the gross and microscopic anatomies of the various parts of the human skeleton.

nase II)-driven long-term potentiation of bone formation as neurons respond to high-frequency stimulation with long-term potentiation of synaptic transmission (Bowe et al., 2004; Mason, 2004; Skerry and Suva. 2003; Spencer and Genever, 2003; Spencer et al., 2004; Szczesniak et al., 2005). In other words bone cells carve their loading experience in bone through a mechanism just like the one used by neurons in the brain to drive the long-term synaptic changes that underly memory storage (Mason, 2004; Spencer et al., 2004). Bone cells have to be smart enough to know where they are and adjust their responsiveness to strain accordingly. If not they would label lightly strained skull bones or ear bones such as the stapes as unused and destroy them while piling too much bone onto ribs and lower leg bones that are continuously squeezed by breathing and heel, hoof or paw strikes. The incredible sophistication of the strain memory is illustrated by Turner et al. (2002) who found that the size of the osteogenic response of the cells of rat ulna to mechanical loading depended on the straining history of the part of the bone where they live. The cells in the habitually high-strain distal region were less responsive to loading than cells in the proximal region, which is habitually exposed to much less strain. In other words, the bone cells are not as stupid as they seem to be. *Incredibly they actually know where they are*!! They know how much strain to expect in order to not overload load-bearing bones or destroy other, seemingly under-loaded, but in fact normally underloaded and essential, bone. Thus it seems that bones, exactly like brains, record their experience, i.e., their loading history, in their structure, for example in the density of bone signaling connections and networks.

Bone seemingly masochistically harbors cells whose job description is brief and destructive—destroy bone! And they are very good at it. Obviously, it must have a good reason for harboring such nasty characters. And of course it does! A load-bearing bone is like a busy interstate highway, which cracks with the constant pounding by traffic and will eventually crumble unless the cracks are detected, dug out and the holes refilled in a timely fashion by road maintenance crews. Bones too have maintenance crews—the bone-remodeling *BMUs* (*Basic Multicellular Units*). A BMU crew is activated every 10 seconds in an adult human bone, and at any time about 35,000,000 of them are at work removing cracks and digging out and replacing about 500 mg of calcium from the skeleton each day. However, the remodeling rate varies widely throughout the adult skeleton according to the level of microfracturing (Robling et al., 2006; R.B. Martin et al., 1998; S. Mori and Burr, 1993) and whether the bone such as the endocortical surface of the ilium or femur is bathed in red hemopoietic (blood cell-making) marrow or fatty yellow marrow (Parfitt, 2002). For example, the remodeling rate in adult human cortical bone can be as low as 2% per year in the distal radius and as high as 50% per year in ilial trabecular bone (Noble and Reeve, 2000). This difference is due to the trabecular bone being weaker (i.e., it has a lower elastic modulus-it takes less force to deform) than cortical bone (R.B. Martin et al., 1998). Because of this it microcracks more easily and ultimately fails and must therefore be remodeled/repaired more often than cortical bone which results in its mineralization being less mature on the average and therefore less hardened than that of cortical bone (R.B. Martin et al., 1998). But as always with bone, things are not so simple (Parfitt, 2002). In cortical bone, remodeling can start anywhere along the so-called Havsersian (osteonal) tunnels running through the bone carrying the dense bones' plumbing (blood vessels) and electrical wiring (nerves), but trabecular remodeling starts only on the surface, therefore the likelihood of remodeling a trabecula drops off the farther it is from the surface. However, trabeculae thicker than 200 m do have a central osteonal canal. This means that the usual claim of a very high trabecular remodeling level (i.e., bone formation rate [BFR]/volume) is an overestimate because it relies only on the surface remodeling level (surface BFR/V) without taking account of the very much lower intra-trabecular remodeling level (Parfitt, 2002).

i a. Why Remodeling?

It appears that this continuous remodeling is characteristic of the big densely "Haversian" bones of long-lived humans, horses, and elephants and was also characteristic of "Haversian" bones of the extinct ornithischian, saurischian and certain therapsid dinosaurs (Bakker, 1986; Chinsamy, 2005; Chinsamy et al., 1998; Enlow and Brown, 1957, 1958; Erickson et al., 2004; Rensberger and Watanabe, 2000; Schweitzer et al., 2005; Stokstad, 2004) but not the tiny non-"Haversian" bones of short-lived mice and rats. The reason for this is simple. Big bones have more flaws than small bones and will therefore have shorter fatigue lives than smaller bones that are equally loaded/unit volume (R.B. Martin, 2003; D. Taylor et al., 1999). R.B. Martin (2003) has given a dramatic example of this size effect. He has estimated that the volumes of the femurs of a shrew, a human and an elephant are in in the relative proportions of 1:900:30,000. If these bone couldn't remodel, the fatigue life of the elephant's femur would be extremely short compared to the shrew's femur when repeatedly subjected to the same physiological strain. In the same paper R.B. Martin gave another example of the relation between the probabilities of failure of equally loaded, equally thick iron wires 1, 2 and 3 units in length. The failure probabilities were 0.63, 0.87, and 0.95 respectively Therefore, the progressively larger and longer bones of the vertebrates that evolved into the big dinosaurs and mammals were intrinsically more prone to continuous fatigue damage. On the other hand despite increasingly heavy musculature the skeletons of these larger animals had to be as light as possible to give their owners the maximum mobility for their size and thus the maximum possible survivability (R.B. Martin, 2003). But there was an obvious and dangerous downside to this failure to increase bone size in step with muscle mass—increasing bone microdamage during physiological activity. The solution to this problem was the increasing use of the remodeling mechanism during normal activity. Thus they had to routinely mobilize BMUs to repair the microcracks continuously appearing at high strain sites in the ribs, and the big leg bones caused by their owners' breathing and heel, hoof and paw strikes (Burr, 1993; Burr et al., 1985; Frost, 1960, 1985; R.B. Martin, 2000a, 2000b, 2003; Mori and Burr, 1993; Tami et al., 2002; Vashishth, 2005; Verborgt et al., 2000; Whitfield et al., 2002b). Actually the continual pulling on the ribs by muscles just during breathing causes them to have more microcracks, and thus the greatest remodeling rate and production of Haversian canals, of all the cortical bone in the body (Frost, 1960)! This of course would explain Enlow and Brown's comment (Enlow and Brown (1958) that "*...if a form does possess Haversian tissue in any part of its skeleton the rib is, with very few exceptions Haversian*". However, the BMU remodeling mechanism was probably available for exploitation at least in our small mammalian ancestors because rat bones which have no osteons and normally no microdamage or remodeling do resort to Haversian remodeling when needed. Thus, Bentolila et al. (1998) found that when the ulnas of 16 rats were cyclically super-fatigue-loaded, 14 of the rats had microcracked ulnas in which resorption cavities appeared by 10 days. The linkage between microcracking and the mobilization of resorption cavity-producing BMUs was clinched by the failure of resorption cavities to appear in 2 of the rats whose ulnas did not microcrack.

But as with all physiological things, we shall see that remodeling can be dangerous when it exceeds a critical level as it does in postmenopausal women because, for example, the remodeling-driven loading of cortical bone with new BMU-produced Haversian canals increases bone porosity and hence fragility (R.B. Martin et al., 1998).

i b. The Road into the Boney Interior—Sloshing Fluids, Cilial Flowmeters, Talking Integrin Moorings, Osteointernet Blogging

i b 1. A Tour of the Boney Interior
The "Haversian" bones are so named because their cortical walls are made up of bundles of individually branching osteons aligned generally along the main stress lines that look like tubular onions each consisting of 4 to 20 layers (or lamellae) of bone wrapped around a central, Haversian canal (named after Clopton Havers (1691). Each tubular onion is separated from surrounding interstitial bone by cement (lower calcium and phosphate) lines that form the boundaries marking the end of bone resorption and new bone formation—the cementing of a new osteon into the old group. The central canal is a conduit for nerves and blood vessels (full-scale arteries and veins in the largest canals but just tubes of flattend endothelial cells in the smallest) from the marrow that convey various signalers to osteoblasts from as far away as the hypothalamus, carry food supplies to, and waste from, the network of osteocytes locked in the lamellae (Cowin, 2000; Kerr, 2000; Laroche, 2002; R.B. Martin et al., 1998; S. Takeda et al., 2002) (Fig. 2). Blood flow in a bone such as the femur is centrifugal (Brookes, 1986). Blood is pumped into a bone such as a femur first through the superior and inferior epiphyseal and

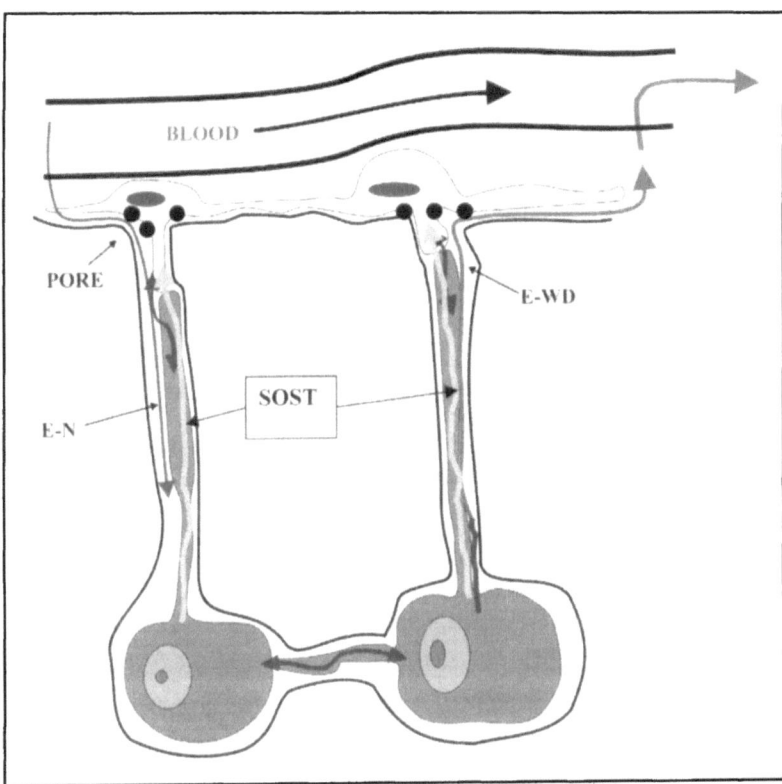

Figure 2. The quiet before the storm. The big bones of large animals such as humans and elephants, unlike the tiny bones of mice and rats which keep growing throughout their short lives, are prone to microcracking at muscle-loading sites and routinely use a mechanism to repair the cracks as soon as possible to prevent the microcracks accumulating and growing up to macrocracks and eventually mechanical failure. (Rats can switch on the same device if the damage is severe enough.) In a normal intact human bone, there is a vast cellular signaling syncytium, the *osteointernet*, which consists of osteocytes in their lacunocanalicular network that opens out onto the surface of a trabecula or the wall of a Haversian or Volkmann canal where it contacts retired osteoblasts known as bone-lining cells—a kind of cellular wallpaper armed with neuron-like glutamate receptors for responding to glutamate (•) release by strained osteocytes. The cloistered osteocytes also produce a protein, SOST (sclerostin; the yellow stream), which prevents the the lining cells from inappropriately starting to make bone. When the network is intact the blood flowing through the canal's blood vessels or sinusoids near the trabeculae has osteoclast precursors sailing peacefully by among the various leucocytes and carries nutrients and oxygen for the network's osteocytes. The nutrients and dissolved oxygen cross over the lining and are sucked through the tiny openings and travel along the "external pathway" of the canaliculi (the canalicular pores) (E-N (nutrient) left side blue line) toward the osteocyte when the network expands. Wastes are pushed along the "external pathway" away from the osteocyte and expelled through the ports and into the blood (E-WD (waste disposal) right side brown line) when the network is squeezed by the alternating expansion and compression of the bone in an activity cycle. The canaliculi are busy places—it has been said that canaliculi can be cleared as many as 2100 times in 24 hours (reviewed by Knothe Tate, 2000). Along with the traffic moving along the external pathway (E-N lines), the osteocyte pushes components along its processes and through gap junctions with other osteocytes and lining cells with waves of contraction of actin microfilaments (nanomuscles) (cell-cell connector lines). Aside from this very busy supply and waste disposal traffic, the lining cells are tranquillized by a stream of signalers (red double-headed arrows) being sent along the internal and external pathways from the normally strained osteocytes. Of course the osteocytes are also busily talking to each other (red double-deaded arrows). A color version of this figure is available online at www.eurekah.com.

metaphyseal arteries, the diaphyseal feeding artery and only then into the endosteal (i.e., internal) surface and then through the Haversian canals (reviewed by Brookes, 1986; Knothe Tate 2000; Laroche 2002). The blood ultimately pulses into the central venous sinus and out of the bone through the superior and inferior epiphyseal and metaphyseal veins and the perforating and nutritive veins (Bullogh, 2001; Glimcher and Kenzora, 1979; Laroche, 2002). The blood is also pumped through the cortex and into the periosteal muscle blood vessels and ultimately to the heart by the pulsing marrow vessels and by the contraction and relaxation of the periosteal muscles (Brookes, 1986). Failure of the periosteal muscle pumps because of nerve damage, limb immobilization, fracture repair, bed rest or old age would slow blood flow through the Haversian system which in turn would stress osteocytes and cause bone loss.

A bone, like the skull, is a rigid box within which the pressure must be kept within certain limits to keep the blood vessels open for the delivery of oxygen and various nutrients to the bone and marrow cells and the carriage of wastes and marrow-hatched blood cells out of the bone (Fig. 2). When the pressure rises above this ceiling level, it causes a painful "bone-migraine" just as increased intracranial pressure causes a migraine "headache". If this happens in the proximal femur because, for example, of the swelling of fat cells responding to alchohol abuse or short-term treatment with high-doses of glucocorticoids, venous channels are squeezed shut, blood stops flowing in the femoral head which becomes necrotic, i.e., dies (Laroche, 2002).

Since the marrow vascular and extravascular compartments equilibrate rapidly, and since the trabeculae and endosteum are very close to the marrow vessels, their osteocytes and lining cells receive any blood-borne osteoactive agent such as PTH (parathyroid hormone to be discussed at length further on) first, fully and faster than osteocytes in the cortical bone (Knothe Tate, 2001). A cortical osteocyte in a long bone receives the agent only after the blood vessel carrying it enters the cortex and then branches into the local Haversian canal and then travels through the lacunocanalicular network to the osteocyte. So the agent can be substantially depleted before the cortical osteocytes, especially those farthest from their local Haversian capillary, can get their chance to be affected by it. In other words the trabecular bone is a pre-filter or affinity column.

The osteocytes are the most numerous cells in bones. Human cortical bone contains 13,900-19,400 osteocytes per mm^3 (Hazenberg et al. 2006). They are the bones' professional strain monitors that are programmed to respond when the strain is above or below a particular bone's expected range (Burger, 2001; Klein-Nulent et al. 2003). Osteocytes are survivors. They are the fortunate few—the 4 % of the osteoblast team members that built the bone but did not have to commit apoptotic suicide (Potten and Wilson, 2004) after a cortical tunnel or trabecular trench was filled and the big osteoblast crew had to be downsized. They were spared as they were being embedded in new matrix when the matrix metalloproteinase (probably MT1-MMP) they were making converted latent TGF-β into active TGF-β which in turn generated a signal that prevented them from commiting apoptotic suicide (Karsdal et al., 2002). But in payment for surviving they had to put away their bone-making tools and shrink down from plump osteoblasts into lacuna-locked cells radiating chemical and strain-sensing dendrites. As a part of the retooling process they began "dendritizing" by activating the E11/Gp38 dendrite driver to send out 40 to 60 processes which were at first oriented toward the mineralization front, but some of which then oriented toward their provisioning and waste-disposing blood vessels when they were finally surrounded by mineralized matrix (Bonewald, 2005). Each of the processes (dendrites) also used MT1-MMP to tunnel through the matrix and plug its tip into a gap junctional link with another osteocyte or lining lining cell in the bone's "wiring transmission system" or, as I prefer to call it, the "osteointernet" (Bonewald, 2005; Hazenberg, et al., 2006; Marotti , 2000).The immensity of the osteointernet is indicated by an awesome 500,000 to1.2 million osteocyte processes coursing through a single cubic mm of human cortical bone—that means a lot of neuron-like blogging for such deceptively inert rocklike structures. Wonderful methacrylate-divinlybenzene replicas of the awesomely interlacing lacunocanalicular cable network of a young man's mandibular bone's osteointernet can be seen in Figures 6a and 6b of Atkinson and Hallsworth (1983).

One of their important new jobs when osteoblasts become osteocytes is to make and secrete a 64-kDa osteoblast-inhibiting protein called sclerostin (SOST) (Fig. 2). But why would preosteocytes do this (see Figs. 8, 23, 27)? The answer is that this is part of an important homeostatic mechanism that prevents excess bone growth by making adjacent osteoblasts stop working and then prevents the lining cells to which they are linked in mature bone from inappropriately starting to make bone—in other words to put the brakes on osteogenesis when enough is enough. The importance of this braking mechanism is indicated by the bone overgrowth when it fails in people with disabled SOSTgenes (Ott, 2005; van Bezooijen et al., 2005b). It is also indicated by the fact that the level of remodeling is correlated to the number of viable, hence SOST-producing, osteocytes—the fewer the osteocytes the higher the remodeling rate (Metz et al., 2003). In bone biopsies from such functionally SOST-less people there is an excessive number of active osteoblasts, increased bone formation as indicated by wider spacing of double tetracycline labeling fronts. But this otherwise excess osteoid is normally mineralized. In other words, normally working osteoblasts are not being turned off when they should be by SOST from the underlying new bone.

The mature osteocytes are directly coupled to other osteocytes and and the post-osteoblast surface bone-lining cells as well as less directly to the blood vessels and nerves running through the Haversian and Volkmann's canals (Burger, 2001; Marotti, 2000; R.B. Martin et al., 1998; Moss and Cowin, 1997; L. Wang et al., 2000) (Fig. 2). They live, monk-like, in tiny cells called lacunae and extend their 40 to 60 processes through narrow canals known as canaliculi to plug directly into the other members of the internet with so-called gap junctions which by directly joining the cells'cytoplasms form the "internal pathway" (Fig. 2). The canaliculi end in tiny openings or ports in the lining cell-covered osteonal canal wall. The processes may also set up gap junctions with the lining cells. High molecular weight nutrients from the osteonal canals' blood vessels are pushed through the ports and along the extracellular channel, the "external pathway", in the canaliculi and wastes are sucked out of them into the osteonal lumen and ultimately the blood by the cycles of compression and expansion of the network during various body movements (Knothe-Tate, 2001, 2003; L. Wang et al., 2000) (Fig. 2). In trabecular bone, the canalicular ports also open onto lining cell layers, but these lie directly on the vascular elements (sinusoids) of the bone marrow which are the equivalents of the blood vessels of the cortical bone's osteonal plumbing system except that any injected agent such as a growth factor or hormone arrives there before entering the cortical osteonal plumbing system (e.g., Knothe Tate, 2001, 2003; L. Wang et al., 2000).

The mature osteocytes' processes tend to be directed perpendicularly to the bone surface, i.e., the vascular surface, which means that their prime communication is with lining cells and moreover some processes can extend beyond the bone surface into the adjacent blood vessels and in the case of adjacent marrow to send information about loading directly to nests of early osteoblast and osteoclast progenitors (Kamioka et al., 2001; Kelin-Nulent et al., 2003; Palumbo et al., 1990). In other words they are actually parts of a giant cellular syncytium gated by large complete gap junctional pores made of connexin 43. But the processes running along the canaliculi are also studded with functional hemi- or half-gap junctions through which the cell can release signaling components such as ATP and PGE_2 in response to shear stress from sloshing canalicular fluid (Bonewald, 2005). Molecules that are too big to pass through the full or hemi-gap junctions can diffuse along or, in an actively bending bone, be swept along in the sloshing extracellular fluid, via the external pathway at a rate depending on canalicular diameter and pushing and sucking intensities (Knothe-Tate, 2001; Sorki et al., 2004; Tami et al., 2002; L. Wang et al., 2000). But the efficiency of the nutrient and factor delivery and waste removal falls off with the distance of the osteocytes from the osteonal or marrow canalicular ports. Therefore these distant cells are more likely to be vulnerable to stress and less responsive to any injected or swallowed factor (L. Wang et al., 2000). However, it must be noted that osteocytes do try to compensate for this long distance (or "remoteness) factor by having more processes and canaliculi than those nearest the Haversian canal much as trees increase their root systems to get enough water and nutrients in dry regions (Knothe Tate, 2001).

Let's now follow molecules such as one of the potent bone-building (anabolic) PTHs (that will figure so prominently further on) as they are pushed from a syringe in the skin and arrive at the nutrient artery of a femur using Knothe Tate's (Knothe Tate, 2003) road map. The molecules in the marrow blood vessels first see the cells lining the trabeculae where some will stick and stimulate while others pass through into the trabeculae. Others are carried by the capillaries into the Haversian channels of the cortical bone. The PTHs are pushed through the vascular walls under a pulsing head of pressure as cyclic bone loading squeezes the bone marrow and its blood vessels. The peptides pass through the layer of bone-lining cells that serve as endosteal and osteonal wall "tiles", regulate the movements of Ca^{2+} and K^+ into and out of the canalicular fluid, and can be recruited to provide osteoblasts for crack-repairing BMUs. The fluid carrying the PTHs then passes through the 500-600-nm canalicular ports. The interlamellar matrix is also a molecular sieve, a slow pathway, through which, for example, the 3-10 kDa PTH fragments (but not molecules larger than 40 to 70 kDa) can diffuse. The fluid then flows along the space between the osteocyte surface and the canaliculus wall. The space is not empty—it contains collagens and proteoglycans that impede the PTH progression. However, in the case of the PTHs, they can help their progression along the canaliculus by increasing the porosity (i.e., penetrability) of the collagen-proteoglycans by inducing osteocytes to line the lacunocanalicular walls with hyaluronan, a macroporous (big-pored) molecule (Midura et al., 2003). The osteocye doesn't just lie there. It also pushes the fluid along its canaliculi with its pulsing micovilli (Knothe Tate, 2003). When the fluid and its peptide passengers approach the lacuna containing the osteocyte's cell body armed with its sensory cilium that we will talk about a little later, they hit a kind of gate or valve formed by the cell processes which favors outflow rather than flow into the lacuna (Knothe Tate, 2003).

While the cells and their fluid-filled lacunocanalicular network make up only about 1% of the bone volume, the total surface area of these tiny pipelines is immense! In an adult male skeleton it is 1200 m^2 while the surface areas of the Haversian and Volkmann's canals and trabeculae are only 3.2 and 9 m^2 respectively (Johnson, 1966; R.B. Martin et al., 1999). Thus, bone is really a stiff, dense, fluid-filled, calciferous sponge. And the rate of flow of fluids and their cargoes into, through, and out of the sponge is determined by how often and how hard the sponge is squeezed by breathing in the case of the ribs and by walking, running and other activities (Burger et al., 2003). This immense throbbing osteocyte internet is the network of cables and modems with which they tell each other and lining cells about strains and microcracks and organize the appropriate responses.

Of course, digging and over-patching of *undamaged* bone is prevented by signals from the cyclic squeezing of the sponges by the owner's breathing and various other movements. The resulting inhibitory signals, especially SOST (to be discussed in much more detail in Chapter 5 iii f), are sent from the osteocytes though the osteointernet via gap junctions to lining cells on the bone surface, between the lining cells to the adjacent bone vascular endothelial cells (which sport for things to come an important peculiarity of endothelial cells—PTH receptors), to the osteonal blood vessel or even to the marrow to talk to osteoclast precursors and osteoprogenitors (R.B.Martin, 2000a,b, 2002; Klein-Nulent et al., 2003; Marotti, 1996; Streeten and Brandi,1990; van Bezooijen et al., 2005a, 2005b; Whitfield et al., 2002) (Fig. 2).

i b 2. Sloshing Fluids, Flowmetering Cilia, Talking Integrin Moorings and Osteointernet Blogging

A microcrack or an orthopedic implant screw cuts osteointernet connections and produces a layer of dying cells bordering the crack which triggers BMU formation by stopping the flow of inhibitory signals, particularly SOST, from the osteocytes to the lining cells and cuts off the flow of fluid and the oxygen and essential nutrients it contains to surrounding cells (Dodd et al., 1999; Marotti, 1996; Noble, 2000; Noble and Reeve, 2000; Skerry, 1999; Schaffler, 2000; Tami et al., 2002) (Fig. 3). The squealing osteocytes surrounding the microcrack self-destruct by making the deadly, apoptosis-driving Bax protein (Potten and Wilson, 2004) and things that attract "vulturing" osteoclasts, but the "kamikaze"wave of apoptotic damage is stopped from going too far beyond the crack by the more distant cells raising an anti-apoptosis "firewall" by making the anti-apoptosis Bcl-2 protein (Verborgt et al., 2002) (Fig. 4).

The normal squeezing, twisting and stretching of the bones of an active skeleton particularly in the hip and distal leg bones by heel striking and in the ribs by breathing, pushes and sucks the fluid in the osteocytes' lacunas and canaliculi back and forth, into and out of the canaliculi's pores in the osteonal walls to produce pulsing shear forces on the cells and their processes (Cowin, 1999; Burr et al., 2002; Ehrlich and Lanyon, 2002; Gooch and Tennant, 1997; Klein-Nulent et al., 2003; Kufahl and Saha, 1990; Knothe-Tate, 2001, 2003; Knothe-Tate et al., 2004; Robling et al., 2006; Sorkin et al., 2004; L. Wang et al., 1999, 2000; You et al., 2001). According to Burger et al. (2003), Knothe Tate (2003), Knoth Tate et al. (1998, 2000, 2004), Piekarski and Munro (1977), Smit et al. (2002) and L. Wang et al. (2000) strain pulses in long bones such as tibia and femur from heel strikes during the walking cycle or in the ribs caused by muscle pulling during breathing cycles cause fluid to flow through the 3-D lacunar-canalicular network to the bone surface and back again—it sloshes back and forth. On the other hand, static strain is a one-shot suck or push with a change in position without the further pumping action needed to drive osteocyte signaling—the fluid in the lacunas and canaliculi must slosh back and forth within a certain frequency range to keep the osteocytes informed of the level of bone function and signaling appropriately to its fellows in the osteointernet (Burr et al., 2002; Ehrlich and Lanyon, 2002; Gooch and Tennant, 1997).

But there is an interesting aspect of strain cycles that affects osteogenic signaling in long bones (Gross et al., 2004). Srinivasan and Gross (2000), using the mid-diaphysis of the avian wing ulna, and murine tibia reported that the outflow rates are maximum during the first load cycle in which fluid is squeezed out of the osteocyte lacunae into the canaliculi. When the load is lifted, the fluid is sucked back into the lacunae. But if the next loading happens before the lacunae are refilled, there is less fluid to be squeezed into the canaliculi. If this continues, fluid flows, and with them osteocyte signaling, drop off to produce a steady state equilibrium. This means that if fluid pulsing and optimal signaling are to be maintained at an optimal level there must be rest periods between the loadings to allow the lacunae to be fully refilled (Gross et al., 2004; Robling et al., 2006). Srinivasan et al. (2002, 2003) confirmed this by measuring the effects of cyclic loading with and without rest periods on bone formation in avian ulna and the tibias of young, adult and aged mice. In all cases inserting rest periods between load cycles enhanced the osteogenicity of low-level loading protocols. For example, 5 consecutive days of 100 uninterruprted low-level loading cycles did not affect the tibial periosteal bone formation rate, but putting 10 seconds of rest between every 10 cycles increased the signaling so that the periosteal bone formation rate increased 8-fold (Sirinvasan et al., 2002).

How do osteocytes sense fluid sloshing? Donahue et al. (2001, 2003) for example have shown that fluid-flow triggers Ca^{2+} oscillations in rat osteoblastic cells. One thing they, like most other non-proliferating cells, have is a strain-sensing gadget that could produce such oscillations (see www.wadsworth.org/BMS/SCBlinks/Cilia1.html for a long list of cells and publications describing it) which was first noticed in almost every tissue by Zimmermann in

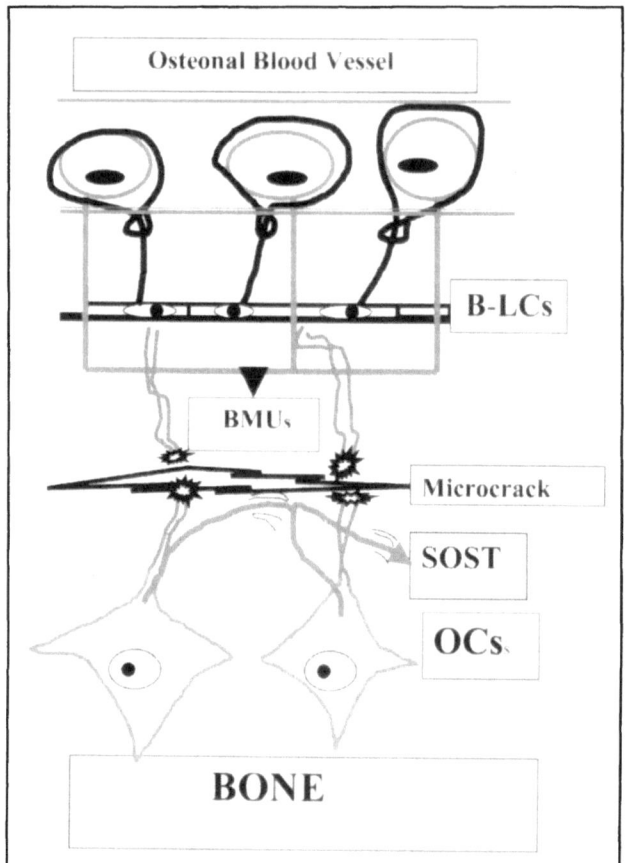

Figure 3. The storm hits with a hard muscle pull or heel strike ! A microcrack slices through the dense tangle of canalicular communication lines and cuts off the local osteointernet. It is through these lines that the osteocytes got their nutrients and oxygen, got rid of their wastes, and sent streams of strain-generated information to the retired post-osteoblasts, the **B-LCs**, lining the osteonal (or trabecular) surface. The crack also blocks the lining cell-restraining **SOS** from reaching its targets. The severing of communications causes the osteocytes (**OCs**) to begin running out of fuel, asphyxiating and drowning in their own wastes. The stressed osteocytes trigger the self-destructive apoptosis mechanism and in the process send distress and then dying cell scents that cause the blood vessel lining cells to velcroize and thus lasso appropriately addressed osteoclast precursors passing by in the circulation. These lassoed cells then invade the damaged patch and fuse with each other to form the first wave of osteoclast diggers of a new BMU. One of the things that attracts these and subsequent osteoclast precursors and the vulturing mature osteoclasts is the "scent" of the phosphatidyl serine-decorated membranes of the dying osteocytes' dendrites poking out of the canalicular pores in the osteonal or trabecular wall.

1898 but until now (C.G. Jensen et al., 2004; Wheatley, 2005; Whitfield, 2003a; Xiao et al., 2006) has either not been noticed or has been ignored by the Bone World. This is the non-motile *primary* (or *solitary*) *cilium* which, unlike a motile cilium with 9 peripheral microtubule doublets, each with attached ciliary dynein motors to move them, and 2 central singlet microtubules, consists of 9 peripheral doublets but no central tubules (Alieva et al., 1999; Alvarez-Buylla et al., 2001; Cameron, 1972; Cohen and Meininger, 1987; Hagiwara et al., 2004; Hellio de Graverand et al., 2001; Matthews and Martin, 1971; Mitchell, 2006;

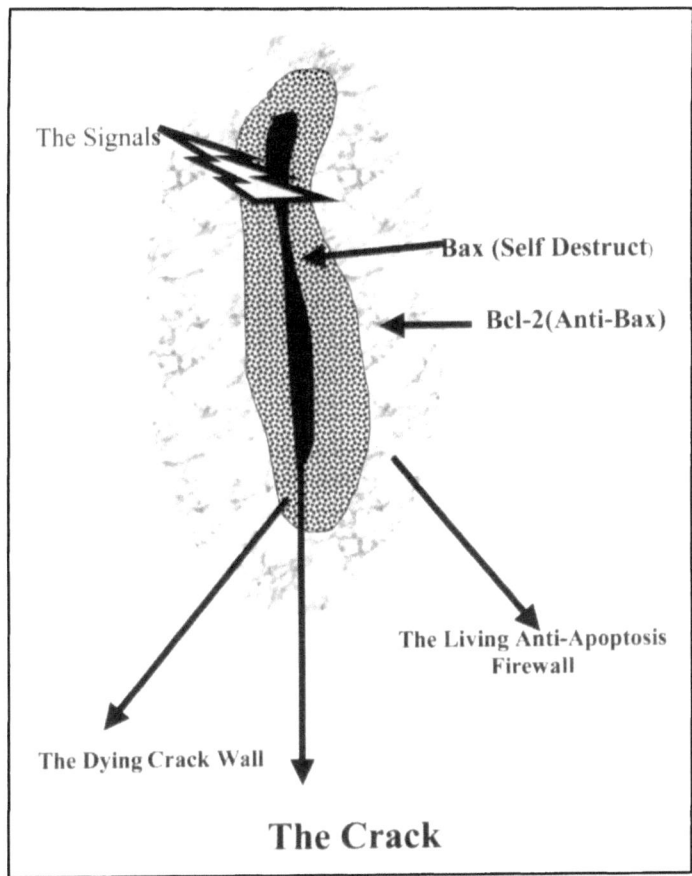

The Signals

Bax (Self Destruct)

Bcl-2(Anti-Bax)

The Living Anti-Apoptosis Firewall

The Dying Crack Wall

The Crack

Figure 4. Because of ruptured supply lines, the starving and suffocating osteocytes bordering a microcrack start the self destructive apoptotic mechanism which is driven in part by the increasingly expressed Bax protein. However, to stop the death zone from spreading a living anti-apoptosis firewall is set up when signals from the dying osteocytes induce outlying osteocytes to turn on their anti-apoptosis Bcl-2 and anti-Bax genes.

Nauli and Zhou, 2004; Pazour and Whitman, 2003; Poole et al., 1985, 1997, 2001; Praetorius and Spring, 2001, 2003a, 2003b, 2005; Satir and Christensen, 2007; Schwartz et al., 1997; Sincla and Reiter, 2006; Tonna and Lampen, 1972; Wheatley, 2005; Wheatley et al., 1996; Zimmermann, 1898). It is the cell's eyes, nose and flowmeter. It grows from the mother centriole of the mother-daughter pair of centrioles embedded in the cell's microtubule-organizing centrosome lying beside, and orienting, the nucleus with the Golgi apparatus. Of course the cilium is at its most glorious in the light-sensing retinal rods and cones. Murine MC3T3-E1 calvarial preosteoblasts, osteoblasts, osteocytes and their chondrocyte cousins also have them (Doxsey, 2001; Malone et al., 2006; Matthews and Martin, 1971; Poole et al., 1997, 2001; Scherft and Daems, 1967; Takaoki et al.,2005; Tonna and Lampen, 1972; Xiao et al. 2006). The primary cilium is a retractable antenna covered by the cell membrane containing stretch-activatable Ca^{2+} channels and studded with various receptors, for example those for serotonin (the 5-HT$_6$ receptor) and somatostatin (the SST$_3$ receptor) in neurons and PDGFRa receptors in quiescent, serum-starved NIH3T3 murine fibroblasts and the

Figure 5. Most cells, chondrocytes, murine MC3T3 preosteoblasts, osteoblasts and osteocytes included have a strain-sensing device known as the *primary or (solitary)* cilium that extends into the extracellular fluid and matrix from the mother-daughter pair of centrioles in a cell's microtubule-organizing centrosome lying beside the nucleus. A striking example of one of these cilia is illustrated here. This one, which belongs to a Ptk kangaroo rat's kidney epithelial cell, looks a lot like a car aerial sporting a decorative ball and is at least the cell's fluid flowmeter, but it might also be equipped with receptors to detect various hormones and other things flowing by in the fluid. The bending back and forth of such a cilium by sloshing extracellular fluid, for example perhaps in an osteocyte's lacuna (Whitfield, 2003a), sends a stream of signals into and through gap junctionally connected cells like flashes of light through a fiberoptic network (Nauli et al., 2003; Praetorius and Spring, 2001,2003, 2005; see also Fig. 6). The frequency and intensity of the signal pulses depend on the frequency and extent of bending of the cilium. This beautiful photograph was generously sent to me by Dr. Sam Bowser of the Wadsworth Center, Albany, NY.

hedgehog (Hh) receptor Smo (probably coupled to Smo's companion Ptch [patched]) we will meet later (Corbit et al., 2005; Fuchs and Schwark, 2004; Huangfu and Anderson, 2005; Pazour and Whitman, 2003; Praetorius and Spring, 2005; Schneider, L. et al., 2005; Whitfield, 2004). A striking example provided by Dr. Sam Bowser of a primary cilium sticking out of a PtK kangaroo rat kidney epithelial cell, looking like an ornate automobile aerial and working like the sensor-covered tusk sticking out of the male narwhal's upper jaw (Holden, 2005) can be seen in Figure 5. An osteocyte might also use its cilium to assess the levels of key components in its surroundings and measure fluid sloshing caused by heel strikes and tell other cells in the network about them (Whitfield, 2003a).

The first clue as to how osteocytes in a femur, for example, might use their primary cilia to translate fluid sloshing in their lacunae into Ca^{2+} oscillations was found in MDCK dog or in mouse embryo kidney cells (Nauli et al., 2003; Praetorius and Spring, 2001,2003a, 2003b,2005; Singla and Reiter, 2006; Whitfield, 2003a). Bending the kidney cell's 8-μm cilium lights up the cell by causing Ca^{2+} to surge into the cell and then sending a Ca^{2+}-triggered signaling wave traveling though gap junctions into the cell's neighbors (Nauli et al., 2003; Nauli and Zhou, 2004; Praetorius and Spring, 2000, 2003a, 2003b, 2005; Singla and Reiter, 2006; Shiba et al., 2005) (Fig. 6). If the osteocyte's cilium is indeed a flowmeter, then the volume and frequency of the signaling would depend on how much and how often the cilium is bent by the extracellular fluid as happens in kidney tubule cells. And it certainly is in the kidney (Praetorius and Spring, 2001, 2005; Singla and Reiter, 2006) and MC3T3-E1 murine preosteoblasts (Takaoki et al., 2005)! Therefore, the cilium would be a device for

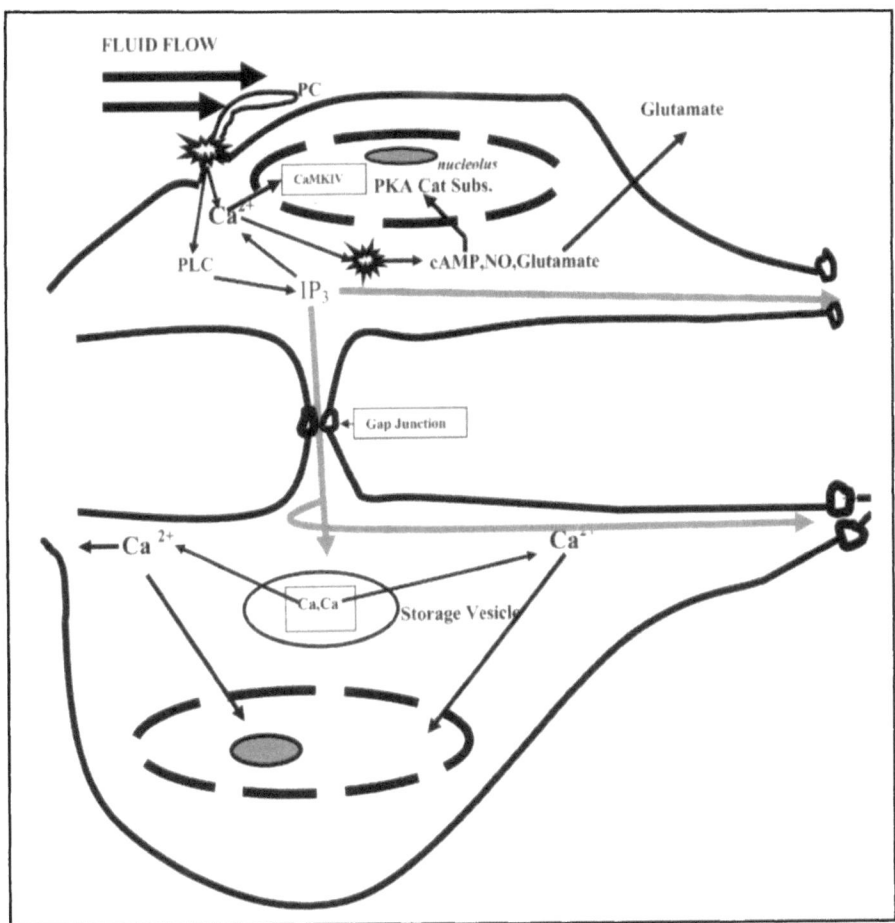

Figure 6. How the bending of a femoral or rib (for example) osteocyte's primary cilium by fluid sloshing back and forth in the cell's lacuna can send signals streaming through the osteointernet when the bone's owner is walking, running, or breathing. The bending of the cilium pries open Ca^{2+} channels in the cilium's membrane through which surges a bolus of Ca^{2+}. This Ca^{2+} bolus "lights up" the cell. It triggers a shower of events such as the stimulation of nuclear CaMKIV (Ca^{2+}•calmodulin–dependent protein kinase IV), a surge of cyclic AMP-dependent protein kinase catalytic subunits into the nucleus which stimulates various gene activities, the stimulation of NO (nitric oxide) by Ca^{2+}•CaM-responsive eNO-synthase, and the release of glutamate to stimulate cells such as preosteoblasts and osteoblasts waiting at the canalicular ports in the osteonal or trabecular wall that have glutamate receptors (Fig. 2). The Ca^{2+}-triggered signal is sent through gap junctions to adjacent cells by the IP_3 produced by the chopping up of the membrane phospholipid PIP_2 by PLC activated by the ciliary stretching. The Ca^{2+} bolus cannot pass into other cells because Ca^{2+} slams gap junctions shut. However, the IP_3 takes over and moves through the junctions and indirectly transmits the Ca^{2+} signal from the bending cilium by easily passing through the gap junctions and triggering the release of Ca^{2+} from the neighbors' calcium-storage vesicles. The reader must not forget that the signaling from the cilium on the cell body in the lacuna collaborate with matricrine (i.e., extracellular matrix-driven) signals from the tugging of the integrins anchoring the cell's processes to the walls of the canaliculi.

measuring the sloshing of extracellular lacunar fluid caused by the squeezing and stretching of bone during a walking or running cycle for example. Indeed, Takaoki et al. (2005) have found that bending the MC3T3-E1 preosteoblast's cilium by fluid flow causes the cell to switch on its *c-fos* gene. And Malone et al. (2006) have further shown that the MC3T3 cell's primary cilium triggers Ca^{2+} signaling and stimulates phosphorylation/activation of the ERK1/2 signaling protein kinase in a flowing medium. Silencing the cilium by disabling the polaris (Tg737) gene suppressed the responses to flowing medium. Thus, the bending cilium is a kind of mechanoresponsive signal lamp that sends flashes of light through the osteocyte fiberoptic network to lining cells on the bone surface and beyond switching on genes and mobilizing various resources (Fig. 6).

How could bending a cilium send a signal spreading from an osteocyte in its lacuna in the depths of femoral cortical bone through a lacunocanalicular labyrinth to cells lining the osteonal canal or endosteal surface? The answer has been found in kidney cells. A kidney cell's primary cilium has in its membrane a G-protein-coupled mechanosensor, PC-1 (polycystin-1), that is tied by its C-tail to the C-tail of PC-2 (polycystin-2), a cation channel that is 4- to 5-fold more permeable to Ca^{2+} than to K^+ or Na^+, to form PC-1•PC-2 complexes (Al-Bhalal and Akhbar, 2005; Delmas, 2004; Singla and Reiter, 2006; Stayner and Zhou, 2001). Bending the cilium with a micropipette or by increasing culture fluid perfusion rate, for example from 2 to 8 μl/second, stretches PC-1 which pulls open the attached PC-2 to let Ca^{2+} flow into the cell (Nauli et al., 2003; Nauli and Zhou, 2004; Praetorius and Spring, 2001, 2003a, 2003b, 2005; Singla and Reiter, 2006). The result of this is a surge of Ca^{2+} into the cilium and stimulation of phospholipase C that releases IP_3 (inositol-1,3,5 trisphosphate) from the cilial membrane's phosphoinositides (Fig. 6). The IP_3 in turn releases Ca^{2+} from internal stores in MDCK kidney cells (Praetorius and Spring, 2001, 2003a, 2003b, 2005; Singla and Reiter, 2006) (Fig. 6), but not in mouse embryo kidney cells (Nauli et al., 2003; Nauli and Zhou, 2004;). However, in both the dog and mouse cells, the result is a cascade of events spreading through the cytoplasm and nucleus and then into the neighboring cells. But, surprising as it may seem, it can't be the little Ca^{2+}ions that move through the big gap junctions because they cause gap junctions to slam shut. Instead it is the Ca^{2+}-releasing IP_3 in MDCK cells and something else in mouse embryo kidney cells that pass though gap junctions (Nauli et al., 2003; Nauli and Zhou, 2004; Praetorius and Spring, 2001, 2003, 2005) (Fig. 6). Thus, bending a primary cilium armed with what appears to be just one PC-1•PC-2 complex triggers a spreading wave of Ca^{2+}- and Ca^{2+}•CaM (calmodulin)-triggered events such as bursts of adenylyl cyclase activity, NO (nitric oxide) production, nuclear CaMKIV (calmodulin-dependent kinase IV) activity and gene activations first in the cilium's owner and then in its neighbors (Nauli et al., 2003; Nauli and Zhou, 2004; Praetorius and Spring, 2001, 2003a, 2003b, 2005; Singla and Reiter, 2006) (Fig. 6). Since osteocytes also have these little antennae, the sloshing canalicular fluid in their lacunae in an active skeleton (Tami et al., 2002) very likely keeps the antennae bending back and forth and sending streams of Ca^{2+}-triggered signals through the network that would keep the members of the network humming with bone-maintaining or growth-inducing activity (Whitfield, 2003a).

Of course, disrupting the kidney cell's cilium eliminates its ability to sense changes in extracellular fluid flow rate (Nauli and Zhou, 2004; Praetorius and Spring, 2003a, 2005; Singla and Reiter, 2006). The horrendous consequence of this sensory deprivation in both humans and mice is polycystic kidney disease (PKD) caused by disrupting the kidney tubule cells' gene for PC-1 or PC-2 (Calvet, 2002; Pazour and Witman, 2003; Singla and Reiter, 2006; Yoder et al., 2002). It seems that the stream of signals from the cilia drives the tubule cells to terminally differentiate and lock down their proliferation machinery (Al-Bhalal and Akhbar, 2005; Singla and Reiter, 2006). But without these cilia and their signaling, the cells cannot terminally differentiate and consequently proliferate inappropriately to form cysts, which expand over time and eventually destroy the kidney (Al-Alal and Akhbar, 2005). Since osteoblasts and osteocytes also have primary cilia, severe skeletal malformations might be expected to accompany PKD. And this is precisely what happens when the mouse's PC-1-encoding *Pkd1* gene is disabled (Boulter et al., 2001; W. Lu et al., 2001). Moreover, there are reports of multiple skeletal limb malformations (polydactyly, syndactyly, tibial agenesis, club foot) and ischiospinal dysostosis associated with polycystic kidney disease (Nishimura et al., 2003; Turco et al., 1993, 1994).

While osteocytes use their primary cilium to measure bone strain and tell lining cells about it, osteoblasts use their probably sensor-studded cilia for driving bone formation. That the primary cilium can drive bone formation has now been shown by Xiao et al. (2006). They have shown that the cilium's mechanically stimulable PC1-product of the *Pkd1* gene together with PC-1's attached PC-2 cation channel can drive, indeed is needed for, osteoblast maturation by stimulating the expression of the Cbfa1/Runx2-II gene, the product of which is the master driver of a panel of the major osteoblast-specific genes. Moreover, mice heterozygous for a mutant *Pkd1* gene (i.e., they are *Pkd1*[+/-]) have low trabecular bone volumes, reduced cortical bone thickness and a decreased mineral apposition rate—in other words they are osteopenic without their full complement of active *Pkd1* genes.

Signals from the osteocytes' waving primary cilia would be accompanied especially at higher strains by the matricrine (i.e., extracellular matrix-driven) signaling from tugged integrins and their linked cytoskeleton that attach the cell and its processes to the osteopontin lining the walls of the lacunae and canaliculi like waves cause a ship to pull on its mooring lines (Turner and Robling, 2004). But in this case the cells' mooring lines can "talk" because they are linked to sophisticated signaling instruments (Denhardt et al., 2002; Miranti and Brugge, 2002). This tugging causes the clustering of integrins with signalers and periodic collection of the actin cytoskeleton into thick, oriented stress fibers with a consequent activation of signaling enzymes on actin-linked cell membrane rafts and mechanosensitive Ca^{2+} channels (Burr et al., 2002; N.X. Chan et al., 2000; Duncan and Misler, 1989). The upshots of this normal pumping and shearing is the appearance of a mob of agents which appear to be needed! There are Ca^{2+} ions surging through mechanosensitive L-type channels; ATP signaling through its $P2X_7$ receptor-Ca^{2+}ion channels; NO from the Ca^{2+}-stimulated NO-synthase; prostaglandins (prostacyclin [PGI_2] and prostaglandin-E_2 [PGE_2]) produced specifically by the mechanosensitive COX-2 (cyclooxygenase-2) and signaling through EP_1 and EP_4 receptors; and glutamate (the well-known excitatory neurotransmitter until recently believed to be used only by central neurons) and serotonin. The members of this incredible mob pass from osteocyte to osteocyte through gap junctions and/or jump from receptor to receptor to keep the bone lining cells from unnecessarily ordering vascular endothelial cells to start the repair process by grabbing passing osteoclast precursors (A.D. Bakker et al., 2003; Bliziotes et al., 2001; N.X. Chan et al., 2000; Chenu, 2002; Duncan and Misler, 1989; Edlich et al., 2001; Ehrlich and Lanyon, 2002; Kufahl and Saha, 1990; Manolagas, 2000; R.B. Martin, 2000; Noble, 2000; Mason, 2004; Noble and Reeve, 2000; Noda et al., 1996; Nomura and Takano-Yamamoto, 2000; Parfitt, 1998; Pavalko et al., 1998; 2003; Ryder and Duncan, 2001; Schaffler, 2000; Skerry, 1999; Skerry and Genever, 2001; A.F. Taylor, 2002; Turner and Robling, 2004; Westbroek et al., 2001).

The signals from bending cilia, stretched cell membranes and matricrine signals from tugged cadherins and integrins turn on osteogenic genes such as the gene for the strongly osteogenic PTHrP cytokine, a main physiological driver of osteoblast differentiation which features prominently further on in our story (X. Chen et al., 2003; T.J. Martin, 2005; Whitfield, 2006b). How? The stimulation of the PTHrP gene is mediated by the opening of stretch-activated TREK-family K^+ channels (X.Chen, C. Macica et al., 2003). The ciliary bending and membrane deformations also open the cilium's PC-2 channel and another mechanosensitive ion channel in the cell membrane known as ENaC (SA-CAT-stretch-activated cation channel) through which Na^+ surges into the cell to depolarize the membrane and thus open L-type VSCCs (voltage-sensitive Ca^{2+} channels) (Kizer et al., 1997; Pavalko et al., 2003). The result of this ionic bombardment is a cascade of Ca^{2+}-driven and other events that create complexes called *mechanosomes* that convert the mechanical deformation into biochemical responses and gene expressions and the generation of osteogenic factors such as PTHrP (Pavalko et al., 2003).

The mechanism goes something like this. The non-receptor tyrosine kinase FAK and cSrc in the tugged integrin-associated focal adhesion plaques are activated and phosphorylate the Crk and p130^{Cas} in Crk•p130^{Cas} complexes in the plaques. The phospho-Tyr-p130^{Cas} attaches to Nmp4/CIZ and kicks Crk out of the complex to form one of the "mechanosomes"—phospho-Tyr-p130^{Cas}•Nmp4/CIZ (Kirsch et al., 2002; Nakamoto et al., 2002; Pavalko et al., 2003; Yamada and Even-Ram, 2002). The activated FAK kinase triggers the Ras $\Rightarrow$ Raf $\Rightarrow$ MEK 1/2 $\Rightarrow$ ERK 1/2 cascade the last of which phosphorylate and thus enhance the activity of the Cbfa1/Runx-2 transcription factor that drives the expression of key osteoblast genes (Lian et al., 2004; Franceschi and Xiao, 2003). Meanwhile the tugged cadherins in a lining cell's adherens attachment junctions with its neighbors also release β-catenins which associate with LEF-1 protein to form a β-catenin•LEF1 mechanosome.

The basic "take-home" message from this complicated cluster of cell surface deformation-triggered events is really quite easy to remember—fluid tugging enhances the activity of the master osteoblast transcription factor Cbfa1/Runx-2 and generates at least two mechanosomes that are fired into the nucleus where they bind to target sites on the nuclear matrix and then together with Cbfa1/Runx-2 grab, bend and twist the promoters of their bone-making target genes into shapes that enable them to be switched on. One of these mechanosensitive genes codes for PTHrP which, as we shall see further on, is a towering figure in osteoblast differentiation (X. Chen et al., 2003; Daifotis et al., 1992; T.J. Martin, 2005; Pirola et al., 1994; Steers et al., 1998; Whitfield, 2006b). The strain-driven maintenance of a steady supply of PTHrP by osteocytes could, for example, facilitate crack repair by promoting the flow of osteoprogenitors into the mature BMU osteoblast pool.

To make the strain-response story more complicated, stretching does something astonishing—it can activate murine MLO-Y4 osteocytic cells' non-genomic (membrane-based) estrogen receptors the signals from which send ERK2 into their nuclei and oppose for example, the initiation of apoptosis by etoposide (Aguirre et al., 2003). In other words strain can activate estrogen receptors *without estrogen* (i.e., ligand-naive receptors)!!

In general terms it seems as if moderate normal body movements keep signals flowing steadily through the extensive osteocyte network (which because of gap junctions is really a giant gated syncytium) from bending cilia. But more strenuous body movements add matricrine signals from cadherin and integrin tugging.

As we shall see in the next section, besides making gene promoter-switchboxes accessible the intercellular chattering caused by mechanical strain looks a lot like neural network crosstalking. Maybe osteocytes could be called "honorary neurons" perhaps without glutamate receptors but able to release it (Szczesniak et al., 2005). An activated neuron sends a wave of Na$^+$ influx-induced membrane depolarization along its axon, which when it hits the axon terminus, it opens Ca^{2+} channels. This triggers the release into the synaptic cleft of packets of a neurotransmitter such as L-glutamate which activates receptors on a post-synaptic neuron. Amazingly, the same Na$^+$/Ca^{2+} shifts caused by the punching and pulling of an osteocyte and the tweaking of its cilium by its owner's heel strikes may also cause the cell to release glutamate which can reach the lining cells that in the rat express a variety of glutamate receptors with the NMDAR2B and GluR2/3 subunits (Mason, 2004; Szczesniak et al., 2005; A.F. Taylor, 2002) (Fig. 2)!

If the battering be too severe it can actually break the local connections of the osteointernet as well as trigger apoptotic self destruction by osteocytes (Noble, 2000; Noble and Reeve, 2000; Skerry, 1999; Schaffler, 2000). The death of the overstrained osteocytes and the loss of SOST signaling would lift the restraints on lining cells as effectively as the microcrack in Figure 3. This would enable the lining cells to start making osteoclastogenesis stimulators such as

M-CSF (Macrophage-Colony-Simulating Factor) and RANKL (Receptor-Activator of NF-κB Ligand) and cut off a source of the principal physiological osteoclastogenesis suppressor, osteoprotegerin (OPG) that the osteocytes and lining cells had been making since its gene was turned on under the influence of the Cbfa1/Runx-2 transcription factor when they became osteoblasts (Atkins et al., 2003; Ducy, 2000; Karsenty, 2000; Komori, 2000; Lian et al., 2004; Noble and Reeve, 2000; Thirunavukkarnasu and Halliday, 2000). Thus are recruited the first, or shall we say pathfinder, BMUs to start fixing the damaged bone.

ii a. Calling the Diggers

"So how can we explain that osteoclasts "know" their way while they are eating through the bone tissue?"—Burger et al. (2003).

As in a highway repair crew, the first on the job in a microcracked bone of a human or some other large animal are the diggers—the big multinuclear osteoclasts. These large cellular syncytia or cellular collectives are armed with a protein shredder (protease) called cathepsin K and the iron-containing TRACP (tartrate-resistant acid phosphatase) in endocytic vesicles that generates ROS (reactive oxygen species) such as ˙OH radicals that collaborate to chop up the demineralized bits of matrix proteins and are heavily loaded with mitochondria to make the large amounts of ATP fuel needed to feed the pump which sprays HCl (hydrochloric acid) to dissolve the bone mineral and operate the garbage disposing transcytosis machinery that loads the matrix debris into the TRACP-containing endocytic vesicles at the digging (apical) end of the cell and carries them to the cell's secretory domain (FSD) at the other (basal) end for emptying (Blair, 1998; Halleen et al., 1999, 2003; Noble, 2000; Noble and Reeve, 2000; Skerry, 1999; Stenbeck, 2002; Troen, 2003;Whitfield et al., 1998b). And one of the first responses of osteocytes to the strain around the microcrack is to make osteopontin to help attract the diggers and to glue them to the damaged patch (Mazzali et al., 2002; Nomura and Takano-Yamamoto, 2000). But while there is a lot to find out about the how the distressed osteocytes summon these diggers to the damaged patch it appears to be the "smell" of osteocytes' corpses that does it—a kind of "microcrack pheromone" package.

Osteocytes can be induced to self destruct and in the process trigger BMU activation either by very low or by very high strain, but they function best when the strain on a hip joint for example is within a range of frequencies and strengths produced by the large muscle forces driving normal body movements such walking or running cycles or the incessant pulling on the ribs by the muscles of breathing (R.B. Martin et al., 1998; Noble and Reeve, 2000). The osteocytes may be prevented from self-destructing by *basal* amounts of NO made probably by Ca^{2+} pulses triggered by the waving of their cilia in the fluid sloshing in their lacunae during normal movements (Klein-Nulent et al., 2003; Whitfield, 2003a) (Fig. 6). The pulsing strain pumps blood and extracellular fluid throughout the network of blood vessels and the osteocytes' lacunocanalicular network. This delivers food and oxygen to the osteocytes and flushes out their waste. If the pumping weakens or stops because of immobilization or the lack of gravity, the oxygen level in the bone drops and starving, oxygen-deprived osteocytes, drowning in their own uncollected wastes respond suicidally by making osteopontin which brings on the osteoclasts to get rid of the unused bone and themselves with it (Arnett et al., 2003; Dodd et al., 1999; Gros et al., 2001, 2005; Mazzali et al., 2002). On the other hand, if the matrix strain rises above a certain level, as happens at the tip of a microcrack where it can be 15 times above average (e.g., 30,000 microstrain as opposed to the normal 2,000-3,000 microstrain)(Nocolella and Lankford, 2002), osteocytes around the crack self destructively start making the pro-apoptosis Bax protein (Potten and Wilson, 2004) probably because of being torn from the walls of their cubicles and the cutting off of the flow of nutrients and oxygen by the severing of their canalicular and dendritic lifelines (Tami et al., 2002; Verborgt et al., 2000) (Fig. 4). However, osteocytes further away from the crack respond to a less severe interruption of nutrient deliveries and prevent the damage from spreading by erecting an anti-apoptosis "firewall" of the anti-apoptosis Bax protein (Verborgt et al., 2000) (Fig. 4).

The sea-level's 20% O_2 is excessive (it's hyperoxic) and the level in tissues is kept far below this level (Lane, 2005). So as expected there is a very low oxygen tension corresponding to 4-7% O_2 in the bone marrow and bone osteoprogenitor cells optimally generate more osteoblasts and bone in an atmosphere containing 5% O_2 instead of the usual 20% O_2 (Lennon et al., 2001). However, lowering the O_2 to 2% reduces immature osteoblast proliferation, maturation and the the formation of mineralized nodules (Utting et al., 2006). But let's have a look at what happens to gene expressions when osteocytes' oxygen supply is further reduced and they rapidly become abnormally hypoxic by the failure of the bone fluid pumping action when it stops because of disuse or the severing of canaliculi by a microcrack (Arnett et al., 2003; Dodd et al., 1999; Gross et al., 2001). All of our cells can respond to an oxygen shortage by turning on or turning up two dozen or more so-called hypoxia-inducible genes, the products of which are meant to maintain at least a minimum supply of ATP fuel and other components for at least a short-term survival. When an osteocyte's nutrient-supplying and waste-eliminating pumps are working normally, the components of the hypoxia-inducible transcription factor, HIF-1α and HIF-1β, are constantly being made, but HIF-1α has an oxygen-dependent degradation domain (ODD) that is ubiquitinated and thus dumped into the cell's proteaseome shredder when there is enough oxygen because the oxygen drives the hydroxylation of the domain's key proline residues. The ODD with its hydroxylated prolines sticks into a pocket in the von Hippel-Lindau factor which causes the von Hippel-Landau factor to ubiquitinate the HIF-1α which marks it for delivery to the shredder (Bruick, 2003; L.E. Huang and Bunn, 2003; Latchman, 2004; Marx, 2004; Wenger, 2002). But when the oxygen supply is cut off, HIF-1α's ODD is no longer hydroxylated by the oxygen-driven hydroxylase, there is no hydroxylated proline to be shoved into the von Hippel-Landau factor's pocket, and HIF-1a is thus not ubiquitinated and shredded. HIF-1α then accumulates and migrates into the nucleus where it combines with HIF-1β to form HIF-1α•HIF-1β transcription factor complexes. One of the incoming HIF-1α's aparagine residues has also not been hydroxylated and because of this the aparagine can bind to the p300 co-activator on the promoter-switch boxes of two dozen or more hypoxia-inducible genes. The p300 can then interact with components of the basal RNA polymerse II gene-transcriber complex or recruiting the polymerase transcriber to these genes's promoters (Latchman, 2004). Thus the now stabilized HIF-1α•HIF-1β transcription factor activates the genes for the glucose transporter-1 (GLUT1) and components of the glycolytic machinery such as phosphofructose kinase I, phosphoglycerate kinase and lactate dehydrogenase A, that will increase the glucose uptake and the anaerobic production of ATP fuel from this additional glucose to try to compensate, albeit less efficiently, for the lack of oxygen-driven mitochondrial fuel production (Ebert, et al., 1995; Gross et al., 2001; L.E. Huang and Bunn, 2003; Schipani et al., 2001). HIF-1α•HIF-1β also stimulates the expression of VEGF in an attempt to drive new blood vessel formation but also stimulates osteoclast migration and activity (Henrikssen et al., 2003; Yamaka et al., 2003). However, another gene, the gene for the anti-apoptosis Bcl-2 protein is turned down or off and the gene for the pro-apoptosis Nip3 protein is turned up. This means that the struggling osteocyte will be able to hang on only for a short time by making more ATP by the glycolytic mechanism, but it is now threatened by the reappearance of oxygen which would cause a massive flow of electrons from accumulated lactate and Krebs cycle intermediates which the mitochondrial respiratory chain cannot fully process and produces an overflow of electrons which leak from respiratory complex I and produce a lethal spray of ROS (reactive oxygen species). Of course the glycolytic mechanism is less efficient at making ATP fuel than the oxygen-driven mitochondrial ATP-making machinery so the osteocytes need more glucose and the glycolytic end product pyruvate will be reduced to lactate instead of being processed into acetylCoA and fed into the mitochondrial machinery. The accumulating lactate increases the cellular acidity (i.e., drops the pH). Since the glucose supply is reduced to a trickle or completely cut off, the osteocyte will soon run out of fuel unless the supply lines are reopened perhaps by the resumption of loading but not if the fuel lines have been cut. To make matters worse a lethal apoptosis-triggering leakage of agents such as cytochrome-c from the mitochondria can no longer be stopped by the Bcl-2 protein and the cell finally, with the help of the apoptogenic p53 protein, does the "honorable" thing and kills itself. This is the honorable way to go, because an apoptosing cell does not perforate its membrane and release things that would kill any innocent bystanders in the osteointernet.

The death of the osteocytes around the crack cuts off signalers such as the SOST protein (see Chapter 5 iii f; Keller and Kneissel, 2005; Poole and Reeve, 2005; Poole et al., 2005; Sevetson et al., 2004; Sutherland et al., 2004; van Bezooijen et al., 2004, 2005a, 2005b) they were sending to the local lining cells in the osteonal central canals to prevent them from reverting to osteoblasts or calling up BMU digging crews (R.B. Martin 2000a, 2000b; Marotti, 1996). The local lining cells alerted by factors emitted from the microcrack could tell immature osteoblastic stromal cells in adjacent bone marrow to make osteoclast recruiters such as M-CSF (macrophage colony-stimulating factor) and RANKL (receptor activator of NF-κB ligand). It is only the *immature* osteoblastic bone marrow stromal cells (they will lose this ability as they mature) that can put RANKL ligands out on their surfaces to bind to preosteoclasts' RANK receptors and stimulate them to differentiate into bone-digging mature osteoclasts. But the microcrack-injured osteocytes can also make and dump M-CSF and RANKL into the adjacent bone marrow or regional blood stream (Heino, 2005). The M-CSF and RANKL and other factors seeping into the blood from the injured bone patch would attract these developing osteoclast precursors to the site where vascular endothelial cells start grabbing them from the passing blood and the M-CSF and RANKL stimulate their development into mature osteoclasts (Fig. 3).

At this point we must try to see what controls the ability of immature osteoblastic cells in the marrow to make the RANKL (mature osteoblasts can't do this) needed to drive the development of osteoclast precursors (Atkins et al., 2003; Kitazawa et al., 1999; G.P. Thomas et al., 2001). As we shall learn further on, the immature stromal preosteoblast owes its promising "osteoblastness" to expressing the master gene transcription factor Cbfa1/Runx-2 we met above. This remarkable factor has a NLS (nuclear localization sequence) that enables it to be shipped into the nucleus and a NMTS (nuclear matrix targetting sequence) that enables it to attach to sites where transcription machinery is located (Lian et al., 2004). At these sites Cbfa1/Runx2 serves as a "platform protein" for the formation of transcriptional complexes with 25 or so positive or negative co-regulators (Lian et al., 2004). In other words at the matrix sites it acts as a kind of magnet for co-regulators of key osteoblastic gene regulators. When it arrives on the matrix it looks for genes with the so called OSE2 (osteoblast-specific 2) AACCACA nucleotide sequences in their promoters (i.e.,their "on-off switches") (Otto et al., 2003). Two of these genes code for collagen 1 and RANKL. But while the RANKL gene does have OSE2 in its promoter (Kitazawa et al., 1999) and switching it on requires Cbfa1/Runx-2 expression, it does *not* in fact depend directly on Cbfa1/Runx-2 being on the promoter (O'Brien et al., 2002). Instead, switching on the RANKL gene requires that the immature osteoblastic stromal cell have integrin receptors to bind to a collagenous matrix, the making of which directly depends on Cbfa1/Runx-2 expression (Lian et al., 2004; O'Brien et al., 2002). The signals from the collagen-bound integrins generate the ERK 1/2 kinase activity which phosphorylates and enhances Cbfa1/Runx-2 activity (Franceschi and Xiao, 2003; Lian et al., 2004).

The upshot of all of this is the arrival and arming of BMU diggers at the damaged site to dig out the microcrack and prepare the site for the making of a new canal by osteoblasts (Marotti, 1996). As the osteoclasts are tunneling through the wall of a Haversian canal in cortical bone or digging a trench on or tunneling into a trabecula, the bone's owner is walking, playing tennis, running, breathing etc.. The tunnel being dug in cortical bone has a *cutting front* or *cone*, an intermediate *resting* or *pause zone* and a *closing* or *filling tail* along which is a gradient of strains produced by pulse-loading by walking, running etc.(Smit et al., 2002). Smit et al. (2002) have analyzed the distribution of strain along the advancing tunnel. It is important to remember for things to come later that low strain at the cutting front stops the lacunocanalicular extracellular fluid sloshing, which, of course, to the ever-alert osteocytes means "*bone-not-in-use*"and accordingly tells the diggers to get rid of it while further along the tunnel the strain rises and generates signals which mean "*bone-in-heavy-use*" and thus calls for osteoblasts to make more bone to reduce strain (Klein-Nulent et al., 2003; Smit et al., 2002). More of this later.

ii b. Calling Diggers with the Neuron-Like Signaling Machinery of Osteocytes, Osteoblasts and Osteoclasts

"It is clear that neurotransmitters have profound effects on bone, influencing the differentiation, proliferation, activity and apoptosis of osteoblasts and osteoclasts."—Spencer et al. (2004)

You may recall that bone cells have memory and the ability to share information with each other through extensive networks. Well, it looks as if it runs much deeper than expected. It seems that the major neurotransmitters, such as glutamate and serotonin, may be major osteoctye ⇒ osteoblast "osteotransmitters" as well as nerve ⇒ lining cell "neuro-osteotransmitters" in the osteonal canals that help initiate the response to loading and microcracking (Bliziotes et al., 2001; Chenu, 2002; Hinoi et al., 2004; Skerry and Genever, 2001; Serre et al., 1999; Skerry and Taylor, 2001; Spencer et al., 2004; Szczesniak et al., 2005; Westbroek et al., 2001). This raises the possibility of there being osteosynapses similar to neuronal synapses and T-lymphocye-dendritic (Langerhans) cell synapses (Dustin and Colman, 2002). But so far no such synapses have yet been seen, although osteocyte processes have been found terminating "suggestively" on osteoblasts during bone formation (Menton et al., 1984).

Osteocytes, with their primary cilia flowmeters (incidentally neurons also have primary cilia [Whitfield, 2004]) waving back and forth in the sloshing extracellular fluid, are not glutamate targets. They don't have glutamate receptors, but they make and release it to signal other cells and they have GLAST(glutamate aspartate transporter)/EAAT1(excitatory amino acid transporter 1) transporters to reaccumulate it in response to Ca^{2+} surging through the ciliary PC-2 channels opened by the stretched PC-1 mechanosensors (Mason, 2004; Nauli et al., 2003; Praetorius and Spring, 2001, 2003, 2005; Skerry and Genever, 2001; Skerry and Taylor, 2001; Szczesniak et al., 2005; Whitfield, 2003a) (Fig. 6). In other words, they can send, but not receive, glutamate signals. But osteoblasts/lining cells and their precursors can receive glutamate signals from nerve terminals and osteocytes, because osteoblast precursors and primary rat osteoblasts have AMPA/kainic acid and neuron-like NMDA "ionotropic" (i.e., ion-passing) receptor/channels and mature osteoblasts have both of these plus the G-protein-coupled, metabotropic glutamate receptors that are related to the CaRs (Ca^{2+}-sensing receptors) and like the CaRs would likely be activated by the extremely high Ca^{2+} concentrations in the osteoclast excavation sites (Brown and MacLeod, 2001; Chattopadhyay and Brown, 2003; Gu et al., 2002; Jensen et al., 2002). Not only do osteoblasts have NMDA receptors, but amazingly what subunits these receptors have depends on the location of the osteoblast—calvarial (skull) osteoblasts' NMDA receptors/channels have NR2A, NR2B, NR2D subunits but no NR2C subunits while femoral osteoblasts' NMDA receptor channels have NR2C subunits (Itzstein et al., 2001). Moreover, as Laketic-Ljubojevic et al. (1999) have reported, glutamate triggers Ca^{2+} transients in osteoblastic cells and these cells also have what were once believed to be brain-specific Na^+-dependent inorganic phosphate transporters that are involved in the receptor-triggered release of glutamate from VGLUT storage vesicles just as in neurons (Hinoi et al., 2001, 2004). Mature osteoblasts also have GLAST (EAAT 1) transporters needed to recharge their glutamate stores,to clear their surroundings of released glutamate to avoid prolonged signaling from their glutamate receptors and prevent cell death caused by excessively prolonged glutamate signaling, and to maintain a high glutamate signal-to-noise ratio exactly like the same transporters in the glial cells enclosing and servicing the neuronal synapses (Carmignoto, 2000; Chenu, 2002; Fields and Stevens-Graham, 2002; Hinoi et al., 2002, 2004 ; Mason, 2004; Skerry and Taylor, 2001; A.F. Taylor, 2002). Indeed this is particularly important for bone cells because of the large amount of glutamic acid circulating in the blood which would otherwise saturate the receptors and "jam" signaling (Hinoi et al., 2002, 2003). To recharge glutamate stores maybe they and osteocytes do the same thing as glial cells in the brain. They might sweep up glutamate released by neighboring osteocytes or nerves with their GLAST transporters, convert it to glutamine and send it back to the osteocytes or the Haversian nerves' axon terminals to be reconverted to glutamate and refill the secretory vesicles (Chenu, 2002).

These receptors are not mere osteoblast/lining cell or preosteoblast ornaments. Inhibiting the AMPA/kainate and NMDA receptor signaling reduces the expression of glutamate receptors, glutamate production and differentiation of osteoblasts/lining cells and causes precursor cells to become adipocytes rather than osteoblasts (Dobson and Skerry, 2000; Skerry and Genever, 2001; Szczesniak et al., 2005; Taylor et al., 2000). All of this evokes the image of a network of nerves, precursor cells, osteoblasts, osteocytes and maybe lining cells talking to each other in fluent glutamate!

At this point we should sit down for a moment and try to put these facts together into a model of how strain might trigger bone formation. When a bone is strained and its osteocytic lacunae and canaliculi are squeezed, its osteocytes will be signaled perhaps by their primary cilium to release glutamate (Fig. 6) which will travel through the canaliculi to the osteoblastic bone-lining cells with their array of glutamate receptors (Fig. 2) (Szczesniak et al., 2005). The signals from these receptors will contribute to their releasing of the SOST brakes on the reversion of the bone-lining cells to bone-making osteoblasts.

O.K. you may say, but what has all of this to do with "*CALLING THE DIGGERS*", the title of this section? While a strain-induced glutamate surge from osteocytes stimulates osteoblast generation it also stimulates osteoclast generation via the baby osteoclasts' wet nurses—immature bone marrow osteoblastic cells (Atkins et al. 2003; Kitazawa et al., 1999; G.P. Thomas et al., 2001). It stimulates the immature, osteoblastic marrow stromal cells to put RANKL on their surfaces while mature, matrix-making osteoblasts cannot make RANKL apparently because its gene's promoter switch box is locked shut by having its CpG sequences methylated at some point during maturation (Atkins et al., 2003; Corral et al., 1998; Kitazawa et al., 1999; G.P.Thomas et al., 2001). RANKL is the ligand for the osteoclast precursors' RANK (Receptor Activator of NF-κB transactivator) receptors, the signals from which stimulate osteoclast differentiation (A. Taylor et al., 2000). And the osteoclasts' glutamate-activated NMDA receptor/channels also stimulate differentiation via the NF-κB transactivator exactly as do the signals from RANKL-activated RANK receptors, (Black et al., 2002, 2003; Chenu, 2002; Espinosa et al., 1999; Hinoi et al., 2004; Itzstein et al., 2000; Laketic-Ljubojevic et al., 1999; Mentaverri et al., 2003; Merele et al., 2003; Patton et al., 1998; Peet et al., 1999; Skerry and Genever, 2001; Szczesniak et al., 2005). Osteoclasts seem to need the Ca^{2+} flowing through these activated NMDA receptor/channels in order to mature and later to assemble the squid's suction cup-like actin ring needed to seal off the excavation site (Itzstein et al., 2000; Mason et al., 1997). And the osteoclasts, short-lived cells that they are, also need the Ca^{2+} surges to stimulate their NOS-1 (or nNOS) nitric oxide synthase to make the NO they need to hold off self destructive apoptosis (Mentaverri et al., 2003). Indeed, blocking the NMDA receptors with MK801 or DEP (±-1-[1,2-diphenylethyl]piperidine) causes the cellular Ca^{2+} to drop which induces osteoclast apoptosis, a response that can be reversed by the NO donor S-nitroso-N-acetyl-D-penicillamine (SNAP) (Mentaverri et al., 2003). (But on the other hand NO strongly appears to prevent *osteoblastic cells* from stimulating osteoclast generation-driving RANKL expression and to stimulate them to make the anti-osteclastogenic OPG instead [Turner and Robling, 2004]).

iii. The Arrival at the Work Site

Signals such as ATP, glutamate, NO, PGE_2 (prostaglandin E_2), osteopontin, PGI_2 (prostaglandin I_2) and VEGF (vascular endothelial cell growth factor) from the overstrained, hypoxic and self-destructing (apoptosing) osteocytes with their normal supply lines cut off in a microcracked patch of cortical bone and transmitted to the cells lining the walls of the nearest Haversian or Volkmann's canal start the remodeling/repair job by causing the VEGF-stimulated endothelial cells of a blood vessel to sprout a loop and by dumping osteoclast–mobilizing cytokines such as M-CSF into the blood to stimulate the osteoclast nurseries in the nearest available hemopoietic red marrow sites and to produce potent osteoprogenitor stimulators such as BMP-2 amd BMP-7 (Osteogenic protein-1; OP-1) (Bouletreau et al., 2002; Carano and Filvaroff, 2003). The flow of blood from the arterioles into the budding capillaries is increased by the vasodilating NO and the membrane-associated guanylyl cyclase and guanylyl cyclase-dependent protein kinase (PKG) it stimulates (Krainock and Murphy, 2000; Lincoln et al., 1997; Zabel et al., 2002; Zaragoza et al., 2002). The NO is initially made in the strained osteocytes from L-arginine by the nitric oxide synthase stimulated by Ca^{2+} flowing through the cells' stretch-activated Ca^{2+}-channels and then sustained by the vascular endothelial cells' nitric oxide synthase (eNOS) turned on by the increased shear strain from the increased blood flow in what will be a new Haversian canal and vascular pipelines running through it (reviewed by Krainock and Murphy, 2000; Lincoln et al., 1997; R.B. Martin et al., 1998; Whitfield et al., 1998b). The cells at the leading edge of the vascular loop switch on a set of genes the products of which make the cells' surfaces selectively sticky—"velcroize" them—to snatch appropriately addressed pre-osteoclasts from the passing blood (Parfitt, 1998, 2000a, 2000b, 2002; Ruoslathi and Rayotte, 2000; Springer, 1994) (Fig. 3).

A BMU's diggers can only come out of a blood vessel near the microdamage site, i.e., the "work site" (Parfitt, 2004). A model for getting osteoclasts to the work site has been proposed by Parfitt (1998, 2000a, 2000b, 2002). According to this model all *appropriately addressed* pre-osteoclasts, be they summoned to cortical bone or to trabecular bone, are plucked from the blood circulating respectively through the Haverserian canals or the marrow sinusoids. When the preosteoclasts among the peripheral blood mononuclear cells sailing along in the blood "smell" the osteocytes' distress emissions and touch the altered blood vessel walls near the damaged site they, like neutrophils sailing into an infected tissue battle zone (see Sompayrac [2003] for a wonderful description of this " *rolling-sniffing-stop-exit*" sequence), start putting selectin-binding (selectin ligand) proteins out on their surfaces. After several passes through the region they ultimately become sufficiently and selectively "velcroized" by putting enough selectin ligands (CD44s) on their surfaces to be grabbed rather loosely by the selectin on the blood vessels' lining (i.e., endothelial) cells (Fig. 3). This slows them down and they start rolling along the vessel wall "sniffing" for damage scents. One of the more important attractants is the increasingly loud signaling from the cells' CaRs by the rising Ca^{2+} level as they approach the excavation site. When they find them, they put integrin hooks out on their surfaces to grab hold of ICAMs (intercellular adhesion moleculse) on the endothelial cell surfaces like descending aircrafts' tail hooks grabbing the arresting wires on an aircraft carrier's deck. This stops them and they are lured from the blood vessel and into the bone by various chemoattractants such as the MCP-1 chemokine released by the stressed osteocytes and lining cells (look ahead to Fig. 41).

The pre-osteoclasts, their generation strongly stimulated by hypoxia are then attracted by the osteopontin and VEGF from the hypoxic osteocytes (Arnett et al., 2003; Dodd et al., 1999; Gross et al., 2001, 2005; Henriksen et al., 2003; Utting et al., 2006; Yamakawa et al., 2003), squeeze between the vascular loop's cells to leave the blood and fuse with others to form large active multinuclear (5 or more nuclei) osteoclasts each of which digs for the next 2-3 weeks. These pathfinding osteoclasts start tunneling through the damaged patch at a speed of about 25-40 μm/day. The osteoclasts first dig into the wall of the osteonal canal at a right angle and then turn to tunnel parallel to the osteonal columns. As they tunnel along they are followed by the vascular loop with new pre-osteoclasts squeezing out of its tip along with the nutrients and the oxygen they will need to do their job. Actually there are only about 10 osteoclasts in a BMU (R.B.Martin et al., 1998), but each osteoclast is a fusion of many individuals. Indeed an osteoclast is a very strange and often a very large beast—a sort of cellular collective in which nuclei actually come and go (e.g., for about 11 days in dog osteoclasts) while the collective with its ever-changing nuclei keeps tunneling ahead (R.B. Martin et al., 1998).

But what, besides VEGF, keeps the osteoclasts tunneling through a microcracked patch like "truffle hound" pigs after mushrooms? What are they looking for? What are the "mushrooms"? Remember that osteocytes are driven to self destruct by apoptosis when a microcrack cuts off their supply lines and they are hit by the huge strain spreading from the tip of the crack (Fig. 4). Moreover, you may recall that according to Klein-Nulent et al. (2003) and Smit et al.(2002) once the osteoclasts have started tunneling, the fluid sloshing of the nutrient-supplying/waste disposing mechanism stops at the tip of the cutting cone. This means that the osteocytes need oxygen. Thus, even the osteocytes that were not immediately cut off or hyperstrained by the microcrack are also gasping and may start making inducible CaR-dependently activated NOS-2 (Dal Pra et al., 2005) which in turn makes *excessive* amounts of NO that could induce neighbors to kill themselves. But of all things, the most important for answering this question is that deep down in their tiny souls osteoclasts are "professional" macrophages with a nano-vulture's taste for apoptotic carrion.

The answer to what drives osteoclast tunneling might go something like this, starting with an apoptosing osteocyte. The killer caspase 3 in an apoptosing osteocyte stimulates Ca^{2+}-independent phospholipase A2 which causes the cell to make lysophosphatidylcholine (LPC) and dump it out into the canalicular channel, which, when it seeps out of the mouth of the channel into the cutting cone, excites the osteoclast's tracking and digging machinery by activating the LPC receptors, GPR4 and G2A, on the osteoclast's surface just as it would on the surface of any macrophgage (Grimsley and Ravichandran, 2003). The dying osteocye also inactivates the "flippase" that has been keeping phosphatidylserine on the inner leaflet of the cell membrane but now activates a nonspecific bidirectional phospholipid "scramblase" that loads phosphatidylserine onto patches or scaffolds on the membrane's outer leaflet (deCathelineau and Henson, 2000; Grimsley and Ravichandran, 2003). Along with this, annexin I is translocated from the cytosol to the outer surface where it associates with the phosphatidylserine patches to produce an *"eat-me!"* sign. This is what the relentlessly digging osteoclast, driven by a rising "scent" of LPC, bumps into at the head of the trail. Then its PSRs (phosphatidylserine receptors) bind to the osteocyte corpse's phosphatidylserine•annexin I complexes. The signals from the vulturing osteoclast's PSRs trigger the mechanism with which the osteoclast "eats" the osteocye corpse by a macropinocytosis-like process that deCathelineau and Henson (2003) have called *efferocytosis* from the Latin prefix *effero* meaning to carry to the grave.

Things are different in trabecular (cancellous bone) and the endocortical surfaces facing the bone marrow (Fig. 1). There, the osteoclasts dig trenches instead of tunnels. Something remarkable happens when the lining cells on a trabecula are uncoupled from the underlying osteocytes. A kind of blister—the "bone-remodeling compartment (BRC)—forms on the trabecular surface (Hauge et al., 2001; Parfitt, 2001)! It appears that the aroused lining cells start making collagenase to remove the matrix cover and lift off the surface to form a kind of tent over the future work site. This blister forms a pseudoblood vessel with a wall of osteoprogenitors and lining cells instead of true CD34-bearing endothelial cells (Hauge et al., 2001). The blister vessel plugs into the local marrow blood sinusoids to bring pre-osteoclasts to the worksite. At first sight it might be assumed that the trabeculae and endocortical bone would enjoy a direct supply of pre-osteoclasts from the red marrow nursery. But while this may be true in children, the marrow in most peripheral bones (except, for example, lumbar vertebrae [Skripitz et al., 2000a)]) in adults is non-hemopoietic fatty yellow marrow so preosteoclasts must be shipped in via the blood, sometimes from distant places such as the residual islands of hemopoietic marrow in the upper femurs and the red marrow in the bones of the central skeleton including the ilium.

iv. Osteoclasts and the Dead Bone Puzzle

Before moving on I must mention the "Dead Bone Puzzle" which is important for those needing bone grafts. When the bone in the head of a femur is killed by having its blood supply atraumatically or traumatically cut off, the bone is not immediately destroyed in the living body. In fact, cadaver bone makes a biomechanically strong and stable allograft for replacing damaged bone. The basic structure of the dead bone is mechanically strong and potentially permanent! But if the dead bone comes in contact with living bone, blood vessels, macrophages and osteoclasts invade the dead bone and cause it to collapse by trying to dig out the dead bone and replace it with new living bone (Glimcher and Kenzora, 1979). The problem here for the graft recipient is the dangerously long time gap between the osteoclasts' rapid removal of the dead bone and the much slower filling of the excavations with new matrix by osteoblasts and the slowly completed mineralization of the new matrix. During this time gap the weakened dead bone is prone to collapse from muscle-induced strain.

v. The Fillers

"Normally, bone cells respond to mechanical loading by increasing their metabolism, activating genes, producing growth factors, and synthesizing bone matrix"— Ehrlich and Lanyon (2002).

"Osteoblasts don't do anything by themselves; they do their job in vivo as sheets of cells, they're all connected"— A. Caplan (in Davies, 2000, p.255).

v a. What Summons the Fillers?

As the the vulturing "truffle hound" osteoclasts are tunneling through the cortical bone looking for dying osteocytes and in the process digging out the microcrack and a large patch of bone around it by spraying the protein-shredding cathepsin K and mineral-dissolving HCl onto the bone, they release a pack of factors (e.g., bone morphogenic protein [BMP]-2, fibroblast growth factor (FGF)-2, insulin-like growth factors [IGFs]-I and –II, IGF-binding protein [IGFBP]-5, TGF-βs) that were deposited in the matrix 2 to 5 years earlier by the osteoblasts of an ancient BMU. It is widely assumed in the Bone Community that these liberated factors, particularly the FGF-2, and the IGFs, form an immediately accessible store of osteoblast generators that can be used by a subsequent BMU to locally increase osteoblast generation by stimulating osteoprogenitor proliferation to start new bone growth to fill the excavation. The TGF-βs would assist by stimulating ostoblastic stromal cells via Cbfa1/Runx-2 and TGF receptor-activated Smad proteins to stop further osteoclast maturation by stimulating OPG expression and reciprocally reducing RANKL expression (Takai et al., 1998; Thirunavukkarasu et al., 2000, 2001; Troen, 2003). However, they may just be litter from a past repair job—the remnants of the autocrine/paracrine factors that the past osteoblasts were using to stimulate themselves—that may be destroyed by the osteoclasts' cathepsin K and acid spray.

But the osteoclasts also leave a trail of chemokine droppings that attract osteoblastic cells. This cytokine is Mim-1 (Myb-Induced Myeloid protein 1) (Falany et al., 2001; Ponomareva et al., 2002; Troen, 2003). It is an acetyl transferase (Allen and Hebbes, 2003) that can lure pre-osteoblasts to the excavation site and expose their mature osteoblast-specific genes' promoters ("switch boxes") to the master osteoblast controller Cbfa1/Runx2's by acetylating histones and thus decondensing chromatin. (Ponomareva et al., 2002)

Serotonin may also be added to the new matrix by osteoblasts for release by the diggers in a future BMU. Osteoblasts and periosteal fibroblastic osteoblast precursors express serotonin receptors and have the machinery to release and reaccumulate it (Bliziotes et al., 2001; Westbroek et al., 2001). And Chopra and Anastassiades (1992) have reported that one of the matrix components—bone sialoprotein (BSP)—binds serotonin, receptors for which might be on the osteoblast primary cilium as they are in neurons in certain parts of the brain (Brailov et al., 2000; Hamon et al., 1999). Although what serotonin does in bone is unknown, it may another of the several factors that stimulate the proliferation of serotonin-receptor-bearing osteoprogenitor cells which in turn generate crack-filling osteoblasts (Westbroek et al., 2001).

A major player in the repair process is Ca^{2+}. A lot of Ca^{2+} (as much as 40 mM) is released and pumped out of the hole when the osteoclasts dissolve the bone mineral (hydroxyapatite) by directing a stream of H^+ (protons) into the hole with their massively expressed vacuolar H^+-ATPase and Cl^- counterions though Cl^- channels in the ruffled border (Blair, 1998; Stenbeck, 2002)— as I said a couple of paragraphs ago, they actually spray the bone mineral with hydrochloric acid (HCl)! They, like roving HCl-secreting gastric parietal cells, ingeniously get the protons for the HCl from water by first using their carbonic anhydrase to make H^+ $(HCO_3)^-$ (carbonic acid) by combining CO_2 with water, then they use their membrane $(HCO_3)^-/Cl^-$ exchanger to pump out the $(HCO_3)^-$ through their upper surfaces away from the hole and bring in Cl^- in exchange for the $(HCO_3)^-$ and finally pump the H^+ (protons) out with the electrogenic ATPase which in order to maintain electroneutrality pulls Cl^- hrough its channels into the hole and onto the hydroxy apatite. The resulting plume of Ca^{2+} is a pivotal two-way switch—it is an '*off*' switch for the osteoclasts and an '*on*' switch for osteoprogenitors coming behind them.

When the Ca^{2+} concentration in a tunnel or trench gets high enough, it activates Ca^{2+} sensors on the osteoclast's apical H^+- and protease-pumping finger-like ruffles sticking into the hole (Blair et al., 2002; Kameda et al.,1998; T. Yamaguchi, 2003; Zaidi et al., 1999). The resulting Ca^{2+} surges and signals cause the osteoclast to pull its acid-dripping fingers out of the hole, pull up the actin sealing ring it had put around the hole and glide over to another part of the cracked patch and start digging again. But if it cannot find another place to attach itself by its integrins in time to generate the matricrine survival signals needed to hold off apoptosis, it will self-destruct— a drastic response to unemployment and homelessness known as anoikis or homelessness-induced apoptosis (Frisch and Screaton, 2001; Lorget et al., 2000; Sakai et al., 2000; Stupack and Cheresh, 2002).

By contrast the increasingly sustained loud signaling from the Ca^{2+}-sensing receptors (which may or may not be identical the parathyroid gland cells' CaR) on the oncoming osteoblast precursors as they feel their way along the surface with their integrins (Brown and MacLeod, 2001; Chattopadhyay et al., 2004; Dvorak et al., 2004; Farzaneh-Far et al., 2000; Hinson et al., 1997; Pi et al., 1997; Tfelt-Hansen and Brown, 2005; T. Yamaguchi et al., 1998a,b, 2000, 2001) directs the cells to the high-Ca^{2+} excavation site, stimulates their proliferation by prolonged activation of the MAP kinase kinase ERK signal pathway and induces them to express BMP-2 and–4, which, in turn, stimulates the expression of Cbfa1/ Runx-2, the master osteoblast-specifying transcription factor that starts the post-proliferative stages of the bone-making process (Brown and MacLeod, 2001; Canalis et al., 2003;

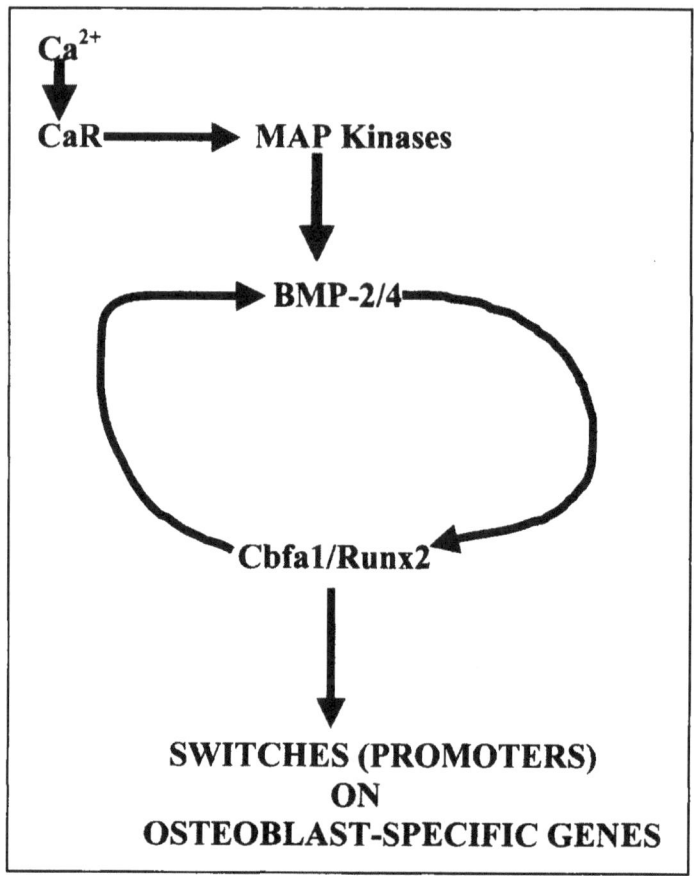

Figure 7. The signals from the CaRs (Ca²⁺-sensing receptors) sticking out of the cell membrane triggers the expression of BMPs (bone morphogenic proteins) which drive the development of osteoblasts by switching on the gene for Cbfa1/Runx-2 which in turn configures the promoters of various osteoblast-specific genes for activation by specific transcription factors. However, Cbfa1/Runx-2 also stimulates BMP expression thereby setting up a self-sustaining maturation cycle.

Chattopadhyay et al., 2004; Dvorak et al., 2004; Huang et al., 2001; Nakade et al., 2001; T. Yamaguch, 2003; T. Yamaguchi et al., 1998a,b, 2000, 2001) (Fig. 7). And the Cbfa1/Runx-2 in turn feeds back to further stimulate BMPs-2 and-4 because their promoters have Cbfa1/Runx-2 sites (Canalis et al., 2003) (Fig. 7). The signals from the Ca²⁺-sensing receptor may also stimulate the expression of PTHrP (parathyroid hormone-related protein), which, as we will see further on, is one of the pivotal osteoblast maturation drivers (MacLeod et al., 2002; T.J. Martin, 2005).

　　As the osteoclast-attracting osteocyte killing stops and the load–driven sloshing resumes, the Ca²⁺ sensor-driven stimulation of ERK (extracellular signal regulated kinases)1/2 in the cells stimulates the expression of eNOS and the production of NO (J. Rubin et al., 2003), the gaseous signaler which diffuses out of the cells to prevent local osteocytes from killing themselves and becoming osteoclast-attracting corpses. This NO also stimulates the enzyme, guanylyl cyclase, in nearby vascular cells by attaching to the enzyme's heme group (Lincoln et al., 1997). The cyclic GMP produced by the cyclase relaxes arterioles, the overall effect of which is to increase the delivery of blood-borne nutrients through the advancing blood capillaries to feed the filler crews (Lincoln et al., 1997). The NO-stimulated cyclic GMP in the osteoblastic cells also stimulates the cells' PKG (protein kinase G) which in turn may, like the Ca^{2+} sensor signals, also activate the ERKs 1/2 which in turn activate the cFos/Jun (AP-1) transcription factor complex and the various matrix-related genes it switches on (Pfeischrifter et al., 2001; Zaragoza et al., 2002). And this in turn also shuts down the expression of RANKL which turns off osteoclast generation (Rubin et al., 2003)

v b. Where Do the Fillers Come from and Where Do They Go?

This is hopefully all very interesting, but where do the osteoblasts come from to make a new Haversian canal surrounded by concentric layers of bone? They come from a group of constantly proliferating spindle-shaped mesenchymal cells that follow the osteocyte cadaver-hunting osteoclasts along the walls of the cutting cone driven by the various mitogenic and chemotactic factors mixed with the large amounts of Ca^{2+} and phosphate released from the bone by the osteoclasts to make a tasty osteogenic "soup" (Bianco et al., 1993; M.J.Doherty and Canfield, 1999; R.B. Martin et al., 1998; Parfitt, 2000a, 2000b, 2001; Schaffler, 2000). The band of proliferating mesenchymal cells advancing along the tunnel wall throw off preosteoblasts that dismantle their mitogenic machinery and become osteoblasts which start depositing layers or lamellae of bone around the tunnel walls. Unlike their progenitors, the new osteoblasts stay in place to lay down concentric 15-mm thick strips of bone around the tunnel wall while the proliferating progenitors move on behind the osteoclasts throwing off new bands of osteoblasts as they go. The result of this is the layering of lamellae around the new blood vessels and nerves.

The question remains as to whether one generation or band of osteoblasts fills the tunnel (except of course the vascular "right-of way") taking rest breaks between each pair of lamellae or whether several generations of osteoblasts are needed do the job which would mean that osteoblasts are still being generated by progenitors on the new surfaces. Probably several osteoblast generations are needed for the refill job (Martin et al., 1998). As the refilling progresses, the rate of formation drops for at least two reasons. First, as the lamellar build-up approaches the central blood vessel, the shrinking space somehow interferes with osteoblast differentiation. Second the progressive drop in stresses with the increasing lamellar build up results in a fading of the strain signals needed for osteoblast differentiation that were at their peak in the microcrack at the head of the cutting cone (R.B. Martin et al., 1998).

At a repair site on the trabecular or endosteal surface, the osteoblasts come from the preosteoblasts from the adjacent marrow and the lining cells that make up the canopy of the repair "blister" mentioned above (i.e., the BRC) canopy (Hauge et al., 2001).

If any of the "legacy" BMP-2 liberated by the osteoclasts from the matrix has escaped being shredded by the osteoclasts' proteases, it would kick osteoprogenitors along the osteoblast differentiation pathway (Chen et al., 1998). They would then be stimulated to start working by the signals from other receptors which include the CaRs in the caveolar pouches in their cell membranes (Brown and MacLeod, 200; Kifor et al., 1998) and the type 1 PTH/PTHrP (PTHR1) receptors (Aubin 1998, 2000, 2001; Aubin and Triffitt, 2002; McCauley et al., 1996), the expression of which shoots up about 10-fold during the transition from mature osteoprogenitor to preosteoblast.

But wait that is not all! I have missed something very important made by vascular endothelial cells of the blood vessels accompanying the osteoprogentiors. That something is endothelin-1 which binds to, and activates, the osteoprogenitors' endothelin-A receptors (Borcsok et al., 1998; Guise and Mohammad, 2004; Mohammad and Guise, 2003; Kasperk et al., 1997; Mohammad et al., 2003; Veillette and von Schroeder, 2004; von

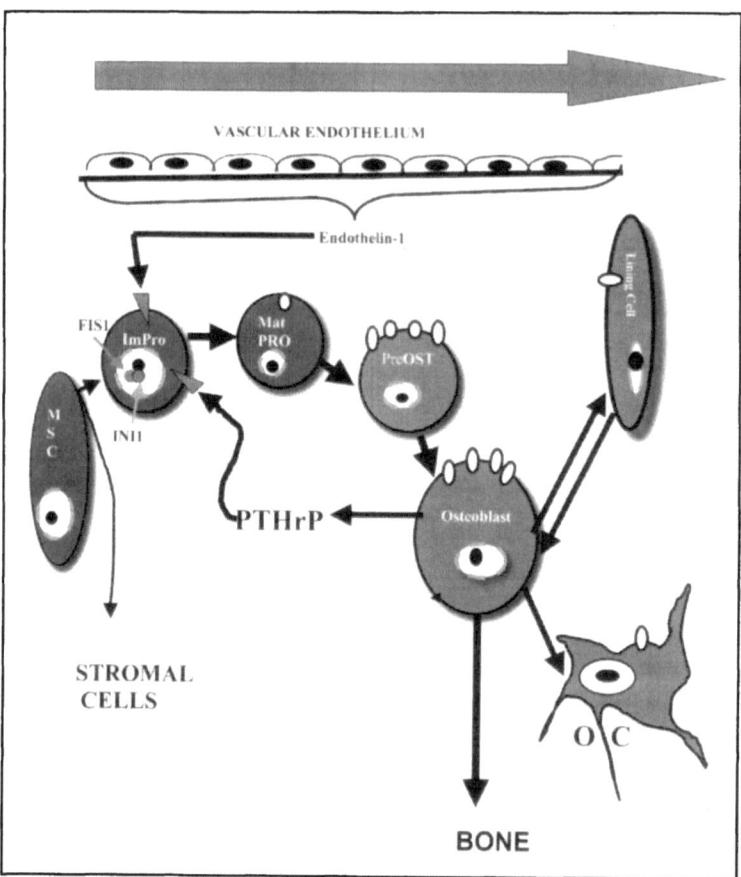

Figure 8. The coupling of vascular budding to osteogenesis—the endothelin-PTHrP cycle. Endothelial cells produce endothelin-1 which binds to endothelin A receptors (**red triangles**) on Imm Pro cells and stimulates the proliferation and maturation of these transit-amplifying osteoprogenitor cells. When the progeny of these stimulated progenitors become osteoblasts, they make and secrete PTHrP which can also stimulate Imm Pro cells' endothelin A receptors with its N-terminal 6-9 (Leu-Met-Asp-Lys) region. This establishes an osteogenic cycle until it is stopped by the newly generated mature osteocytes making SOST brakes. **Imm Pro**, immature ostoprogenitor cell; **MSC**, multipotenet mesenchymal stem cell; **OC**, osteocyte; **PreOst**, preosteoblast. A color version of this figure is available online at www.eurekah.com.

Schroeder et al., 2003) (Fig. 8). Specifically, very low concentrations (e.g., 10^{-10} M) of endothelin-1 from the endothelial cells of the new blood vessels following osteoclasts can stimulate the endothelin-A receptors on osteoprogenitors with its 6-9 (LMDK; Leu-Met-Asp-Lys) region and drive osteoblast production. But this is not all to this mechanism! We will see further on that the 8-11 region of the osteoblast maturation-driving PTHrP when it starts being made by maturing osteoblasts can also stimulate the endothelin A receptor and with this further promotes pre-osteoblast proliferation, differentiation and bone formation (Mohammad et al., 2003; von Schroeder et al., 2003) (Fig. 8). *Thus, endothelin-1 and its endothelin-A receptor is a major device for BMUs to couple blood vessel extension to new bone formation when refilling an excavation site..*

Signaling by glutamate from the dying or hypoxic osteocytes or from invading or adjacent nerves is also needed. However, the osteoblasts also release glutamate using exactly the same vesicular release machinery as neurons (Bhangu et al., 2001). As we learned above, the osteoblasts are also equipped with neuron-like NMDA-type glutamate receptor/channels (Gu et al.,2002; Hinoi et al., 2003; Skerry and Taylor, 2001; Spencer et al., 2004), AMPA/Kainate ionotropic receptor channels as well as the G-protein-coupled metabotropic mGluR1b receptors with their 7 transmembrane a-helices which belong to the same GPCR (G-protein-coupled receptor) family as the Ca^{2+} receptors (Gu and Publicover, 2000; Hinoi et al., 2001; Skerry and Taylor, 2001; Spencer et al., 2004). Moreover, they have Na^+-dependent GLAST (EAA1) transporters to suck the glutamate back into the cell just as they do in neurons to keep the external glutamate low which keeps the signaling noise down and prevents excessively extended receptor activation by released glutamate (Chenu, 2002; Mason et al., 1997). The activated mGluRs, like activated Ca^{2+} sensors and the related PTHR1 receptors, trigger a burst of PLC (phospholipase C) activity which triggers a prompt release of Ca^{2+} from internal stores followed by the flow of Ca^{2+} through opened membrane channels and a surge of PKCs (protein kinase Cs) activity (Gu and Publicover, 2000).

The importance of activated NMDA receptor/channels at a critical point in the osteoblast maturation program for driving post-proliferative osteoblast differentiation by stimulating the expression and possibly the transport of the master osteoblast differentiation driver Cbfa1/Runx-2 to its target genes has been demonstrated by Hinoi et al (2003) using freshly isolated rat calvarial osteoblastic cells. Thus, NMDA receptor/channel inhibitors such as MK801 interfere with the binding of Cbfa1/Runx-2 to its DNA targets and thus block the expression of key osteoblast genes such as alkaline phosphatase and osteocalcin and ultimately the arrival of the cell at the stage where it accumulates Ca^{2+} for matrix mineralization (Hinoi et al., 2003). The NMDA receptor/channels on the calvarial osteoprogenitors in Hinoi et al's and others' experiments (A.F. Taylor, 2002) must have been continuously stimulated by the 500M glutamate in the a-MEM culture medium with no relief being possible from the GLAST transporters as would be the case in a bone. (*This incidentally suggests the extremely important and as yet unappreciated possibility that glutamate has been a covert signaler for osteoblasts in standard culture media.*) In a bone, the osteoblastic cells' NMDA receptor/channels would only be stimulated by glutamate boluses from glutamatergic nerves responding to strain or to glutamate released by damaged and dying osteocytes in microcracks.

Despite this, there are some nay-sayers about the importance of glutamate as an osteogenic osteotransmitter. Although Dobson and Skerry (2000) and Taylor et al. (2000) have found that inhibiting the glutamate receptors reduce bone formation in vivo and in vitro, Gray et al. (2001) have claimed that high doses of NMDA receptor/channel inhibitors did not affect bone formation by cultured rat osteoblasts and moreover bone formation in GLAST transporter knock-out mice was normal. But Skerry et al. (2001) have shot back with a list of serious shortcomings of Gray et al.'s GLAST knockout mice which I will not repeat here. Osteoblasts also have AMPA/kainate receptors that also activate glutamate receptor/channels (Skerry and Genever, 2001) and could take over from blocked NMDA receptor/channels, and we do not know whether GLAST activity and glutamate re-uptake are as critical for bone making as they are for neurotransmission. One of the most serious errors in Gray et al.'s experiments is their use of *competitive* NMDA receptor/channel antagonists. These antagonists simply could not compete with the high levels of glutamate in culture media that support the growth of osteoblastic cells (A.F. Taylor, 2002).

The upshot of all of this is that osteoblast differentiation and osteocyte functions, such as the sensing and memorizing of local strain pulse frequency, involves glutamate and to do their job these cells use exactly the same signaling machinery and transmitters as central neurons (Skerry and Taylor, 2001; Spencer et al., 2004; Turner et al., 2002). In other words it looks as if we have been looking at "osteoneurons"—what a fantastic change in our view of these cells and the smart bones they make!!

These unexpected neural similarities are not restricted to glutamate receptors and transporters! Active osteoblasts also make BDNF (brain-derived neurotrophic factor) and the TrkB receptors needed to bind it (Yamashiro et al., 2001). Of course you might say that this BDNF is simply used to stimulate the innervation of new bone. However, the osteoblasts' TrkB receptors mean that *they too* use their BDNF to stimulate themselves and their neighbors for some aspect of bone making.

Before the osteoblasts arrive on the scene to start filling in the trench or tunnel, the lining cells must sweep up the litter left by the untidy osteoclasts on the resorption cavity floor (Everts et al., 2002). There are collagen "bristles" sticking out of the cavity floor. The lining cells then move onto the surface to give it a good shave and make a smooth surface upon which they slap a layer of osteopontin to glue the new collagen that will be made by the incoming osteoblasts (Everts et al.,2002; Mazzali et al., 2002; McKee and Nanci, 1996). As we shall see below, the residual phosphate from the dissolved bone mineral in the resorption cavity could be the stimulator of this first burst of osteopontin expression (Beck, 2003; Beck and Knecht, 2003; Beck et al., 2000, 2003). These lining cell janitors are in fact reversibly retired osteoblasts who when stimulated by agents such as PTHs can revert to full osteoblasthood to give a first wave of osteoblasts to "kindle" bone building.

As mentioned above, the strained osteocytes in the microcracked site make and release NO gas, but it cannot save those Bax-makers closest to "ground zero" (Fig. 4). But it increases arteriolar dilation by stimulating the vascular smooth muscle cells' soluble guanylyl cyclase and cyclic GMP production which in turn drop the contraction-driving Ca^{2+} signaling (Tiyyagura et al., 2004). And it stimulates the movement of osteoclast precursors into the cortical tunnels and trabecular and endocortical trabecular blisters, and it also stimulates the migration of osteoblasts into the site (Afzal et al., 2000).

While there were as few as 10 osteoclasts doing the digging, there may be hundreds of osteoblasts working in a tunnel or trench (R.B. Martin et al., 1998). Despite their numbers, the osteoblasts take about 4-8 times longer to fill the cortical tunnels and trabecular "blister"-covered trenches than the osteoclasts took to dig them. Therefore, at any moment in the microfracture sites, a mature bone has holes that are being dug or have been recently dug by the microcrack-repairing BMUs—these holes together make up the *remodeling space* with slowly mineralizing, hence initially weak, nascent bone. As the osteoblasts are slowly filling these holes with new factor-loaded matrix ('osteoid'), some of them will be trapped in it like insects in amber as it is gradually mineralized with apatite-like Ca-phosphate. When the large amount of Ca^{2+} in the excavation site has been used by the osteoblasts to mineralize the new matrix and the filling has been finished at the endocortical and trabecular worksites, the blister canopy collapses and the old lining cells together with new ones from a lucky few osteoblast survivors connect to the entrapped osteocytes and cover the repaired patch (Hauge et al., 2001). Walled up in their matrix-lined cells they reduce their PTHR1 receptor density (Aubin 1998, 2000,2001; Aubin and Triffitt, 2002) and become osteocytes that connect to the osteointernet and start sending messages from their bending cilia and tugged and twisted integrins to the overlying lining cells and from them to adjacent marrow stromal cells and Haversian blood vessels to again summon BMUs to repair damage when needed or mobilize Ca^{2+}if needed to restore the circulating level (Marotti, 1996; Martin, 2000a, 200b, 2002; Hauge et al., 2001; Skerry, 1999; Whitfield et al., 1998b; Yellowley et al., 2000).

When the new patch is finally in place 3-9 months later and the osteocytes start sending the signals needed to prevent lining cells from unnecessarily summoning osteoclast precursors from the blood vessels running through the cortical Haversian canals or from forming a BMU-activating blister on endosteal or trabecular surfaces, the members of the last osteoblast crew are now out of work—they have become redundant! Those that have failed to find a free space onto which to attach and become lining cells or osteocytes to keep their surface-sensing integrins emitting their matricrine survival-promoting signals trigger apoptosis and self-destruct like the unemployed osteoclasts before them (Frisch and Screaton, 2001; Stupack and Cheresh, 2002; Whitfield et al., 1998b, 2000a). However, the doomed, self destructing osteoblasts make one last contribution to bone formation, specifically mineralization, by dumping alkaline phosphatase along with Ca-apatite crystals into the new matrix in vesicles released from their blebbing surfaces (Farley and Stilt-Coffing, 2001). Some of this alkaline phosphatase will get into the blood to become one of the serum markers that tell the outside World of bone being made and osteoblasts dying .

Most of the self-destructing (apoptosing) cells in bone are located in microcrack sites being repaired by BMUs. The large amount of inorganic phosphate (Pi) released along with Ca^{2+} from the apatite crystals dissolved by the osteoclasts (Gupta et al., 1996) may increase the chance of osteoblasts triggering apoptosis unless they are attached and protected by survival factors such as Bcl-2 (Allen et al., 2002; Meleti et al., 2000). The reason for this is likely to be the Na^+ gradient-powered pumping of P_i into the osteoblasts by their Pit 1 and 2 type III Na^+/ P_i transporters, particularly the Pit 1 transporter that they selectively upregulate during maturation for matrix mineralization (Adams et al., 2001; Meleti et al., 2000; Nielsen et al., 2001 Takeda et al., 1999). When the phosphate reaches a critical level, it would be carried into the mitochondria by the Pi^-/OH^- exchanger and trigger the assembly of the huge mitochondrial permeability transition pore through which cytochrome-c, AIF (apoptosis-inducing factor), procaspase-9 and Diablo leak into the cytoplasm from the intermembrane space. The appropriately named and deadly Diablo neutralizes a group of caspase inhibitors and cytochrome-c forms a complex with APAF-1, which causes caspase-9 to auto-activate and trigger a cytocidal cascade of so-called "executioner" caspase protein shredders (Crompton, 2000; Finkel, 2001). (This ability of P_i to greatly promote apoptogenesis was first noticed by me nearly 40 years ago, when I was searching for a way to enhance the inducibility of apoptosis by irradiated thymic lymphocytes and human peripheral blood lymphocytes for an ultra-sensitive radiation bio-dosimeter [e.g., Whitfield et al., 1967]). Meleti et al. (2000) have shown that inhibiting the transporter with low concentrations of phosphonoformic acid (PFA) prevents Pi from killing primary human osteoblasts and Mansfield et al. (2001) have shown that PFA also prevents Pi from triggering apoptosis in chondrocytes. It follows from this that inhibiting the Na^+/Pi transporter might prolong the lifetime of osteoblasts by shielding them from the high Pi in their workplace. However, as we shall see later on, this could be dangerously counterproductive because the transporter is also a key player in osteoid mineralization.

The apoptogenic action of Pi in cultured human osteoblasts from bone fragments and murine MC3T3-E1 preosteoblasts is greatly enhanced by a small, itself harmless, increase (0.1-1.0 mM) in the Ca^{2+} concentration in the culture medium (Adams et al., 2001). Indeed it has been shown that both proliferation and the phosphate transporter in osteoblastic cells are stimulated by Ca $^{2+}$ (Schmid et al., 1998). Therefore, it is likely that the high Ca^{2+} concentrations in the excavation sites would stimulate the proliferation of osteoprogenitors as well as promote Pi-induced apoptogenesis of both them and mature post-proliferative osteoblasts unless the cells are protected by some anti-apoptosis protein such as autocrine and paracrine IGF-I, the expression of which is stimulated by the signals from the osteoblasts' PTHR1 receptors activated by the PTHrP that is normally expressed alongside the receptors (Aubin, 1998, 2000, 2001; Aubin and Triffitt, 2002) or by an injected PTH (Allen et al., 2002) (Fig. 9). It seems likely that the burst of phospholipase-C activity caused by activated CaRs would produce a surge of IP_3 from membrane phospholipid breakdown and a highly localized release of Ca^{2+} (Ca^{2+} "hotspots") from IP_3 receptor-bearing endoplasmic reticulum Ca^{2+} stores lying within or beside clusters of mito-

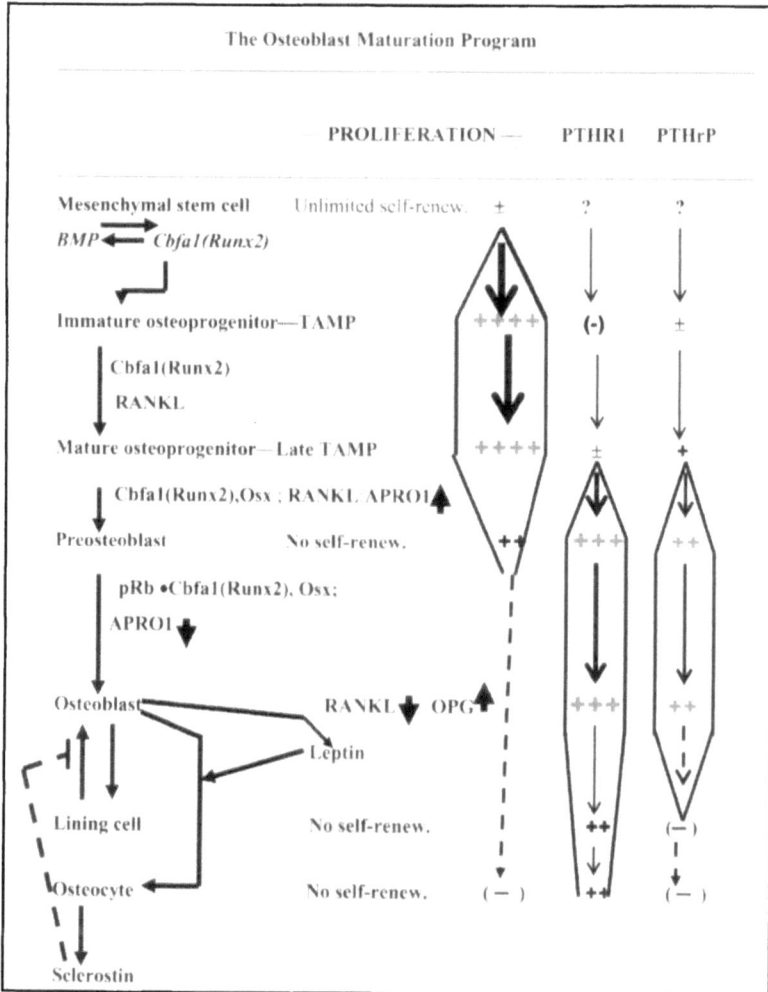

Figure 9. The relation between proliferative activity and PTHR1 and PTHrP expressions at the various stages of osteoblastic differentiation, using the information provided by Aubin (2000, 2001) and Aubin and Triffitt (2002) as the foundation of the scheme. The appearance and disappearance of the various other things as the differentiation progresses are discussed in the text. The message from this figure establishes the fact that to understand a PTH's bone-building action it is necessary to know that only the terminally differentiated, permanently proliferatively shut-down mature osteoblasts are the maximal expressors of PTHR1 receptors and neither they nor post-osteoblastic osteocytes and bone-lining cells can be induced to start proliferating. **TAMP**, transit amplifying

chondria (Hajnóczky et al., 2000, 2003a, 2003b; Mannella, 2000; Pozzan et al., 2002; Rodlan, 2003). This released Ca^{2+} would enhance Pi action by jumping on the Ca^{2+} uniporters of the mitochondria clustered around the mouths of the IP_3 –opened channels in the endoplasmic reticulum and riding the uniporters into the mitochondria to help Pi stimulate the formation of permeability transition pores for cytochrome c to leak into the cytoplasm to trigger the lethal apoptogenic caspase cascade unless the cell is sufficiently protected by a protein such as Bcl-2 (Hajnóczky et al., 2000, 2003).

Finally, osteoblasts are recruited in load-bearing mature human bones only for remodeling BMUs to repair microcracks. But the osteoblast recruitment and osteoclast activity are not coupled in growing bone. Clusters of osteoblasts operate independently from osteoclasts in the growing, so-called *modeling*, bones of rat pups and human children (Frost, 1997; R.B. Martin et al., 1998; Pugsley and Selye, 1933; Selye, 1932). And as we shall see below PTHs can stimulate a massive layering of osteoblasts on trabeculae and bone formation *without a prior activation of osteoclasts*. Indeed growing mutant mice, which cannot generate functional osteoclasts, become osteopetrotic ("marble-boned") due to unopposed bone building by osteoblasts; and mutant mice that cannot generate functional osteoblasts become the opposite, osteoporotic, due to the unopposed osteoclast activity (Karsenty, 1999; Whitfield et al., 1998b).

v c. Heterotopic Osteogenesis I: Deadly Boney Blood Vessels

"The emerging view is that plaque calcification represents a meeting of bone biology with chronic plaque inflammation."—T.M. Doherty et al (2003a)

"…we see ossification declare itself in precisely the same manner as when an osteophyte forms on the surface of bone.....the osteophytes of the inner table of the skull…follow the same course of development as the ossifying plates of the internal coat of the aorta and even of the veins …"—R. Virchow (1863/1971, p.408).

At this point we should look at some very dangerous amateur osteoblasts which don't live in the skeleton, but instead build bone where they shouldn't in aortic atherosclerotic plaques, coronary arteries, heart valves and in arteries and in veins that cause limb amputations in diabetics (Doherty et al., 2004; Vattikuti and Towler, 2004). These are the so-called CVCs (calcifying vascular cells), that are derived from arterial SMCs (smooth muscle cells) and vascular pericytes (Collett and Canfield, 2005; Doherty et al., 2004; Vattikutti and Towler, 2004). There are also the migratory adventitial myofibroblasts that generate the medial CVCs (Shao et al., 2005). Obviously these cells also would be expected to be stimulated to make bone in the walls of blood vessels by potent bone anabolics such as the PTHs, a frightening possibility that will be addressed in Chapter 5 ix and schematically represented in Figure 41.

Before going on, it is important to examine some of the functional consequences of vascular bone. Aortic ossification increases the risk of myocardial infarction, heart failure and death by impairing coronary blood flow (Boström and Demer, 2000). Coronary blood flow is normally driven by the diastolic elastic recoiling (collapse) of the aorta from its systolic ballooning which, like a second heart operating by the "Windkessel" effect, pumps blood into the the coronary vascular network. The rigid wall of a bone-lined aorta cannot recoil—it won't "balloon" like a normal aorta during systole and then pump the residual blood into the coronary vessels by elastically recoiling or collapsing (Boström and Demer, 2000; O'Rourke et al., 2002).

Another problem created by ossified vascular plaques are the biomechanically unstable interfaces or seams between hard ossified and soft regions of the plaque. These interfaces are sites of high shear stress that in a pulsing blood vessel tear and rupture the plaque which blocks the vessel or releases pieces and clots that can block distant blood vessels. The risk of rupture should therefore rise to a peak along with the number of plaques and the extent of the hard/soft plaque interfaces but then crashes when the rigid plaques coalesce into a seamless sheet of bone (Abedin et al., 2004). Therefore the risk of plaque rupture is biphasic and the coalescence of ossified plaques into a boney sheet may actually stabilize a blood vessel. This would be why Hunt et al. (2002) found that patients with heavily calcified carotid plaques had fewer strokes and transient ischemic attacks (TIAs). In other word a few ossified plaques might rupture in one of your carotids and cause a stroke or TIA. But the good news is that if you survive until there is a lot of plaques that are big enough to fuse and eliminate the strain-prone interfaces, you will have a tough, armor-plated carotid blood vessels, However, because of your now recoilless aorta the blood flow through your coronary blood vessel will be falling toward infarct and heart attack.

There are two kinds of vascular ossification mechanism. The first resembles the cartilage-independent, intramembranous bone formation and operates in heart valves and the medial vascular ossification of types 1 and 2 diabetics (Doherty et al., 2004; Vattikutti and Towler, 2004). The second is atherosclerotic ossification which resembles perfectly normal endochonral bone formation. It involves a typical preliminary formation of cartilage which is chewed up by osteoclasts and replaced by the bone-making vascular smooth muscle cells in osteoblast attire (Doherty et al., 2004; Fitzpatrick et al., 2003; Vattikutti and Towler, 2004). Endochondral bone formation is discussed in much more detail in Chapter 5v.

We will focus on atherosclerotic ossification. The atherosclerosis story, in a coronary artery for example, begins with injury-induced accumulation of low density lipoprotein complexes (LDLs) in an arterial wall where their lipids are oxidized and their apolipoproteins are glycated (Libby, 2002; Steinberg, 2002; Stocker and Keaney, 2004). The modified LDLs send distress signals that cause endothelial cells to display molecules such as E-selectin that snare passing monocytes and T-lymphocytes (as described above in Chapter 2 iii (p. 24) for the trapping of preosteoclasts in bone blood vessels) which collect in the vessel wall. There the monocytes mature into macrophages which, while gorging on the altered LDLs to become fat globule-loaded "foam cells", produce pro-inflammatory cytokines that stimulate the T-cells and vascular smooth muscle cells (VSMCs) (look ahead to Fig. 41). The excited VSMCs proliferate, crawl under the endothelium lining and build a relatively fragile matrix canopy over the fulsome nest of foam cells and T-cells. The swelling mass of foam cells in this nest make enough matrix metalloproteinases to cut through the cover and cause an artery-blocking clot or embolism.

But of interest here are the responses of potential osteoprogenitors in the wall of such blood vessels that are the CVCs and VSMCs. The plaque endothelial cells, stirred up by proinflammatory cytokines such as IL-1β and TNF-α make factors which target the CVCs and the VSMCs. The VSMCs respond with dramatic phenotypic changes (Walsh and Takahashi, 2001). They downregulate transcription factors such as GATA-6 and gax that were holding them in a proliferatively quiescent, differentiated state and start expressing BTEB2 and Egr-1 transcription factors that drive proliferative activation. They reduce the expression of the SM1 and SM2 myosin heavy chain isoforms needed for differenetiated function and start expressing the non-muscle Smemb isoform of a more primitive state (Walsh and Takahashi, 2001). These cells are pushed into osteoblast maturation by BMP-2 and BMP-4 made by the inflammation-factors-driven endothelial cells and lay down real bone that bone-plates nests of foam cells—they literally make armored blood vessels (Bunting, 1906; (Doherty et al., 2002, 2003a, 2003b, 2004; Fukui et al., 2003; Jakoby and Semenkovich, 2000; Virchow 1863/1971) (look ahead to Fig. 41)!

The BMP-2 from the endothelial cells is likely the principal driver of vascular ossification (Hruska et al., 2005; Shao et al., 2005). According to the model of Shao et al. (2005) for diabetic ossification, the BMP-2 causes adventitial myofibroblasts to express Msx2 which in turn causes the cells to make and secrete Wnt glycoproteins (which will be discussed in far more detail in Chapter 5 iiif) (Nusse, 2005; Westendorf et al., 2004) (look ahead to Fig. 27). The Msx2 also downregulates a factor called Dkk-1 (Dickkopf, German for thick head because when injected into a four-cell frog embryo it produces embryos with big heads; Nusse, 2001,2005). A secreted Wnt binds to both the single-pass LRP (low-density lipoprotein (LDL)-receptor-like protein) and to the 7-transmembrane-pass Frizzled receptor and the acti-vated receptors enable β-catenin surge into the nucleus and stimulate the expression of a large cluster of target genes the products of which trigger the osteogenic cascade in the CVCs and VSMCs (Nusse, 2001,2005).

But the stirred up endothelium and VSMCs also express endothelin-1-converting enzyme and make endothelin-1 (Corti et al., 2003; Fei et al., 2000; Iwasa et al., 2001; Minamino et al., 1997). If the VSMCs also express the endothelin A receptor, the endothelin-1 could also stimulate them to proliferate, osteoblastically differentiate and make bone (Börcsök et al., 1998; Guise and Mohammad, 2004; Mohammad and Guise, 2003; Kasperk et al., 1997; Mohammad et al., 2003; Veillette and von Schroeder, 2004; von Schroeder et al., 2003) (Fig. 8).

But there is another cell-driven mechanism by which blood vessels are first mineralized which in turn triggers the trans-differentiation of VSMCs into osteoblasts and ossification. This mechanism is triggered by the high inorganic phosphate concentration in the bloods of atherosclerotics, diabetics and uremics and the circulating high phosphate and Ca^{2+} levels (i.e., high calcium X phosphate product) in ESRD (end-stage renal disease) (Shanahan, 2005). If the VSMCs are damaged by loading up with high Ca^{2+} and phosphate, they will die apoptotically and release calcium-phosphate-loaded apopotic bodies while other cells may avoid dying by releasing calcium-phosphate-loaded vesicles like those released by mineralizing osteoblasts (Shanahan, 2005). If these vesicles are not promptly swept up they will spill out their basic calcium phosphate and directly mineral-plate the vascular elastic lamina and collagen fibrils.

This release of calcium-phosphates can also trigger the trans-differentiation of surrounding undamaged VSMCs into functioning osteoblasts which start building bone. When the VSMCs are exposed to high phosphate concentrations they shed their muscular identity and start behaving like osteoblasts (Jono et al., 2000; Steitz et al., 2001). They lose their specific smooth muscle lineage markers and α-actin and then retool themselves for making bone just like normal osteoblasts (Stein et al., 1996). They express the master osteoblast transcription factor Cbfa1/Runx-2, make type I collagen, alkaline phosphatase and then ostocalcin and osteopontin when mineralizing the matured matrix (Doherty et al., 2002, 2003a, 2003b, 2004; Jono et al., 2000; Steitz et al., 2001). A rather worrying, but yet very strong, support for what follows further on in our discussion in Chapter 5 of PTH's cyclic AMP(cyclic adenosine 3',5' monophosphate)-triggered bone-building action is the ability of a short exposure to cyclic AMP to cause VSMC-CVCs to stop proliferating and assume a classical cuboidal osteoblast shape and osteoblast functions (Tintutt et al., 1998).

Presumably the danger of vascular ossification should be far less in osteoporotics who are losing skeletal bone. They should also lose vascular bone, shouldn't they? But they don't! Elderly osteoporotic post-menopausal women are surprising prone to ossifying their arteries while deossifying their skeletons—the "calcification paradox" (Demer, 2000; Demer and Tintut, 2003; Iba et al., 2004; Parhami et al., 1997; Rubin and Silverberg, 2004; Schultz et al., 2004; Wallin et al., 2001). Indeed according to Schultz et al (2004) *"aortic calcifications are a strong predictor of low bone density and fragility fractures"*. A murine model of this ossifying blood vessels-resorbing skeleton paradox has been provided by Bucay et al.(1998, 2006) with OPG knockout mice which, while developing severe trabecular and cortical bone porosity, thinning their parietal skull bones and fracturing their bones, at the same time ossify their aortae and renal arteries. Parhami et al.(1997) have suggested that the reason for the "calcification paradox" is that sub-endothelial minimally modified LDLs actually stimulate the re-differentiation of arterial VSMCs into osteovascular CVCs, but they inhibit the differentiation of skeletal preosteoblasts.

v d. Heterotopic Osteogenesis II: FOP—Rampaging Osteogenesis and Second Skeletons

"These people aren't just forming little bones here and there. They are forming a whole extra skeleton. It doesn't necessarily look like the first one , but that's what it is…It behaves like normal bone—if it bears weight it gets denser, and if it doesn't bear weight, it becomes

osteoporotic. If you break it, it heals, just like a normal fracture. It even contains marrow. It's normal in every way except one: it shouldn't be there. "— F. Kaplan (quoted by T. Maeder in The Atlantic Monthly online, February 1998, p.6).

A person of either sex or any race or from any part of the World who has the gene for FOP (fibrodysplasia ossificans progressive) seems normal at birth except for short great toes without skin creases because of having only one instead of two phalanges (Conner and Evans, 1982; Kaplan and Smith, 1997; Zasloff, 1980). While these ominous toes are of course noticed by parents they are not regarded as indicating anything serious until catastrophic osteogenesis begins later in the child's neck and back (Connor and Evans, 1982; Kaplan et al., 2004; Maeder, 1998; McCarthy and Sundaram, 2005; Shore et al., 2006).

In general terms this condition is characterized by the replacement of soft connective tissue by mature bone without changes in the normal serum calcium and phosphate levels. It may begin as a large soft highly vascularized fibroproliferative growth involving tendons, ligaments fascia and skeletal muscle that is immediately assumed to a be tumor and accordingly biopsied. But this shouldn't be done because FOP tissues are like loaded osteoguns that fire in response to any trivial injury by causing skeletal muscle, VSMCs and other mesenchymally derived cells to retool themselves to start making bone—a crazy effort to produce protective armor plating. Axial, cranial and proximal parts of the body ossify first followed by ventral appendicular caudal and more distal parts of the body (Kaplan et al., 2004).The spreading bone formation in response to biopsies, dental procedures, falls, surgical efforts to remove excess bone nodules, intramuscular injections, and even routine immunization injections eventually result in all of the major joints, including the jaws, freezing and ultimately the conversion of the still vivdly conscious person into a living, frozen boney statue able to survive only on a liquid or semi-solid diet (Kaplan et al., 2004; Lanchoney et al., 1995; Maeder, 1998).

What causes this disastrous osteogenic responsiveness to even the most trivial injuries? It is an awesome demonstration of the ability of BMPs to trigger osteogenesis. It is driven by a hyperproduced, though perfectly normal, BMP-4 (Kaplan and Shore, 1998; Kaplan et al., 2004; Shafritz et al., 1996). But the problem is that hyperproduced BMP-4 in a FOP person has a reduced ability to limit itself by triggering the expression of its antagonists gremlin and noggin as happens in normal people (Gazzero et al., 1998; Kaplan et al., 2004; Pereira et al., 2000a, 2000b).

The very recently discovered core cause of FOP is a mutant type I BMP receptor, ACVR1(Shore et al., 2006). The gene for ACVR1 is on chromosome 2q 23-24 (Shore et al., 2006). BMP receptors are activated when a BMP such as BMP-4 binds to the type I receptor chain which causes the types I and II receptor chains to pair (dimerize) (Massague, 1998; Seebald et al., 2004). The type II chain is an always active S/T (serine/threonin) protein kinase which when it mates with the type I chain phosphorylates the type I chain's "G-S" (glycine-serine) region. This activates the type I chain S/T protein kinase which phosphorylates associated Smad proteins. The different phosphoSmads then move into the nucleus where they turn on the Cbfa1/Runx2 and other osteoblastic driver genes. Other targets of the Smads are the promoter-switchboxes of the BMP-4 gene and the genes encoding the BMP-4 antagonists gremlin and noggin (Gazzero et al., 1998; Pereira et al., 2000a, 2000b). The mutant FOP ACVR1-receptor chain has undergone a G→A nucleotide shift in its "G-S" region which must somehow distort Smad signaling so that the gremlin and noggin stops cannot be put on. .

It seems likely that the trigger of a spreading osteogenic explosion is started by inflammatory response cells such as lymphocytes and mast cells as well as local muscle cells all expressing large amounts of BMP-4 at the site of a perhaps unnoticed trivial injury (Kaplan et al., 2004). This triggering by inflammatory cells is really just the same as the triggering of vascular ossification by the proinflammatory cytokines from foam cells in an injured blood vessel wall which drive overlying vascular endothelial cells to produce osteogenesis-triggering BMP-2 (Hruska et al., 2005) (look ahead to Fig. 41). The local flood of BMP-4 in the FOP tissue stimulates a subpopulation of skeletal muscle cells with osteoprecursor potential (Bosch et al., 2000), the VSMCs of rapidly proliferating blood vessels and fibroblastic cells to start building the tumor-like soft preosseous growth and finally bone. But why this wildly inappropriate osteogenic rampage? The response of a normal person, for example to injection of a vaccine, is initially the same, but with only a very limited. indeed unnoticed, amount of swelling and reddening. The problem is that the FOP cells' excessive BMP-4 production triggers an accelerating autoexpression via the mutant ACVR1 receptors which cannot be stopped because of an impaired production of gremlin and noggin. In other words, BMP-4 starts an un-braked buildup of itself, osteoblast expression and bone formation.

vi. The Decline and Fall of Bone Strength and BMU Efficiency but a Rise in Remodeling Rate and Cortical Porosity with Age

"While the ape heel is solid with thick cortical bone, the human heel is puffed up and covered with only a paper-thin layer of cortical bone; the rest is thin lattice-like cancellous bone. This enlargement of cancellous bone is pronounced not just in the heel, but in all the main joints of our lower limbs—hip, ankle, knee—and has likely marked the skeleton of our ancestors since they first got upright; it has been found in the joints of 3.5 million-year-old hominin human fossils from Ethiopia.
The greater volume of bone is an advantage for dissipatring the stresses delivered by normal bipedal gait. However, it is not without cost: the redistribution in our bones from cortical to cancellous means that humans have much more surface exposure of their skeletal tissue. This results in an accelerated rate of mineral loss—or osteopenia—as we age, which may eventually lead to osteoporosisand hip and vertebral fractures."—Bruce Latimer quoted by Ackerman (2006).

Among the many unpleasant consequences of aging from our ancient *Australopithecus* ancestors to to-day are bone loss (dropping bone mineral density [BMD]), increasing vulnerability to bone fracturing and ultimately osteoporosis. It is accompanied by changes in collagen crosslinking and a decline in the quality of bone mineral (Vahshishth, 2005). In the cortex of the aging bone the osteocyte lacunar density drops in a stepwise manner and when it drops below about $600/mm^2$ there is a sharp rise in microcrack accumulation (Vahshishth, 2005). Cancellous architecture also weakens as trabecular separation and anisotropy increase, trabecular interconnection drops and trabeculae become thinner. These declines are associated with mesenchymal stem cells being less able to express their osteogenic genes including the gene for the anti-osteoclast OPG (Gimble et al., 2006). For example, aged mesenchymal stromal stem cells are less inclined to express osteoblastc pathway-driving Cbfa-1/ Runx-2 and Dlx5 transcription factors and genes for collagen I and osteocalcin while being more inclined to express the adipocyte-specific osteoblast-specific gene-suppressing PPARγ transcription factor (Moerman et al., 2004). Therefore, with advancing age the number of adipocytes in bones increases while bone formation drops—the marrow cavities fill with fat (Gimble et al., 2006). And this fattening of marrow cavities increases more in osteoporotics (Gimble et al., 2006).

With advancing age, yellowing, fattening marrow, and decades of microcrack remodeling/ repair, the density of osteons and their Haversian canals in the cortical bone rises with the replacement of primary bone with secondary and beyond bone. In other words, the bone with its yellowing marrow becomes increasingly "moth-eaten". Indeed fatigue-induced microcracks accumulate with age and in the ribs of 50-60year-old women, for example, they occur more frequently in increasing amounts of interstitial bone where the osteocyte lacunae are significantly less frequent and the microcracks are much longer than in osteons (Qiu et al., 2005). This microcrack-accumulating interstitial bone rises as osteon diameter drops while osteon

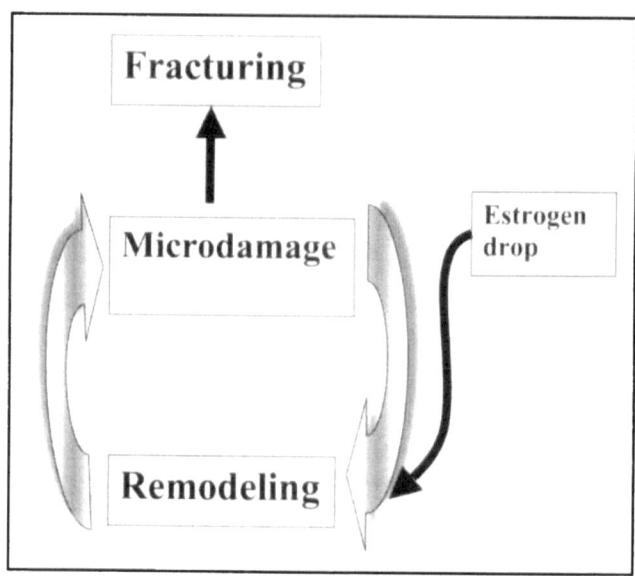

Figure 10. The vicious cycle of remodeling ⇔ microdamage ⇒ fracturing which is the engine that drives postmenopausal bones to osteoporotic hyperfragility and fracturing. With advancing age remodeling and the remodeling hole-refilling deficit rise which increases microarchitectural deterioration, fragility and microdamage at load-bearing sites. The signals from the damaged sites increase remodeling. When the estrogen level drops during menopause, the cycle accelerates and the microdamage level and weakening escalates toward "spontaneous" fracturing by the impacts of normal muscle pulling. Breaking the cycle by reducing remodeling with osteoclast-killing anti-catabolics reduces microdamage, permits the *existing* remodeling holes that have been dug out of damaged patches to be more or less refilled with more prolongedly mineralized, hence stronger, bone matrix. This reduces the probability of re-fracturing. But the prevalent microaschitecural deterioration and attendant fragility can only be reversed by an anabolic drug such a PTH that can stimulate osteoblast generation and bone-making activity.

density (in a human femur, for example) rises from about $3/mm^2$ at 5 years of age to about $25/mm^2$ at 70 years of age (Kerley, 1965). According to Laval-Jeanet et al. (1983) the humeral porosity of women increases from 4% at 42-49 years of age to 7.5% at 60-69, 12% at 80-89 years of age and 14% at 90+ years of age while the corresponding porosities in men are 4.5 %, 7%,10.5% and 11.5% respectively. Since osteons with their channels are effectively hollow (i.e., like worm holes or bubbles in Swiss cheese), the unavoidable price of microcrack repair is increasing cortical porosity and microarchitectural weakening of the bone (R.B. Martin et al., 1998). And of course with this weakening comes a higher frequency of microdamage and remodeling at prime loading sites—a vicious cycle has begun operating (Fig. 10)! In other words, in a bipedal human the repair of microdamage in its cancellous-enriched bones (Latimer and Ackerman, 2006) increases the risk of fracture especially since the bones are now being used increasingly far beyond their designed shelf-life.

Contributing to this structural bone weakening and loss with advancing age is the fact that our BMUs are not as good at patching as interstate highway repair crews. (But to be fair to the BMUs when there has been too much cracking the highway repair crews can tear up the whole road and replace it, which obviously is not an option for an old patched-up skeleton.) First, the availability of osteoblasts for BMUs drops with advancing age as the stromal cells become more inclined to express the adipocyte pathway-driving PPARγ transcription factor and less inclined to express the osteoblastc pathway-driving Cbfa-1/ Runx-2 and Dlx5 transcription factors and number of osteoprogenitor cells in the bone marrow declines (Moerman et al., 2004; Nishida et al., 1999). Osteoblasts working on the walls of the osteon tunnels in endocortical bone and in the trabecular trenches in the cancellous (trabecular) compartment of bone do not completely refill the osteoclasts' excavations, but periosteal osteoblasts do tend to overfill the holes (Eriksen, 1994; Eriksen et al., 1994; Frost, 1997). Therefore, with advancing age the cortical shells become thinner as the endocortical and trabecular parts of the bones waste away (R.B. Martin et al., 1998). Fortunately the periosteal overfilling increases the diameters of load-bearing bones (such as the femur) which somewhat compensates for the overall thinning and loss by resisting an increase in the bone's vulnerability to bending and breaking (Einhorn, 1996). But the cortex of the femoral neck becomes thinner without increasing in diameter because there are no periosteal BMUs (Einhorn, 1996). This combination of a thinning, increasingly porous cortical shell with low quality collagen and increasingly anisotropic, disconnected and thinning trabecular struts and plates in the cancellous (trabecular) compartment without an increase in the neck's diameter makes the aging hip especially vulnerable to bending and breaking by the huge loads, as much as 5 times the total body weight, that are constantly being put on it during every walking cycle (Einhorn, 1996; R.B. Martin et al., 1998).

The amount of bone removed by BMUs depends on the number of pre-osteoclasts that can be grabbed from the passing blood (Fig. 3) as well as the lifespans of the osteoclasts into which they fuse (Manolagas, 2000; Parfitt, 2000a). When the signaling from a repaired patch stops, i.e., when the " smell" of dying osteocytes fades, osteoclast recruitment also stops and so should the digging. However, depending on the number of osteoclasts in the excavation, less focused digging may go on for some time after the signaling has stopped (Parfitt, 1998, 2000a, 2000b). Indeed once a BMU starts tunneling, it tends to overshoot (R.B. Martin et al., 1998). Therefore, any anticatabolic agent that lowers the number of osteoclasts by reducing their recruitment or killing them can reduce or even stop resorption and remodeling. But as is true for so many good things there is a "catch". Reducing or killing osteoclasts impedes the removal of microcracks. This would result in a limited amount of bone growth as the unaffected osteoblasts continue filling the existing holes without having to contend with osteoclasts digging more holes faster than they can be filled. When they have finished, the osteoblast crews disband and obligingly self destruct, except for those 4% or so who are lucky enough to avoid the call to downsize and become osteocytes or lining cells. But now the owner with suppressed BMU responsivenss to microcracking continues to move and crack bone, the accumulating microarchitecural deformation of which can no longer easily be repaired.

The amount of new bone put into the osteoclast excavations is a function of the number of mature osteoblasts and how long they can work. Therefore, bone formation beyond the amount removed by osteoclasts can be stimulated by agents that increase the number and active lifespan of PTHR1 receptor-expressing mature osteoblasts (Dempster, 1997; Manolagas, 2000) (Figs. 9, 23). As we shall now see, definitely the PTHs, probably leptin and possibly lipophilic statins, are just such *anabolic* bone-builders some of which, like the PTHs can stimulate bone growth without compromising crack repair.

CHAPTER 3

Menopause and Bone Loss

i. Leptin, Fat, Brains and Bones

So far it has seemed that estrogen is the primus inter pares of an ever-growing number of agents that control bone growth and strength in both women and, perhaps surprisingly, men (Baylink et al, 1999; Klein, 1999; Stevenson and Lindsay, 1999; Vanderschueren et al., 2000). Now it appears that bones are also the direct and indirect targets of one of the the principal operators of the mechanism that was originally discovered managing the white fat energy reserves in mice (Ducy et al., 2000a,b; Fleet, 2000; Himms-Hagen, 1999; Karsenty, 2000b; S. Takeda et al., 2003) (Fig. 11).

i a. What Is Leptin?

"The leptin endocrine system is like a dynamic puzzle. As more additional pieces of the puzzle are found, more perplexities arise and more extra pieces are needed."—Jean Himms-Hagen (1999).

"Leptin: the voice of the adipose tissue"—W.F. Blum (1997).

At first it was white fat cells in mice and then a lot of other cells, including brain cells and osteoblasts, were found to make a helical cytokine hormone, belonging to the IL (interleukin)-6 family of cytokines, the non-glycosylated,167-amino acid, 16.7-kDa leptin (from the Greek word λεπoξ meaning thin) (Ahima and Flier, 2000; Bradley et al., 2001; Friedman, 2000; Frhhbeck, 2006; Grasso et al., 1999; Gordeladze et al., 2001, 2002a, 2002b; Himms-Hagen, 1999; Otero et al., 2005; Wiesner et al., 1999; F. Zhang et al., 2005). The gene for leptin is on chromosome 6 in the mouse and chromosome 7 in humans. The production of leptin by fat cells is a function of their fat load. This means that the amount of circulating leptin tells the neurons in the hypothalamic arcuate nucleus the size of the fat stores—it's a kind of fuel guage needle. From the mouse model, leptin got the reputation of being a satiety signaler—a "Fat-O-Stat"—an adipokine that inhibits the secretion of the orexigenic (eating-stimulator) neuropeptide Y (NPY) by arcuate neurons (Ahima and Flier, 2000; J. Friedman, 2000; Himms-Hagen, 1999; Pedrazzini et al., 2003). If the fat load is at the body's optimal set point, the circulating leptin holds NPY secretion from the arcuate nucleus at the appropriate level. However, if the fat load and with it leptin production drop below a critical limit, NPY secretion, and eating, rise to refill the fat fuel tanks and raise the circulating leptin concentration which puts the brakes on eating.

But there is a problem with this popular mouse story. In her hard-hitting review, Himms-Hagen (1999) has presented a convincing case for leptin *not* being a significant satiety hormone in humans. As she has pointed out the very name leptin which was of course coined for the mouse model is wrong for humans—it is not a "Fat-O-Stat" in humans. Leptin's main role in humans is to serve as an energy monitor, the loss of which, when food becomes scarce enough to critically deplete the leptin-generating fat stores, sets off alarm bells that unleash a set of responses that try to maintain energy reserves without counterproductively reducing

Growing Bone, Second Edition, by James F. Whitfield. ©2007 Landes Bioscience.

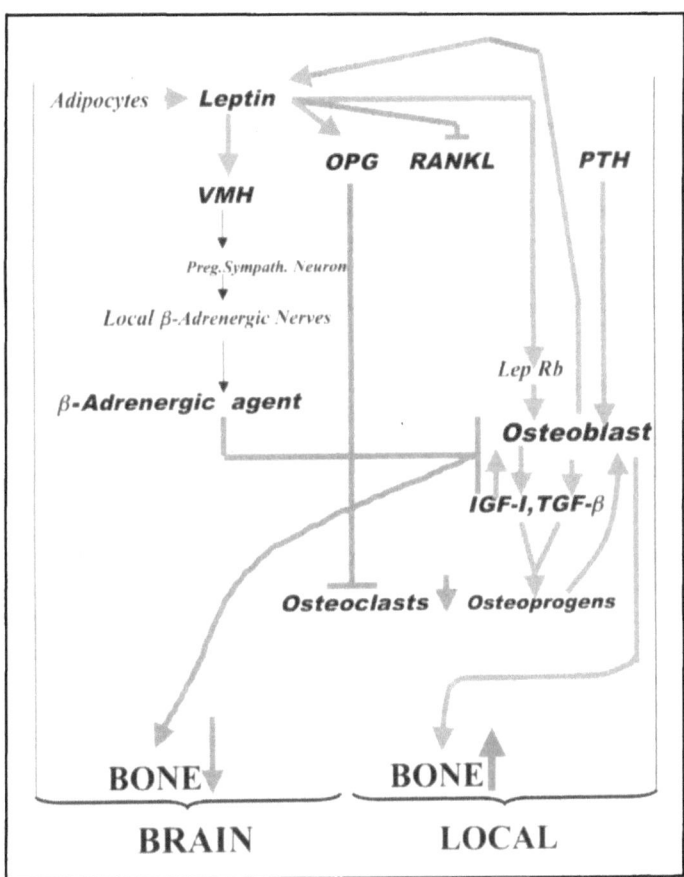

Figure 11. The overall restraining of bone growth from the brain by leptin via local, hypothalamic nuclei-dependent β-adrenergic agents and locally the direct positive, brain-independent control of bone growth by leptin and its receptors. Leptin from fat cells (see also Fig. 33), osteoblasts and even brain cells, or leptin that is injected intracerebrovascularly stimulates receptor-bearing neurons of the VMH, ventromedial hypothalamic nuclei. These neurons project to the preganglionic sympathetic (**Preg. Sympath**) neurons of the lateral cell columns of the spinal cord which in turn project to the local β-adrenergic nerves in the bones which fire their β-adrenergic transmitters at osteoblasts which have β2-adrenergic receptors on their surfaces. Details of the brain and local control mechanisms may be found in the text. LepRb, the large completely functional leptin receptor; **Osteoprogens**, osteoprogenitors.

activity (e.g., looking for food) as happens in rodents, but as in rodents to shut down ovarian cycling to avoid a pregnancy and a voracious fetus that would quickly consume any remaining energy reserves.

But leptin doesn't just monitor fuel tank guages and couple the levels to ovarian cycling. It also stimulates sympathetic activity that may be responsible for the deadly hypertension of obesity and injecting it into the brain can activate sympathetic nerves in various regions of the body *including the rat hindlimb* (Canatan et al., 2004; Rahmouni and Haynes, 2002, 2004). This driving of sympathetic nerve activity is separable from the hypothalamus-based metabolic effects, but of course sympathetic activity drives the burning of fat to extract the energy needed to try to replenish the fat stores (Rahmouni and Haynes, 2002). And most important for our story—it is turning out to be a strikingly versatile cytokine that can also indirectly inhibit bone growth by stimulating sympathetic nerve activity but at the same time stimulate bone growth by directly targeting bone cells..

i b. What Does Leptin Do to Bone?

i b 1. It Can Indirectly Block Bone Formation via the Hypothalamus

It isn't the arcuate nucleus through which leptin operates on bone from the brain. Karsenty's group have found the boney consequences of leptin stimulating the neurons of the ventral medial hypothalamic nuclei (VMHN) which projects to the sympathetic nervous system (S. Takeda et al., 2003). The hypothalamus is really the head nucleus of the autonomic nervous system and leptin keeps the β-adrenergic nervous system functioning by stimulating VMHN neurons which project to the preganglionic sympathetic neurons of the lateral cell columns of the spinal cord. And directly tweaking the VMHN neurons by squirting leptin into the third cerebral ventricle of a mouse affects the beast's bones. But I am getting ahead of myself.

Of course bones are affected by leptin's stimulation of GnRH (gonadotropin-releasing hormone) secretion and the estrogen-dependence of the *Lep* gene (originally called the *Ob* gene for the obese mouse in which it was discovered) expression (Ahima and Flier, 2000; Bann et al., 1999; Chu et al., 1999; J. Friedman, 2000; Himms-Hagen, 1999). Therefore, when either the *Lep(Ob)* gene or the LepRb receptor (Sweeny, 2002) is disabled the animal will have reduced sympathetic activity and thus be hypotensive (Mark et al., 1999). And it will also make too much cortisol and too little estrogen and therefore should be osteopenic. This should be very bad for bones!

But instead of being severely osteopenic as expected, $Ob(Lep)^{-/-}$ obese mice that cannot make leptin and $Db^{-/-}$ mice with disabled LepR receptors have a *high* vertebral bone mass (HBM) (Ducy et al., 2000a,b; Karsenty, 2000b; S. Takeda et al., 2003). Since the lack of estrogen increases osteoclast production, they do have more osteoclasts than normal lean animals but only *a normal number of hyperactive osteoblasts* that override the diggers by making twice as much bone matrix as the osteoblasts in lean mice (Ducy et al., 2000a,b; Karsenty, 2000b).

This supernormal bone production, despite the soaring osteoclast population, is due to a hypothalamic response to the lack of leptin production throughout the body in $Ob(Lep)^{-/-}$ obese mice or the incompetence of leptin receptors in $Db^{-/-}$ obese mice, and it can be stopped in $Ob(Lep)^{-/-}$ mice (which have functional leptin receptors) by injecting leptin into the animals' cerebral ventricles (Ducy et al., 2000a,b; Karsenty, 2000b). Moreover, bone loss can be caused by intracerebroventricular leptin injection into normally boned lean mice—leptin seems to cause causes the release from the brain of an osteoblast suppressor. Perhaps the osteoblast hyperactivity is due to the dramatic surge of NPY gene expression in the leptin-lacking $Ob(Lep)^{-/-}$ mouse's hypothalamus (Wilding et al., 1993). But no—cerebrally injected NPY does the same thing as leptin—it causes bone loss (Ducy et al., 2000; Karsenty, 2000)!

Thus, it seems that in mice a set-point level of leptin might stimulate the neurons in some hypothalamic nucleus to produce a hypothalamic osteoblast inhibitory factor, a HOBIF, that somehow holds osteoblast activity at an optimal level (Ducy et al., 2000a,b: Karsenty, 2000b; Whitfield, 2002b; Whitfield et al., 2002) (Fig. 11). It appeared at one point that the nNOS (neural Nitric Oxide Synthase) might be part of this brain-based leptin signaling mechanism that dampens osteoblast activity, because knocking out the nNOS gene causes large increases in bone mass (van't Hof et al., 2002b). But this is not the HOBIF. At any rate without signals from leptin-activated receptors on hypothalamic neurons, there is no osteoblast-restraining HOBIF production and bone growth climbs along with the food consumption and body weight (Fig. 11).

But is all of this just much to do about nothing?! The bone growth in a fat mouse could simply be driven by the escalating loading by the massive increase in body weight and the consequent operation of the body's "mechanostat" to make more bone to reduce the strain as we have learned in the last chapters. But the bone mass starts rising *before* the weight rises. Furthermore, it also rises in A-ZIP/F-1 mice that do not have the leptin-making white-fat adipocytes without which the animals cannot gain weight (Ducy et al., 2000a; Karsenty, 2000b). But reader should beware of this model—all of the other sources of leptin must still be functioning in the A-ZIP/F-1 mice and as we shall see further on this bone growth may not be due to a lack of leptin and HOBIF.

The Karsenty group (S. Takeda et al., 2002, 2003) now knows what HOBIF is. First it is not a circulating agent because when the circulations of two high-bone-mass $Db^{-/-}$ mice are connected (parabiotic pairing) and one receives intracerebral ventricle leptin only the injected mouse loses bone—the factor does not spread via the circulation (S.Takeda et al., 2002, 2003). Second, selectively destroying a normal mouse's ventromedial hypothalamic nuclei (VMH) with goldthioglucose, but not destroying the arcuate nuclei with monosodium glutamate, drives the bone mass up to the high level in $Db^{-/-}$ mice and prevents the bone mass from dropping after intracerebroventricular leptin injection (S. Takeda et al., 2002, 2003). Third, knocking out dopamine β-hydroxylase, and with it the ability of β-adrenergic nerves in the bones to make epinephrine and nor-epinephrine, does not increase body weight, but it does increase bone mass and blunts the bone-reducing response to intracerebroventicularly leptin (S. Takeda et al., 2002, 2003). Therefore, HOBIF is a non-circulating, i.e., within-bone, product of dopamine β-hydroxylase in the bones' nerves.

In mice, and probably humans, leptin operates on bone independently from the appetite and body weight controlling hypothalamic neurons of the arcuate nuclei (which as noted above can be destroyed without affecting bone mass) by stimulating the sympathetic nervous system via the LepRb receptor-bearing cells of the ventromedial hypothalamic nucleus (Funahashi et al., 2003; S. Takeda et al., 2002, 2003), which increase catecholamine secretion by peripheral nerves (S. Takeda et al., 2002, 2003). It appears that it is the VMHN-dependent non-myelinated sympathetic nerves in the bone that terminate near the trabeculae and follow the osteoblasts filling osteonal tunnels and locally bombard these osteoblasts with β-adrenergic agonists. And osteoblasts are certainly well-armed with β1- and/or β2-adrenoreceptors (Kellenberger et al., 1998; Majeska et al., 1992; Moore et al., 1993; Spencer et al., 2004; Takeuchi et al., 2000; Togari et al., 1997). For example in my experience the β-adrenergic agonist isoproterenol very strongly stimulates adenylyl cyclase activity in ROS 17/2 rat osteoblasts. Activation of β-adrenoreceptors increases the expression of the pro-osteoclastogenic RANKL in MC3T3-E1 preosteoblasts and mouse bone marrow cells and chemical sympathectomy inhibits preosteoclast differentiation and impairs bone resorption in adult rats (Cherruau et al., 1999; Spencer et al., 2004; Takeuchi et al., 2000). Isoproterenol, acting via the β2 adrenergic receptor, increases vertebral bone turnover, decreases femoral trabecular density and volume and substantially reduces femoral endocortical and trabecular surfaces in mice (Ke et al., 2004). Knocking out β-adrenergic receptors in male and female mice significantly increases vertebral and distal femoral trabecular bone volume (1.3 and 3.5-fold respectively) and femoral failure load and this increased bone formation is accompanied by *a doubling of the circulating leptin level* (Dhillon et al., 2004). Eleftriou et al (2005) have also shown that mice lacking the β-adrenergic receptor gene (*Adrb2*) have increased trabecular bone mass and are resistant to OVX-induced bone loss in part because of the inability of osteoblasts to make the osteoclastogenic RANKL. Also, we know that β-adrenergic stimulators such as clenbuterol and salbutamol decrease bone mass in leptin-deficient mice and female Wistar rats by reducing bone cell proliferation, that a β-blocker such as propranolol increases bone mass in intact and ovariectomized wild-type mice, that propanolol reduces the bone loss due to mechanical unloading in tail-suspended rats, and that taking β-adrenergic blockers alone or along with thiazide diuretics reduces the fracture incidence (Bonnet et al., 2003; Lavasseur et al., 2003; Pasco et al., 2004; Pierroz et al., 2004;

Schlienger et al., 2004; S. Takeda et al., 2002, 2003). Interestingly, in mice, proproanolol only increases osteoblast number and surface of trabecular bone while hPTH-(1-34) increases both cortical and trabecular bone formation with propranolol enhancing the potent osteogenicity of PTH in ovariectomized mice (Pierroz et al., 2004). However, Reid et al (2005) have reported that use of β-blockers is not strongly correlated to a decreased risk of hip fracture and the bone mineral density at the hip.

To summarize—injecting leptin into the third ventricle of the brain of a mouse stimulates the sympathetic nervous sytem and the adrenergic nerves in the bones via the lepRb-receptor-bearing cells of the ventromedial hypothalamic nuclei and this raises the level of β-adrenergic transmitters in the bone that slow the proliferation of osteoprogenitor cells (certainly *not* the terminally post-proliferative osteoblasts as claimed by S. Takeda et al. [2002]) (Fig. 11). But there is something inconsistent with their earlier paper. According to Ducy et al. (2000a) the high bone mass in the $Ob(Lep)^{-/-}$ is *not* due to there being more osteoblasts making bone as claimed by S. Takeda et al but to the same number of osteoblasts making more bone.

On the basis of leptin's indirect hypothalamically mediated osteoblast-suppressing action, Karsenty (2000b) has called it the osteoblast-suppressing equivalent of the osteoclast-suppressing OPG. *But it isn't* (Gordeladze et al., 2001; Reseland and Gordeladze, 2002; Gordeladze et al., 2002a, 2002b; Reid and Comish, 2004; Whitfield, 2001, 2002)!! While the primary adult mouse osteoblasts in the experiments of Ducy et al (2000a) apparently did not express the fully signaling, long isoform Lep receptor—LepRb (Funahashi et al., 2003; Sweeny, 2002; Zabeau et al., 2003)—and thus could not respond directly to leptin, others have found that mouse bone cells can *directly and positively* respond to the cytokine both in culture and, most importantly, *in the mouse*. MC3T3-E1 mouse preosteoblasts and MCC-5 mouse chondrocytes both have LepRb receptors, make leptin and dump it out into the culture medium (Kume et al., 2002). Liu et al. (1997) have reported that *non-cerebrally injected leptin stimulates endocortical bone formation in obese mice*. Steppan et al. (1998, 2000) have also reported the results of experiments showing that intraperitoneally injected, instead of intracerebrally injected, leptin *directly and potently stimulates* bone growth in $Ob(Lep)^{-/-}$ mice. Yamasaki et al (2003) have also reported that *ob/ob* mouse head condylar cartilage cells have the long LepRb receptor isoform and that intraperitoneally injected leptin stimulated the recovery of maxillo-facial bone length. And could a direct action of the leptin be responsible for the high bone mass in A-ZIP/F-1 mice which do not have white fat cells, but must have all of the other leptin makers?

As could have been expected from these other reports of leptin's bone-stimulating action in mice it is now becoming clear that the Karsenty story is misleading because of an inadequate examination of their $Ob(Lep)^{-/-}$ mice and a failure to pay attention to the earlier reports of Steppan et al (1998, 2000). But Hamrick et al (2004) have now properly looked at $Ob(Lep)^{-/-}$ mice. It turns out that Steppan et al (1998, 2000) were right for the appendicular skeleton! Leptin lack does indeed shorten femurs and reduce BMD, bone mineral content, cortical thickness, trabecular thickness and trabecular volume in the femur. The expression of TGF-β1, an osteoprogenitor stimulator and strong anti-adipogenesis agent, is also reduced which is consequently associated with a massive increase in femoral adipocytes and a drop in the osteocyte population density. However the Karsenty group focused on an increased lumbar vertebral bone mass in their $Ob(Lep)^{-/-}$ mice. And according to Hamrick et al. (2004) they were also right! Hamrick et al. found increased bone mineral content, BMD and trabecular bone volume in the lumbar vertebrae of their $Ob(Lep)^{-/-}$ mice. The $Ob(Lep)^{-/-}$ lumber vertebrae were also longer and broader than in normal lean mice. This suggested to Hamrick et al. that the chondrocytes of the vertebral growth plates respond differently from the femoral growth plate— maybe they are more sensitive to, or more controlled by, leptin /VMNH-dependent β-adrenergic activity. The loss of long bone (appendicular skeleton) mass is likely due in part to severely reduced loading because of a significant drop in muscle (e.g., quadriceps) mass as well as the basic inactivity of such obese beasts. As discussed in detail in Chapter 1 this would stress osteocytes and ultimately trigger their self-destruction and osteoclast scavenging of the cadavers by restricting the transcanalicular provision of nutrients and clearance of waste. And of course another contributor would be the lack of leptin's direct anabolic action discovered in mice by Liu et al (1997) and Steppan et al (1998, 2000). Therefore all that remains of the Karsenty mouse story is the ability of an intracerebroventricular shot of leptin to stimulate the VMHN-neurons and through them the preganglionic sympathetic neurons of the lateral cell columns of the spinal cord and the bones to which they project.

i b 2. But Leptin Directly Stimulates Bone Formation

Cornish et al. (2001) have reported that leptin directly stimulates fetal rat osteoblasts as strongly as IGF-I and increases bone strength in mature male mice by 20% or more. And Picheret et al. (2003) have shown that obese female and male Zuker *fa/fa* rats with their disabled LepR receptors that cannot respond to the large amounts of leptin pouring out of their fat cells have significantly *lower* femoral bone density and osteoblast activity (as indicated by a significantly lower blood osteocalcin concentration) than lean control rats. This is supported by Tamasi et al. (2003) who have reported that in these rats tibial trabecular bone volume, trabecular number and trabecular thickness are also *lower* than in normal Zuker rats although the formation rates in the normal and *fa/fa* rats were not different. Moreover, Maor et al. (2002) and Gat-Yablonski et al. (2004) have reported that leptin stimulates endochondral bone formation in cultured murine mandibular condyle and humeral growth plate by causing the cells to express and secrete IGF-I which, as we shall see further on, is also stimulated by PTHs.

Leptin is also one of the drivers of endochondral bone formation in the mouse. And Dumond et al.(2003) have found that leptin stimulates rat chondrocyes to grow and make cartilage by making IFG-I and TGF-β1. Leptin also stimulates the proliferation and differentiation of human articular as well as rabbit growth plate chondrocytes (Figenschau et al., 2001; R. Nakajima et al., 2003). Large amounts of the cytokine are made specifically by growth-plate hypertrophic chondrocytes (see Chapter 5v to find out what these cells do) near invading capillaries and osteoblasts in the primary spongiosa extending from the growth plate of long bones (Kerr, 1999). Here we see leptin doing its thing as an angiogenic factor (Boulomie et al., 1998; Cao et al., 2001; Goetze et al., 2002; Park et al., 2001; Sierra-Honigmann et al., 1998). It seems that hypertrophic chondrocytes far from blood vessels do not make leptin but the hypertrophs adjacent to marrow blood vessels make leptin and use it to get the blood vessels' endothelial cells to start moving, proliferating and making matrix metalloproteinases-2 and –9 which they use as nanodrills for tunneling through the collagenous matrix to deliver blood-borne osteoprogenitor cells and other essentials to the construction sites for vascularization and ossification (Kume et al., 2002).

Morroni et al. (2004) regard the osteogenically inactive bone stromal cells, osteocytes and bone-lining cells as the permanent members of the BBCS, the bone basic cellular system. The actively bone-making osteoblasts are not in the group because they are transient and either die after making bone or become osteocytes or bone-lining cells. It appears that in normal human and rat bones only the osteogenically inactive osteocytes and bone-lining cells make leptin while the active osteoblasts do not (Morroni et al., 2004). However, it seems that isolation and cultivation in vitro causes human osteoblasts to make leptin (Reseland et al., 2001).

Freshly isolated human osteoblasts (e.g., from trabecular bone in the head of the femur or the iliac crest), normal rat osteoblasts and ROS 17/2.8 rat osteoblastic osteosarcoma (bone cancer) cells express the all-important big, signaling LepRb receptors (Bassilana et al., 2000; Enjuanes et al., 2002; Evans et al., 2001; Gordeladze et al., 2001; Y.-J. Lee et al., 2002; Reseland and Gordeladze, 2002a, 2002b; Reseland, et al., 2002; Steppan et al., 1999, 2000). Signals from these receptors obviously must stimulate bone growth because Thomas et al.

(2001) have reported that intraperitoneally infusing human leptin prevents disuse ("pseudo-microgravity")-induced bone loss in tail-suspended rats. And Burguera et al (2001) and Gordeladze et al. (2002a, 2002b) have shown that leptin promotes rat bone growth by suppressing the osteoclast-promoting RANKL expression and stimulating osteoclast-suppressing OPG expression in osteoblastic stromal cells (Fig. 11). Actually Lamghari et al (2006) have shown that while 12 ng of murine recombinant leptin /ml of culture medium stimulate MC3T3-E1 murine preosteoblasts to express and secrete RANKL, higher concentrations such as 24 ng/ml completely suppress RANKL expression without affecting OPG expression. Iwaniec et al. (1998) have reported that low concentrations (1-100 ng/ml of medium) of leptin increase bone nodule formation in human bone cultures. Leptin also makes human marrow stromal cells differentiate into osteoblasts instead of adipocytes (Thomas et al., 1999), and it stimulates the proliferation of cultured (i.e., probably incompletely matured simply because normal mature osteoblasts cannot proliferate) human osteoblastic cells and causes human marrow stromal osteoprogenitor cells to express alkaline phosphatase, collagen I, osteocalcin and to mineralize matrix (Gordeladze et al., 2002a, 2002b; Evans et al., 2001; Thomas et al., 1999). Leptin also enhances the expression of the genes for major osteoblast differentiation driver, the master Cbfa1/Runx-2 gene transactivator, as well as alkaline phosphatase and osteocalcin and it protects the human osteoblasts from self destructing by suppressing the expression of the proapoptotic Bax protein and stimulating the expression of anti-apoptotic Bcl-2 protein (Gordeladze et al. 2002a, 2002b). Leptin also facilitates the transition of cultured osteoblasts into preosteocytes (Gordeladze et al. 2002). Leptin's ability to facilitate the osteoblast $\Rightarrow$ osteocyte transition and stimulate matrix mineralization by human osteoprogenitors fits in nicely with the fact that that the LepRb receptor-bearing human osteoblasts start making and secreting autocrine (self-stimulating)/paracrine (neighbor-stimulatimg) leptin when they are either in the mineralization or early osteocyte stage (Reseland et al., 2001) (Fig. 11.).

Most important for our story is the very exciting possibility that *signals from type 1 PTH/PTHrP receptors* (the so-called PTHR1s we will soon be getting to know rather intimately) *stimulate the expression and secretion of leptin which, from the accumulating evidence, could at least partly mediate PTH's stimulation of bone growth* (Fig. 11). This has been suggested by Torday et al. (2002)'s report that PTHrP stimulates the leptin gene, leptin production and leptin secretion in lung lipofibroblasts and by Gordeladze et al. (2002a)'s report that the effects of leptin on human osteoblasts mimic those of hPTH-(1-84).

The amazing story emerging from all of this goes something like this. Leptin is expressed first in mesenchymal stem cells to steer them onto the osteoblastic pathway, but this stops in proliferating osteoprogenitors. Then it starts again, after proliferation stops, and PTHrP and PTHR1 receptors appear and osteoblasts have matured, in order to help drive matrix deposition and mineralization. The leptin from the osteoblasts stimulates more osteoprogenitors to advance along the walls of the osteoclast tunnels and trenches and stimulates the activity and increase the longevity of their osteoblast friends and neighbors.

i b 3. A Summary of the Different Impacts on Bone of Leptin and Its Hypothalamic Associates

To some people leptin has caused a major Paradigm Shift in the World View of the control of bone growth. Thus, osteoblasts are controlled negatively by the body-wide leptin / hypothalamic VMH-promoted β-adrenergic nerve activity. But there really has been no Paradigm Shift! The osteoblasts are actually directly and positively controlled peripherally by leptin itself coming from white fat adipocytes, local yellow marrow adipocytes and most importantly from the late-stage osteoblasts themselves (Coen, 2004; Laharrague et al.,1998; Reid et al., 2006; Reseland et al., 2001, 2002; S. Takeda et al., 2003; Whitfield, 2002b) (Fig. 11). As we have seen with the stimulation of bone growth by peripherally injected leptin, the positive actions of the cytokine override any negative hypothalamically driven effects. Any doubts about peripheral leptin's direct positive osteo-actions has been addressed by people with common (as opposed to the much rarer genetic (e.g., $Ob[Lep]^{-/-}$)) obesity who make a lot of circulating peptide that cannot get into the brain through the blood-brain barrier. However, these large amounts of the peripheral cytokine in these people do *not* inhibit bone formation; there is in fact a positive, instead of negative, correlation of bone density to fat mass!

But how does all of this interesting stuff fit into a post-menopausal story? The postmenopausal estrogen drop might turn down both of the leptin mechanisms (Brann et al. 1999; Chu et al., 1999) which would cause a rise in weight and help promote the osteoclast population explosion caused by the lack of estrogen. Thus, for example, in the rat model ovariectomy silences $Ob(Lep)$ gene expression and leptin consequently vanishes from their circulation (Chu et al., 1999). And the animals get hungry and heavier while their trabecular bone is being chewed up by osteoclasts and their marrow is getting fattier (Whitfield et al., 1995). The leptin drop and the resulting surge of NPY expression in the hypothalamus cause the hunger and weight gain. At first sight this seems to conflict with the amount of circulating leptin expression being positively correlated to the body fat load—leptin should rise. But $Ob(Lep)$ gene expression is *estrogen-dependent* and giving estrogen to the OVXed rats prevents the leptin drop, overeating and weight gain (Chu et al., 1999). Therefore, without enough estrogen to maintain the $Ob(Lep)$ gene expression leptin virtually vanishes from the circulation by 7 weeks after OVX, but the increasing accumulation of white fat in its swelling adipocytes eventually overrides the estrogen lack, leptin production rebounds and its level soars above the sham OVX level between 9 and 13 weeks after the operation (Chu et al., 1999). Therefore, the bones of an overeating OVXed rat do not at first have to face the debilitating double whammy of leptin-induced, VMH-driven β-adrenergic-suppressed osteoblast activity as well as an osteoclast feeding frenzy, but the direct stimulation of osteoblasts by the eventual leptin surge should slow or stop the bone loss. Alternatively any positive action of the leptin surge might be counterbalanced by an adrenergic surge. It follows from this that it would be important to find out whether a leptin loss might also affect bone loss in postmenopausal women and whether leptin injection would stimulate bone growth in these women as it does in rats. Indeeed Blain et al. (2002) have concluded that leptin may limit the excessive bone reorption in postmenopausal women. Moreover, Weiss et al (2006) have found that there is a positive relation between serum leptin levels and BMD with a negative relation between leptin levels and bone turnover in older women but not men. And Ke et al (2001, 2002) have anticipated leptin's admission into the exclusive *Growing Bone Club* by patenting it and its analogs for treating osteoporosis and all other conditions causing bone loss and for accelerating fracture healing.

But the hypothalamus has a lot of jobs to do and there seems to be another player in the brain-and-bones saga—NPY's Y2 receptor. Why do we think so? Well, if the Y2 receptor gene is selectively knocked out in mouse hypothalamic neurons, osteoblasts make more matrix just as they do in Ob(Lep)$^{-/-}$ mice (Baldock et al., 2002; Herzog, 2002)—signals from the Y2 receptors in hypothalamic neurons somehow lead downstream to the reduction of osteoblast activity. And of course this why Ducy et al. (2000a) found that directly injecting NPY into the cerebral ventricles caused bones to *lose* mass in their Ob(Lep)$^{-/-}$ mice just as did leptin injection. Thus, when the leptin level is "normal", NPY secretion and Y2 receptor signaling are minimal or suppressed. But any possible upsurge of osteoblast activity caused by shutting down NPY-Y2 receptor activity is prevented by leptin-driven VMH-mediated β-adrenergic secretion. However, the direct osteogenic action of leptin would still be working in the bones.

But things are complicated by the fact that NPY also directly targets osteoblasts. Indeed osteoblasts have receptors for the NPY released into the bone by the same sympathetic nerves that are driven by leptin via the VMHN neurons (Bjurholm, 1991; Ekblad et al., 1984; Togari et al., 1997). But we don't yet know how NPY receptors signals directly affect osteoblast activity. All we know at the moment is that NPY inhibits the abilities of noradrenaline and PTH to stimulate adenylyl cyclase in osteoblastic cells (Bjurholm, 1991). However, it is clear that there is a complex control system operating at several levels running from the brain and spinal cord to the bones and osteoblast surfaces to control bone formation.

But before moving on to other things, it must be clearly understood that as matters stand today, leptin is a direct bone anabolic agent which may indirectly, but only under conditions such as sustained anxiety stress, limit bone growth and cause actual bone loss by increasing local bone β-adrenergic activity via the VMHN neurons (Reid and Comish, 2004; Whitfield, 2002b). Thus, it would seem to be unreasonable to try to treat osteoporotic patients without cardiovascular disorders with potent β-blockers as suggested by S.Takeda et al. (2003). But Pasco et al (2004) have presented data indicating that women with cardiovascular disease who have been taking such drugs do have higher bone mineral densities and fewer fractures.

ii. How Does an Estrogen Loss Cause a Bone-Wasting Osteoclast Population Explosion?

The ability of osteocytes to know when the strain exceeds a site-appropriate level and to react to this by signaling through the osteointernet the need to make and release various factors such as PGE_2, PGI_2, IGFs, and TGF-βs as well as our new friend leptin to generate and mobilze osteoblasts to make more bone to bring the strain to a subthreshold level depends on estrogen (Ehrlich and Lanyon, 2002; Turner and Robling, 2004). But as the estrogen level starts falling, estrogen receptor-α (ERα) signals start fading as must leptin on the approach to menopause, the osteocytes become less responsive to strain, i.e., they become less able to feel the sloshing of the canalicular fluid. This results in fewer calls for resources to maintian bone and increasing uncompensated strain from muscle pulling and limb impacts with a preferential loss of minimally strained bone. But along with the osteocytes becoming "strain-deaf", the estrogen drop also causes an upsurge of osteoclast generation.

However, the vigilant reader will challenge this blanket assertion that the signals from estrogen receptors fade as the estrogen level drops because *strain by itself can activate estrogen receptors without estrogen.* (Aguirre et al., 2003). Thus I must now qualify this by saying simply that estrogen seems to be replaceable in some cells such as osteocytes by strain.

Osteoclast progenitors respond to estrogen with their conventional "genomic" estrogen receptors which targets the chromosomes to activate target genes with so-called EREs (estrogen response elements) in their "signal boxes" as well as through what appear to be "non-genomic (genotropic)" receptors that are translated into protein from the same primary transcript as the genomic receptors and fire signals into the cell from little caves (caveolae) or pouches in the plasma membrane (Chambliss et al., 2000; Coleman and Smith, 2001; Falkenstein et al., 2000; Fionelli et al., 1996; Kelly and Levine, 2001; Kim et al., 1999; Krishnan et al., 2001; Levin, 2000; Manolagas et al., 2002; Morley et al., 1992; Razandi et al., 1999, 2002; Segars and Driggers, 2002; Toran-Allerand, 2000; C.S.Watson, 2003). There is yet another estrogen receptor, GPR30, which is a typical transmembrane G-protein-coupled receptor that responds to estradiol-17β by mobilizing intracellular Ca^{2+} and triggering the synthesis of PIP3 in the nucleus (Hewitt et al., 2005; Revankar et al., 2005). But this is an usual G-protein-coupled receptor because it is located on endoplasmic reticulum membranes instead of the plasma membrane and responds to estrogen only after the steroid passes into the cell through the cell membrane (Hewitt et al., 2005; Revankar et al., 2005). Only time will tell whether this receptor has any role in osteoclast production.

Signals from these receptors in a young woman might limit the size of her pre-osteoclast population by a TGF-βs-mediated killing of the cells by apoptosis (Manolagas, 2000; Whitfield et al., 1998b; Zecchi-Orlandini et al., 1999) (Figs. 12 and 13). Therefore, as the estrogen level

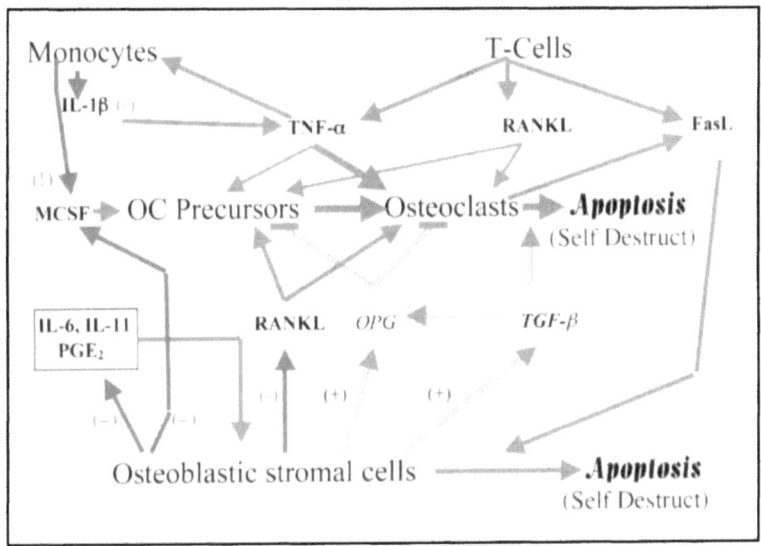

Figure 12. How estrogen and things made by T-lymphocytes control osteoclast generation and bone turnover. The (–) symbols indicate that estrogen inhibits expression of the indicated factor such as MCSF (macrophage colony-stimulating factor) and that lowering the estrogen level releases the expression and production of the factor. It is also shown that T-cells make RANKL and TNFα that stimulate osteoclast generation and FasL (Fas ligand) that causes ostoblastic cells to self destruct by switching on the apoptotic mechanism.

falls with approaching menopause, less TGF-β is made and osteoclast precursors live longer to make bigger osteoclast crews for more extensively and deeply digging BMUs (Fig. 13). It is also likely that the lack of estrogen could cause a drop of caveolin production and loss of osteoblastic cells' caveolae which would normally hold together and restrain groups of signaling enzymes (Chambliss et al., 2000; Couet et al., 2001; Okamoto et al., 1998; Razandi et al., 1999, 2002; Razani and Lisanti, 2001; Razani et al., 1999, 2001; Solomon et al., 2000). This would contribute to the surge of osteoclastogenic signaling.

The number of osteoclast progenitors in the microenvironment created by marrow stromal cells also rises because of a surge into the stromal microenvironment of factors such as interleukin (IL)-1β and tumor necrosis factor (TNF)-α from the IL-1β-stimulated marrow monocytes and perhaps most importantly TNF-α from an increasing population of T-lymphocytes (Cenci et al., 2000; Hofbauer et al., 2000; Jilka, 1998; Katagiri and Takahashi, 2002; Kobayashi et al., 2000; Martin et al., 1998; Pilbeam et al., 1997; Roggia et al., 2001) (Figs. 12 and 13). Indeed in mice, TNF-α from a growing, OVX-induced population of marrow T-lymphocytes is required for the estrogen drop to cause bone loss (Roggia et al., 2001). This TNF-α stimulates marrow stromal cells to make M-CSF that puts committed macrophage-family members on the road to osteoclasts (Hofbauer et al., 2000; Roggia et al., 2001). In the estrogen-rich youthful koric (from the Greek for "youth") bones, the responsiveness of the osteoclast progenitors to the TNF-α-inducing IL-1β is blunted by the estrogen-promoted, functionless type 2 IL-1 decoy receptors on their surfaces that compete with the functional receptors for IL-1β (Hofbauer et al., 2000; Manolagas, 2000; Pilbeam et al., 1997; Suda et al., 1999; Sunyer et al., 1999) (Figs.12 and 13). But as estrogen declines, so do the decoy receptors and IL-1β is no longer diverted from driving the development of the osteoclast progenitors, which it can do by stimulating the expression of the transcription factor NF-κB (Xing et al., 2003). The estrogen-deprived osteoblastic stromal cells also start making the potent osteoclast

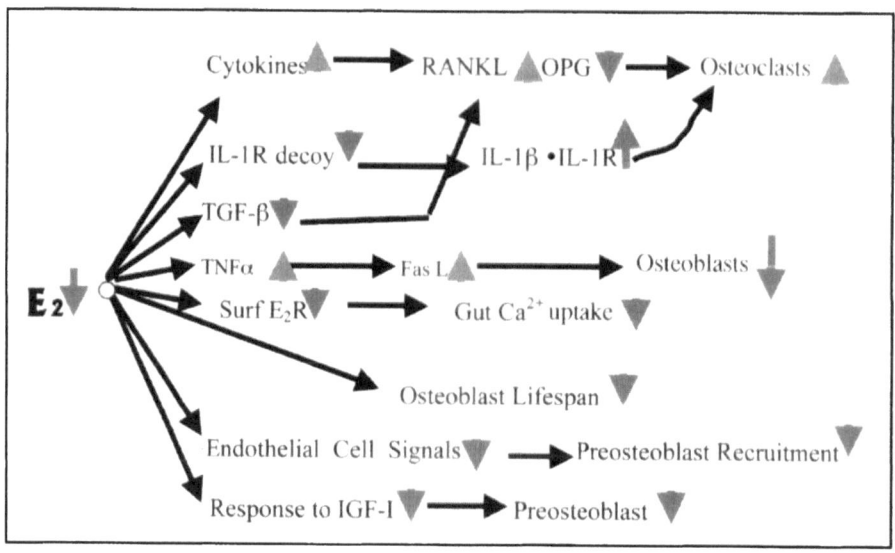

Figure 13. A summary of some of the many of the anti-osteogenic consequences of an estrogen crash on bone.

differentiation stimulators, IL-6 and IL-11 that have been suppressed by estrogen (Jilka, 1998; Manolagas, 2000; Whitfield et al., 1998b) (Fig. 12).

These several interleukins, prostaglandins and TNF-α focus on immature osteoblastic marrow stromal cells to persuade them to make more RANKL and put it out on their surfaces (Atkins et al., 2003; Aubin and Bonnelye, 2000; Blair et al., 2002; Greenfield et al., 1999; Hofbauer et al., 2000; Karsenty, 1999; Suda et al., 1994) (Figs.12 and 13). However, the TNF-α also tends to increase osteoblast apoptosis and at the same can itself stimulate the differentiation of osteoclast precursors by activating TNFR1 and TNFR2 receptors and the receptors'-associated TRAF(TNF-receptor-associated factor)2 factors which switch on osteoclast-differentiating genes via the NF-κB transcription factor (Hill et al., 1997; Katagiri and Takahashi, 2002). In fact TNF-α is actually a better stimulator of osteoclast activity than differentiation from precursors (Fuller et al., 2002). Moreover, TNF-α can very strongly synergize with RANKL, so that even a slight increase in TNF-α that is too small to increase osteoclast *generation* in the presence of normal levels of the differentiation-driving RANKL would greatly increase osteoclast *activity* without an increase in osteoclast numbers (Fuller et al., 2002).

The RANKL surfacing on immature osteoblastic marrow stromal cells binds to and activates the differentiation-stimulating RANK receptors on pre-osteoclasts when they dock on the stromal cells (Atkins et al., 2003; Aubin and Bonnelye, 2000; Hofbauer et al., 2000; Suda et al., 1999). (However, cell-cell contact is *not absolutely necessary* because free RANKL is also *secreted* by osteoblastic cells and other cells, such as activated T-lymphocytes, which stimulate the generation of the osteoclasts that for example dig holes in rheumatoid arthritic joints [Fig. 12.; Karsenty, 1999; Kong et al., 1999].) The RANKL-activated RANK receptors on osteoclast precurors and beyond drive maturation, bone-digging activity and survival via receptor-associated TRAF proteins that activate transcription factors such as AP-1, NFAT and NF-κB that switch on the genes for osteoclast differentiation (Katagiri and Takahashi, 2002; Z.H. Lee and H.-H. Kim, 2003). The TRAFs-mediated RANK signals also stir up a swarm of other kinases that turn on the digging machinery and hold-off the self-destruct apoptogenic mechanism (Z.H. Lee and H.-H. Kim, 2003).

The ability of osteoblastic precursor cells to make RANKL falls as the *rankl* gene's promoter-switch box is locked shut by methylation of its CpG domains while the gene for the anti-osteoclast *opg* gene stays active as the progeny differentiate into mature bone-making osteoblasts—a wise stratagem to protect the new bone they are making (Atkins et al., 2003; Gori et al., 2000; Kitazawa et al., 1999; G.P. Thomas et al., 2001). To understand the control of osteoclast differentiation we must know that the expressions of the genes for RANKL and OPG in the individual cultured cells are logically linked by a reciprocal control mechanism, which will be discussed in Chapter 5. Suffice it to know here that when one gene's expression is turned up the other's expression is turned down (e.g., Ma et al., 2001; Yamagishi et al., 2001).

Estrogen controls the OPG/RANKL expression ratio, which determines the level of osteoclast generation. Thus, a young woman's estrogen stimulates the production of TGF-βs which, beside directly stimulating osteoclast apoptosis, increases the osteoblastic stromal cells' OPG/RANKL expression ratio which probably reflects the relative fractions of RANKL-expressing immature mature cells and OPG-expressing mature cells in the marrow stroma and restrains osteoclast generation (Atkins et al., 2003; Aubin and Bonnelye, 2000; Hofbauer et al., 1999, 2000; Murakami et al., 1998; Takai et al., 1998) (Fig. 12). And estrogen also hits osteoclast production directly by preventing the RANKL-triggered RANK signals in osteoclast progenitors in the marrow from driving differentiation by blocking the AP-1 (c-Fos•c-Jun) transcription factor complex's activity by lowering the expression and activity of the c-Jun component (Shevde et al., 2000). OPG is another decoy receptor, like the non-signaling type 2 IL-1 receptor that is designed to restrain osteoclast production. OPG is actually a floating piece of the RANK receptor that is made by mature osteoblasts and binds to, and thus covers up, RANKL molecules sticking out of the surfaces of immature osteoblastic stromal cells which prevents RANKL from binding to the differentiation-driving RANK receptors on osteoclast precursors (Aubin and Bonnelye, 2000; Hofbauer et al., 1999, 2000; Makhluf et al., 2000; Moonga et al., 1998; Murakami et al., 1998; Oyajobi et al., 2001; Takai et al., 1998).

It is important to note at this point that knocking out the OPG gene in mice produces a severe osteoporosis. While endogenous OPG is not needed for embryonic bone formation, it is very much needed for the postnatal bone maintenance (Bucay et al., 2006). Indeed in the OPG$^{-/-}$ mouse trabecular bone is almost gone in femurs, humerus and tibias by 1-2 months .And there is increased cortical porosity as indicated by an increased number of cortical blood vessels by about 1 month. Since OPG specifically brakes the terminal stages of osteoclast development (Simonet et al., 1997),this massive bone destruction in OPG$^{-/-}$ mice is due, not to an increased production of osteoclast precursors, but to increased precursor maturation.

The estrogen decline doesn't only increase the numbers of longer-lived, bone-excavating osteoclasts. The increased number of osteoclasts associated with immature stromal osteoblastic cells and cytokine-stimulated T-lymphocytes probably produce soluble and/or membrane-bound FasL (Fas ligand or binder), which cause osteoblastic cells with Fas receptors on their surfaces to self destruct apoptotically (Kawakami et al., 1997, 1998) (Fig. 12).

The TGF-β stores in the bone matrix are also depleted (revewed by Jilka, 1998 and Whitfield et al., 1998b). Therefore there is even less of it that could evade the proteases and reduce osteoclast generation and kill mature osteoclasts when they dig it out of the matrix (Jilka, 1998). And there is also less of it to attract immature osteoblasts into the excavation sites, although TGF-β receptor signals may be less important for this than the migration-stimulating signals from the osteoblastic cells' Ca^{2+}-sensing receptors stimulated by the huge amount of Ca^{2+} in the excavation sites (Brown amd MacLeod, 2001; T.Yamaguchi et al., 1998).

Since the endothelial cells of the capillary bud following behind the trench-digging and tunneling osteoclasts express estrogen receptors and secrete an entire set of osteoprogenitor-stimulating factors such as BMP-2, endothelin-1 and OP-1 (Bouletreau et al., 2002; Brandi et al., 1993; Carano and Filvaroff, 2003; Mohammed et al., 2003; Soerensen and Eriksen, 1998;

von Schroeder et al., 2003), a lack of estrogen receptor signaling could impair the ability of the capillary's endothelial cells to attract and stimulate the development of osteoprogenitors which would reduce the size of the BMU osteoblast filler crews (Fig. 13). The impaired ability to snare blood-borne BMU team members is specifically due to the endothelial cells' surface selectin expression being estrogen dependent (Q. Chen et al., 2004). Thus, the loss of selectin caused by an estrogen crash cripples the ability of the cells to catch passing blood-borne preosteoclasts and osteoprogenitors and start assembling BMUs.

Since estrogen also lengthens the working lives of osteoblasts and the microcrack-detecting osteocytes by suppressing their apoptotic self-destruct mechanism, the estrogen loss unleashes this mechanism and reduces the number of osteocytes (Manolagas, 2000; Manolagas et al., 1999; Tomkinson et al., 1998). This loss of osteocytes would reduce the inhibitory signaling through the osteocyte-lining cell network, which would be equivalent to severing the signaling network by microcracks (R.B. Martin, 2000, 2002). This generalized pseudo-microcrack signaling would increase BMU activation and therefore bone turnover and the remodeling space—functionally it would be equivalent to a widespread microcracking like the shattering of an automobile windshield.

Perhaps counterintuitively such BMU activation should, and in my experience does, increase the number of osteoblasts. However, the hole-filling efforts of these osteoblasts can't keep up with the hole-digging zeal of the large osteoclast crews!

Under these low estrogen conditions, osteoblast working life and activities such as laying down bone matrix and secreting various factors may also be reduced, and the BMU replacement deficit thus increased, by the loss of signals from the non-genomic estrogen receptors such as those found on the surfaces of cultured female rat osteoblasts (Falkenstein et al., 2000; Lieberherr et al., 1993; Manolagas, 1999). These surface receptors appear either to be the products of the gene for the nucleus-seeking, gene-activating ERα genomic receptor that have been modified to stay at the cell surface and fire Ca^{2+} and cyclic AMP signals into the cell when activated by 17β-estradiol (Chambliss et al., 2000; Falkenstein et al., 2000; Fiorelli et al., 1998; Kelly and Levin, 2001; Lieberherr et al., 1993; Razandi et al., 1999; C.S.Watson, 2003; Whitfield et al., 1998b). Or they might be the entirely unrelated ER-X receptors (Toran-Allerand, 2000) or the "γ-adrenergic" receptors that are activated by dopamine, epinephrine and norepinephrine as well as 17β-estradiol and various xenoestrogens (Benten et al., 2001; Nadal et al., 1998, 2000).

The loss of Ca^{2+} influx-driving signals from these non-genomic estrogen receptors on the surfaces of intestinal epithelial cells may further reduce intestinal Ca^{2+} uptake and increase the bone loss due to the vitamin D deficiency that commonly develops in older persons (Chen and Kalu, 1998; Doolan et al., 2000; Picotto et al., 1999) (Fig. 13). Adding to this vitamin deficiency may be a drop in the estrogen-dependent expression of the 1α,25-dihydroxyvitamin D_3 receptor that drives Ca^{2+} uptake and transcellular transport to the blood by colon epithelial cells (Schwartz et al., 2000).

In summary, all of this boils down to the menopausal estrogen decline producing more BMUs with bigger osteoclast and smaller osteoblast crews that cannot fill the larger number of deeper holes dug by the swollen osteoclast crews. The results are weakening bones and increasing remodeling, for example because of the escalating number of osteonal "worm holes" and poorly mineralized matrix in remodeled cortical bone, and a dramatic upsurge of microcracking and remodeling (Bouxsein, 1999). But the BMU crews answering the calls are getting ever less osteoblastic cell power. Thus the vicious microdamage-driven remodeling cycle of aging is accelerated (Fig. 10). The cause of osteoporosis is basically escalating remodeling by osteoclast-overmanned, osteoblast-undermanned BMUs, just as road repair crews with over-manned, over-zealously digging teams and under-manned filling teams would produce an increasingly fragile, crack-prone road.

iii. How to Stop (or at Least Slow) Menopausal Bone Loss

To properly stop menopausal bone weakening and fracturing, it is necessary to break the vicious remodeling⇔microdamage cycle (Fig. 10). Ideally to do this we should restore the optimal composition of BMUs by reducing the osteoclast and increasing the osteoblast crews. This would reduce the number of osteonal "worm holes" produced during the repair of microdamaged cortical bone and lower the amount of cutting and perforating of trabecular struts, especially the critical horizontal struts, and plates by osteoclasts. It would also prolong the mineralization and hence strengthen new bone. Therefore as said by Parfitt (2004) "...only small reductions in remodeling (and hence only small doses of anti-remodeling agents) may confer a large therapeutic benefit".

Obviously bone loss can be stopped by giving the post-menopausal woman estrogens or one of the partially estrogen-mimicking SERMs (Selective Estrogen Receptors Modulators) such as raloxifene (Eli Lilly's Evista™) (Roe et al., 2000). But this does not cause a sustained increase in bone formation—it works by reducing remodeling. Giving one of these, raloxifene, to OVXed rats does not affect the expression of bone formation-related genes such as alkaline phosphatase, biglycan, collagen I, IGF-II, collagen-binding protein, and osteocalcin in their femurs, but it actually *reduces* the expression of decorin and IGF-I (Onyia et al., 2002). On the other hand it reduces osteoclast generation by reducing the expression of CSF-1(colony-stimulating factor-1).

Bone loss can also be reduced or even stopped with one of the many anti-catabolic bisphosphonates such as alendronate (Merck & Co.'s Fosamax™) (Bone et al., 2004; Fleisch, 1997, 1998), or either oral or nasally sprayed calcitonin peptide that shuts down osteoclast activity by activating receptors on the cells of the osteoclast lineage (Roodman, 1996; Suda et al., 1999).

The bisphosphonates are Ca-apatite-loving, bone-binding analogs of pyrophosphate (P-O-P) in which the oxygen has been replaced by a carbon atom, the so-called geminal carbon, which makes them non-metabolizable. They preferentially attach to osteoclast excavation sites and attack the working mature osteoclast crews (Rogers, 2003). Then upon the arrival of osteoblasts where the bone mineral is exposed, they are buried in the new bone. There they lie in wait, unchanged for months to years, until the poisonous bone patch is once again cracked and marked for removal by osteoclasts. The osteoclasts inadvertently release the deadly bisphosphonates and suicidally gobble them up along with the other debris they want to expel from the hole by picking them up by receptor-mediated endocytosis or pinocytosis and carrying them from their apical acid-protease-dripping fingers poking into the holes to their upper basal surfaces by a process called "transcytosis" (Stenback, 2002). (The terms apical (upper end) and basal (bottom end) may be a bit confusing here because osteoclasts stick their apical "jaws" into the bone and their basal tails in the "air".) Because the bisphosphonates lock onto the exposed mineral surfaces of osteoclast-excavating microcrack repair sites, and since trabecular bone reputedly has the highest turnover rate (though see Parfitt [2002]), the bisphosphonates are best able to prevent resorption and remodeling and increase the strength of trabeculae-rich bones such as vertebrae and femoral trochanter which have the largest numbers of attractive remodeling excavation sites at any given time (Rodan and Reszka, 2002).

The bisphosphonates can persuade osteoclasts to self destruct à la apoptosis in one of two ways (Amin et al., 1992; Benford et al., 1999, 2001; Reszka et al., 1999; Rodan and Reszka, 2002; Rogers, 2003; Rogers et al., 2000; van Beek et al., 1999) (Fig. 14). The nitrogen-containing bisphosphonates (N-BPs) such as alendronate (Fosamax™), ibandronate, incandronate, risedronate and zoledronate disable and kill osteoclasts as well as lower cholesterol just like like the cholesterol-lowering statins that you will be meeting in Chapter 8 (Rogers, 2003). Both the N-BPs, directly, and statins, indirectly, inhibit FPP (farnesyl pyrophosphate) synthase (Coxon and Rogers, 2003; Dunford et al., 2002; Luckman et al., 1998; Rodan and Reszka, 2002; Rogers, 2003; van Beek et al., 1999) (Fig. 14). This prevents the

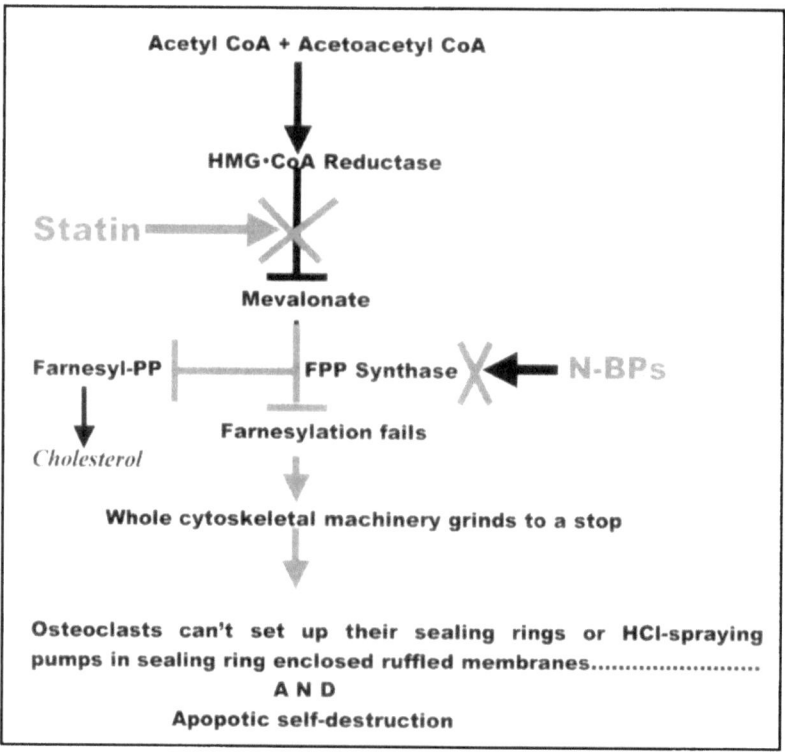

Figure 14. The ways statins and N(nitrogen)-bisphosphonates (N-BPs) disable and kill osteoclasts. See Figure 15 and the text for further details of how N-BPs and other bisphosphonates kill osteoclasts.

isoprenylation by farnesyl-PP or geranylgeranyl-PP tranferase that small GTPases—Cdc 42, Rac, the Rabs, Rases and Rhos—need to anchor themselves to cell membrane scaffolds and rafts and cytoskeleton-shaping devices. Thus, for example, on its way to the cell membrane Ras is first isoprenylated by having farnesyl-PP diphosphate tacked onto its C-terminal cysteine residue by farnesyl-PP transferase in the cytosol, then it is delivered to the endoplasmic reticulum and from there to the Golgi membrane system where it is also palmitoylated and either plugged into a new scaffold and sent to the cell membrane or separately plugged into the cell membrane (Bertiaume, 2002). If farnesylation fails, the Cdc 42, Rabs, Racs, Rases, and Rhos will not be plugged into signaling scaffolds (Coxon and Rogers, 2003). This will disrupt and disable signaling rafts, scaffolds and their attached actin structures, effects that will reverberate throughout the cell to prevent the formation and operation of the osteoclasts' sealing ring and the proton pumps in the ruffled membrane that sprays the mineral-dissolving HCl and matrix-shredding proteases onto the bone. All of this metabolic mayhem can be prevented simply by feeding the cells geranylgeraniol, an analog of geranylgeranylpyrophosphate that can pass through the cell membrane (Coxon and Rogers, 2003; Benford et al., 1999, 2001; Fisher et al., 1999; Reszka et al., 1999).

Let's summarize what we have learned so far. A mature osteoclast, the specific target of an N-BP, does it job with its cytoskeletal machinery. It lays down an actin ring—a bit like a micro-octopus's suction cup. Inside the ring it sets up its ruffled border to form a kind of giant exteriorized, bone-dissolving lysosome loaded with molecular hoses for spraying the apatite-dissolving HCl and matrix-chewing enzymes for digging the hole (Stenbeck, 2002). These cytoskeletal creations are driven and shaped by the various isoprenylated small GTP-ases (Davies, 2005). When a N-BP sits down on a bone patch and is swallowed (by fluid phase endocytosis) by the osteoclast the whole cytoskeletal machinery grinds to a stop as the cell runs out of its supply of isoprenylated GTPases. So the first effect of a N-BP is to seek out and disable mature osteoclasts wherever they are digging. But something even more sinister follows the shutdown.

One of the effects of the N-BPs towers above the others—the stimulation of the caspase-3 protease that will kill the osteoclast unless blocked in time by an inhibitor such as zVAD-fmk or SB-281277 (Benton et al., 1999, 2001; Reszka et al., 1999). Caspase 3 slices the 50 kDa Mst 1(Krs2) kinase into a regulatory C-terminal fragment and an active 34 kDa kinase fragment, which in turn activates more caspases (Fig. 15; Reszka et al., 1999a).

The non-nitrogen-containing bisphosphonates (clodronate, etidronate, tiludronate) do not disable the osteoclasts by inhibiting FPP, but they combine with ATP to form toxic ATP analogs (e.g., the forgettably named adenosine-5'[β,γ-dichloromethylene]triphosphate) that kill the cell (Benford et al., 2001; Frith et al., 1997, 2001; Reszka et al., 1999; Rogers, 2003) (Fig. 15). Tiludronate can also directly disable the diggers by inhibiting their mineral-dissolving, protease-activating acid-spraying proton pump (David et al., 1996).

But alendronate (Fosamax™) unfortunately does not leave osteoblasts alone in rats. Far from it—it actually *turns down* the expression of genes for bone formation in the proximal femoral metaphysis of ovariectomized rats (Onyia et al., 2002)! The inhibited genes include the ones for alkaline phosphatase, biglycan, collagen I, decorin, IGFs-I and II, collagen-binding protein, and osteocalcin. This strong suppression of osteoblast genes explains the strong suppression of bone formation activity by alendronate in rats (Onyia et al., 2002). And this happens not only in rats. Black et al. (2003), Finkelstein et al. (2002) and Neer et al., 2002) have found that alendronate also inhibits the osteogenesis serum marker alkaline phosphatase and prevents hPTH-(1-34)(Forteo™) and hPTH-(1-84) from stimulating it and bone growth in hip and vertebrae in osteoporotic men and postmenopausal women.

But as I said at the start of this section, alendronate does appear to do something more positive to osteoblasts and osteocytes. It can protect osteoblasts and osteocytes from apoptotic self destruction. The cytoplasms of these cells are normally directly interconnected in a functional syncytium by gap junctions which are pipes made of end-to-end joined halves ("half-pipes") or connexons, each of which is a bundle of six, 4-pass (tetraspanning) transmembrane proteins, such as Cx (connexin) 43, through which passes a 1.5-μm channel through which ions (except Ca^{2+} which causes the channel to slam shut) and a wide variety of small molecules such as cyclic nucleotides and ATP can pass (Goodenough and Paul, 2003). But these cells also have uncoupled, or extra-junctional (EJ), "half-pipe" connexons that independently stick out of the cell and serve as non-specific, large channels which slam shut when the external Ca^{2+} concentration rises above a critical level (Goodenough and Paul, 2003). It seems that alendronate operates by binding to and opening these "1/2 pipes", then passing through them into the cell and activating the cSrc protein-tyrosine kinase which triggers an apoptosis-blocking MAP/ERK kinase cascade that closes the pipes by phosphorylateing their Cx43 proteins (Goodenough and Paul, 2003).

Because these antiresorptives (or antiremodelers) suppress osteoclasts and may also reduce osteoblast action, they only cause at the most a limited amount of growth in mature human bones as the osteoblasts continue filling in the microcrack repair holes (the remodeling space) which account for 10% or more of the current total bone volume at any given moment (Dempster, 1995, 1997; Fleisch, 1997, 1998). But this is the limit—the gains in bone mineral density (BMD) in lumbar spine caused by optimal doses of alendronate, etidonrate, HRT, raloxifene and risedronate range only from 2.6 to 6.8 % and from 1.0 to 4.8 %in the femoral neck by 3 years (Pearson and Miller, 2002).

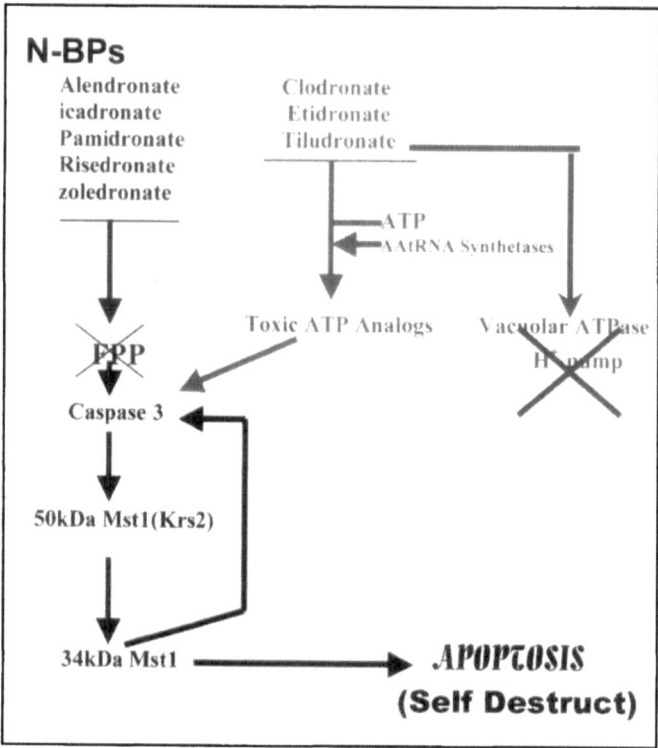

Figure 15. The different mechanisms by which the different bisphosphonates drive osteoclasts to self destruction by causing them to switch on their apoptosis mechanism.

Because they stop osteoclast development and therefore micocrack-triggered remodeling, bisphosphates such as alendronate at least temporarily reduce fracture rates by prolonging the so-called secondary mineralization (i.e., by hardening the remaining bones and increasing the bone mineralization density level and therefore strength) but they do so without affecting trabecular thickness or number, i.e., *without increasing the amount of bone tissue* (Boivin and Meunier, 2002; Boivin, et al., 2000; T.J. Martin, 2004; Pearson and Miller, 2002). They may also improve hole-filling by making osteoblasts live and work longer (Allen et al., 2002; Manolagas, 2000) by preventing them from self-destructing à la apoptosis by sucking up the large amounts of phosphate in the osteoblast excavations with their Pi transporters (Mansfield et al., 2001). Nevertheless, they can't raise bone mass above the normal level in rats or reduce the fracture risk in the mature human skeleton by more than 50% of the baseline risk (Rosen and Bilezekian, 2000; Seeman, 2001). However, a wrong impression of anabolic ability can be obtained by extrapolating the responses to antiresorptives of the continuously growing bones of mice and rats to the non-growing mature microcracking and remodeling human bones, where, because of osteoblast generation's tight coupling to osteoclast activity, the reduction of osteoclast generation ultimately reduces osteoblast generation. Thus, rodents with their always growing, minimally microcracking bones treated with antiresorptives are like mutant mice, which cannot produce functional osteoclasts and become osteopetrotic as their contiuously and independently generated osteoblasts continue their unopposed bone making (Karsenty, 1999).

There is a potentially very serious problem with bisphosphonate therapy. Microcracking (i.e., microcrack density) increases exponentilally with age in bones such as the femur, probably because declining bone mass increases strain on the remaining bone and thus microcracking (R.B. Martin et al., 1998; Schaffler et al., 1995). Therefore, murdering mature osteoclasts and preventing their replacement with one of the anti-catabolic agents (to use the new term recommended by Riggs and Parfitt, 2005) might be counterproductive because it will impede or even prevent the activation of BMUs to repair the rising microcracks, which could with time lead to a resurgence of fracturing, the seriousness of which would be mitigated of course by the weakening muscle-imposed strain on the aging body by its weakening muscles (Mashiba et al., 2001; R.B. Martin et al., 1998). That this is a real concern for a bone-remodeling animal is indicated by Iwata et al (2006) who treated beagle dogs for 1 year with a clinical or a 5x clinical dose of alendronate or risedronate, removed their L5 lumbar vertebrae and subjected them to 100,000 cycles of loads corresponding to 100-300% of body weight. The two doses of alendronate increased the microcrack density 2.4- and 3.1-fold respectively but risedronate increased microdamge less than alendronate. Large increases in microdamage initiation in beagle dogs' lumbar vertebrae by alendronate and a smaller increase by risedronate has been reported by Allen et al. (2006). The reason for this difference between alendronate and risedronate is unknown, but alendronate does interfere with the expression of osteoblast-specifc genes (Onyia et al., 2002).

CHAPTER 4

The Amazing Bone-Anabolic PTHs

i. Anabolics—The Answers to the Osteoporotics' Prayers

Clearly the antiresorptives are far from being the 'Holy Grails' of osteoporosis therapy although they do break the vicious cyle of escalating remodeling and microdamage (Fig. 10). Since by the time of their first fracture, the bones of osteoporotic postmenopausal women have undergone considerable microarchitectural deterioration (e.g., increased cortical osteonal porosity [R.B. Martin et al., 1998; Sietsema, 1995]) and have lost about 30% of their bone mass because of escalating microfracturing and resulting remodeling activity, it is essential to find something that can increase bone mass a lot more than the 6-10% which is the most that can be got from a 3-year treatment with the various anti-catabolics. The ideal drug for treating osteoporosis and accelerating fracture healing in both women and men (without accelerated remodeling?) would be a true *anabolic* that directly stimulates osteoblast production which actually makes strong new bone instead of just permitting the unopposed filling of existing remodeling holes by murdering osteoclasts and lowering the remodeling activity to a "normal" or even subnormal level.

In osteopenic ovariectomized animal models such as rats or monkeys such a true osteoblast-stimulating anabolic would push the bone mass well above the normal, sham-operated level without interfering with microcrack repair, which otherwise could lead to an accumulation of microdamage as may happen with long-term treatment with anti-catabolics such as bisphosphonates (Mashiba et al., 2001). It would largely bypass the BMU remodeling/repair cycle by directly and strongly stimulating bone formation by directly available or mobilizable osteoblasts. It would directly stimulate a buildup of longer-lived, more apoptosis-resistant osteoblasts, which would restore the mechanical strength of bones such as vertebrae and femurs by thickening their remaining trabeculae and thus strengthen their connectivity (Dempster, 1997; Manolagas, 2000).

Another property expected of an anbolic agent is a much greater stimulation of trabecular and endosteal growth than cortical growth because growth can only be started on bone surfaces and the total trabecular surface in an adult human male is 9 m^2 while the total surface area of the Haversian and Volkmann's canals inside the cortex are only about 3.2 m^2 (Johnson, 1966; R.B. Martin et al., 1999). However, as I pointed out in Chapter 2i the greatest total surface area, 1200 m^2, is in the 3-D osteocyte lacunar-canalicular network, but it is obviously inaccessible—it is stuffed with cell bodies and processes (Johnson, 1966; R.B. Martin et al., 1999).

As we shall now see, the PTHs—the big native parathyroid hormone and certain of its small fragments and at least one of the fragments' lactamized (cyclic) analogs—are the first and currently the foremost of the bone anabolics (Whitfield et al., 1998b, 2002a, 2002b, 2002c).

ii. The Ancient Origins of PTH and Its Receptor

The ancestor of the modern receptor for PTH and the PTH-like N-terminal 13 aminoacids of PTHrP (PTH-related protein) which we will soon be meeting belong to an ancient family that probably appeared hundreds of millions of years ago in the first animals in the Precambrian seas.

Growing Bone, Second Edition, by James F. Whitfield. ©2007 Landes Bioscience.

The descendents of the Mother Receptor are the modern receptors for calcitonin, corticotrophin-releasing hormone, gastric-inhibitory protein, glucagons, glucagons-like peptide-1, growth hormone-releasing hormone, an insect (the tobacco hornworm *Manduca sexta*) diuretic hormone, insect (*Drosophila*) calcitonin-like hormone and vasopressin-like hormone, pituitary adenylyl cyclase-activating protein, secretin, vasoactive intestinal peptide (Brody and Cravchik, 2000; Whitfield et al., 1998b; Wong, 2003). This family is an example of Nature using a basic design to do many different jobs. Thus, although these receptors have different amino acid sequences, they have similar higher order structures. Indeed, they are so alike that changing only one amino acid enables the secretin receptor to be activated by PTH and switching the equivalent amino acid in the PTH/PTHrP receptor to the secretin receptor's amino acid enables the receptor to be activated by secretin!

One of major events, if not the major event in the "Cambrian Explosion" was the arrival of Ca^{2+} on the scene. Ca^{2+}, its receptors, channels and signal transduction mechanisms improved sense organ signaling to drive the newly emerging muscles. And Ca^{2+} made body armor, diverse weaponry and even eye "lenses" for the increasingly mobile animals. The fore-runners of PTHrP and its receptors were probably involved in this revolution. The antiquity of the PTH family and their receptors is suggested by the ability of mammalian PTH to increase the Ca^{2+} concentration in crustacean hemolymph by mobilizing the ion from the exoskeleton just they do from our endoskeletons, by the presence of a molecule in the neuroendocrine cells of the pond snail that cross-reacts with anti-mammalian PTH antibodies, and by the ability of PTH to stimulate Ca^{2+} influx into snail neurons (reviewed in Whitfield et al., 1998b).

Is it possible that the Family was emerging in the armored trilobites and the heavily armored and armed *Anomalocaris*? Certainly our Cambrian ancestors, the notochord-bearing *Myllokunmingia fengjiaoa* and *Pikaia* (from Chengjian China [Xian-Guang et al., 2004] and Alberta Canada, respectively [Long, 1995]) must have been able to swim 500,000 years ago in the warm Ca^{2+}-rich (ca. 10 mM) Panthalassic and Iapetus oceans because their Precambrian ancestor(s) had found a way to avoid lethal hypercalcemia from Ca^{2+} flowing in through their gills, guts and other portals. The solution was a proto-calcitonin which was made and secreted by the first ultimobranchial bodies that later became the thyroid parafollicular cells of the terrestrial animals and the installation of calcitonin-specific offspring of the ancient mother receptor on the surfaces of gill and proto-kidney cells (Whitfield et al., 1998b).

Another very ancient anti-hypercalcemia invention was the stanniocalcin (STC) glycoprotein which was designed to prevent excess Ca^{2+} from getting into at least one kind of invertebrate (freshwater leeches) and through fish gills (Gerritsen and Wagner, 2005; Tanega et al., 2004). The circulating Ca^{2+} level in fish is measured by Ca^{2+}-sensing receptors (closely related to mammalian cells' Ca^{2+} receptors) on the cells of the kidney-attached corpuscles of Stannius (Radman et al., 2002). When the Ca^{2+} level rises above the 1.2 mM set point the cells release STC which reduces Ca^{2+} uptake by the gill cells and gut epithelial cells (Radman et al., 2002; Gerritsen and Wagner, 2005). As we shall see below STC is also involved in controlling Ca^{2+} uptake in mammalian cells and is involved in skeletogenesis (Gerritsen and Wagner, 2005).

But the problem with Ca^{2+} changed radically for some fishes during the Devonian period when the rapidly evolving lobe fins, already equipped with Ca^{2+}-sening receptors to maintain their circulating Ca^{2+} set point concentration and sporting their new many-toed limbs instead of fins, left the ocean and invaded weedy coastal lagoons and river deltas (Zimmer, 1998). Eventually they invaded fresh water and crawled out onto dry land. Then the challenge switched from preventing lethal *hyper*calcemia to preventing equally lethal *hypo*calcemia because it is harder to get this essential ion in fresh water and even harder on dry land (Whitfield and Chakravarthry, 2001). They had already pre-equipped themselves for invading dry land and losing the protection of "anti-gravity" underwater buoyancy by arming themselves with dramatically strengthened endoskeletal ribs and vertebrae made of true "mesodermal cellular calcified bone" (Clack, 2005; Smith and Hall, 1990). These pre-adaptations enabled them to breathe without being crushed by their own weight like modern beached whales as well as to serve as large Ca^{2+} storehouses (bones have 99% of the body's calcium) from which the ion could be withdrawn when needed. But they also had to redesign their kidneys so that they could dump water without losing Ca. Around this time calcitonin was fired from its job as the prime controller of Ca^{2+} levels. Its place was taken by PTH.

The two great engines of evolution have been gene duplication and gene fusion the first of which gives the opportunity for safely getting novel mutant genes (i.e., new "ideas") from the duplicates while keeping the original gene information and letting the duplicates mutate freely and safely. The second of which produces novel multi-modular proteins with new functions. Fusions would put once individually controlled genes or gene duplicates under a single command of one promoter. But where did PTH come from? To try to answer this we must travel back maybe 540 million years to Western Canada and the Cambrian Burgess Shale protovertebrate *Pikaia* or further back to China and the Early Cambrian Chengjiang protovertebrate *Myllokunmingia fengjiaoa* (Xian-Guang et al.,2004)

The first PTH was probably what we now know as the N-terminal part of PTH and PTHrP. The gene for this peptide duplicated, mutated and perhaps fused with others to make the string of different factors which is produced by brain cells and and various other organs, but, unlike the current mammalian PTHrP which doesn't circulate except in cancer patients, is a true hormone that normally circulates in the blood of fish from lampreys and sharks to sea bream (*Sparus aurata*) and Puffer fish (*Fugu rubripes*) (Danks et al., 1998; Guerreiro et al., 2001; Flanagen et al., 2000; Ingleton, 2002; Ingleton et al., 2002; Power et al., 2000; Redruello et al., 2005; Trivett et al., 2002). The ancestral hormone was probably designed first to drive the development of many different tissues of the early beasts in the Precambrian seas just as its descendent does in modern beasts. It was meant to drive Ca^{2+}-dependent epithelial differentiation in various organs, later the formation of tough mineralized dermal structures such as the ancient jawless fishes' boney head shields, dermal denticles (skin teeth) and their then new internal cartilagenous skeletons (Carter and Beaupré,2001) It also used its N-terminal region in kidneys, gills, brains and nerves to control Ca^{2+} and phosphate levels in the very high, indeed the dangerously high-Ca^{2+} (e.g., 10mM) seawater.

But PTHrP has another major role in the calcium economy of some modern fish by targeting their scales which are really neural crest-derived dermal bones complete with osteoblasts and osteoclasts (Sire and Huysseume, 2003). In a bony fish such as the sea bream (*Sparus aratus*), the scales, like our bones, are calcium storehouses the cells of which are studded with the type 1 PTH/PTHrP receptors which, like ours, when activated by binding PTHrP sends signals that trigger the formation of osteoclasts that mobilize the scales' calcium (Rotllant et al., 2005).

The original PTHrP might have been a much slimmer molecule than the modern mammalian molecule (Ingleton, 2002). Indeed judging from the modern fish PTHrPs it was rather like the mammalian PTH. It had the same N-terminus, nuclear transport and RNA-binding domains as the "modern" mammalian PTHrP (Ingleton, 2002). But it did not have the mammalian "osteostatin" domain with the 107-111 (TRSAW) sequence. On the other hand the mammalian molecule does not have the modern fish and frog PTHRPs' functionally mysterious 38-54 domains (Ingleton, 2002). The ancient PTHrP's gene was located beside the thymopoietin gene and the mammals have kept their PTHrP gene there during the subsequent hundreds of millions of years of evolution. This enduring linkage seems to be due to thymus epitelial cells and parathyroid gland cells being derived from the same group of pharyngeal endodermal cells in the embryonic third and fourth branchial pouches (Gordon et al., 2001; Günther et al., 2000; Su et al., 2001).

Then something momentous happened in a certain line of fish which determined our destiny—a something needed to avoid a lethal, Ca^{2+} shortage it any fish were foolish enough to try to leave the Ca^{2+}-rich ocean for Ca^{2+}-poor fresh water or dry land. If this had not happened we would not be here. These fish built parathyroid glands from their gill buds (Okabe and Graham,2004).Until then these fish, like others, used the product of their *gcm2 gene* (one of the two known homologs of the *Drosophila's* *"glial cells missing"*, *gcm*, gene [Ghnter and Karsenty, 2005; J.Kim et al.,1998]) to make their gills. But the new parathyroid glands in the gill buds were Ca^{2+}-sensing machines equipped with CaRs (Ca^{2+}-sensing receptors) connected to machinery for secreting PTH which the fish had been making but now localized to the new gland. The parathyroid glands appeared when cells driven by the *Hoxa3* gene started expressing the *gcm* 2 gene in the ancient water-drenched gill slit region, specifically the third and fourth pharyngeal pouches. There the analge was marked out by the collaboration of pharyngeal pouch endodermal cells with neural crest cells from the adjacent overlying hindbrain rhombomeres (Ghnter and Karsenty, 2005). From the beginning the *gcm2*-expressing parathyroid cells make PTH, but so can thymus cells. The parathyroid cells originate beside the thymus gland anlage in the gill buds which is demarcated by the expression of the *foxn* 1 gene instead of the *gcm2* gene (Ghnter and Karsenty, 2005). This *gcm2*-expressing part of the initially adjacent thymus and PTH-producing parathyroid anlagen in the third and fourth pharyngeal pouches was the forerunner of the encapsulated parathyroid gland of terrestrial vertebrates (Gordon et al., 2001; Su et al., 2001). Echoes of their ancient origin parathyroid glands from this same gill slit region shared with the thymus primordium are the facts that although knocking out *gcm2* in mice eliminates the parathyroid glands, the thymic epithelial cells can take over and make PTH and, of course, the continuing location of the mammalian PTHrP gene beside the thymopoietin gene (Günther et al., 2000).

With time, the emerging pharyngeal PTH became the circulating Ca^{2+} and phosphate managing hormone. Big PTHrP then lost its original side-job of being the circulating Ca^{2+} controller, but it kept its many other jobs of being a multipurpose, widely expressed and acting autocrine, paracrine, "intracrine" and nuclear gene-activaing polyprotein—a non-circulating string of *cytokines* instead of a single circulating hormone—that mediates various mesenchymally driven epithelial differentiation programs, endochondral bone formation, smooth muscle relaxation, placental Ca^{2+} transport, and neural Ca^{2+} channel repression (Chatterjee et al., 2002; Macica and Broadus, 2003; Nissenson, 2000; Re, 2002a, 2002b; Re and Cook, 2005; Whitfield and Chakravarthy, 2001). In fact the various add-ons (gene fusions and/or mutant duplicates of the original gene) to the big new multi-modular PTHrP molecule, marked out by 8 endoproteolytic consensus sites, could be scissored out of the cytokine chain by processing proteases and then used separately to activate different receptors and stimulate the functions of the cells expressing them. On the other hand, the smaller, new PTH, with its potent N-terminal nose and dragging a functioning but seemingly unessential, tail took over as the circulating hormonal controller of the body's all-important Ca^{2+} and phosphate levels via the bone reserves, kidney tubules and gut epithelium. But it could also functionally overlap with big PTHrP's ancient N-terminal domain because it could still target cells with the PTHR1 receptor which had been used by the original PTHrP.

Thus, when the first Devonian vertebrates invaded dry land 350,000,000 years ago they already had parathyroid glands, the cells of which had the by then ancient Ca^{2+} sensors/receptors housed in the tiny flask-shaped caveolae (little caves or pouches) in their surface membranes with which they could monitor the blood Ca^{2+} concentration (Kifor et al., 1998, 2002, 2003). If the Ca^{2+}-driven signals from these sensors should start fading, the cells try to restore the volume of sensor signaling by releasing enough PTH to raise the circulating Ca^{2+} concentration by increasing the outflow of phosphate, stemming the outflow of Ca^{2+} through the kidney, increasing Ca^{2+} uptake through the gut, and commanding the members of the osteocyte internet and bone lining cells to pull Ca^{2+} out of the bone stores if the other two fail to do the job (Kifor et al., 2002, 2003).

iii. PTHs-Induce Bone Growth and Fracture Mending in Non-Human Animals

iii a. The Discovery

In 1929, the Gospel-Carved-in-Stone or perhaps the Gospel-carved-in-Bone was that PTH's job in bone was to extract Ca^{2+} and dump it into the blood when needed, for example, to mobilize Ca^{2+} for fetal bone building. But Bauer et al. (1929) got a big surprise when they treated female rats with the only available PTH (a bovine parathyroid gland extract from Eli Lilly) to assess the role of trabecular bone as a readily resorbable, accessible Ca^{2+} reserve, but instead they saw—*bone growth*! But they didn't know what to make of it and didn't follow it up. There was an excellent reason for their reluctance to believe that PTH could be osteogenic. Even the intermittent injections that we now know strongly stimulate bone growth still stimulate osteoclasts with their acid hoses and cathepsin K shovels, but their destructive digging can be overridden by the bone-building arm provided *the injections are sufficiently far apart*. However, if the PTH is injected at close intervals or continuously infused into mice or rats the response is net bone resorption instead of net growth (e.g., Shneider et al., 2003; Tam et al., 1982).

With all due respects to Bauer et al, the bone-growing PTHs story was really kick-started in Paris by a group who boldly went where no one had gone before—*they suggested that a young boy's ultimately lethal bone overgrowth (osteopetrosis) was due to the PTH secreted by his parathyroid tumor* (Péhu et al., 1931)! Hans Selye was fascinated and then promptly fathered the field of PTHs-As-Anabolics when he and his colleague L. Pugsley announced that daily injections of small doses of the Eli Lilly extract ("Parathormone") into rat pups caused a massive pile up of bone-making (osteogenic) osteoblasts on the surfaces of the trabeculae in cancellous bone just as had happened in the unfortunate French boy (Selye, 1932; Pugsley and Selye, 1933). The Selye experiment gave birth to a paradox—PTH was supposed to be a potent bone destroyer, not a potent bone builder. As Riordan (1987) says on page 31 of his book "The Hunting of the Quark"—"*A paradox is a sure sign that nature is trying to tell us something completrely new and different about herself*".

iii b. The PTHs—Who Are They? What Do They Do in Non-Human Animals?

At this point we should remember from Chapter 2 i that when something like a PTH is injected it first enters the bone via the marrow blood vessels which means that the cells in the cortical shell will see it only after the PTH receptor-bearing cells in the extensive surfaces of the trabeculae and endosteum have had the first grab at it. Therefore a recurring theme will be the PTHs' apparent "trabeculophilia"—its ability to stimulate trabecular bone growth more than cortical bone growth.

An incredible thing about the dramatic accumulation of osteoblasts induced by the parathyroid extract in Selye's rat pups was that it happened *directly without being primed by osteoclasts as happens in human BMUs*—there was a direct stimulation of osteoblast precursors in the little beasts' rapidly growing bones which we now know would have been loaded with PTH receptor-loaded, bone-building osteoblasts! But who would have believed it? In those days people had to use a bovine parathyroid gland extract. The extract undoubtedly contained intact bovine PTH-(1-84) and some of its various proteolytic fragments, but it almost certainly contained other things (Bringhurst, 2002). And it took almost 40 years for the bone people to get around to proving that it was indeed the PTHs (Kalu et al., 1970; Walker 1971)!! Since then, the basic features of PTH's osteogenic action described by Selye and Selye and Pugsley have been repeated by the results of scores of experiments on cynomolgus monkeys, dogs, ferrets, rabbits and sheep, but on literally thousands of OVXed rats, using a synthetic piece of PTH, h(human)PTH-(1-34), that has been wrongly believed for decades to be fully bioactive (comprehensively reviewed by Dempster et al. [1993] and Whifield et al. [1998a, 2002a, 2002b, 2002c, 2006a, 2006b]).

Actually hPTH-(1-34) is *not* fully bioactive—the rest of the native hormone is not just inert filler as has been generally believed! It now appears that following the gene duplication that put PTH and PTHrP on their separate evolutionary pathways, PTH acquired an additional target and functions. In the mid-1980s T.M. Murray and his group in Toronto Canada published almost completely ignored (by me included) evidence for there being another receptor for a region of the native hormone's C-tail (McKee et al., 1985; Rao et al., 1985). Then, in 1998, Erdmann et al. (1998) reported that the big native PTH's residues 73-76 form the core of a region that activates a receptor, the signals from which cause Ca^{2+} to surge into growth plate chondrocytes. Divieti et al (2001, 2002a, 2002b) have also found more evidence of this receptor which they have named the CPTHR(C-terminal PTH receptor), that sees C-terminal PTH fragments. The binding affinity is maximum (20-30 nM) for hPTH-(24-84), but drops 25-fold when residues 25-27 are missing, i.e., the affinity for hPTH-(28-84) is 500-700 nM (Bringhurst, 2002). This receptor is apparently maximally expressed by the spidery osteocytes, i.e., the locked-in-bone cells with extended dendrites that make mRNAs for gap junctions' connexin 43 and the mineralization-restraining osteocalcin but not the osteoblast-specific alkaline phosphatase or Cbfa1/Runx-2) (Bringhurst, 2002; Divieti et al., 2001). And the CPTHR signals promote the formation of gap junctions which are such an important part of the intercellular connections in the osteocyte osteointernet signaling syncytium that registers strain and microfracture (Burger, 2001; D'Ippolito, et al., 2002). If the big PTH should somehow reach the osteocytes in their lacunocanalicular hideaways, the signals from this receptor would cause the osteocytes to turn on their self-destruct apoptogenic mechanism (Bringhurst, 2002; Divieti, 2005; Divieti et al., 2001). But this CPTHR-driven osteocyte killing is counterbalanced by the reduction of osteoclast generation by the activation of the CPTHRs on osteoclast progenitors (Divieti et al., 2002a, 2002b). However, the differentiating osteoclasts can be protected from lethal CPTHR signaling by having their CPTHR expression suppressed when the differentiating osteoclasts bind to immature osteoblastic stromal cells' PTH-stimulated RANKLs (Divieti et al., 2002b). Thus, the intact PTH-(1-84), unlike hPTH-(1-34), has many targets—preosteoblasts, osteoblasts, osteocytes and of course the many other cells in the body that have the N-terminal PTH-binding PTHR1 receptors as well as osteocytes and osteoclast precursors that have CPTHRs. The clinical meaning of this additional signaling may never be known because of the demise of hPTH-(1-84) (PREOS™) at the hands of the USFDA and the decision of its producer NPS Pharmaceuticals, Inc. in June 2006.

The intact PTH that has not clamped onto its target cells, is promptly chopped into mid-molecule and C-terminal, but not N-terminal, fragments in the liver by Kupffer cells—the half-life of circulating exogenous or endogenous intact hPTH-(1-84) is just 2 minutes (Bringhurst, 2002, 2003; Bringhurst et al., 1988; Kronenberg et al., 2001). N-terminal fragments of the big mother PTH have half-lives of only about 5 minutes and seem to be selectively destroyed, but the mid-molecule and C-terminal fragments have much longer half-lives of 24 to 36 hours (Bringhurst, 2002; Bringhurst et al., 1988; Kronenberg et al., 2001). One of the long-lived, large circulating C-terminal fragments is hPTH-(7-84) (Kronenberg et al., 2001), which is directly released from the parathyroid glands (Yamashita et al., 2003). It can promote osteocyte apoptosis and impair osteoclastogenesis (Bringhurst, 2002; Divieti et al., 2002). However, as we shall see a bolus of normally rare N-terminal PTH fragments by subcutaneously injecting the hPTH-(1-34) fragment can obviously override any destructive, apoptogenic action of CPTHR signaling from the circulating large C-terminal breakdown products of the big native PTH and strongly stimulate bone formation.

One of the effects of CPTHR signaling is the stimulation of the proliferation of human vascular endothelial cells by a Ca^{2+}-triggered, CAM kinase II-driven mechanism (Isales et al. (2004). A fascinating feature of the interactions of the hPTH-(1-84)'s two receptor-targeting regions in the endothelial cells is the ability of hPTH-(1-84) to stimulate the expression of "only" 166 genes and downregulate the expression of "only" 56 genes presumably by activating PTHR1 and CPTHR receptors while just activating the CPTHR receptor with hPTH-(53-84) stimulates 922 genes and downregulates 699 genes!! Evidently the signals from PTHR1 somehow interferes with most of the gene responses to CPTHR signals.

These observations suggest that the selective destruction of N-terminal fragments (e.g., amino acids 1-34) of the native PTH-(1-84) could be meant to prevent circulating PTH from interfering with differentiation mechanisms driven in the cells of various tissues including bones by signals from the cells' PTHR1 receptors activated by the hPTH-(1-34)-like N-terminal domain of locally produced and secreted PTHrP the ancient differentiation driver.

The native PTH is a string of 84 aminoacids (reviewed by Morley et al. 1999, Whitfield, 2006 b; Whitfield et al., 1998a). Injecting one small dose (e.g., 1-50 nmoles/100 g of body weight) of the native hPTH-(1-84) (PREOS™) or certain of its N-terminal fragments ([Leu^{27}]*cyclo*(Glu^{22}-Lys^{26})hPTH-(1-28)NH$_2$ (mini-C), hPTH-(1-30)NH$_2$, hPTH-(1-31)NH$_2$ (Ostabolin™), [Leu^{27}]*cyclo*(Glu^{22}-Lys^{26})hPTH-(1-31)NH$_2$ (Ostabolin-C™), b(bovine)PTH-(1-34), hPTH-(1-34)NH$_2$, recombinant hPTH-(1-34) (Forteo™) [Leu^{27}]*cyclo*(Glu^{22}-Lys^{26})hPTH-(1-34)NH$_2$, recombinant [Asp^{35}]hPTH-(1-35), hPTH-(1-36), or hPTH-(1-38)) into OVXed mice and rats or orchidectomized (castrated) male rats each day stops the bone loss, stimulates cortical bone growth, acts synergistically with external loading to increase cortical bone strength, and dramatically raises the bone mineral density (BMD) and mean trabecular thickness in trabeculae-rich bones such as the distal femur, jaw, tibia and vertebrae or subcutaneous ectopic "ossicles" *as much as 2 or more times above the level in normal or still intact sham-OVXed or orchidectomized (i.e., operated on but the ovaries or testicles are not removed) animals* (Table 1; Akhter et al., 2001; Andreassen et al., 1999; Arita et al., 2002; Barry et al., 2002; Brommage et al.,1999; and Lindsay, 1998; Dempster, 1997, 2000; Dempster et al, 1993; Dobnig and Turner, 1995; Gabet et al., 2003; Gasser, 1997; Gomez et al., 2004; Hagino et al., 1999; Hodsman et al., 1999a, 1999b; Jerome et al., 1994, 1999, 2001; Jilka et al., 1999; Kawane et al., 2002; Kneissel et al., 2001; Kostenuik et al., 1999,2000,2001; Lane et al.,1995; Leaffer et al., 1995; Liang et al., 1999; Ma et al., 2000; Mohan et al., 2000; Manolagas, 2000; Morley et al., 1999, 2001a,b; Mosekilde, 1997; Mosekilde et al., 1997; M. Nakajima et al., 2000; Nishida et al., 1994; Opas et al., 2000; Rixon et al., 1994; Samnegard et al., 2991; Schneider et al., 2003; Shirota et al., 2003; Sogaard et al., 1997; Strein, 1994; Sung et al., 2000; Tam et al,. 1982; Tormanoff et al., 1997; Valenta et al., 2005; von Stechow et al., 2003;

Table 1. The anabolic/osteogenic PTHs

---

rhPTH-(1-84)
hPTH-(1-38)
hPTH-(1-36)
rhPTH-(1-34) (Forteo™ [NA]; Forsteo™[Eur])
rhPTH-(1-31)NH$_2$
hPTH-(1-31)NH$_2$ (Ostabolin™)
[Leu^{27}]cyclo(Glu^{22}-Lys^{26})hPTH-(1-31)NH$_2$ (Ostabolin-C™)
hPTH-(1-30)NH$_2$
[Leu^{27}]cyclo(Glu^{22}-Lys^{26})hPTH-(1-28)NH$_2$ (mini-C)

Walker, 1971; Whitfield et al., 1995, 1996b, 1997a,b,c,d, 1998a,b,c, 1999a,b, 2000a,b,c,d, 2001, 2002a,b,c; Wronski and Li, 1997; Wronski et al., 2001).

Starting daily subcutaneous injections of 30 or 75 μg of hPTH-(1-34)/kg of body weight into a young female rat when she is only 5-8 weeks old and continuing for the rest of her life will produce big, grossly abnormal bones (Sato et al., 2002). The PTH injections will keep stimulating bone growth throughout rat's life. The injections will push the growth of bones such as the femur far above normal values, indeed so far that the marrow cavities will be nearly filled with bone as in Pehu et al's little boy, the difference between cortical and trabecular bone will disappear in a kind of fusion and the vertebrae and femora will be greatly stiffened. An example of the immense PTH-induced trabecular thickening in a rat femur may be seen in Figure 16E,F. Moreover, although PTH normally stimulates only the growth of lining cell-covered, marrow-bathed endosteal and trabecular bone (Andreassen and Oxlund, 2000) the lifelong massive stimulation will eventually include periosteal apposition. But, as we shall see further on this relentless lifelong battering by daily PTH injections into Fisher rats and the excessive amounts of PTH constantly being made in humans with primary

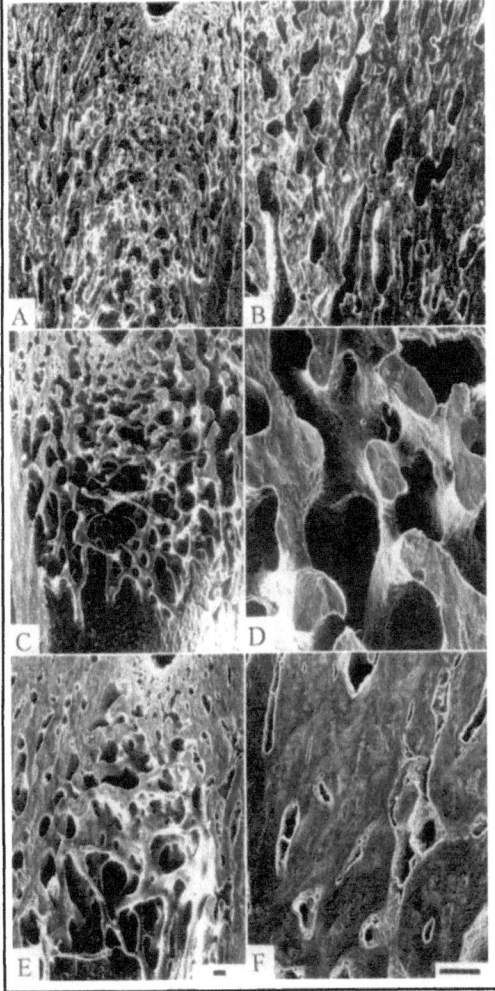

Figure 16. A striking demonstration of the potent ability of hPTH-(1-31)NH₂ (Ostabolin™) to very strongly stimulate trabecular growth in rat distal femurs. The samples were taken at the end of 6 weeks of single daily subcutaneous injections of 0.8 nmole of hPTH-(1-31)NH₂/100 g of body weight that were started 2 weeks after ovariectomy (OVX). A,B) Vehicle-treated, sham-OVXed rat; C,D) An OVXed rat; E,F) the OVXed rat received 6 weeks of daily injections of 0.8 nmole of the PTH starting 2 weeks after OVX. These are scanning electron micrographs of demineralized bone obtained with a Philips SEM-505. The bar for the A, C, E column represents 30 μm and the bar for the B, D, F column represents 100 μm.

hyperparathyroidism had a deadly consequence in Lilly's preclinical trials of rhPTH-(1-34) (Forteo™)—osteostarcomas (Vahle et al., 2002, 2004).

According to the results of experiments on rats, humans and cultured human and rat bone cells, the PTHs owe their ability to stimulate bone growth by dramatically stimulating the proliferation of marrow osteoprogenitor cells followed by the layering of osteoblasts on quiescent bone surfaces without a preliminary appearance of, and signaling from, osteoclasts exactly as Selye reported in 1932 (Hodsman and Steer, 1993; Kostenuik et al., 1999). Indeed a contribution of osteoprogenitor proliferation to this layering is indicated by PTH-induced stimulation of DNA synthesis in the the tibias of female rats, a response that is nearly doubled by mechanically loading the tibias (Barry et al., 2002). But at least in rats, but maybe not in mice, there is also a direct and reversible stimulation of the 'reversion' of retired osteoblasts—the quiescent lining cells covering the osteonal walls, the endosteum and trabeculae—to active osteoblasts, an increase in osteoblast activity, and an increase in the osteoblast lifespan by preventing apoptotic self-destruction (Dobnig and Turner, 1995; Hock, 1999; Manolagas, 2000; Jilka et al., 1999, 2004; Leaffer et al., 1995).

The PTHs are most potently anabolic for the bones of the fast-growing fetal and newborn animals simply (Walker, 1971) because the marrow is still red and loaded with stem cells and osteopogenitor cells and the growing bones are covered with gangs of mature osteoblasts bristling with PTHR1 receptors (Fermore and Skerry, 1995). Because the growing bones are in this super-responsive state, PTHs can so greatly stimulate trabecular bone growth that it fills the marrow cavities and drives hematopoiesis back into fetal pre-marrow sites such as spleen and liver in these young animals (and, as we shall soon see, in at least one historically very significant young French boy) (Walker, 1971).

Of course, this is why Selye (1932) found that the PTH in the Lilly bovine parathyroid extract so greatly stimulated osteoblast accumulation and bone growth in his rat pups without first stimulating osteoclast activity. As fatty yellow marrow replaces red marrow (Bianco and Riminucci, 1998) and bone growth slows, but does not stop, with age so does the responsiveness of bone to PTH in rats (Walker, 1971). In big animals such as humans with no longer growing microcracking-prone bones at their loading sites (R.B. Martin, 2002), the availability of PTH receptor-bearing preosteoblasts and mature osteoblasts for an *immediate* or *first* response to PTH injection in the mature skeleton is a function of the number of responsive bone-lining cells (Dobnig and Turner, 1995) and BMUs which is linked to osteoclast activation (Manolagas, 2000).

None of the PTHs can restore the original bone structure. In rats, they cannot generate new trabeculae de novo, increase the trabecular number, or make broken trabeculae rejoin if they have drifted too far apart (Dempster, 2000; Jerome et al., 2001; Lane et al., 1995; Whitfield et al., 1998b). Nor, it seems, can hPTH-(1-34) stop or reverse the drop in the number of central metaphyseal trabeculae in the distal femur when its injections are started 2 weeks after OVX, because it stimulates osteoclast generation as well as osteoblast bone-building (Whitfield et al., 1998b). In other words while the OVX- and PTH-(1-34)-stimulated osteoclasts are busily destroying central trabeculae in the femur, PTH-stimulated osteoblasts are thickening the more lateral, load-transmitting trabeculae (reviewed by Whitfield et al., 1998b) (Fig. 1). The awesome, marrow-obliterating trabecular growth that can be stimulated by a potent osteogen such as hPTH-(1-31)NH$_2$ (Ostabolin™) is shown in Figure 16. The net result of this is an increased trabecular thickness (but not number) and increased total cancellous volume and a disproportionate strengthening of the OVX rat's load-bearing bones (Dempster et al., 1997; Whitfield et al., 1998b,1998c) (Figs. 1 and 17). However, [Leu^{27}]$cyclo$(Glu^{22}-Lys^{26})hPTH-(1-31)NH$_2$ (Ostabolin C™) does not allow the same loss of trabeculae as does hPTH-(1-34)(Forteo™) (Whitfield et al., 1998c). This adds to the other observations that intermittent injections of hPTH-(1-31)NH$_2$ either do not cause, or cause far less, resorption than hPTH-(1-34) in the bones of humans and mice (Fraher et al., 1999; Mohan et al., 2000). This has been strikingly confirmed by the demonstration in preclinical trials on cynomolgous (*Macaca fascicularis*) monkey that even intermittent injections of very high doses of [Leu^{27}]*cyclo*(Glu^{22}-Lys^{26})hPTH-(1-31)NH$_2$ (Ostabolin-C^{TM}) did not cause hypercalcemia and in fact seemed actually to *reduce* osteoclast activity while at the same time strongly stimulating bone growth (Jolette et al., 2003, 2005)! The failure of [Leu^{27}]*cyclo*(Glu^{22}-Lys^{26})hPTH-(1-31)NH$_2$ (Ostabolin C^{TM})-injected, OVXed rats to lose as many trabeculae as the untreated OVXed or hPTH-(1-34) (Forteo™)-treated rats and the apparent ability of the peptide to reduce osteoclast activity in monkeys could be explained by [Leu^{27}]*cyclo*(Glu^{22}-Lys^{26})hPTH-(1-31)NH$_2$ (Ostabolin C^{TM}) preferentially stimulating the maximally receptor-loaded mature osteoblasts, which according to Atkins et al. (2003) express OPG-instead of RANKL as do immature osteoblastic cells (look ahead to Fig. 35).

Daily injections of hPTH-(1-34) also stimulate the layering of new bone primarily on the inner cortical tunnels and trabecular trenches and thus raise the bone mass and bone strength in ovariectomized non-human primates such as *Macaca fascicularis* cynomolgous monkeys above the normal level in sham-operated monkys (Turner et al., 2001; Sato et al., 2000). But this PTH does something in OVXed monkeys that it can't do in rats—it *increases the number of trabeculae* (Jerome et al., 2001). But it does not do this by making new trabeculae from scratch. Instead, it does this by increasing the thickness of existing trabeculae, which are then sliced in two by osteoclastic 'tunneling' when they reach a certain size (Jerome et al., 1994, 2001). Trabecular tunneling is thus a primate's ingenious way of restoring the original trabecular thickness and number without making new trabeculae from scratch.

Without estrogens, the new bone in a PTH-treated OVXed rat should be, and apparently is in some cases, as much a victim of overdigging BMUs as was the old bone, when the PTH injections are stopped (reviewed by Whitfield et al., 1998b, 2000c,d). But according to the results of most of the relevant experiments on rats, the new "PTH bone" be it endocortical or trabecular, can be effectively protected by an anti-catabolics such as a bisphosphonate, calcitonin, or an estrogen (Samnegård et al., 2001; reviewed by Whitfield et al., 1998b, 2000c, 2000d). However, the new bone's need for protection would depend on its rate of turnover (i.e., the size and activity of the osteoclast population). Thus, for example, Mosekilde et al. (1997) found that the new "PTH bone" in the sluggishly turning over bones of aged osteopenic OVXed rats did not need protection by the bisphosphonate, risedronate. And the positive effect of a 12-month treatment with PTH-(1-34) on the cancellous bone of femoral neck, iliac crest and vertebrae of OVXed cynomolgus monkeys was maintained for 3-6 months after the injections were stopped (Jerome et al., 2001). But new 'PTH bone' can face another threat. If it should have grown beyond mechanical needs, it will be under-strained which sounds the alarm of redundancy from a growing number of dying osteoclasts and then it will be destroyed by a flock of vulturing osteoclasts (Dempster et al., 1997).

Co-treatment with an antiresorptive to block the resorption component theoretically should enhance the effectiveness of the PTH. But the maximum increase in trabecular and cortical bone mass in rats treated with an optimal dose of hPTH-(1-34) or hPTH-(1-38) alone is the same as in rats co-treated with PTH and a bisphosphonate, calcitonin or an estrogen (Hodsman et al., 1999a,b). However, co-treatment with the SERM raloxifene reduces the dose of the PTH fragment needed to get this maximum effect. But Boyce et al. (1996) found that co-treating beagles with hPTH-(1-34) and a bisphosphonate, risedronate, was osteogenically better than treating with hPTH-(1-34) alone. On the other hand Delmas et al. (1995) found that a bisphosphonate, tiludronate, eliminated the otherwise strong bone-building action of hPTH-(1-34) in old sheep with their slowly remodeling, hence sparsely BMUed, bones. Evidently the only source of PTH receptor-bearing preosteoblasts and mature osteoblasts in these old beasts were BMUs. So getting rid of BMUs with the bisphosphonate eliminated the waves of osteoclasts needed to kindle an osteogenic response. In other words, there was no prior prompting from osteoclasts, no osteoblasts, few PTH targets to start bone growth.

OPG, the physiological inhibitor of osteoclast generation, should potently enhance PTH osteogenicity by selectively stopping the PTH from stimulating osteoclast generation without affecting the stimulation of osteoblast accumulation and activity (Capparelli et al., 2003; Hofbauer et al., 2000; Kostenuik and Shaloub, 2001; Kostenuik et al., 2000a,b). Nor would it have any of the side effects of the other less selective artificial antiresorptives. This expectation has been confirmed with young growing rats by Kostenuik and his colleagues (Kostenuik and Shaloub, 2001; Kostenuik et al., 2001). They OVXed 3-month-old rats and began treating them with PTH and/or OPG 15 months later. OVX significantly reduced the bone mineral density (BMD) in the distal femurs and lumbar vertebrae. Subcutaneously injecting hPTH-(1-34) (0.08 mg/kg) three times a week for 5 1/2 months into the aging rats *increased* the osteoclast-gnawed surface by 50%, but it also significantly increased the BMD in the distal femurs and lumbar vertebrae. Subcutaneously injecting recombinant OPG (10 mg/kg of body weight) stopped OVX from increasing osteoclast activity. The femoral and vertebral mineral density then rose as the unaffected osteoblasts continued making bone. But the OPG was less effective than the PTH injections in increasing the BMD. Injecting the recombinant OPG along with the PTH prevented the PTH from increasing osteoclast activity without affecting its osteogenic activity and consequently caused a greater increase in BMD than did the injections of either peptide alone. However, Valenta et al (2005) have now shown that PTH-OPG combination therapy did not further increase the large increase in bone volume caused by PTH alone in aged OVXed rats treated when they were between 18 and 23.5 months-old. Regardless of whether the rats were young and growing or very old, these observations again underscore Selye's original observations on rat pups given daily small doses of bovine PTH extract that PTH does not need a prior activation of osteoclasts to stimulate bone growth (i.e., it stimulated bone growth despite an essentially total osteoclast shut down) and that OPG strongly enhances a PTH's bone-building action. PTH only needs enough receptor-bearing preosteoblasts and osteoblasts to start the process.

It is important for managing postmenopausal bone loss to know that OPG can prolongedly inhibit bone resorption in postmenopausal women. One subcutaneous injection of an OPG dose such as 3.0 mg/kg of body weight reduced the urinary levels of collagen (i.e., bone) breakdown products such as N-terminal telopeptides by as much as 80% by the 4th day after injection and were still down by as much as 14% as long as 6 weeks after the injection (Bekker et al., 2001). This prolonged suppression of osteoclast activity reduced the blood Ca^{2+} concentration. This drop turned off the parathyroid glands' Ca^{2+}-sensors, which caused a prolonged elevation of the circulating PTH concentration (from a mean normal 40 pg/ml to about 80 pg/ml at 5 days and still about 55 pg/ml by 30 days) to try to bring the blood Ca^{2+} concentration back to the normal level. Of course the hormone couldn't do this by stimulating the osteoclasts to chew up more bone so it had to rely on reducing phosphate uptake by proximal kidney tubules and increasing the Ca^{2+} uptake by distal kidney tubules which resulted in a 80% drop in the urinary Ca^{2+} concentration by 5 days , a 50% drop by 20 days and a lingering 10% drop by 30 days.

As expected, the osteogenic PTHs accelerate fracture mending in both normal and osteoporotic bones. Indeed, a fracture enhances the effectiveness of PTH by triggering a shower of BMUs with their PTH receptor-bearing, hence directly PTH-stimulable, mature working osteoblasts (R.B. Martin et al., 1998). For example, daily subcutaneous injections of 200 μg of hPTH-(1-34)/kg of body weight increased both the ultimate load that could be tolerated *before* breaking and the callus volume of rat tibial fractures by as much as 75% and 95% respectively over the control values by 20 days after fracturing and by 175% and 72% respectively over the controls during the next 20 days (Andreassen et al., 1999). A similar acceleration of fracture mending by Ostabolin-C™ ([Leu27]*cyclo*(Glu22-Lys26)hPTH-(1-31)NH2) has been reported by Andreassen et al. (2003). Kim and Jahng (1999) have also reported that a daily injection of recombinant hPTH-(1-84) for 30 days significantly accelerated the mending of surgically produced bilateral tibial shaft fractures in OVXed rats. A.Nakajima et al. (2002) have found that daily injections of 10 μg of hPTH-(1-34)/kg of body weight into male Sprague-Dawley rats with unilateral femoral fractures stimulated the probably IGF-I-mediated proliferation of sub-periosteal osteoprogenitor cells in the callus as indicated by the fraction of cells expressing PCNA (proliferating cell nuclear antigen), the DNApolymerase-δ processivity clamp (Whitfield and Chakravarthy, 2001). By 28 days the PTH injections had increased BMC (bone mineral content) 61%, BMD 46% and load to failure 32% more than in control vehicle (the solution in which the PTH was dissolved)-treated rats. By 42 days, these values had risen respectively 119%, 74% and 55% above those in the control untreated rats. Alkhiary et al (2003) have reported a similar enhancement of femoral fracture healing in male Sprague-Dawley rats. Finally, Seebach et al. (2004) have reported that hPTH-(1-34) (60 μg/kg subcutaneously injected every second day) significantly increased distraction-induced osteogenesis (as indicated by increased ultimate load, stiffness, total regenerate callus volume, callus BMC and bone density) in the progressively widened gap between the surgically cut ends of femurs in 3-months-old male Sprague-Dawley rats.

In a different and very elegant way to accelerate fracture mending, Bonadio and colleagues implanted a degradable GAM (Gene Activated collagen Matrix) sponge loaded with a plasmid containing DNA encoding hPTH-(1-34) into surgical breaks in rat femurs as well as beagle femurs and tibias that were so wide that they normally would have have healed only very slowly or not at all (Bonadio, 2000; Bonadio et al., 1999; Fang et al., 1996; Goldstein and Bonadio, 1998). But the fibroblasts in the fractures' granulation tissue picked up the plasmid DNA and briefly became mini-bioreactors making and secreting hPTH-(1-34) that dramatically stimulated fracture healing. Implanting a second GAM with DNA coding for BMP-4 (bone morphogenic protein-4) along with the PTH-GAM accelerated mending even more.

A fracture destroys blood vessels creates a clot and the resulting hypoxia stimulates the production of VEGF and the formation of new blood vessels. These new blood vessels invading the fracture site secrete the potent osteoprgenitor stimulators BMP-2, endothelin-1and OP-1 (Bouletreau et al., 2002; Carano and Filvaroff, 2003) (Fig. 8). And the bone endothelial cells differ from other endothelial cells by having PTH receptors (Streeten and Brandi, 1990), which likely play a role in fracture healing and the enhancement of fracture healing by injected PTH.

PTHs can also dramatically accelerate bone formation in and around an implant—an ability of immense orthopedic importance. Indeed screwing an implant into a bone, like a fracture, should make more PTH receptor-bearing PTH targets and therefore increase the primary response to a PTH in humans by driving squadrons of osteoblast-inducing BMUs out in all directions from the screwholes (Davies, 2003; R.B. Martin et al., 1998). Skripitz et al (2000a, 2000b) showed the implant-fixing power by inserting perforated hollow titanium chambers into the cortices of the proximal tibias of male Sprague-Dawley rats and followed the penetration of endosteal cells into the chambers and the growth of bone in the chambers. In untreated rats by 6 weeks the bone in the chamber had been hollowed out by osteoclasts to form a marrow space with few trabeculae, but in rats which had been injected daily and subcutaneously with hPTH-(1-34) the chamber was filled with thick trabecular struts and plates. Then Skriptz and Aspenberg (2001a, 2001b) showed that subcutaneously injecting hPTH-(1-34) only on Mondays, Wednesdays and Fridays stimulated the fixation of stainless steel screw implants to rat tibias. In just 2 weeks, the PTH injections so strongly stimulated the growth of dense fibrous bone-anchoring tissue around the screws that it took twice as much force to pull them out of the tibias than to pull the implants out of the control rats' tibias. Shirota et al. (2003) found that hPTH-(1-34) reversed the thinning of bone around titanium screw implants on ovariectomized rats. More recently M. Allen et al.(2003) have reported that [Leu^{27}]*cyclo*(Glu^{22}-Lys^{26})hPTH-(1-31)NH$_2$ (Ostabolin-C™) is significantly better than hPTH-(1-34) (Forteo™) at stimulating the growth of anchoring bone around a methacrylate pin in rat tibias. Since then Skripitz et al (2005) have reported that intermittent injections of hPTH-(1-34)OH into male Wistar rats increased the contact of bone with a stainless steel tibial implant about 3-fold by 2 weeks and 5-fold by 4 weeks, but the peptide increased the contact of bone with rougher polymethylmethacrylate rods as much as 6.9 fold by 2 weeks and had reached a maximum contact by 4 weeks. These data clearly indicated the tremendous potential of the PTHs for increasing the micro-interlocking of bone with implants, particularly rough-surfaced implants. Even more recently Gabet et al. (2006) have tested the ability of intermittent injections (particularly 25 and 75 μg/kg /day) of hPTH-(1-34) to strikingly enhance the osteointegration and anchorage of titanium screws horizontally inserted into the proximal tibial metaphyses with reduced trabecular bone caused by 7 weeks of pre-exposure to testosterone deprivation in gonadectomized rats. The question answered here is that PTH could promote the implantation of prostheses in osteoporotic patients. All of these findings herald a new era for people needing new joints because a long lifetime of an implant needs at the outset a lot of new bone to build up around and cling to the implant surface without spaces that could fill up with wear debris (Davies, 2000).

So far we have been considering the effects of injected PTHs. What would happen if we blinded the parathyroid cells to circulating Ca^{2+} by disabling their CaRs? The cells would interpret the CaR silence as being due to a large drop in the blood Ca^{2+} concentration that urgently warrants PTH release. NPS 2143 is a CaR silencer (a "calcilytic"), which when given orally in a dose of 100 μmol/kg causes the PTH concentration in the blood of OVXed Sprague-Dawley rats to shoot up from 25 pg/ml to about 110 pg/ml in 30 minutes and then stay there for at least 4 hours (Gowen et al., 2000; Hebert, 2006; Nemeth and Fox, 2003). This prolonged, indeed continuous, release of endogenous PTH dramatically increases both bone turnover and bone formation in the proximal tibia, but without a net bone loss or gain or without significantly affecting the OVX-induced trabecular crash. However, bone mineral density as well as trabecular thickness and area, but of course not trabecular number, can be increased by giving 17β-estradiol along with the calcilytic to suppress osteoclasts and resorption . But the NPS 2143/estrogen combination is still nowhere nearly as osteogenic as the subcutaneous injections of 10 nmoles/kg of hPTH-(1-34)NH$_2$ or [Leu^{27}]*cyclo*(Glu^{22}-Lys^{26})hPTH-(1-31)NH$_2$ (Ostabolin C™) into OVXed rats (Whitfield et al., 1998b). But maybe the osteogenicity of the calcilytics can be improved by shortening their circulating half-life and therefore the duration of the PTH surge, which would shift the balance from resorption to bone-building. Remember the Principal Golden Rule for PTHs—*they are most effective when given in well-separated brief boluses.* It appears that such "quick PTH rise-quick PTH fall"-inducing calcilytics have been developed by Avery et al. (2005) at Bristol-Myers Squibb Company. One of these CaR inhibitors, "Compound 1", stimulates PTH release in rats with an EC$_{50}$ of 0.4 μm and, for example, when given orally at 50 mg/kg of body weight causes the circulating PTH concentration to rise 3-fold and then fall back to normal with a half-time of 2 hours. It must now be determined whether compound 1 can stimulate bone growth as effectively as an injected PTH.

While it is clear that exogenous PTHs have strongly stimulated bone growth in every one of many animal models, the question remains as to whether normal circulating PTH affects bone growth and maintenance. According to Chow et al. (1998) the answer is a resounding—Yes! First they found that removing the thyroid-parathyroid complexes from young female Wistar rats dramatically reduced bone growth in the caudal vertebrae. And they also showed that circulating PTH is required for the mechanical responsiveness. Mechanical stimulation of the caudal vertebrae in normal rats induced osteogenesis. But the vertebrae in the thyroparathyroidectomized rats could not respond to the mechanical stimulation.

Then there is the delightful story of PTH and bears. The reader has been bombarded with the fact that unloaded bones are treated as expensive and useless toys which are destroyed by gangs of osteoclasts. But what about bears who are inactive for 6 months while hibernating ? Surely their bones must be devastated by swarms of osteoclasts because of disuse osteoporosis. But no! In fact bone resorption is balanced by bone formation during hibernation. It appears that these clever beasts know a lot about osteomaintenance and actually increase the circulating level of PTH during and directly after hibernating (Donahue et al., 2006).

Finally, according to rats PTHs should be added to dentists' toolboxes. They prevent periodontitis-associated bone loss. This has been shown with a rat model in which periodontitis is induced in a mandibular first molar by submarginally placing a cotton ligature (Barros et al., 2003). In untreated control rats, the ligature causes a peri-gingingival accumulation of inflammatory cells producing proinflammatory cytokines, osteoclasts and consequently bone loss. Injecting 40 µg of hPTH-(1-34)/kg of body weight 3 times a week for 4 weeks prevented periodontitis and bone loss (Barros et al., 2003).

iv. PTHs-Induced Bone Growth in Humans

It was Bauer et al.'s 1929 passing mention of the Lilly bovine parathyroid extract's unexpected stimulation of bone formation in rats, but much more the case of a 8-year-old osteopetrotic boy who was killed by the aplastic anemia caused by wildly growing bone filling his marrow cavities and sending his hematopoiesis back to its fetal sites (Péhu et al., 1931), that prompted Selye's dramatic demonstration in 1932 of the awesome osteoblast-generating, osteogenic action of the Eli Lilly bovine parathyroid extract in rat pups that was to be repeated hundreds of times in various animals. The thing that riveted Selye's attention to the French paper was the authors' iconoclastic suggestion that the cause of the unfortunate boy's lethal bone growth was the abnormal amount of PTH pouring out of the boy's parathyroid adenoma—a PTH-producing tumor. But as Enlow and Brown warned 46 years ago: *"A great many experimental studies, conclusions, and generalizations concerning bone tissue are based on observations of laboratory animals which do not possess typical human-like bone tissue"* (Enlow and Brown, 1958) and the rat is one of these!. However, with the advantage of 2006 hindsight PTH's osteogenicity was already staring the Bone Community in its collective face—postmenopausal women with *mild* (and *mild* is the key word) primary hyperparathyroidism do not lose trabecular bone mass. Indeed, these women have more trabecular bone packets with greater wall width, higher bone apposition rates and active formation periods than their healthy postmenopausal peers (Dempster et al., 1999).

The main reasons for the long delay in extending the animal results to humans were the unshakable belief in the then current paradigm plus the lack of an affordable, pure PTH. Forty-four years later the affordable synthetic peptide, hPTH-(1-34) was born. It was destined to be become Forteo™ the first anabolic to reach the clinic. (Andreatta et al., 1973; Potts et al., 1971; Tregear et al., 1974) (Table 2). It was given the name *teriparatide* (an abbreviation of tetratriaconta parathyroid peptide). Its appearance on the market sparked efforts to find out whether PTH is as osteogenic in humans as it is in animals.

To start the story of PTHs and humans, let's look at the rise and fall of the hPTH-(1-34) fragment in the circulation after it is injected subcutaneously. For example, Lindsay et al. (1993) have reported that injecting 25 µg of this peptide once a day into women with osteoporosis (20 µg is the currently recommended daily dose for treating osteoporosis) caused a peak of circulating fragment by 30 minutes later at a level which averaged 10-times the basal level followed by an exponential drop in the level with a mean half-time of 75 minutes. Of course this brief surge of the fragment caused the level of circulating endogenous, native hPTH-(1-84) to drop immediately by 35% as the parathyroid gland cells responded to the sudden loading of the blood with normally very rare N-terminal fragments. All of the many studies, starting in 1976, using

Table 2. *The clinical studies and trials that led to the clinical acceptance of PTHs for treating osteoporosis*

Author	PTH (dose)	Duration	Co-Therapy	Primary Endpoint
Reeve (1976)	500 IU hPTH-(1-34)	6 months	none	histology
Reeve (1976)	500-2000 IU hPTH-(1-34)	1-6 months	none	calcium balance
Reeve (1980)	500 IU hPTH-(1-34)	6-24 months	none	histology
Hesp (1981)	500 IU hPTH-(1-34)	12 months	none	BMD
Reeve (1981)	500 IU hPTH-(1-34)	6 months	none	calcium balance
Slovik (1981)	450-750 IU hPTH-(1-34)	1 month	none	calcium balance
Slovik (1986)	400-500 IU hPTH-(1-34)	12 months	calcitriol	BMD
Reeve (1987)	1000-1500 IU hPTH-91-34)	12 months	calcitriol	BMD
Neer (1987)	400-500 IU hPTH-(1-34)	18-24 months	calcitriol	BMD
Hesch (1989)	750 IU hPTH-(1-38)	14 months	calcitonin	BMD
Reeve (1990)	500 IU hPTH-(1-34)	12 months	estrogen	BMD
Hodsman (1990)	400 IU hPTH-(1-38)	6 months	calcitonin	biochemistry
Reeve (1991)	500 IU hPTH-(1-34)	12 months	estrogen	calcium balance
Hodsman (1991)	400 IU hPTH-(1-38)	6 months	calcitonin	BMD
Neer (1991)	400-500 IU hPTH-(1-34)	12-24 months	calcitriol	BMD
Bradbeer (1992)	450-750 IU hPTH-(1-34)	6-12 months	estrogen	histology
Hodsman (1993)	800 IU hPTH-(1-34)	3 months	calcitonin	markers
Reeve (1993)	500 IU hPTH-(1-34)	12 months	estrogen	BMD
Finklestein (1994)	40 μg hPTH-(1-34)	6 months	naferelin	BMD
Lindsay (1995)	400 IU hPTH-(1-34)	36 months	estrogen	BMD
Sone (1995)	20 IU hPTH-(1-34)	6 months	none	BMD
Hodsman (1997)	800 IU hPTH-(1-34)	24 months	calcitonin	BMD
Lindsay (1997)	25 μg hPTH-(1-34)	36 months	estrogen	BMD
Lindsay (1998)	50-100 μg rhPTH-(1-84)	12 months	none	BMD
Fujita (1999)	50-200 IU hPTH-(1-34)	12 months	none	BMD
Cann (1999)	400 IU hPTH-(1-34)	24 months	none	BMC
Roe (1999)	400 IU hPTH-(1-34)	24 months	estrogen, calcitriol	BMD
Rittmaster (2000)	rhPTH-(1-84)	12 months	none	BMD
Cosman (2001)	400IU hPTH-(1-34)	36 months	estrogen	BMD
Neer (2001)	20 or 40 μg hPTH-(1-34)	19 months	none	decreased fractures
Bilezikian (2001)	400IU hPTH-(1-34)	18 months	calcitriol	BMD, markers
Dempster (2001)	400IU hPTH-(1-34)	18 months (men) 36 months (women)	calcitriol	increased cortical width and trabecular connectivity
Kurland (2000)	400 IU hPTH-(1-34)	18 months (men)	none	lumbar bone mass and femoral BMD
Orwoll (2003)	20, 40 μg hPTH-(1-34) Forteo™	11 months (men)	none	lumber spine, proximal femur BMD
Hodsman (2003)	100 μg h hPTH-(1-84)	12 months (women)	none	lumbar spine BMD

this fragment have shown that such a daily bolus of a PTH can potently stimulate bone formation in humans as it does in dogs, mice monkeys and rats.

Now that we have an example of the pharmacokinetics of a PTH injection, let's go back to the beginning of the human story. In 1976 Reeve et al. (1976a, 1976b) reported that single daily injections of 100 μg/kg of hPTH-(1-34) stimulated iliac trabecular bone growth, but they did not affect femoral BMD. In a subsequent uncontrolled, multicenter study, they injected 16 osteoporotic women and 5 osteoporotic men once each day with 50 to 100 μg/kg of hPTH-(1-34) for 6 months to 2 years (Reeve et al., 1980). The treatment produced a significant amount of new, apparently normally mineralized iliac lamellar trabecular bone.

Since 1980, the results of more than 20 studies have been published (reviewed by Cosman and Lindsay, 1999; Dempster et al., 1993; Hodsman et al., 1997; Meunier, 1999; Netelenbos, 1998; Whitfield et al., 1998b, 2000b 2000d; summarized in Table 2), but I really need to present only the more recent examples. By far the most often used, but inadequate measure, of a PTH's osteogenic effectiveness in these human studies has been the increase it causes in the overall BMD as determined by dual-energy X-radiation absorptiometry (DEXA) or the much better quantitative computed tomography (QCT) with which the cortical and cancellous (trabecular) compartments can be separately measured (Genant et al., 1982). Remember that in reading what follows that despite its shortcomings a low BMD generally does indicate a higher fracture risk (Faulkner, 2000).

However, before continuing I must also note Heaney's warning that BMD (areal density) is the "misbegotten surrogate of bone mass" (Heaney, 2003). Actually as discussed above the level of remodeling activity is a better indicator of bone fragility than bone mass or BMD.

In Lindsay et al.'s 3-year study, single daily subcutaneous injections of 25 μg of hPTH-(1-34) increased the vertebral BMD by 12.8% in 27 postmenopausal osteoporotic women who also received estrogen, which by itself did not affect the BMD in the 25 women of the control group (Lindsay et al., 1997). As expected, the rise in the vertebral BMD was accompanied by a drop in vertebral fracturing which was indicated by the PTH-treated women not losing as much radiographically measured vertebral height as the untreated women. Just as in the animal models, the increases were lowest in bones with smaller cancellous (trabecular) compartments such as hip (4.4%) and forearm (1.0%). The whole body BMD increased by 8%.

In a follow-up to this study, Cosman et al. (2001a) reported that the BMD did not change during the 3-year treatment and for 1 year afterwards in the control group receiving HRT (hormone [estrogen] replacement therapy). In the PTH-HRT group, the serum markers of both bone formation (bone-specific alkaline phosphatase) and resorption (urinary crosslinked collagen N-telopeptides) peaked at 6 months and returned to baseline levels by 30 months. During this period, the high cancellous spinal bone mass increased 13.4 %, the lower cancellouis hip bone mass increased 4.4 % and total body bone mass increased 3.7%. The bone masses did not change significantly during the first year after stopping the PTH injections while continuing HRT. The PTH-HRT treatment dramatically and significantly reduced the number of "incident" (in other words "spontaneous" or muscle-pull or bending-crush) vertebral fractures during the follow-up year. The frequency of incident fractures in the PTH-HRT group was reduced to as little as 0% or 25% of the frequency in the HRT-only group depending on which height reduction (their measure of vertebral crushing) 'cut-off point' was used.

A larger increase in vertebral BMD was seen by Hesch et al. (1989) in a smaller group of 13 osteoporotic women. After 14 months of single daily subcutaneous injections of 54 μg (about 700-750 Units of activity) of hPTH-(1-38) the mean vertebral BMD had increased by 20%. This large response might have been due to the post-PTH prevention of a resumption of resorption by the nasally sprayed anti-osteoclast calcitonin that was also given to the women.

In a randomized, controlled 2-year trial, the first "cyclical therapy" trial, with 30 osteoporotic women receiving 6 treatment cycles each one of which consisted of 1 month (28 days) of single daily subcutaneous injections of 800 U (or about 60 µg)/day of hPTH-(1-34) followed by a 3-month rest period (to restore full anabolic responsiveness to the PTH for the next round of injections) there was a 10.1% increase in the BMD of lumbar vertebrae, a non-significant 2.4% increase in the femoral neck BMD and a 80% drop in the vertebral fracture incidence (Hodsman et al., 1997). However, in this study treating the osteoporotic women with calcitonin during the rest periods did not enhance PTH's osteogenicity.

Rittmaster et al. (2000) have reported the results of a study in which 75 postmenopausal women were treated for 1 year with daily injections of 50, 70 or 100 µg of recombinant hPTH-(1-84) and then with a daily dose of 10 mg of the anti-catabolic alendronate for the next year. This treatment stopped any post-PTH cortical bone loss and increased the spinal bone mass by 14% as compared to 6.9 to 9.2% increase in patients without post-PTH alendronate treatment.

In a much larger study Fujita et al. (1999) treated 220 osteoporotic patients with weekly subcutaneous doses of 50 U (3.7 µg; 0.8 nmole), 100 U (7.4 µg; 1.6 nmole), or 200 U (14.8 µg; 3.2 nmole) of hPTH-(1-34) for 48 weeks. The highest dose caused a significant 8.1% increase in the lumbar vertebral BMD, without affecting cortical thickness or metacarpal BMD. The increase in vertebral BMD was accompanied by a 30-40% reduction in "backache".

In another study (Roe et al. 1999) osteoporotic women injected themselves with a placebo or 400 U (about 30 µg) of hPTH-(1-34) subcutaneously once each day for 2 years and also took an oral estrogen (0.625 mg Premarin/day) with or without methoxyprogesterone. They all received 800 U of vitamin D and 1500 mg of calcium/day. By the end of 2 years the overall (cortical plus trabecular) lumbar vertebral BMD in the PTH/estrogen-treated group was a dramatic 28.3% higher than in the placebo/estrogen-treated control group. The femoral neck BMD in the PTH/estrogen-treated group was 10.8% higher than the femoral neck BMD in the placebo/estrogen control group. The density of the trabecular (cancellous) compartments of the L1 and L2 vertebrae in the PTH/estrogen-treated women (as measured selectively by QCT) had increased even more dramatically—it had increased by 74% during the 2 years while the mean density in the vertebrae of the placebo/estrogen-treated controls had *dropped* 2.1%.

At the end of 1998, Lindsay et al. (1998) summarized the results of a one-year, Phase II, multicenter (18 centers in Canada and the USA), placebo-controlled, double-blind study of the effects of recombinant rhPTH-(1-84) (i.e., NPS Pharmaceuticals Inc.'s *now discontinued* PREOS™) on BMD and serum indicators of bone resorption and formation in 217 postmenopausal osteoporotic women between 50 and 75 (average 64.5) years of age. The results of the prior Phase I study had indicated that one subcutaneous injection of 0.02 to 5.0 µg of PREOS™/kg into healthy postmenopausal women did not produce a frank hypercalcemia even at the highest dose (though you will learn further on that this has turned out not to be the case and excessive hypercalcemia has prevented FDA approval and caused NPS Pharmaceuticals to discontinue further development of the drug at the end of its Phase III trial) which indicated that the hormone was safe and well-tolerated in this dose range (Schweitert et al., 1997). The apparent failure to cause hypercalcemia could be due to the killing of osteoclasts by the stimulation of their CPTHR receptors for a C-terminal domain of the large native PTH (Divieti et al., 2002).The Phase II patients were given single daily injections of 50, 75, or 100 µg of hormone (~ 0.8 to 1.6 µg/kg). The largest dose increased the vertebral BMD by 6.9% while the densities of the vertebrae in the placebo-treated women did not change. However, the BMD *dropped* in the arms and legs by 0.3% (75 µg) and 0.9% (100 µg) respectively. The serum levels of both formation markers (bone-specific alkaline phosphatase, osteocalcin) and resorption markers (deoxypyridinolines, N-terminal cross-linked collagen peptides) increased 100 to 200%. There were no serious side effects nor did the patients make antibodies to the recombinant human hormone. The authors assumed that the drops in arm and leg BMDs were only transient and that PTH appeared to be a promising novel treatment for osteoporosis.

Transient though they might have been, the drops in the BMDs of the arms and legs of Lindsay et al's patients are reminiscent of the results of an earlier study by Neer et al. (1991) that nearly kicked the PTHs off the list of credible therapeutics for osteoporosis. As expected, giving hPTH-(1-34) and calcitriol (1α,25(OH)$_2$vitamin D$_3$) to 15 osteoporotic women for 1 to 2 years increased the lumbar vertebral (i.e., trabecular) BMD by a whopping 32%! But at the same time there was a net 5.7% *drop* in the radial BMD, which was what set off the alarm bells. The great fear was that although PTH can reduce the unpleasant vertebral crushing, it might increase the far more crippling hip fracturing. To explain this worrisome result they suggested that the PTH somehow stole cortical bone to make trabecular bone. Fortunately for the PTHs this '*Cortical Steal*' hypothesis as it was called has faded away although there is a good reason why there is often a transient drop in cortical BMD. So the common, though not universal (e.g., Lindsay et al.'s 1998 study) experience has been virtually the same as the experience with non-human bones—PTHs strongly stimulate trabecular bone growth but either do not affect or less strongly stimulate cortical bone growth and increase cortical porosity as they do in experimental animals (Strietsma, 1995) . To understand this remember that it is the trabecular cells that get the first grab at an injected PTH coming into the bone through the nutrient blood vessels.

Then came the reports that propelled old hPTH-(1-34) into clinics and the skins of osteoporotics. Neer et al (2001) and Zanchetta et al. (2003) have reported the effects on bone geometry and fracturing of daily subcutaneous injections of a placebo, or 20 or 40 μg of Eli Lilly's recombinant hPTH-(1-34) (Forteo™) into 1637 post-menopausal women who had already suffered vertebral fractures. These numbers were from this venerable, 33-years-old peptide's phase III clinical trial. The experimental details responsible for this Grand Entry of the first of the Anabolics into the Osteoporosis World can be found in the Ly333334 (Teriparatide Injection) Briefing Document (2001) from Lilly Research Laboratories.The relative risk of vertebral fracturing was reduced to 0.35 (95% CI 0.22-0.55) in women receiving daily injections of 20 μg of the peptide and to 0.31 (95% CI 0.19-0.50) in women receiving daily injections of 40 μg of the peptide. The relative risk of non-vertebral fracturing was 0.47 in women receiving 20 μg of the PTH (95% CI 0.25-0.88) and the same 0.46 in women receiving 40 μg (95% CI 0.25-0.86). Both doses of the PTH increased the BMD of hip and spine by 2 to 4%. While there was no drop in the BMD of the distal 1/3 of the radius in the women treated with 20 μg of hPTH-(1-34) (Forteo™) there was a small drop in the radial BMD in the women treated with 40 μg during the first year. However, according to Zanchetta et al. (2003), using pQCT instead DEXA, hPTH-(1-34) actually significantly increased the BMD and area of the distal radius.There were: increased porosity on the inner region of the distal radius where the mechanical stress is low; resorption of fully mineralized endocortical bone; and a compensatory deposition of periosteal bone.

Also the "PTH" bone in the osteoporotic women turned out to be immature with respect to its mineralization (Paschalis et al., 2005). And this has also been reported by Misof et al (2003) who additionally found that despite its immature mineralization the new "PTH" matrix is normal at the microstructural and nanostructural levels. This means that "PTH" bone has lower matrix mineralization and collagen cross-link ratio at periosteal, endosteal and trabecular bone surfaces than the mature bones in placeo control women. In other words the new "PTH" bone had an abundance of divalent cross-links while the stronger mature bone had an abundance of trivalent cross links—the new bone had the characteristics of weaker young bone. The new "PTH bone needs time to minerally mature and this could be helped by post-PTH treatment with a bisphosphonate such as alendronate which over the next months would significantly increase the degree of mineralization (Boivin and Meunier, 2002; Boivin, et al., 2000; Roschger et al., 2001). Also this initial mineralization deficit could have been due to PTH stimulating the expression of matrix Gla and osteopontin inhibitors of mineralization.

Of course, this effect would be much greater if the PTH were continuously administered (Gopalakrishnan et al., 2005).

Nevertheless, hPTH-(1-34), Lilly's Forteo™, improves bone geometry as indicated by increased axial and polar cross-sectional moments of inertia, decreases the risk of wrist, vertebral and non-vertebral fracturing, increases femoral and vertebral BMDs, and is well tolerated although there was some (29%) hypercalcemia with 40 μg, which, because it causes such unpleasant things as muscle weakness, fatigue, nausea, vomiting, kidney stones, limited the dose to 20 μg. Indeed according to Lane et al. (2003) a daily injection of 40 μg of hPTH-(1-34) caused the circulating level of the osteoclastogenic soluble RANKL to peak broadly around 250% of the baseline value between the 3rd and 9th month and the osteoclastogenic IL-6 to peak at about 150% of the baseline value at the 3rd month. Obviously the new potently anabolic PTHs such as oral Ostabolin™ or Ostabolin-C™ (Jolette et al., 2003, 2005) that apparently do not cause hypercalcemia even at very high doses will have a considerable clinical advantage over Forteo™ when they complete their clinical trials.

A comparison of what hPTH-(1-34) (Forteo™) treatment does to cortical and trabecular bone has been provided by Cann et al. (1999). They used 3D-QCT to separately track the hPTH-changes in the density and mass of the cortical and trabecular (cancellous) parts of the proximal femurs of the osteoporotic women in Roe et al (1999)'s study. By the end of the second year the PTH injections had increased the spinal BMD by a very substantial 74% and the BMD of the trabecular compartments of the proximal femur from 10.5% in the trochanter to 12.4% in the whole hip. On the other hand, the Density-Dropping Demon reappeared here again—the cortical BMD *dropped* by less than 3.5% during the first year. But, as in Lindsay et al.'s 1998 study, the drop then stopped. So were Neer et al.(1991) right? Yes, but PTH treatment is certainly not hazardous for femoral necks. Despite the small drop in cortical bone *density*, the use of 3D-QCT enabled Cann et al. to see that the cortical bone *mass* in the proximal femurs had actually *risen* 10.9% in the trochanter and by an awesome 20.7% in the femoral neck by the end of the second year. These large increases in cortical mass took place on the endosteum (just as Andreassen and Oxlund [2000] found in the femurs of hPTH-(1-34)-treated OVXed rats) because the endosteum unlike the periosteum is both bathed in marrow and covered with PTH receptor-bearing lining cells, which like the equally marrow-bathed trabecular lining cells, can throw PTH-induced growth factors directly at osteoblastic progenitors in the adjacent marrow stroma and may revert to active osteoblasts as sen in rats (Dobnig and Turner, 1995; Kostenuik et al., 1999; Pun et al., 2001). Thus, as expected in a mature skeleton, PTH rapidly stimulated osteoclast generation by RANKL-producing immature PTHR1 (the common PTH receptor)-poor osteoblastic stromal cells, which would reduce the cortical density by increasing the porosity (i.e., the number of osteoclast-dug tunnels; Sietsema, 1995). But this was far more than balanced by a delayed, large production of endocortical bone by increased numbers of osteoblasts resulting from the PTH's three osteoblast-generating actions soon to be unveiled in the next chapter.

The same changes have been seen by Burr et al. (2001) in OVXed cynomolgus monkeys (*Macaca fascicularis*), that received one daily subcutaneous injection of 1 or 5 μg of hPTH-(1-34)/kg for 12 or 18 months. The injections increased the cortical porosity especially in the inner third of the mid-diaphysis of the left humerus where it was increased 5 to 16 times above the porosity in the sham-operated monkeys. Surely this must have reduced the bone's mechanical strength. But it did not! The strength actually increased because while the osteoclasts were digging more holes in the endocortical zone the PTH-driven osteoblasts were busily increasing the mineralizing surface 11-fold more than in the controls and making bone 4 times faster than their peers in control monkeys.

Sato et al (2004) have also seen the same thing in 9-year-old, OVXed cynomolgous monkeys that had been subcutaneously injected daily for 18 months with 1 or 5 μg of hPTH-(1-34)

(Forteo™). The PTH increased osteon activation, osteonal bone formation and cortical width and area, but it also increased bone turnover and cortical porosity. However, again the increased bone growth more then compensated for the increased porosity and the bone strength (as indicated by the ultimate load to failure) ranged above the weakened OXV value and the normal sham value.

A similar conclusion was reached by Hirano et al. (2000) from a study of the effects of an osteogenic dose of hPTH-(1-34) on cortical bone porosity and bending rigidity of rabbit tibias. They found that here too the PTH greatly increased the porosity mostly on the endocortical surface, but this was more than offset by the formation of new endocortical bone as well as periosteal bone, which *increased* the tibial bending rigidity. The same coincidence of bone loss and large bone building has been seen by Whitfield et al. (1998b) in OVXed rats where hPTH-(1-34) injections did not stop the estrogen deprived, stimulated osteoclasts from destroying the femoral central metaphyseal trabeculae but they increased thickness of the surviving laterally sited trabeculae 1.65 times.

At this point you might ask whether preventing the PTH-induced surge of osteoclasts boring holes in the cortex would increase the potency of the anabolic response to a PTH. And the answer is almost certainly "Yes" if one uses the potent, osteoclast maturation/activation inhibitor osteoprotegerin or better the much safer AMG 162 human monoclonal anti-RANKL antibody ("denosumab" Bekker et al., 2004; Kostenuik et al., 2001; Lacey, 2006; McClung et al., 2004; Simonet et al., 1997; Smith et al., 2003) instead of the bisphosphonate alendronate (Fosamax™) which does kill osteoclasts but unfortunately as we shall see below also *interferes* with the expression of osteoblast-specific genes (Onyia et al., 2002).

Accumulating mature osteoblasts should release an indicator of their presence such as osteocalcin into the blood. Indeed this is what happens in PTH-treated osteoporotic humans. The hPTH-(1-34) injections in the study of Lindsay et al. (1997) caused the serum osteocalcin content to rise for the first 6 months after which it slowly subsided. Cosman et al. (1998, 2000) found that the PTH injections in these experiments caused the serum osteocalcin and pro-collagen I-C-terminal propeptide (released during collagen fiber processing and assembly) to rise during the first 4 to 6 weeks, which indicated a rapid build-up of osteoblasts. There should also have been evidence of PTH-stimulated osteoclasts. But the osteoclast response, as indicated by the release of cross-linked N-telopeptides from osteoclast-cleaved collagen, was not detectable until 6 months and then dropped to baseline values by 24 months. In other words there was an "anabolic window" which is the basis for Hodsman et al.'s cyclical therapy— first open the anabolic window ⇒ then close it before the catabolic machinery can effectively start ⇒ re-open the anabolic window when the catabolism subsides …and so on.

Reeve et al. (2001) reported the effects of treating severely osteoporotic women with hPTH-(1-34) or hPTH-(1-38). They found that the PTHs improved the "body calcium balance", "impressively" increased the spinal BMD and stimulated smaller increases in proximal femur and radius. They concluded *"hPTH or comparable PTH receptor activators remain the most promising anabolic treatment for osteoporosis currently under clinical evaluation …"*.

In 2005, Cosman et al. (2005) have reported the results of a study on 126 osteoporotic women which confirms that using cyclical therapy could revolutionize the PTH treatment. Normally, an osteoporotic patient self-injects 20 µg of hPTH-(1-34) (Forteo™) once daily for no more than 2 years. But Cosman et al (2005) shifted to Hodsman et al's (1997) *"3-months-on-3 months-off"* cyclic protocol. The first reason for cyclic treatment is based on the now-ancient fact (discovered by Selye (1932) with his rat pups receiving the PTHs in the old bovine parathyroid extract) that a daily injection of intact hPTH-(1-84) or hPTH(1-34) (teriperatide/Forteo™) first stimulates the appearance of osteoblasts and bone formation (i.e., the pure bone formation phase; the anabolic window) and then the build-up of osteoclasts and resorption/remodeling catches up by 6 months (Hodsman and Steer, 1993; Hodsman et al., 1993,

2003). And we have seen a dramatic example of this in sexually mature Sprague-Dawley rats when hPTH-(1-34) was injected daily in a dose of 2.0 nanomoles/100 g of body weight for the first two weeks after OVX. In these animals, the number of osteoblasts shot up 2.2-fold over the number in sham-operated animals while the number of osteoclasts actually dropped. And as expected Cosman et al. (2005) found that bone resorption increased more in the women receiving continuing daily injections of the PTH than in the cyclically treated women. So if there is a sufficiently long rest period between injections, osteoclast production and resorption would not have the chance to catch up. The second reason is that if cyclic treatment works at least as well as continuing daily injections, there would be less syringe needle stress for the patient. And last, but not least, the treatment could be only half as expensive/year.

Cosman et al's results, like those of Hodsman et al. (1997), are encouraging. When hPTH-(1-34)(Forteo™) was "on" for 3 months the serum markers of bone formation rose, dropped when it was "off" for the next 3 months, but then rose robustly when turned on again during the next 3 months. By the end of 15 months of such cycling there was the same approximately 5.4-6.1 per cent increase in the lumbar vertebral BMD that had been stimu-lated by either the costly continuing daily treatment or the cheaper cyclic treatment. But even greater savings might be possible if the once-weekly injection regimes of Fujita et al. (1999) and Miki et al. (2004) should prove to be biomechanically effective in further trials. Fujita et al. (1999) reported that injecting 100 or 200 units of hPTH-(1-34) (Forteo)™ once a week for 48 weeks significantly increased spinal BMD by 3.6 and 8.1 % respectively in osteoporotic patients. Miki et al. (2004) reported that weekly subcutaneous injections of 100 units of PTH-(1-34) for 48 weeks into one man and 9 women with primary osteoporosis increased lumbar BMD and increased the trabecular bone volume and improved trabecular microstruc-ture in iliac crest.

Parenthetically cyclic therapy is also effective in mice as well as humans. Iida-Klein et al. (2005) have found that cyclic therapy with hPTH-(1-34) (1 week on, 1 week off for 7 weeks) was as osteogenic for the lumbar vertebrae of 20-week-old female C57BL/J6 mice as were daily injections.

At the end of 2003, Hodsman et al. (2003) reported the effects of 12 months of treatment with NPS Therapeutics's now discontinued (June 2006) recombinant full-length hPTH-(1-84). Postmenopausal women self-injected 50, 75 or 100 µg of the peptide every morning for 12 months and swallowed one or two $CaCO_3$ (500 mg elemental calcium) tablets and 400 IU of vitamin D along with it. The most effective dose was 100 µg, which by 3 months had increased the lumbar (i.e., L1-L4) spine BMD by a significant 2% and then 7.8 % by 12 months. The latter number was actually underestimated because of a 2% areal increase by 12 months. There were also transient dose-dependent incidences of hypocalcemia which the authors did not consider particularly worrying. At the 25th meeting of the American Society for Bone and Mineral Research in September 2003, Paul Miller (from the University of Colorado in Denver) also summarized (without providing a record of his talk) some of the results coming out of this full length PTH's Phase III trial. The results he cited were got from 50 post-menopausal women who injected themselves subcutaneously every day with 100 µg of the PTH which in terms of molecular weight was approximately equivalent to the 20 µg of hPTH-(1-34) (Forteo™) that osteoporotic women are now allowed to self-inject each day. And the anabolic effectiveness of this full length PTH is about the same as that of hPTH-(1-34) (Forteo™). But with all due respect to Hodsman et al. (2003) there is a very good reason not to use this PTH—as many as 22% of the women suffered from one or more hypercalcemic experiences. So why would a physician prescribe PREOS™ when the patient could get exactly the same anabolic punch/nmole with hPTH-(1-34) (Forteo™) that is already on the pharmacy shelves or with the even smaller osteogenic PTHs that are in the pipeline (Table 1)? Indeed in March 2006, the USFDA has required NPS Therapeutics to go back and re-trial their dosage of PREOS™ to eliminate the excessive hpercalcemia (MorganStanley, 2006). Then in June 2006 NPS announced that it was stopping further development of the protein, despite the fact that the peptide has been approved by CHMP (Committee for Medicinal Products for Human Use of the European Medicines Agency) under the name of Preotact™ for sale in the European Union.

Glucocorticoid therapy, like menopausing, wandering around in space, or lying immobilized in bed can also cause osteoporosis. It does this by inhibiting osteoblastogenesis (apparently by suppressing Cbfa1/Runx-2 and Cbfa1/Runx-2-stimulated TGF-β type I receptor expression [Chang et al.1998; Ducy, 2000]), promoting osteoblast and osteocyte apoptosis and inhibiting intestinal Ca^{2+} uptake (Lane et al. 2000; Weinstein et al., 1998). Injecting a PTH can overcome the glucorticoid action and stimulate bone growth. This is shown in a recent experiment involving 51 women (mean age of 63 years) receiving glucocorticoids and estrogens (Lane et al. 2000). Twenty-eight injected themselves with 40 μg (400 U) of hPTH-(1-34) (Forteo™) once each day for 12 months and continued taking estrogen while 23 just continued taking estrogen. Here again as expected PTH preferentially stimulated growth of cancellous (trabeculae-rich) bone. By 12 months the PTH injections had increased the trabecular *mass* (as measured by QCT) mainly in cancellous lumbar vertebrae by about 35%, which continued rising to 45% during the next 12 months as the PTH-enhanced remodeling space (remember that PTH also increases osteoclast generation and thus excavations) continued to be filled up after the injections were stopped. By 24 months the overall BMD (as measured by dual energy X-radiation absorptiometry instead of QCT), which is only a function of the extent of mineralization and the amount of remodeling space in the increasingly massive vertebral bone, had increased by only 12.6 %. The hip mass rose by 4.7% over the 24-month period. However, the treatment did not affect the cortical bone mass in the forearm.

Rehman et al (2003) have also found that treating women with glucocorticoid-induced osteoporosis for 1 year with hPTH-(1-34) (Forteo™) plus hormone-replacement therapy (HRT) increased the cross-sectional areas of their L1 and L2 vertebrae by 4.8% over the areas in patients given HRT only. hPTH-(1-34) (Forteo™) also increased the compressive strength of the vertebrae by 200% over the base line levels while HRT alone had no effect.

Glucocorticoid treatment along with alchohol consumption is also a major cause of atraumatic osteonecrosis along with bone-killing traumas and over-fast bubbling of blood nitrogen (caisson disease) all of which variously occlude or sever bone blood vessels. The mechanism of this is still obscure, but it seems to involve the hypertrophy of bone marrow adipocytes which cause the marrow to swell just as adipocyte hypertrophy causes the "buffalo hump" and "moon facies" of steroid-treated patients (Boss and Misselevich, 2003; Richardson, 2003). Since the marrow like the brain is held within a rigid bony box and cannot expand, such swelling causes the intramedullary pressure to rise, squeeze the intramedullary veins and eventually cause ischemia and infarction (Richardson, 2003). The vascular occlusion can be aggravated by blockage by fatty embolisms resulting from an elevated ratio of LDL (low density lipoproteins)/HDL (high density lipoproteins).

Adipocyte expansion is not the only cause of glucocorticoid-induced osteonecrosis. Glucocorticoids also increase the expression by vascular smooth muscle cells of the potent vasoconstrictor endothelin 1,which would cause a cytocidal medullary ischemia (Börcsök et al., 1998o; Drescher et al., 2004).

The femoral head is the most vulnerable to necrosis because it has no effective collateral circulation. While dead bone by itself is as strong and as per-sistent as any plastic replacement, the necrotic head is caused to collapse by osteoclasts when they dig out the dead cortical bone in preparation for receiving new living bone from the slowly working osteoblast. But now there is a dangerous biomechanical hiatus! There is an uncompensated loss of subchondral bone and joint cartilage as the osteoclasts' attack spills over the bone-cartilage border (Glimcher, 1999; Glimcher and Kenzora, 1979; Richardson, 2003).

Obviously glucocorticoid-induced osteoporosis must contribute to this kind of necrosis and traumatic osteonecrosis and its prevention by the PTHs might restrain osteonecrosis. But could the PTHs alo help to restore already necrotic bone? To do this it would be necessary first to stimulate an invasion of vascular buds from adjacent live bone. To do this the PTH would have to stimulate the expression of angiogens such as VEGF which stimulates osteoblast migration and differentiation along with angiogenesis (Midy and Plouet, 1994). And indeed it could (Carter et al., 2000; Esbrit et al., 2000; Turner et al., 2003; Wang et al., 1996). The budding blood vessels would deliver blood-borne preosteoclasts and osteoprogenitor cells to the dead bone and the PTH boluses would collaborate with the BMP-2, endothelin-1, leptin and VEGF from the PTHR1 receptor-bearing bone vascular endothelial cells to suppress osteoclastogenesis and stimulate the osteoprogenitor cells to proliferate, differentiate into osteoblasts and lay new live bone on the dead trabeculae.

But one must be careful! The marrow spaces in the dead bone are invaded and filled by mesenchymal cells which mature into osteoblasts when they contact the surfaces of the dead trabeculae. They produce trabeculae with new bone envelopes and still-dead cores. The dead cores are then invaded by blood vessels and osteoclasts which resorb the dead cores which osteoblasts fill with new bone. While everything is fine with trabecular bone, things can be very different in the femoral head's cortical bone. There, osteoclast SWAT teams rather than osteoblasts are first one the scene and do what they are programmed to do—rapidly start digging out the dead bone. However, the later arriving osteoblasts work much more slowly while the subchondral bone in the femoral head is being progressively destroyed and weakened by the osteoclasts which can also attack the joint cartilage. This difference in the times of arrival and the speeds of osteoclast excavation and osteoblast filling of subchondral cortical bone can end in the collapse of the femoral head!! So when using a PTH one can, and must, stop osteoclasts from rapidly destroying cortical bone by using osteoclast-killing anti-catabolics that do not hamper the osteoblasts too much if at all (Glimcher, 1999; Hofstaetter et al., 2004; Kim et al., 2004). In other words for femoral subchondral bone timing is everything!!!

PTH can also treat idiopathic (a fancy word for "cause is unknown") osteoporosis in middle-aged men (Kurland et al., 2000). Unlike postmenopausal osteoporosis in women, which is associated with high bone turnover due to osteoclastic overactivity, male idiopathic osteoporosis is associated with low bone turnover and thus is not as amenable to the antic-catabolics designed to suppress postmenopausal women's demonic bone diggers. Kurland et al. (2000) reported the results of an experiment on 23 men (50 ± 1.9 years of age) with idiopathic osteoporosis. They gave 10 of the men a daily subcutaneous injection of 400 IU (about 30 µg of hPTH-(1-34)(Forteo™) for 18 months. While the lumbar spine bone mass did not change in the 13 placebo-treated control patients, it significantly increased by 13.5% in the PTH-treated patients. The PTH treatment also significantly increased the femoral neck BMD by 2.9%. It also increased bone turnover markers such as serum osteocalcin (growth) and urinary N-telopeptides (resorption) by 300 to 400% without causing hypercalcemia. More recently Orwoll et al. (2003) reported the effects of 11 months of daily injections of 20 or 40 µg of rhPTH-(1-34)(Forteo™) on bone BMD in men aged 35-85 with lumbar spine, or proximal femur (neck or total hip) having BMDs which were 2 or more standard deviations below the average for young healthy men. At the end of the 11months the BMDs of spinal and femoral neck BMDs were significantly increased above the values in placebo-treated men. The spinal BMDs were respectively 5.9 and 9.0% higher in men given 20 and 40 µg hPTH-(1-34)/day than in the placeo-treated men. The corresponding increases in the PTH-treated mens' femoral necks were 0.6 and 0.9%.

In July 2001, the FDA panel assessing Lilly's recombinant (i.e., bacterially made) hPTH-(1-34) (Forteo™) concluded that this now venerable fragment clearly increases bone mineral density (though remember that "PTH" bone is young bone [Paschalis et alk., 2005]), improves bone microarchitecture, prevents fractures by a unique mechanism of action, changes the paradigm for the treatment of osteoporosis and offers benefits that cannot be had with the current antiresorptives (Center for Drug evaluation, 2001). In November 2002 this fragment was approved by the USFDA for treating osteoporosis in both women and men. Thus did the aging teriparatide metamorphose into Forteo™ (in North America) or Forsteo™ (in Europe) (Table 1) and was sold to osteopototics to inject themselves subcutaneously each day in a dose of 20 µg for no more than 2 years.

A very important question at this point is—*how does the human skeletal response to Lilly's hPTH-(1-34) (Forteo™) compare with the response to the Merck-Frosst's bisphosphonate alendronate (Fosamax)?* The appropriately named Body et al. (2002) treated 73 postmenopausal osteoporotic women with 10 mg of oral alendronate plus subcutaneously self-injected placebo (instead of PTH) and oral calcium (1000 mg) and vitamin D (400-1200 IU) each day for 14 months. Another 73 postmenopausal osteoporotic women injected themselves subcutaneously once daily with 40 µg of hPTH-(1-34) (Forteo™) and swallowed an oral placebo (instead of alendronate) as well as calcium and vitamin D supplements for the same period of time. The greater effectiveness of PTH appeared as early as 1 month. By 14 months alendronate had increased the BMD of the lumbar spine by only 5.6% while hPTH-(1-34) had increased it by 12.2% (p <0.001). The PTH also increased the femoral neck BMD by about 6% while alendronate increased it by only about 2%. And by 14 weeks PTH had increased total hip BMD by about 4.5% while alendronate had increased it by about 2.5%. And the non-vertebral (ankle, foot, radius, ribs, toe) fracture incidence was 4.1% in the PTH-treated group compared with 13% in the alendronate-treated group. Interestingly, the PTH promptly began increasing bone alkaline phosphatase, an indicator of bone growth, by about 50 to 100%, while alendronate equally promptly caused the alkaline phosphatase level to *drop by 50%*. This would be expected because of alendronate's ability to inhibit the genes for bone formation such as alkaline phosphatase, collagen I, and IGFs-I and II in the proximal femoral metaphyses of OVXed rats (Onyia et al., 2002). Also there was a small drop (about 4%) in the BMD of the distal 1/3 of the radius in the PTH-treated group which was expected from other studies in humans, monkeys and rabbits, was probably transient, and would have been biomechanically insignificant (Burr et al., 2001; Cann et al., 1999; Hirano et al., 2000; Neer et al., 1991, 2001). Clearly, as expected the anabolic hPTH-(1-34) (Forteo™) is much better than the anti-catabolic alendronate at increasing BMD at various skeletal sites and decreasing non-vertebral fractures in humans (D.M. Black et al., 2003; Finkelstain et al., 2003).

Alendronate co-treatment actually blunts the anabolic actions on hips and vertebrae of daily injections of 40 µg of hPTH-(1-34) (Forteo™) into osteoporotic men (Finkelstein et al., 2003) and 100 µg of hPTH-(1-84) into osteoporotic women (D.M. Black et al., 2003). The reason for this has been found in rat femurs by Onyia et al. (2002)—alendronate prevents hPTH-(1-34) (Forteo™) from stimulating key bone-growth-driving genes

Finally an equally important question is—*Does prior exposure to the anticatabolic bisphosphonates or SERMs compromise the anabolic action of a subsequent treatment with hPTH-(1-34) (Forteo™)?* This is very important because nearly all osteoporotic patients presenting for PTH treatment will already have been or still are taking one of these antiresorptives. Ma et al (2003) have reported that prior treatment of OVXed Sprague-Dawley rats for 10 months with alendronate (Fosamax™), 17α-ethinyl estradiol, or raloxifene before switching to hPTH-(1-34) (Forteo™) did not affect the responsiveness of the rats' skeletons to the PTH. But this result would be expected because the bones of continuously growing rats retain a pool of PTH-responsive osteoblastic cells and bone-lining cells. However, this might not be the case with old humans who normally have more microcracking bone and thus osteoblasts-containing BMUs but in whom an anti-catabolic-induced lack of BMUs and microcrack repair would blunt the response to a PTH just as happened in Delmas et al.'s (1995) anti-catabolic-treated old sheep.

And this appears to be the case for the potent alendronate (Fosamax™). Ettinger et al. (2004) measured the effects of prior treatment with alendronate (Fosamax™) or raloxifene on the anabolic action of daily injections of hPTH-(1-34) (Forteo™) in postmenopausal, 60-87 year-old women for 18 months after stopping the anti-catabolic bisphosphonate treatment. Briefly, raloxifene pretreatment did not affect the subsequent ability of the PTH peptide to increase the BMD in hip and spine, but the rise in the BMDs in the alendronate-pretreated women was delayed and peaked at lower levels than in the raloxifene-pretreated women. The authors suggested that the nearly complete suppression of remodeling by alendronate pretreatment in contrast to the less complete suppression by raloxifene pretreatment would more severely reduce the pool of PTHR1-bearing target osteoblasts for stimulation by the PTH injections in the alendronate-pretreated women as probably happened in Delmas et al.'s anti-catabolic-treated old sheep. And indeed Cosman et al (1998) had reported 6 years earlier the results of a trial in which bone resorption in the patients happened to be less suppressed by alendronate than in Ettinger's patients and PTHwas then able to rapidly increase BMD.

v. Rat versus Human—Problems for Extrapolation and Prediction

The rat has been the standard pre-clinical model for assessing the osteogenicities of the PTHs instead of the much more relevant, but far more expensive, cynomolgous monkey. But there are very important differences between the osteogenic and carcinogenic actions of the peptides in rats and humans.

Rats do not completely stop growing as do humans. Consequently they maintain a pool of osteoprogenitors and PTHR1 receptor-bearing osteoblasts and the post-osteoblastic bone-lining cells. After humans stop growing they reduce their PTH target bone cells to only those in remodeling/microcrack-repairing BMUs and probably the post-osteoblastic lining cells. Therefore, rat bones are much more responsive to PTHs which can double the bone growth in osteopenic OVXed rats but cannot raise the human bone growth up to the normal level. Another possible consequence of this difference as we shall see is the ability of Forteo™ to induce osteosarcomas in rats.

CHAPTER 5

How Might PTHs Stimulate Bone Growth?

"...I shall need a herd of elephants, I thought, and a wilderness of spiders, desperately referring to the animals that are reputed longest lived and most multitudinously eyed, to cope with all of this. I should need claws of steel and beak of brass to penetrate the husk. How shall I ever find the grain of truth embedded in this mass of paper?"—Virginia Woolf (1975).

"The whole system is controlled by a vastly complex "signaling" system. Membrane-bound receptors themselves do not float freely in the membrane, but are tied to scaffolding proteins. This locates them precisely with with respect to the other proteins ....in the cascades with which they interact."—John Smythies (2002; p. 2-3) speaking of neuronal synaptic signaling but applicable to bone.

i a. The Starter Gun: The PTHR1 Receptor

"...the receptor protein sticks its molecular equivalent of head and arms out to catch passing...molecules and keeps its foot on the molecular gears inside the cell."—L.H. Caporale (2002; p. 139).

"...it is getting more and more complex and we are frequently overloaded with information. In addition, we have moved from absolute fidelity (one ligand, one receptor, one G protein, one effector, etc.) to almost complete promiscuity (multiple ligands, receptors, G proteins, effectors, and all interacting among them)."—J. Váquez-Prado et al (2003; p. 556).

To tackle the formidable job of understanding how the PTHs stimulate bone growth in humans, rodents and other animals (Fig. 17) we must know where and how things start. What signals do they send into their target cells via the PTHR1 receptors to trigger osteogenesis? However, before going on I must warn the reader that most people call the receptor *PTH1R*. Since there is no type 1 PTH (i.e., PTH1) but there is a type 1 PTH receptor, I use what surely must be the more accurate name for the receptor—PTHR1 (i.e., PTH's receptor No.1).

But aside from the semantics, how do I know that the PTHR1 receptor is the molecular gun which the PTHs use to fire their osteogenic bullets into the bone cells? To show this, Calvi et al. (2001),Schipani et al. (2002) and Kuznetsov et al. (2004) coupled a mutant human PTHR1 gene (HKrk-H223R) encoding a permanently switched-on receptor to the mouse α(I) collagen promoter in order to restrict the gene's expression to the osteoblasts in a "transgenic" mouse otherwise there would have been widespread chaos in the many other PTHR1-expressing tissues. The receptor encoded by this gene was "frozen" in an active configuration—it didn't need a PTH to activate it. The switched-on PTHR1s in the osteoblasts, like normal PTH-activated PTHR1s, increased endosteal bone formation, caused an accumulation of inter-trabecular pre-osteoblastic marrow stromal cells, eliminated the appearance of other components of the skeletal marrow stromal cell population in favor of a build-up of hyperactive trabecular osteoblasts in both proximal tibia and cranial bones and dramatically increase trabecular bone volume. As must happen with PTH infusion or, as in this case, non-stop PTHR1 signaling, there was also a large increase in cortical and trabecular osteoclasts, a lack of periosteal growth, increased cortical porosity and a resulting drop in cortical bone mass. However, there was no bone loss and the trabecular volume was even greater in doubly-mutant mice making collagenase-3-resistant α(I) collagen along with the switched-on PTHR1 (Schipani et al., 2002). In these resistant beasts, the diggers could'nt reduce the bone build up by chopping up the collagen matrix.

The PTHR1 receptor (Figs. 18 and 19) is one of the 1200 to 1300 members of the huge family of G-protein-coupled (GPCR) or 7TM (7 transmembrane helical segments) receptor superfamily and belongs to the secretin receptor-like family B (Karnik et al., 2003; Schoneberg et al., Schwartz et al., 2006; 2002; Wong, 2003). It has the familial 7 transmembrane α-helices (i.e., like barrel staves inserted through the membrane) and a middle-size extracellular "nose" or N-terminus of 160 aminoacids compared, for example, to the extracellular 360 aminoacids of the parathyroid gland cells' CaR (Ca^{2+} receptor; [Brown and MacLeod, 200]) (Bockaert et al., 2002; Hoare and Usdin, 2001; Hoare et al., 2001; Schoneberg et al., 2002; Whitfield et al.1998a). PTH-(1-34) dispenses most of its binding energy when it attaches the 15-31 region of its C-terminal tail first to the receptor's N-terminus sticking out of the cell from the first barrel stave and then sticking its N-terminal "nose" onto the Leu^{232}-Lys^{240} region of the receptor's second barrel stave (TM-2) (Barbier, et al. 2006; Chorev, 2002; Gensure et al., 2002; Hoare and Usdin, 2000; Hoare et al. 2001) (Fig. 18). The key players in this interaction are Arg^{233} and Lys^{240} and replacing one or both of them with alanine prevents the interaction.

According to Karnik et al. (2003) when a PTH finds and then settles down on the receptor's N-terminus it nudges the extracellular loop 2 connecting staves 4 and 5 (Fig. 19). This nudge stretches the S-S bond linking outer segments of staves III and IV which causes the intracytoplasmic segments of the staves to flare open to let a GDP•Gs three-piece complex (i.e., a heterotrimer) to get at the intracellular loops and the C-tail sticking out of the bottom of the barrel. According to Wong (2003), the shifting staves cause the three intracellular loops and C-terminal tail to make a pocket that grabs an adjacent G-protein dangling from the membrane. The G-protein is probably hanging beside and somehow linked to the receptor in the inactive GDP•Gα-$\beta\gamma$ form initially tethered by the $\beta\gamma$ complex to the cell membrane (Petsko and Ringe, 2004). When the incoming PTH pushes the receptor staves and causes their inner segments to flare open, the GDP•Gα-$\beta\gamma$ complex attaches to the C-tail and is pulled into the "pocket" formed by the intracellular loops, where its GDP is pulled off and replaced by a GTP to form active GTP•Gα which is separated from the still C-tail-bound $\beta\gamma$ pair (Mahon et al., 2006). The active GTP•Gα is then changed back to inactive GDP•G either slowly by its own GTPase activity or much faster by an associated GAP (GTPase activating protein) protein and reassociates with $\beta\gamma$. Of course the final signal mix that is fired into the cells is determined by whether the receptor tail is attached to a NHERF and what other signalers are also riding on the NHERF (Fig. 18).

The above description is based on Schwartz et al. (2006)'s global "Toggle Switch Model" of the ligand-induced intracytoplasmic flaring of GPCR/7TM receptor barrel staves. According to this model the "homing" PTH (like the various other large and small GPCR/7TM receptor ligands) triggers the inward pivoting of TM-6 around its proline 15 hinge weak point to approach TM-7 which pulls and locks the outer ends of the staves together and pushes their intracytoplasmic segments apart (Fig. 19).The spreading of the inner (cytoplasmic) segments exposes the binding sites for adjacent G-proteins and arrestin (Schwartz et al. 2006) (Fig. 19).

But where are these binding sites? As I said above, a docking site for G-proteins, specifically their $\beta\gamma$ subunits, has recently been found by Mahon et al. (2006). Specifically they found that Gi/o protein heterotrimers bind to a site on the receptor's C-tail which cannot be got at in the inactive receptor because of the more closely clusterd cytoplasmic segments of the transmembrane helices. When the receptor is jiggled by a PTH, the heterotrimer grabs the activated receptor's C-tail with its $\beta\gamma$ subunits. The binding site is located on the juxta-membrane part of

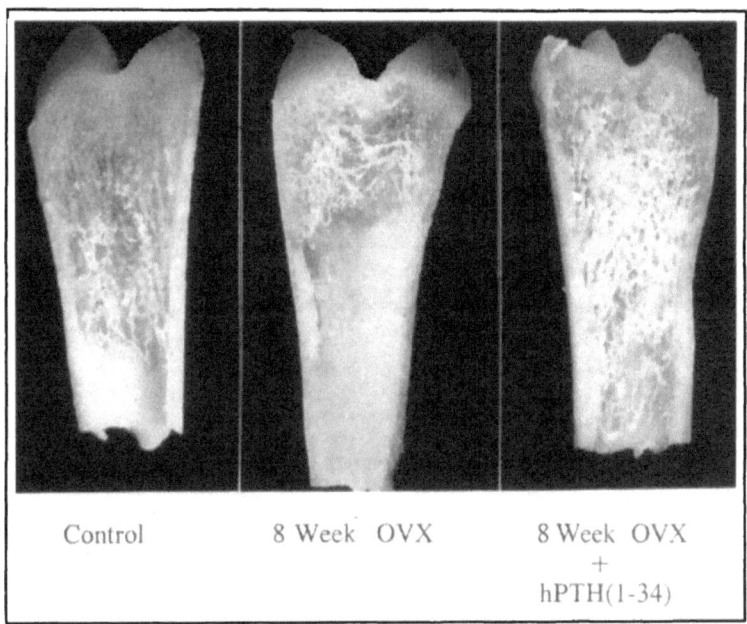

Control   8 Week OVX   8 Week OVX
                                           +
                                       hPTH(1-34)

Figure 17. A demonstration of the dramatic ability of hPTH-(1-34) (Forteo™) to stimulate trabecular bone growth in the rat distal femur. The rats received 6 weeks of daily injections of 1 nmole of the PTH/100 grams of body weight starting 2 weeks after PTH.

the receptor's C-tail between amino-acids 468 and 491. Activating the tail-bound heterotrimer does not release the βγ pair but it disrupts the binding of, and releases, the heterotrimer's Gα subunit that is activated by acquiring a GTP and stimulates or inhibits the adenylyl cyclase and cyclic AMP signaling. As expected, preventing the βγ binding blocks PLCβ1 activation as well as adenylyl cyclase activation. In other words when the PTHR1 is activated and its cytoplasmic helical segments spread apart a heterotrimeric G-protein can reach the 468-491 region upon which to bind its βγ subunits and activate and release the Gα subunit (Fig. 19).

i b. Signal Bullets Fired from the PTHR1 Starter Gun

Depending on the cell, the receptor-flaring binding of hPTH-(1-34)'s N-terminal nose specifically generates active GTP•Gs subunits from the inactive GDP•Gsα-βγ protein complexes. GTP•Gs then stimulates adenylyl cyclase. Adenylyl cyclase is a potent signal amplifier that can make lots of cyclic AMP from ATP. Also, depending on the cell the binding of the PTH's amino-acids 1-20 nose region to the receptor's N-terminus may generate active GTP•Gq/11α subunit from the GDP• Gq/11complex that stimulates phospholipase-Cβ1 (PLC-β1) that in turn chops a minor, but an incredibly potent event-driving, membrane phospholipid with the daunting name, phosphatidylinositol(4,5)bisphosphate (or PIP_2 for short), into diacylglycerols (DAGs) and inositol (1,4,5)trisphosphate (IP_3) (Hoare and Usdin, 2001; Hoare et al., 2001; Morley et al. 1999; reviewed by Bilezekian et al., 2001; Whitfield et al.1998a; Yang et al., 2006) (Fig. 18). The DAGs then stimulate protein kinases such as PKC-δ. But as I will show below there is the 28-32 region of hPTH-(1-34)'s tail that also somehow stimulates membrane-associated PKC-δ without activating PLC-β1 (Yang et al., 2006). It is important to find out how the contact of the PTH's negelected 28-34 region with PTHR1 activates PKCs such as PKC-δ (Fig. 20).

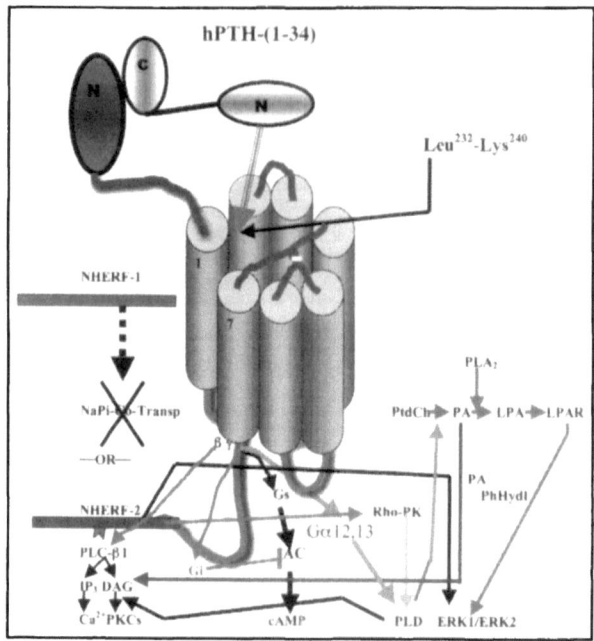

Figure 18. A possible model of the PTHR1 receptor. Cohen and Stewart (2003, p.165) have beautifully described receptors such as PTHR1 as *"proteins laced through the cell membrane"* and that *"pushing a receptor button on the outside rings a chemical bell inside"*. PTHR1 is a barrel with 7 α-helical staves laced through the cell membrane with a N-terminal "nose" and stave-connecting loops sticking out of the cell as well as connecting alternating loops hanging down into the cell's cytoplasm from the bottoms of the staves. And then there is the receptor's *"magic tail"* (Bockaert et al., 2003) hanging out of stave 7 that attaches to a NHERF-1 and/or NHERF-2 scaffold lying beside the receptor and bearing a cluster or network of interacting signal-transmitting enzymes and factors. Depending on what's in them, these clusters can increase the variety of signals that the receptor sends into the cell when the PTH pushes the receptor's button—its N-nose—which causes the receptor to ring the chemical bell by flaring out its intracellular stave segments and rearranging the internal loops and the C-tail to interact with G-proteins hanging from the cell membrane by their βγ subunits and slapping its C-tail on the NHERF scaffold *if there is one*. As indicated in the drawing, the signaling starts when the PTH attaches its C-tail (C) to the receptor's N-"nose" (and perhaps to some of the barrel loops sticking out of the membrane) and then sticks its own nose (N) onto the Leu^{232}-Lys^{240} patch on barrel stave 2. According to the "toggle switch" model of Schwartz et al (2006), the result is an inward pivoting of TM 6 around its Pro^{15} hinge weak point to approach TM7 which pulls and locks the outer ends of the of the staves together and pushes their intracytoplasmic segments apart. The spreading of the inner segments exposes binding sites on the receptor's C-tail for G-proteins and arrestin. When this happens a heterotrimeric G protein can stick its βγ subunits to an exposed site on the receptor's cell membrane-associated C-tail which is exposed by the inner stave flaring. Then the G-protein is activated and its α-subunit separates from the still C-tail bound βγ and can interact with its specific targets. But the signal mix emitted by the activated PTHR1 depends on whether the PTH has a normal or shortened N-"nose" and C-tail and on what scaffolds the cell has on its membrane within reach of the receptor tail as well as what things the cell has loaded onto the scaffolds. As discussed in the text, this receptor may also dimerize with another PTHR1 or even heterodimerizes with another kind of receptor to send a different mixture of signals. Abbreviations: AC, adenylyl cyclase; βγ subunits of Gi G-protein; DAG, diacylglycerol; ERK1/ERK2, extracellular signal-regulated protein kinases also known as MAP kinases; Gi, the AC-inhibiting subunit of Giβγ G-protein heterotrimer; Gs, the AC-stimulating subunit of Gsβγ G-protein heterotrimer; LPA, lysophosphatidic acid; IP$_3$, inositol-1,4,5-trisphosphate; LPAR, lysophosphatidic acid receptor; NaPi-Co-Transp., Na$^+$-phosphate co-transporter; PAPhHydl, phosphatidic acid phosphyhydrolase; PKCs, various PKC isoforms; PLA$_2$, phospholipase A$_2$; PLC-β1, phospholipase-Cβ1; PLD, phospholipase D; PtdCh, phosphatidyl choline; Rho-PK, rho (a G-protein)-controlled protein kinase.

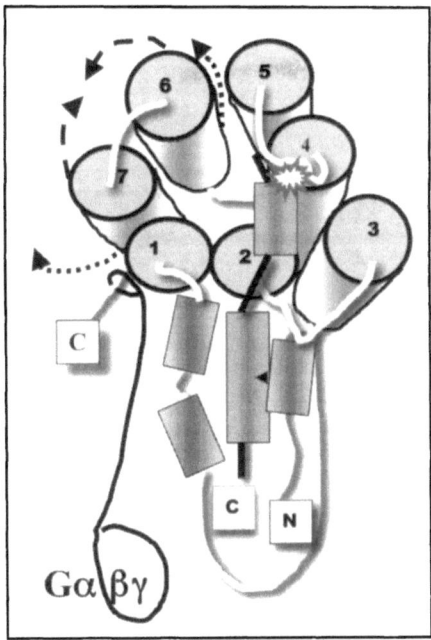

Figure 19. What the PTH-PTHR1 complex might look like from above after the attachment of the PTH to the receptor. The PTH is the two green cylindrical α-helices connected by thick black lines. In this cartoon, the PTH has just landed on the receptor's N-nose and shoves its nose against the receptor's extracellular loop 2 that links staves 4 and 5 which strains the S-S bond linking barrel staves (i.e., α-helices) 3 and 4. The outer segments of staves may also 6 and 7 tip toward each other according to the Schwartz et al (2006) "toggle switch" model. This triggers a flaring open of the internal segments of the staves that opens the way for a GDP•Gs-protein to bind its βy subunits to a now available site on the C-tail where it can somehow be activated by having the GDP removed and a GTP put in its place (Karnik et al., 2003). The GTP•Gs subunit then activates the nearby adenylyl cyclase hanging from the cell membrane (Cooper, 2003). But whether the adenylyl cyclase can be activated will depend on whether the receptor tail is attached to a NHERF scaffold lying to one side and on what other signalers are on the scaffold as in Figure 18.

The PTHR1 signaling must *not* be allowed to continue beyond a certain time if it is to be optimally effective. As we shall see further on, one way of doing this is to simply drag the PTHR1•PTH complex into the cell on a β-arrestin platform, pull the PTH off and destroy it, and put the cleaned up receptor back to work on the surface (Fig. 21). However, the job of being the cell's general GPCR receptor off-switch has been given to RGS-2, a small member of the B/R4 clan of the 26-member family of RGS (regulators of G-protein signaling) proteins (Hollinger and Hepler, 2002; Petsko and Ringe, 2004). PTHs rapidly stimulate RGS-2's gene via adenylyl cyclase/cyclic AMP (but we don't know what happens when PKC-δ, but not adenylyl cyclase and PLC-β1, is stimulated by N-noseless PTHs) and the resulting RGS-2 protein shuts the signaling off by converting active GTP•Gαs and GTP•Gαq to inactive GDP•Gαs and GDP•Gαq as well as by directly inhibitng adenylyl cyclase (Hollinger and Hepler, 2002; Miles et al., 2000; Thirunavukkarasu et al., 2002; Tsingotjidou et al., 2002). However, RGS-2, like RGS-4, another small member of the B/R4 clan, might have a more positive role such as triggering Ca^{2+} oscillations (Hollinger and Hepler, 2002). But it appears that RGS-2 might either require help from β-arrestin to shut the receptor off or it is not really needed for this, because knocking out the β-arrestin-2 gene in mouse osteoblasts prolongs PTH-triggered cyclic AMP signaling which increases cortical (but not trabecular) bone growth (Pierroz et al., 2003).

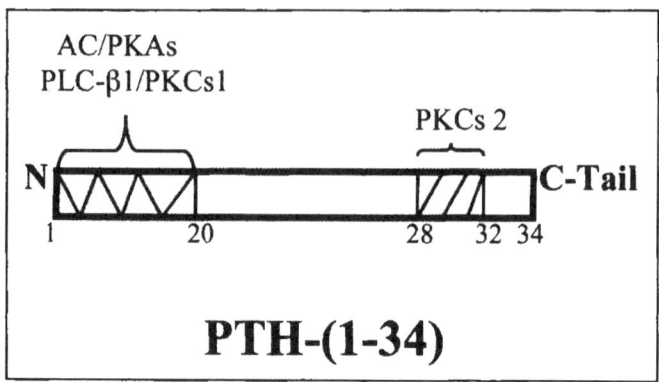

Figure 20. A PTH such as hPTH-(1-34) (Forteo™) has two separable protein kinase C (PKC)-stimulating regions. One region is in the peptide's first 20 residues. It is not coupled to adenylyl cylase activation, but it operates by stimulating phospholipase C-β1 (PLC-β1) which in turn chops phosphatidylinositol-4,5 bisphosphate into inositol-1,4,5 trisphosphate and diacylglycerol which stimulates PKCs such as PKC-δ. The second region is located in the 28-32 region of the molecule and stimulates PKCs by a mechanism that does not operate via PLC-β1. It is via this region that peptides such as hPTH-(13-34) and even hPTH-(28-32) can stimulate membrane-associated PKS activity. It will fascinating to find out how these regions so differently cause the PTHR1 receptor to activate PKCs!

So far the effort to find out how a PTH such as hPTH-(1-34) causes the PTHR1 barrel loops and staves to assume an active G-protein-grabbing state has concentrated on the stimulation of adenylyl cyclase as the measure of activation. But PTHs such as hPTH-(1-34) stimulate two surges of membrane-associated PKC activity, one peaking in the picomolar range well before cyclic AMP production and the second peaking in the nanomolar range along with adenylyl cyclase (Janulis et al., 1993; Jouishomme et al., 1992). Peptides such as hPTH-(3-34), hPTH-(7-34), hPTH-(13-34), hPTH-(28-32), hPTH-(28-42), or hPTHrP-(5-36) which have had their normal N-terminal noses cut off are *very wrongly* called antagonists just because they cannot activate adenylyl cyclase themselves and by binding to the PTHR1 receptor and prevent the N-terminally intact PTHs such as hPTH-(1-34) from doing it. But they can stimulate PKCs as strongly as hPTH-(1-34) in some cells such as ROS 17/2 and UMR106-01 osteoblastic cells (osteoblasts variously express PKCs -α, -β, -δ, -ε, -η, -θ, -ζ, -1/λ !) as well as freshly isolated human foreskin fibroblasts, freshly isolated rat proximal kidney tubules cells, OK opossum kidney cells, and *cyc⁻* S49 murine T-lymphoma cells (Azarani et al., 1995a, 1995b; Chakravarthy et al., 1990; Erclik and Mitchell, 2001; Janulis et al., 1993; Jouishomme et al., 1992; Lichong et al., 1998; Sriussadaporn et al., 1995; Whitfield et al., 1998b, 2001). And bPTH-(3-34) has been shown to stimulate Ca^{2+} uptake by rat thymic lymphocytes (Atkinson et al., 1987). Because of this misconception, there has been a large gap in our understanding of PTHR1 signaling. But the gap is now being filled by the discovery of PTHR1's "*magic tail*" to use Bockaert et al. (2003)'s happy term as well as the confirmation by Yang et al., 2006) of presence of PTH's second, seemingly very high-affinity, PLC-independent PKC-stimulating C-terminal domain (Fig. 20).

Yang et al. (2006) have found that the hPTH-(1-34) peptide has two different domains for stimulating PKCs which correspond to Janulis et al.'s (1993) and Jouishomme et al.'s (1992) responses to picomolar and nanomolar PTH concentrations (Fig. 20). This story actually began when Whitfield, Isaacs et al. (2001) showed that N-terminally truncated PTH peptides, which according to Takasu et al. (1999) do not stimulate PLC-β1 (they did not measure PKCs activity) in HKRKB7 pig cells, significantly increase membrane-associated PKC activity in

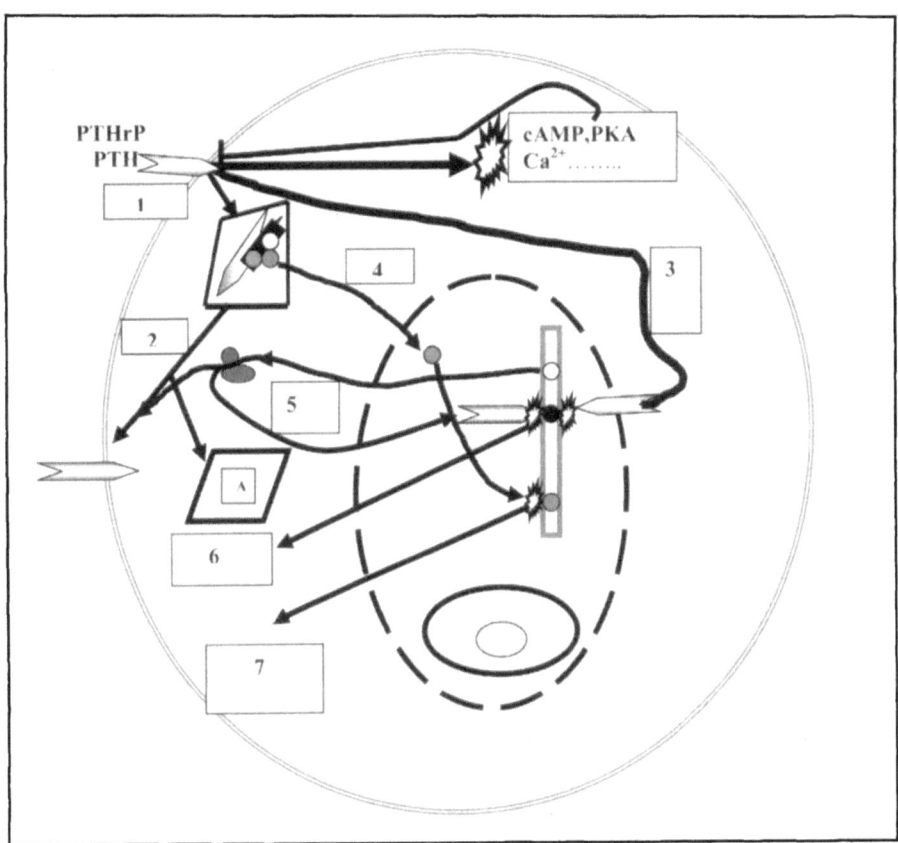

Figure 21. The travels of a PTHR1 (shaded arrows) laced into the membrane of a target cell such as an osteoblast. (1) The receptor is activated by either an endocrine, injected, inhaled, or swallowed pill-borne PTH from the blood or by autocrine/paracrine PTHrP from the cell or its neighbors. The receptor fires cyclic AMP, Ca²⁺ and PKC signals into the cell and the cyclic AMP signals feed back to shut down the receptor. (2) PTHR1 and the attached PTH are attached to β-arrestin -2 scaffold (rectangle) containing a cluster of signalers (circles) and carried into the cell in a clathrin-coated endocytic vesicle where the PTH is destroyed and the receptor is reactivated. The reactivated PTHR1 is returned to the surface ready to go again and the empty β-arrestin -2 scaffold (A) is left behind in the cytoplasm. (3) The receptor might enter the nucleus because it has a NTS (nuclear transport sequence) and activate genes. (4) One of the signalers that had been activated when the β-arrestin-2 scaffold had grabbed PTHR1 is a JNK (c-Jun kinase—one of the colored circles brought in on the arrestin rectangle) which gets into the nucleus and stimulates gene transcription. (5) The PTHR1 gene transcript is translated into PTHR1 in the cytoplasm by the ribosomal translating machinery. The newborn PTHR1 has a NLS with which it can be grabbed by nuclear cargo transport importins and directly pulled into the nucleus instead of being plugged into the membrane. (6,7) Genes are stimulated by PTHR1 or JNK.

these cells just as they do in rat kidney and bone cells provided they have 28-34 regions. Yang et al. showed that when *wt9* murine osteoblasts with 80,000 to 100,000 PTHR1s/cell were exposed to hPTH-(1-34), they activated adenylyl cyclase and also phosphorylated and activated their PKC-δ. [G1,R19]hPTH-(1-28) with an impaired N-terminal PLC-activating domain and without the 29-34 PLC-independent PKC-activating domain caused the cells to activate adenylyl cyclase, but as expected from Takasu et al. (1999), could not activate PLC or phosphorylate and activate PKC-δ. On the other hand, the adenylyl cyclase-activating [G1,R19]hPTH-(1-34), without the functional N-terminal PLC-activation domain (adenylyl cyclase stimulation is not coupled to PLC-β1 activation) did cause the cells to activate PKC-δ because it still had the 29-34 PLC-β1-independent PKC-activating region discovered in my laboratory (Jouishomme et al., 1992). But there is a big problem here—we were surprised to find that [G1,R19]hPTH-(1-28) actually powerfully (as much as 3- to 25-fold!) stimulated membrane-associated PKC activity (which was certainly no artifact because it was inhibited by a specific pan-PKC blocker) in the cells of a primary isolate of human foreskin fibroblasts. Evidently the owner of the PTHR1 receptors has a lot to say about what the receptor can and cannot activate.

But how do N-terminally truncated PTHs persuade PTHR1 receptors to stimulate PKCs without activating PLC-β1? It seems that they can't stimulate adenylyl cyclase and PLC-β1 because they can't get down on the receptor TM2's (barrel stave 2) Arg^{233} and Lys^{240} to bump the extracellular loop 2 and start the adenylyl cyclase-activating stave movements (Gensure et al., 2002) (Fig. 19). Nevertheless they can still bind to TM-2's Leu^{232}-Lys^{240} region with the barrel stave's critical interacting amino acid being Phe^{238} (Gensure et al., 2002). But we don't know how this interaction along with the binding of their C-tails to the receptor's N-terminus could cause the receptor to activate PKCs independently from activating PLC-β1.

The ability of PTH-(1-34)'s 1-20 N-nose to activate PLC-β1 is the result of the PTHR1 receptor, like the β2-AR (β2-adrenergic receptor), having an atypical PDZ-binding domain— Glu-Thr-Val-Met (ETVM) sequence—in its magic C-tail (Fig. 18). When a N-terminally intact PTH binds to a PTHR1 receptor, the receptor's staves flare open and its C-tail attaches its PDZ ligand to the PDZ-II domain in the C-tail of a NHERF-2 (the similar C-tail of β2-adrenergic receptor binds to the PDZ-I domain [Hall, 2004; Hall et al., 1998]) which is the Na+-H+ exchanger's inhibitory regulatory factor-2 (Bockaert et al., 2002, 2003; Harris and Lim, 2001; Mahon and Segre, 2001; Mahon et al., 2002; Voltz et al., 2001). It is this contact with NHERF that fires up PLC-β1 (Fig. 18).

Because of its PDZ patch, PTHR1's C-tail, like a dog's tail, is a sophisticated signaling device. The signaling by the PTHR1 receptor, like several other GPCR receptors, such as the β2-AR, is in some cells mediated by a large particle called a "transducosome" consisting of a core scaffold and several attached and interacting signal components (Kreienkamp, 2002). These transducosomes can be awesomely large and complex—the 5-HT2c receptor's tail, for example, can associate with as many as 15 different proteins (Becamel et al., 2002)! And PTHR1 plugs into a transducosome with its tail. One of the cores of PTHR1's transducosomes is NHERF-2, a kind of scaffold or circuit board attached to the cell membrane, loaded with a set of interconnected signaling enzymes, and is tethered to underlying cytoskeletal actin cables via an "ERM" (ezrin, radixin, moesin) bridge (Harris and Lim, 2001; Kreienkamp, 2003; Mahon and Segre, 2002; Ponting et al., 1997; Voltz et al., 2001). Thus, *if* the cell has NHERF-2s, each of its PTHR1 receptors can be part of a signal-transducing supercluster on the NHERF-2 circuit board perhaps along with another GPCR receptor as well as several membrane-associated G-proteins and other components such as PLC-β1.

The NHERF-2 scaffold in cells such as PS120 fibroblasts grabs PLC-β1 by the tail with its PDZ-I domain and puts the phospholipase within easy reach of the active PTHR1 C-tail, and subunits from the associated membrane-anchored Gi/o or Gq/11 heterotrimers (Mahon et al., 2002; Suh et al., 2001). It appears that when the target cell has NHERF-2 (and it is essential to know that *not all cells do!*) and PTH-activated PTHR1 attaches to the scaffold with its cytoplasmic C-tail's PDZ-II-binding patch, adenylyl cyclase stimulation is *suppressed* but PLC-β1 stimulation by the nose's PKC-1 activation site (Fig. 20) is greatly *amplified.* Why?

The reactions triggered by the activated PTHR1 take place in a cluster of enzymes, G proteins and receptors on the underside of the cell membrane. In PS120 cells, the mechanism seems to be due to the NHERF-2-connected PTHR1 receptor generating active GTP•Gi/oα subunits that find, bind and inhibit adenylyl cyclase and the NHERF-2 complex abetting this cyclase suppression by hindering the coupling of the AC–activating Gs in the pocket formed by the intracellular loops to the adenylyl cyclase. Meanwhile, the βγ subunits released from the Gi/o or Gq/11 complexes stimulate the PLC-β1 sticking to the NHERF-2 circuit board's PDZ-I domain (Mahon et al., 2002).

This means that cells with lots of NHERF-2s will respond to a PTH very differently from those with no NHERF2. Thus, ROS 17/2.8 rat osteosarcoma cells without NHERF-2 respond to PTH-(1-34) with as much as a *30-fold* burst of adenylyl cyclase activity but *no* PLC-β1 activity, but if NHERF-2 be put into these cells the PTH can stimulate PLC-β1 and PKC (Mahon et al., 2002). By contrast hPTH-(1-34) can stimulate PLC activity 3-4 fold and adenylyl cyclase by only 1.2-fold or less in ECV304 cells which have lots of NHERF-2 (and NHERF-1) (Mahon et al., 2002). And then there are the lymphocytes with lots of NHERF-2s that respond to hPTH-(1-34) with a burst of PLC-β1 and membrane-associated PKC activity but with no increase in adenylyl cyclase activity (Mahon et al., 2002; Whitfield et al., 1998b, 1999a). Of course this could explain why PTH cannot directly stimulate adenylyl cyclase in lymphocytes (Whitfield et al., 1971, 1999a).

*The take-home message from all of this is that the PTHR1 is basically an adenylyl cyclase activator, a cyclic AMP pulse generator that must be plugged into a NHERF-2 circuit board if it is to activate PLC-β1 and its downstream effectors such as PKCs and if it also has the PLC-β1-independent PKC 2 activating region. Put in another way, PTHs by themselves are adenylyl cyclase stimulators, but they are PLC/PKC stimulators only when plugged by their nose into a PTHR1/NHERF-2 cluster and/or onto the receptor by the C-terminal PKC 2 region. But there is also a revolutionary postscript to this take-*home message—*N-terminally truncated PTHs can still stimulate PKC, though not adenylyl cyclase or PLC-β1, if thy have the PKC-activating 28-32 region (Fig. 20).*

Another example of the daunting complexity of PTH signaling imposed on it by the transducosome on the NHERF-2 scaffold has been reported by Mahon and Segre (2002). As long as PTHR1 is attached to NHERF-2 in PS120 fibroblasts, hPTH-(1-34) can also stimulate extracellular-regulated kinases (ERKs; Pearson et al., 2001)-1 and -2 within 5 minutes (Fig. 16). Since this stimulation was unaffected by inhibiting PKAs (the protein kinases activated by cyclic AMP) and PKCs, the authors concluded that the ERK pathway must have been activated by a novel mechanism. But it still might be cyclic AMP itself, instead of cyclic AMP-activated PKA, which directly stimulated ERKs via the cyclic AMP•Epac ⇒ GTP•Rap1 ⇒ B-Raf protein kinase pathway to be discussed further on. However, since NHERF-2 attachment prevents adenylyl cyclase stimulation, disconnecting PTHR1 from NHERF-2 should have enabled rather than prevented a direct cyclic AMP stimulation of the ERK kinases.

A model for this action might be what happens when β2-AR receptor sticks to NHERF's PDZ-I domain (Ahn et al., 1999). The activated β2-AR receptor in this cluster causes the dimerization (pairing) and *trans*activation of the EGF receptor. The β2-AR receptor also activates the c-Src tyrosine protein kinase, which causes the β2-AR receptor to cluster with the EGF receptor pair. This huge β2-AR •[EGFR]2 cluster is then put by a β-arrestin phosphoprotein into a tear drop-shaped clathrin-coated pit, the neck of which is grabbed and pinched off by a dynamin "pinchase" and the pinched-off "tear drop" vesicle is delivered into the cytoplasm (Smythies, 2002). But to get to the point—the active c-Src and the still-functioning pair of EGF receptors trigger the ERK1/2 cascade and its many actions (Maudsley et al., 2000; Friedman et al., 2002). So if PTHR1 attached to NHERF-2 can do the same things, this would be how hPTH-(1-34) could stimulate ERK1/2 phosphoylation/activation.

The attachment of the PTHR1 to the NHERF-2 scaffold in some cells means that there might be yet another signal—an internal pH spike—when PTHR1 is activated as happens when the β2-AR is activated. NHERF-2 inhibits the Na$^+$/H$^+$ exchanger when the exchanger is sitting on NHERF's PDZ1 domain. But the activated β_2-AR knocks the exchanger off NHERF's PDZ-I (Barber and Ganz, 1992; Barber et al., 1992; Bockaert et al., 2002). The displaced and rejuvenated exchanger then causes a pH spike in the cell by pumping protons (H^+s) out of the cell. So PTH might also stimulate the Na^+/H^+ exchanger, because when the activated PTHR1 receptor's tail attaches to PDZ-II, though this is not the PDZ-I to which β_2-AR attaches, it might also displace Na^+/H^+ exchanger from its inhibitory attachment to NHERF (Voltz et al., 2001). Indeed PTH fragments do trigger a PKA-mediated extracellular acidification response (ECAR; a squirt of protons out of the cell) in neonatal mouse calvaria and ROS 17/2 rat cells, but an apparently PKC-mediated ECAR in human SaOS-2 cells (Belinsky and Tashjian 2000; Belinsky et al., 1999). But we don't know whether these different cells have different amounts of NHERFs. However, this is probably a "false trail" because proton -squirting by human SaOS-2 osteosarcoma cells cannot be the work of the Na^+/H^+ exchanger because it is not affected by inhibiting the exchanger (Barrett et al., 1997).

Also among the several PTHR1 signal transducers is (PLD), which chops the major membrane phospholipid—phosphatidylcholine (PtdCh)—into choline and phosphatidic acid (PA) which is converted by phosphatidic acid phosphohydrolase into PKCs-stimulating diacylglycerol (Friedman et al., 1999; Radeff et al., 2002; Singh et al., 1999, 2004) (Fig. 18). The PTHR1 signals start the PLD stimulating mechanism not by activating $G\alpha s$ or $G\alpha q$ but by activating $G\alpha 12$ and $G\alpha 13$ proteins (Singh et al., 2004). The phosphatidic acid could also be converted to autocrine (self-stimulating)/paracrine (neighbor-stimulating) LPA (lysophosphatidic acid) by PLA_2 (phospholipase A_2). LPA is the ligand (binder/activator) for G-protein-coupled LPA receptors (LPARs) which can generate various signalers such as the mitogen-activated protein kinases ERK1/2, p38 MAPK, and MSK 1 that phosphorylate the CREB transcription factor like the cyclic AMP-activated PKA and generate small GTPases such as GTP•Rho A (C.-W. Lee et al., 2003; Xie et al., 2002) (Fig. 18). This GTP•RhoA in turn can feed back to activate the PLD-stimulating, RhoPK with a consequent escalation of PKCs (e.g., PKCα-activity by diacylglycerols (Radeff et al., 2002; Xie et al., 2002). We will meet the "Rho Gang" lurking in membrane caveolar scaffolds in Chapter 8 ii when discussing how the erratic statins might stimulate bone formation. This propensity for scaffolding means that RhoPK may be another member of the NHERF-2 signaling gang.

As if this were not enough, there is another NHERF involved in PTH actions! At the 24th meeting of the American Society for Bone and Mineral Research in San Antonio in September 2002 J.M. Mahon announced (without any written record of his talk) that the C-tails of PTHR1 receptors in kidney proximal convoluted tubule cells, specifically OK (opossum kidney) cells, attach to NHERF-1 instead of NHERF-2 (Fig. 18). In these cells PTHR1-activated bursts of ezrin-phosphorylating PKA and/or PKC prevent phosphate re-uptake by causing NHERF-1•type II Na/Pi-co-transporters to be dragged by phospho-ezrin•actin from their caveolae ("little caves") microdomains in the apical surface into the depths of the cell and destroyed (Pfister et al., 1997; Taketani et al., 2003, 2004). Disrupting the gene for NHERF-1 prevents the type II cotransporter from being plugged into the apical cell membranes of kidney proximal convoluted tubule cells (Shenolikar et al., 2002). And Mahon reported that phosphate uptake by mutant OK cells lacking NHERF-1 was unaffected by PTH, but giving the cells back their NHERF-1 restored the ability of PTH to inhibit phosphate uptake. So we are faced with yet another PTHR1-coupled scaffold with more signaling complexes though maybe not in bone cells (Fig. 18).

However, NHERF-1 can also couple PTHR1 activation to stimulation of phospholipase C as does NHERF-2 (Mahon and Segre, 2003). This has been shown using cells of the OKH substrain of opossum kidney cells. Exposing these cells to hPTH-(1-34) stimulates their adenylyl cyclase but without affecting PLC activity, inhibiting phosphate uptake by the Na-phosphate co-transporter or changing the intracellular Ca^{2+} concentration. But putting NHERF-1 into these cells results in the localization of PTHR1 to the NHERF-1 and enables the PTH to stimulate the release of βγ subunit complexes from Go/i which then activate phospholipase-Cβ which in turn opens a type of Ca^{2+} channel exactly as would happen with the NHERF-2•PTHR1 complex in fibroblasts and osteoblasts (Fig. 18).

But sorry to say there could be even more! It now appears that GPCR/7TM receptors do not work alone. So it is likely that PTHR1 receptor homodimerizes (twins with another PTHR1) or far more important *heterodimerizes* (couples with different kinds of receptor). It has long been an article of faith that GPCR receptors like PTHR1 don't do either of these—they were believed to be strong individualists that would never pair with any different receptors such as the EGF receptor and its several kin. But now the prevailing cocept is that GPCR/7TM receptors habitually form stable homodimers and heterodimers during their synthesis in the endoplasmic reticulum (Schwartz et al., 2006). For example the metabotropic glutamate and GABA receptors dimerize (Schwartz et al., 2006). Also there are the adenosine A1 receptor•dopamine 1 receptor heterodimer, the dopamine D2R receptor•somatostatin 5 receptor heterodimer, and the opioid κ receptor•opioid δ pair (Lee et al., 2000; Liebmann, 2004; Rocheville et al., 2000a, 2000b; Váquez-Prado et al., 2003). And of course we must not forget the ability of β2-adrenergic GPCR/7TM receptors to activate the entirely unrelated EGF tyrosine kinase receptors by physically joining them in a heteromultimeric transducosome that produces active EGF dimers (Maudsley et al., 2000)! *If* as now seems likely PTHR1is a dimerizer, its pairings with other family members along with large clusters of various signalers on NHERFs could produce dazzling displays of multi-track signaling fireworks depending on the target cell's make up that could be ignited by PTH and, here is the important thing—*the ligand of the other member of the receptor couple.*

Then there are the RAMPs (receptor activity-modifying proteins) that are variously and generally expressed and serve as partners to GPCRs. When these proteins partner with receptors such as calcitonin receptor-like receptor (CL-R) or the calcitonin receptor (CTR), they can actually change the receptor's specificity (Christopoulos et al., 2003; Conner et al., 2004; Liebmann, 2004; Morfis et al., 2003; Udawela et al., 2004). For example the CL-R(calcitonin receptor-like receptor)+RAMP1 makes a CGRP (calcitonin gene related peptide) receptor, CL-R+RAMP2 or 3 makes an adrenomedullin receptor, or CTR(calcitonin receptor)+ RAMPs make different amylin receptors. The RAMPs are held in the endoplasmic reticulum until their receptor partners are expressed at which time they move to the cell surface with their partners. Our hero PTHR1 partners specifically with RAMP 2 (Christopoulos et al., 2003; Morfis et al., 2003; Udawela et al., 2004). Unfortunately we don't yet know how this association with RAMP2 affects PTHR1 signaling and specificity. But you can be sure that the RAMP 2 content of the cell will be yet another factor controlling them.

In summary, what PTHR1, like other GPCR/7TM receptors, surrounded by a crowd of hangers-on clinging to rafts with widely ramifying actions, is telling us through the fog of ignorance is actually very clear—*there are no standard PTH receptor signals!* There are no solitary signaling PTHR1s. The kinds of partner receptors, RAMPs and membrane scaffolds with particular sets of signal transmitters and transducers provided by the cell determine what signals a PTHR1 (or, for that matter, any other GPCR receptor) gives the cell. But this is not all! There is also the all-important question of what determines the kind of G-protein the activated receptor couples with.

There are several ways that a GPCR receptor like PTHR1 might couple to more than one kind of G-protein and activate different signal mechanisms. There is the *sequential model* in which the activated receptor initiates a signal cascade that in the process switches back and modifies the receptor to activate a different G-protein and start a second cascade (Schoneberg et al., 2002). As an example of this, phosphorylation of a β-adrenergic receptor by PKA can shift its coupling from Gs to Gi (Daaka et al., 1997; Zamah et al., 2002). Likewise, the activated prostacyclin receptor couples first to Gs, which through its Gsα subunit stimulates adenylyl cyclase, cyclic AMP production and the cyclic AMP activates PKA which then turns around and phosphorylates the receptor on its Ser^{357}which enables the receptor to couple to Gi and Gq and trigger other signal cascades (Lawler et al., 2001). Then there is the *parallel model* in which the receptor is in equilibrium with two possible activation states (Schoneberg et al., 2002). Thus, in one state PTHR1 without access to a NHERF2 "chip" would couple to and activate Gs and the Gs would activate adenylyl cyclase, while activated PTHR1 attached to a NHERF2 "chip" would activate GTP•Gi/o or GTP•Gq complexes which would prevent Gs from activating adenylyl cyclase and at the same time stimulate PLC-β1 on the NHERF's PDZ-I domain with the βγ subunits libertated from the GTP•Gi/o or GTP•Gq/11 complexes. Perhaps PTHR1 behaves like the $β_2$-adrenergic receptor which when activated unfurls a motif on its C-tail with which it binds to the appropriate PDZ domain on NHERF-2's tail and thus displaces and releases the Na^+/H^+exchanger from its inhibitory linkage to NHERF (Schoneberg et al., 2002; Voltz et al., 2001). The relative responses of adenylyl cyclase and PLC-β1 to a PTH would depend on how many receptors the cell has and how many of these can attach to NHERF-2.

    In other words, the adenylyl cyclase / PLC-β1 / ??? contents of the primary signal from PTHR1s in a cell and whether bone growth will be started by this signal will depend on the cell's PTHR1:NHERF-2 ratio.

i c. Different PTH Receptors

    Now we really should hurry to catch the bone-forming signal train before all of this complexity becomes mental mush. But I can't resist adding one last bit to PTH receptor complexity which I hope will notbe the proverbial straw that breaks the camel's back and cause the camel to kick the book away. In Section v, I will briefly outline the PTHs' surprising ability to eliminate psoriatic skin lesions that are caused by the dysregulated proliferation and differentiation of epidermal keratinocytes. Keratinocytes, like osteoblasts, can be induced by either hPTH-(1-34) or hPTH-(1-84) to increase their membrane-associated PKC activity and generate a Ca^{2+} transient, but unlike osteoblasts they *will not activate adenylyl cyclase* (Orloff et al., 1995; Whitfield et al., 1992, 1996a). The reason for this difference could be that that they do not have PTHR1 receptors as indicated by their failure to make the gene's 2.3 kb transcript (Hanafin et al., 1995; Orloff et al., 1995; Sharpe et al., 1998). The receptor they make might be smaller than PTHR1— its gene's mRNA transcript seems to be only 1kb instead of 2.3 kb (Orloff et al., 1995). The keratinocyte receptor, which could be called PTHRK, is partially homologous to PTHR1's 155-281 region which includes the immediately juxtamembrane part of the N-terminal extracellular region, the first 2 transmembrane α-helices and the first extracellular loop linking transmembrane α-helices 2 and 3 in the normal receptor (Figs. 18 and 19) (Orloff et al.,1995). Thus there may be another—*pthrk*—gene (or perhaps an alternative transcript of the PTHR1 gene) that progeny of basal keratinocyte stem cells turn on instead of the PTHR1 gene when they become rapidly proliferating transit amplifying cells (Errazahi et al., 1998; Whifield and Chakravarthy, 2001). But whatever this dwarf PTHR1 is, it can bind PTHs, stimulate PLC-β1 and generate inositol -1,4,5 trisphosphate that triggers the release of Ca^{2+} from internal stores.

    But wait! As usual with PTHR!s there is a problem. According to Errazahi et al (2003) keratinocytes from *newborn* rats seem to make functional PTHR1s. But they cannot be normal because they do not respond to bPTH-(1-34) as do other rat cells such as ROS 17/2 cells. Unfortunately these authors also don't say whether PTHrP also stimulated adenylyl cyclase in their keratinocytes. Thus, rat keratinocytes may have a modified PTHR1 which just cannot activate adenylyl cyclase or perhaps is prevented from doing so by associating with NHERF-2 (if keratinocytes do indeed have NHERF-2). So stay tuned to the far too slowly unfolding saga of keratinocytes and their adenylyl cyclase-shunnung PTH/ PTHrP receptor.

Of course, we must not ignore other members of the PTH receptor clan—yes dear reader indeed there are others!There is PTHR2, which, unlike PTHRK, is a potent adenylyl cyclase stimulator designed to be activated by the PTH-related, 39-residue tuberoinfundibular peptide, TIP39, and is expressed most abundantly in the nervous system and to a lesser extent in lung, the pancreatic islets' somatostatin-producing D-cells and placenta (Usdin et al., 2002). However, the human PTHR2, but not the rat PTHR2, can also be strongly activated by PTH but not by PTHrP while TIP39 binds to, but doesn't activate, PTHR1 (Usdin et al., 2002). Then there is the zebrafish's zPTHR3, which shares a 67% of its amino acid sequence with hPTHR1 but has a 50-fold greater affinity for hPTHrP than hPTH (Rubin and Jüppner, 1999). Like zPTHR1, zPTHR3 can activate adenylyl cyclase, but unlike zPTHR1 it cannot stimulate PLC-β1 (i.e., it cannot stimulate IP_3 accumulation) (Rubin and Jüppner, 1999). Then there is a mysterious receptor in the brain (specifically the supraoptic nucleus) that is activated by PTHrP-(1-34) but *not* by hPTH-(1-34) or hPTHrP-(7-34) (Yamamoto et al., 1998). The signals from this receptor stimulate the expression and secretion of arginine vasopressin by the supraoptic nucleus cells (Yamamoto et al., 1998).

Now let's see how activated PTHR1s might turn on the bone-growing machinery.

ii. What Signal Bullets Switch on the Bone-Making Machine?

ii a. What PTHR1 Signal Bullets Target Bone Cells?

If the adenylyl cyclase response is not muffled by a high NHERF-2:PTHR1 ratio, the burst of cyclic AMP synthesis directly activates certain GEFs (guanosine nucleotide exchange factors) to be discussed further on and whatever cyclic AMP-dependent protein kinase isoforms (PKAs) are around as well as the B-Raf/ERK protein kinase system. The diacylglycerols from PLC-β1-induced PIP_2 and PLD-induced PtdCh breakdown activate the available and responsive protein kinase-C isoforms (PKCs), and the IP_3 from PIP_2 opens IP_3 receptors and releases Ca^{2+} from endoplasmic reticulum stores (Fig. 18). These PTH-induced signalers stir up a veritable cyclone of gene expression changes and metabolic events. For example, waiting at the the mouths of some of the endoplasmic reticulum's IP_3-activated Ca^{2+} channels are mitochondria, the Ca^{2+} uniporters or transporters of which carry the Ca^{2+} released from these channels into the mitochondria to fill the cell's ATP fuel tanks by stimulating key components of the ATP-making machinery such as pyruvate dehydrogenase, NAD^+-isocitrate dehydrogenase and 2-oxoglutarate dehydrogenase, and the ATP synthase ATP-maker (Ashby and Tepikin, 2001; Csordas and Hajnoczky, 2003; Dumollard et al., 2006; Hajnoczky, et al., 2000, 2002; Pozzan et al., 2002). Perhaps the most important of these PTH-triggered events is the expression and secretion of FGF-2 and IGF-I which will also send Ca^{2+}, IP_3, PKCs and PLD signals from their receptors attached to their specifc scaffolded superclusters in the membranes of both the PTHR1-bearing producer cells and in the "juvenile" osteoprogenitor cells which are not yet mature enough to have started expressing PTHR1 receptors (Figs. 9 and 23). At this point I must point out that cells with the appropriate receptors can stimulate themselves (cellular masturbation) with growth factors they produce—these are called "autocrine" factors. They can also stimulate appropriately receptored neighboring cells with factors they produce—when they do this they are called "paracrine" factors. So the FGF-2 and IGF-I made by PTH-stimulated osteoblasts are both autocrine and paracrine factors in bone.

If it is cyclic AMP, bursts of cyclic AMP-stimulated PKA activity and the direct targets of cyclic AMP-activated GEFs (guanine nucleotide exchange factors) such as Epacs 1 and 2 (deRooij et al., 2000) that trigger bone growth, and if a PTH fragment could be made that needs only to activate adenylyl cyclase to stimulate bone growth, maybe it could have potentially fewer side effects resulting from PKCs' activity in the many different cells expressing PTHR1 receptors.

Such a fragment could be made if the multi-signaling hPTH-(1-34) could be cut into smaller, still partly functional, pieces each of which stimulates either one or the other signal mechanism. And this we seemed to have been done in the early '90s when PTHR1 signaling seemed to be so direct and simple without the complications of NHERFs and homo- and heterodimerizations.

The potently osteogenic (in OVXed rats and cynomolgous monkeys) Ostabolin™ (hPTH-(1-31)NH$_2$) and Ostabolin C™ ([Leu^{27}]cyclo(Glu^{22}-Lys^{26})hPTH-(1-31)NH$_2$) are as effective adenylyl cyclase stimulators as hPTH-(1-34)NH$_2$ and they still have the N-terminal PLC-β1/PKC-activating region, but unlike hPTH-(1-34)NH$_2$, they do not have the C-terminal PKC-activating region (i.e., amino acids 29-32) that causes a PLC-independent stimulation of PKC (probably PKC-δ) nuclear translocation and activity in cultured non-transformed murine MC3T3-E1 calvarial preosteoblasts, rat ROS 17/2 osteoblast-like osteosarcoma cells, UMR 106-01 rat osteoblast-like osteosarcoma cell, wt9 mouse osteoblasts and primary rat osteoblasts (Barbier et al., 1997; Erclik and Mitchell, 2001; Jouishomme et al., 1992, 1994; Ryder and Duncan, 2001; Singh et al., 1999; Swarthout et al., 2000; Whitfield et al., 1996b, 1999a; Yang et al., 2006) (Fig. 20). Cultured human fetal osteoblasts (hFOB cells) appear to respond to hPTH-(1-31)NH$_2$ in the same way. In these human osteoblasts hPTH-(1-31)NH$_2$ does not stimulate the PKC-dependent expression of TGF-β1 (and therefore presumably not PKC-δ activity) in these human fetal osteoblasts although it stimulates the cyclic AMP-dependent expression of TGF-β2 as strongly as hPTH-(1-34) (Wu and Kumar, 2000).

As might be expected from what the PTHR1 receptor has been trying to tell us, whether such C-terminally truncated fragments can also stimulate PKC activity depends on which cells' membranes the PTHR1 receptors find themselves and are enabled to use their N-terminal PLC-β1/PKC-stimulating region (Fig. 20). For example, as expected hPTH-(1-31)NH$_2$ with a normal N-nose can increase membrane-associated PKC activity in pig kidney cells engineered to put a huge number (i.e., about 950,000) of human PTHR1 receptors on their surfaces and in normal human newborn foreskin fibroblasts which have only a very few (in fact not detectable by competitive binding assay) human PTHR1 receptors (Whitfield et al., 2001). The failure of hPTH-(1-31)NH$_2$ (Ostabolin™) and [Leu^{27}]cyclo(Glu^{22}-Lys^{26})hPTH-(1-31)NH$_2$ (Ostabolin-C™) to increase membrane-associated PKC activity in ROS 17/2 osteoblast-like cells and fetal human osteoblasts means that while PTHR1 receptors on cultured osteoblasts can't activate PLC or PLD, they can activate one or both of these enzymes in densely human-receptored pig kidney cells and very sparsely receptored cultured human foreskin fibroblasts that are differently equipped with the necessary messengers, kinase isoforms and NHERF scaffolds.

At this point the reader may be muttering enough's enough with this horrendous signaling complexity. But there's even more! Even the once seemingly simple and all-important adenylyl cyclase-cyclic AMP mechanism has become formidably complicated (Chin et al., 2002). There is a very large number of possible cell-type-specific combinations different types of cyclic AMP-activated PKA R (regulatory) and C (catalytic subunits in or on the cell membrane or in or around the nucleus. There are cyclic AMP-binding G-protein-activating guanine nucleotide exchange-stimulating factors (GEFs) factors. There are ion channels and transcription factors such as the very important NF-κB that directly bind and respond to cyclic AMP independently from PKA activity. And R sub units can function independently from the catalytic subunits. For example, the cyclic AMP•RIIβ complex can bind to the CREs (cyclic AMP-responsive elements) in the promoters (the transcription switch boxes) of some genes. Therefore the cell's response to the cyclic AMP pulses triggered by activated PTHR1 receptors depends on the whether the cells have NHERF scaffolds, what kinds of cyclic AMP-binding targets are available, and what PKA R (regulatory) and catalytic subunit isoforms the cell has and where they are located in the cell.

GPCRs/7TM Rs such as PTHR1 do not directly activate receptor tyrosine kinase receptors (RTKs). But surprisingly they can indirectly transactivate RTK receptors such as the EGF receptor (Cole, 1999; Daub et al., 1997; Iwamoto and Mekada, 2000; Prenzel et al., 1999)! At first, it seemed that in these cases the EGF receptor was not activated by its EGF ligand, but it now appears that the activated GPCR stimulates a transmembrane metalloproteinase that cuts soluble sHB-EGF (a heparin-binding protein with EGF-like motifs) from membrane-anchored pro-HB-EGF (precursor of heparin-binding EGF) (Iwamoto and Mekada, 2000; Prenzel et al., 1999). The liberated sHB-EGF then activates EGF receptors on its own or adjacent cells. When the PTH/PTHR1-"transactivated" EGF receptors phosphorylate themselves, they become sticky flypaper-like or Velcro-like scaffolds for collecting and clustering the components of the Ras/Raf-1-mechanism that triggers the multipurpose MAP kinase cascade (Cole, 1999; Daub et al., 1997; Pearson et al., 2001). However, we don't know what the Ras/Raf/MAP kinase cascades triggered by PTHR1-transactivated RTKs or whatever other receptor with which PTHR1 might have paired might contribute to osteogenesis. *Nevertheless the important message is that the signals triggered by a PTH could come from both PTHR1 receptors and any transactivatable RTKs the cell might be expressing.*

PTHR1 signaling is far more complicated, busier and far-reaching than originally believed in the simple schemes of the 1980s and early 1990s. After having sent its first signals, the PTH•PTHR1 complex is rapidly silenced and pulled into the cell (endocytosed) with a half-time of 3-5 minutes but then PTHR1 slowly makes its way back to the surface, ready to work again, with a half-time of about 2-4 hours (Chauvin et al., 2001) (Fig. 21). But it appears that if the endocytosis of the PTH•PTHR1 complex is prevented in human embryonic HEK 293 kidney cells by having them express β-arrestin-2's C-terminus along with the normal β-arrestin, the receptor is still switched off when the PTH•PTHR1 complex binds to normal β-arrestin, but the lack of endocytosis does not stop the receptor from being put back to work (Bisello et al., 2002). However, if cells such as osteoblasts have their arrestin gene knocked out, PTH(PTHR1) signaling is sustained (Pierroz, et al., 2003). But under normal conditions the PTH•PTHR1 complex's endocytosis starts when the receptor's C-tail binds to the β-arrestin-2 scaffolding protein (Chauvin et al., 2001; Pouysségur, 2000) (Fig. 21). The PTH•PTHR1•β-arrestin-2 complex is wrapped in a clathrin-coated pit and c-Src is brought into the complex to phosphorylate and switch on the tiny dynamin motors that pull the packaged receptor complex down into the cytoplasm (Ahn et al., 1999; Chauvin et al., 2001; Pouysségur, 2000). By analogy with the type IA angiotensin II receptor, while a combination of RGS-2 and the endocytosis of the PTHR1 receptor stops adenylyl cyclase and PLC-β1 signaling, it might at the same time trigger a second tier of signaling because the β-arrestin-2 scaffold has an RRSLHL (Arg-Arg-Ser-Leu-His-Leu) motif which brings Raf-1, ASK-1, MAP Kinase Kinase-4 and a JNK (c-Jun kinase) together with the PTH•PTHR1 complex (Keenan and Baldassare, 2001; Luttrell et al., 2001; Miller et al., 2001; Pouysségur, 2000). When β-arrestin-2 attaches to the PTHR1 of the PTH•PTHR1 complex, ASK1 could activate MAP Kinase Kinase-4 which in turn activates JNK which is released from the scaffold, surges into the nucleus and phosphorylates and activates other gene transactivators such as c-Jun which forms part of the c-Fos•c-Jun AP-1 gene transcription complex (Keenan and Baldassare, 2001) (Fig. 21). However, the c-Src, which turns on the endocytosis motor may also activate the MAPK mechanism (Mahon and Segre, 2002).

The already challenging tangle of PTHR1 signals (Fig. 21) is increased still further by PTHR1's nuclear localization signal! This means that either newly made PTHR1 might be transported into the nucleus without reaching the cell membrane or after being released from the β-arrestin-2 complex, the endocytosed receptor might go into the nucleus along with JNK to stimulate some as yet unknown nuclear functions (Fig. 21). The surprising possibility of a receptor directly stimulating genes after being transported from the cell surface into the nucleus after binding and activation by its ligand has recently been demonstrated with EGF and the

ErbB-1 and ErbB-4 receptors (Lin et al., 2001; Ni et al., 2001). The EGF receptor, with a strong gene transactivation domain in its C-tail, gets into the nuclei of a variety of cells where it can bind to AT-rich sequences such as the consensus site in the cyclin D1 gene's promoter-switch box (Lin et al., 2001). When the ErbB-4 receptor on a mammary carcinoma cell surface binds its ligand, heregulin, and starts signaling, its cytoplasmic tail is snipped off by γ-secretase (which we will meet further on when talking about statins and Alzheimer's disease) and it travels to the nucleus where it activates a gene(s), the product(s) of which inhibits the growth and stimulates the differentiation of the cell (Ni et al., 2001). However, at second glance this should really not be too surprising because estrogen receptors can be activated on the cell surface and then be carried into the nucleus to stimulate suitably marked genes.

ii b. Do We Know Which PTHR1 Signal Bullets Stimulate Bone Growth?

For our retrospectively very naive effort in the late 1980s and early 1990s to find out whether the osteogenic signal from PTHR1 receptors to start making bone required the stimulation of adenylyl cyclase, membrane-associated PKCs, or both, we needed fragments to stimulate membrane-associated PKC but not adenylyl cyclase. Unlike the C-terminally cropped fragments, no N-terminally truncated PTH fragment can stimulate adenylyl cyclase and that is still true. However, as we learned above the ability of a N-terminally truncated PTH to stimulate PKC activity also depends on the kind of cell expressing the PTHR1 receptors. For example, b(bovine)PTH-(3-34)NH_2, hPTH-(3-34)NH_2, hPTH-(13-34)OH, hPTH-(28-48)OH, hPTH-(8-84)OH and even the tiny hPTH-(28-34)OH, hPTH-(29-32)OH can increase membrane-asociated PKC activity in cultured ROS17/2 osteoblastic cells, UMR-106 rat osteoblastic cells, human fetal osteoblasts, human dermal fibroblasts and primary rat kidney cells, and hPTH-(8-84) can strongly stimulate membrane-associated PKC activity in spleen cells in rats, but they are at best only very weak stimulators of membrane-associated PKC activity in the pig cells with nearly a million human PTHR1 receptors (Fujimori et al., 1992; Jouishomme et al., 1992, 1994; Li Chong et al., 1998; Neugebauer et al., 1995; Swarthout et al., 2001; Whitfield et al., 2001; Wu and Kumar, 2000). This of course could be due to the artificially engineered pig cells having only very little NHERF-2 (M. Mahon, personal communication to F.R. Bringhurst) which is required for PTHR1 to activate PLC and thus stimulate PKC activity. But of course this might only apply to the PLC-β1-activating domain in the peptide's N-nose.

hPTH-(1-31)NH_2(Ostabolin™), its lactam derivative, [Leu^{27}]cyclo[Glu^{22}-Lys^{26}]hPTH-(1-31)NH_2 (Ostabolin-C™), as well as [Leu^{27}]cyclo[Glu^{22}-Lys^{26}]hPTH-(1-28)NH_2, which stimulate AC but not membrane-associated PKC activity in cultured rat and human fetal osteoblasts, strongly stimulate bone growth in OVXed rats (Rixon et al., 1994; Whitfield et al., 1996b, 1997b,1997c,1997d, 1998b,1999c, 2000a,2000b, 2000c). On the other hand, the N-terminally disabled for adenylyl cyclase stimulation (but I don't know about PLC-β1/ PKC activation) 1-desamino-hPTH-(1-34) and the N-terminally truncated hPTH-(28-48) and hPTH-(8-84) that strongly stimulate membrane-associated PKC activity, but not adenylyl cyclase, in cultured osteoblasts and in rats do not stimulate bone growth in the OVXed rats (Rixon et al., 1994; Strein, 1994). Therefore, activating adenylyl cyclase and the cascades it triggers seems to be all a PTH needs to start, but maybe not subsequently drive, bone formation (at least in rats). This is supported by the ability of selective cyclic AMP-elevating, cyclic nucleotide phosphodiesterase inhibitors (Pentoxyfylline and Rolipram) to stimulate bone growth in normal male mice (Kinoshita et al., 2000). But (this nasty, monkey wrench-throwing word again!) Locklin et al. (2003) have found that hPTH-(3-34), which stimulates PKC activity but not adenylyl cyclase or bone formation, nevertheless can stimulate the expression of that master driver of osteoblastic genes—Cbfa1/Runx-2—in cultured C57BL/6 mouse marrow cells as effectively as the cyclase-stimulating, osteogenic hPTH-(1-34) and hPTH-(1-31).

However, adenylyl cyclase stimulation might not be enough to start osteogenesis in ovariectomized rats and mice. MPTH-(1-21) (i.e., [Ala^{3,12},Nl e^8,Gln^{10}, Har^{11},Trp^{14},Arg^{19},Tyr^{21}]rPTH-(1-21)) can bind to PTHR1 and strongly stimulate adenylyl cyclase, but it cannot stimulate bone formation in OVXed mice and rats and was unexpectedly much less potent in affecting alkaline phosphatase activity in human HOB-03-C5 and Saos bone cells (Murrills et al., 2002). Thus, for some reason the cyclic AMP signals from PTHR1 activated by this short PTH can't start osteogenesis. But wait! I have two questionns. Did the fragment still have the N-terminal PKC-activating activity? And did the fragment reach the bones in an active state? Rixon et al. (1994) had to answer the second question when they found that hPTH-(3-34) and hPTH-(13-34) did not stimulate bone formation in OVXed rats. Was this because they couldn't stimulate bone formation or they couldn't reach the bone in an active state? In this case both turned out to be the case. The inability of Rixon et al.'s PTHs to reach their bone targets in the rat was shown by their abilities to strongly stimulate membrane-associated PKC activity as did hPTH-(1-34) when added to spleen lymphocytes suspensions just a few minutes after they were taken from the rat, but they did not increase membrane-associated PKC activity in spleen lymphocytes still in the rat as did hPTH-(1-34). So one could ask Murrills and colleagues whether their little peptide actually reached bone cells?

The failure of small (less than 34 amino acids) N-terminally truncated PTH fragments to stimulate PKC activity in spleen cells in the living breathing rat while strongly stimulating PKC activity in freshly isolated spleen cells suggests the existence of something that sees and trashes such small fragments but ignores a larger N-terminally truncated fragment such as rhPTH-(8-84) or a 34-aminoacid fragment such as 1-desamino-hPTH-(1-34) with its quasi-normal N-terminus. This nemesis of small "nose-less" fragments just might be the 230-kDa protein in rat and human serum that Kukreja et al (1994) found to inactivate bPTH-(7-34)NH$_2$ and hPTHrP-(7-34)NH$_2$, but not hPTHrP-(1-34). As we learned above, the unselected rhPTH-(8-84) would be able to stimulate both osteocytes having the CPTHR (Bringhurst, 2002; Divieti et al., 2001, 2002) and in my experience (Rixon et al., 1994) the PLC-β1/PKC arm of the PTHR1 response. However, one must be careful of this because Bringhurst (2002) has said that hPTH-(7-84) does not bind to PTHR1 receptors. If this were true, it would mean that spleen lymphocytes responded with either the CPTHR or another receptor as suggested by Whitfield et al. (1999a). Remember (if by now you need reminding) that nothing is simple with PTHs and their receptors.

There are reasons to believe that hPTH-(1-31)NH$_2$ (Ostabolin™) may also only stimulate adenylyl cyclase in adult human osteoblasts as it seems to do in cultured fetal human osteoblasts (Wu and Kumar, 2000). If hPTH-(1-31)NH$_2$ were a dual-signaler like hPTH-(1-34) in adult human bones, it should affect adult human bone metabolism exactly like hPTH-(1-34). But it doesn't! While the two fragments equally stimulate adenylyl cyclase in human volunteers, hPTH-(1-31)NH$_2$ is a much weaker stimulator of bone resorption than hPTH-(1-34) as indicated by a failure to cause hypercalcemia and increase urinary collagen breakdown products (Fraher et al., 1999). hPTH-(1-31)NH$_2$ is also as strong a stimulator of adenylyl cyclase and bone growth as hPTH-(1-34) in OVXed mice, but it is only half as strong a stimulator of osteoclast activity as hPTH-(1-34) (Mohan et al., 2000). There seems to be some support for this in the observations of Whitfield et al. (1998b). They found that daily injections of a low dose (0.6 nmole/100 g of body weight) of hPTH-(1-34) into 3-months-old rats, starting 2 weeks after OVX, did *not* reduce the loss of femoral trabeculae (56% fewer bone hits/mm of metaphyseal scan in the treated OVXed rats compared to 52% fewer in the untreated OVXed rats), particularly the central metaphyseal trabeculae, although they raised the mean thickness of the surviving lateral trabeculae by about 1.7 times above the normal level in the sham-operated control rats. On the other hand, while injections of the same dose of Ostabolin C™ ([Leu^{27}]*cyclo*(Glu^{22}-Lys^{26})hPTH-(1-31)NH$_2$) raised the trabecular thickness 1.5 times above the sham-operated value they also significantly reduced the OVX-induced loss of central trabeculae from 56% to 29%. And Jolette et al. (2003, 2005) have found that Ostabolin-C™ can stimulate bone growth while reducing osteoclast activity in cynomologous monkeys. However, when the trabecular number in Whitfield et al.'s rats was allowed to drop for 9 weeks before starting the 6 weeks of daily injections, neither fragment could appreciably stop any further drop although both still strongly increased the mean thickness of the remaining lateral trabeculae to 1.6-1.7 times the mean thickness of the trabeculae in the stimulated femurs of the sham-operated, vehicle-injected control animals. The responses to a higher dose of Ostabolin C™ or hPTH-(1-34) (2 nmoles/100 g of body weight) were identical.

As we shall see further on, this difference between the effects of hPTH-(1-34) and the two C-terminally truncated fragments in humans and rodents just might be due to lower doses of hPTH-(1-31)NH$_2$ (Ostabolin™) and Ostabolin-C™ not having hPTH-(1-34)'s ability to stimulate, low PTHR1-expressing osteoblast progenitors to make the osteoclast differentiation stimulator RANKL and reduce the expression of the osteoclast-inhibiting OPG (Hofbauer et al., 2000; Suda et al., 1999) (see also Fig. 35). Indeed maybe Ostabolin™ and Ostabolin C™ increase the OPG/RANKL expression ratio to a level that actually inhibits OVX-induced osteoclast activation as suggested by the findings of Jolette et al. (2003, 2005) with monkeys.

Before going any further, I must repeat that we do not have the full measure of the signals that the PTHs send into their target bone-making cells in the various provinces of the Real Bone World. For example, what NHERFs do the cells in the various stages of osteoblastic development in a PTH-treated bone have, and what things are plugged into their NHERFs that can be tweaked by activated PTHR1s?? In other words what do the PTHR1-bearing cells have that can be targeted by the primary, osteogenesis-kick-starting cyclic AMP signal?

iii. PTHs Speak to Bones with Two-Pronged Forked Tongues— One Prong Says Make It!

iii a. Generalities

From all that has been said so far, adenylyl cyclase activation is the *sine qua non* for bone building by PTH. Obviously we must find out how the adenylyl cyclase-stimulating (i.e., the N-terminally intact) PTHs and the signals from their PTHR1 receptor stimulate bone growth in order to at least understand how the current PTHs stimulate bone growth and then go back to the drawing board and make even better anabolic drugs for the future. But this is not easy. The first problem is trying to identify the key players in the horde of genes and other things that are hit by PTHR1's cyclic AMP buck-shot. Indeed, the PTHR1 signals with their many target genes, transcription factors, enzymes, enzyme regulators, cytoskeletal components and motors set off a barrage of bone-relevant as well as bone-irrelevant reactions. Within this tangle of events are the few that actually lead to bone formation. The challenge is to find the bone-building needles in this tangled web of responses (Fig. 22).

As a tiny tantalizing taste of things to come, Qiu et al. (2003), starting by characterizing the promoters of 8 "training" genes known to be regulated by PTHs, have identified 31 transcription factors and 80 genes that could be affected *directly* by PTHs. And Onyia et al. (2001a) have found that 158 genes associated with bone growth are stimulated in rat metaphyseal bone by hPTH-(1-34) (Forteo™). Added to these are those in the differently equipped and scaffolded members of the osteoblastic lineage in different parts of the bones that belong to the vast signaling "syncytium" stretching from the cilium-waving strain-sensing osteocytes locked in their lacunae to their dendrites snaking through canaliculi to the retired osteoblasts lining the bone surfaces and from there to the osteoblastic and osteoclastic progenitors and precursors in the bone marrow (Marotti, 1996). We see osteoblastic cells signaling and, amazingly, remembering strain frequencies using exactly the same machinery as neurons in the central nervous system (Mason, 2004; Skerry and Taylor, 2001; Spencer et al., 2004; Turner et al., 2002).

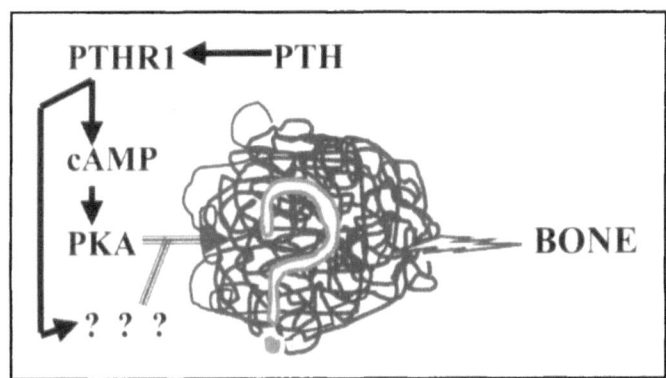

Figure 22. The dense tangle of transmitters, transcription factors, transcripts and translations of the information encoded by the literally hundreds of genes activated by signals from PTH-activated PTHR1 receptors. The stream of signals from the receptors includes a burst of cyclic AMP-dependent protein kinase (PKA) which is needed to start the bone-building machinery. But the signal mix and the cascades it triggers depend on to what other receptors the PTHR1s are linked or hetrodimerized and to what signalers-loaded scaffolds, such as NHERFs -1 and -2, the PTHR1s are attached. The upshot of this is a seemingly ever pullulating mass of candidate drivers hidden among whom are the real ones. The challenge for the future is to separate the driving "wheat" from irrelevant "chaff".

But we are being swamped by an avalanche of reports of relevant but undoubtedly many more irrelevant gene transactivators, hundreds of gene expressions, dozens of enzymes and mechanisms in osteoblastic cells not working in their native bones but that have been ripped from their native signaling "syncytium" and planted on plastic or in cells of established or semi-established lines in which cell cycle controls, PTHrP and PTHR1 expressions and the maturation program have been more or less disconnected. Indeed such dysregulated cells are like crazy computers simultaneously running many normally mutually exclusive programs such as cell cycling and maturation. Therefore, what follows can only be glimpses of the mechanism(s) by which the PTHs cause bones to grow as seen while struggling in the swirling stream of events.

When an adenylyl cyclase-stimulating PTH is injected subcutaneously, or, as may soon be happening, swallowed in a pill or inhaled (Fig. 23), it first meets and is grabbed by trabecular bone cells in the bone marrow as it passes out of the marrow blood vessels and only later and probably more or less depleted does it reach cells in the Haversian canals and the lacunocanalicular network (see Chapter 2 i b; Knothe Tate, 2001). The target cells in the various regions will have their PTHR1s in different superclusters and some cells will have unconventional PTH/PTHrP receptors. But how many target cells with PTHR1 receptors are available to kindle an osteogenic response? This depends on the age of the rat or human. In a growing rat or human there are packs of PTHR1-bearing osteoblasts and osteoclasts working independently to build and shape the growing bones. But when growth stops in humans (and other big-boned animals) there are few or no independently working osteoblasts. The only maximally receptored osteoblasts will be in the crack-repairing BMUs. So there shouldn't be many PTHR1-bearing first responders in middle-aged humans but the increased microcracking of the increasingly fragile bones in aging skeletons should cause the production of more BMUs with working osteoblasts. However, one must not forget the extensive population of bone-lining cells each with a reduced number of PTHR1 receptors, but still responsive to PTH (Dobnig and Turner, 1995). Also if a long bone be broken and an implant screwed into it, BMUs will be activated and fan out from the screwhole in all directions along the diaphysis on their several-millimeter journeys

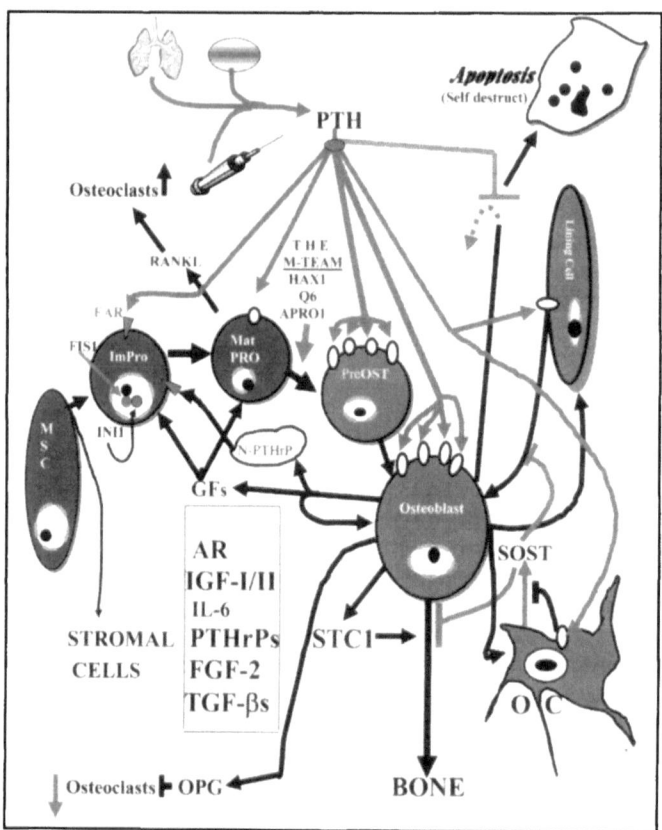

Figure 23. A working model of the osteoblastic maturation and how a subcutaneously injected, swallowd, or inhaled PTH such as hPTH-(1-34) (Forteo™) or one of the new Ostabolin™ family peptides such as hPTH-(1-31)NH$_2$ (Ostabolin™) and [Leu^{27}]cyclo(Glu^{22}-Lys^{26})hPTH-(1-31)NH$_2$ (Ostabolin-C™) might affect it. The blue cells are the proliferatively competent mesenchymal stem cell and transit-amplifying osteoprogenitors and the red cells are the terminally (i.e., permanently) proliferatively shutdown cells in the more advanced stages of osteoblastic differentiation. The white buttons on the cells are PTHR1 receptors and their number indicates the relative densities which peak as the cells enter the preosteoblast and osteoblast stages and then fall if the cells are lucky enough to be chosen to be strain-sensing osteocytes or lining cells after mineralization is finished. Abbreviations: AR, amphiregulin; EAR, endothelin-A receptor; FGF-2, fibroblast growth factor-2, one of the missionary growth factors; FIS1, the mitochondrial fission driver which appears briefly in the nucleus of early osteoprogenitors to bind INI1 to promote the proliferation and progression to maturity of these transit-amplfying cells; IGF-I, insulin-like growth factor-I another missionary growth factor; ImPro, immature osteoprogenitor; INI1, the cell cycle inhibitor in the SWI/SNF complex that is inactivated by FIS1 to drive the proliferation of the transit-amplifying osteoprogenitor cells; MatPro, mature osteoprogenitor; PreOst, preosteoblast; PTHrPs, the cytokines cut out of the PTHrP chain such as PTHrP-(107-111), PTHrP-(107-139), and PTHrP-(8-11) that is virtually identical to endothelin 1's residues 6-9 and thus would trigger osteogenic signals from endothelin A receptors; MSC, mesenchymal stem cell; SOST, sclerostin which is produced by mature osteocytes and reduces osteogenesis by inhibiting osteoblasts and preventing the reversion of lining cells to functioning osteoblasts; STC1, stanniocalcin which is produced by osteoblasts and promotes osteogenesis; TGF-βs, tumor growth factor-βs still more missionary growth factors. The green lines mark the PTH targets. But we must not forget that the exogenous PTH is an intruder. The physiological pusher of the PTHR1 buttons is PTHR1 which is produced by the differentiating preosteoblasts and the mature osteoblasts and stimulates their and their neighbors' PTHR1 receptors and is thus one of the programmed drivers of osteoblastic maturation.

and affect the PTH-responsiveness of whole length of the bone for months afterwards (R.B. Martin et al., 1998).

The potently osteogenic hPTH-(1-34) (Forteo™), but it seems not other equally potent bone-building PTHs such as hPTH-(1-31)NH$_2$ (Ostabolin™) and [Leu^{27}]cyclo[Glu^{22}-Lys^{26}]hPTH-(1-31)NH$_2$ (Ostabolin-C™) (Jolette et al., 2003, 2005), increases osteoclast generation like an estrogen crash and hence the number of holes being dug in the bone at any given moment (i.e., the remodeling space). It stimulates (probably via cyclic AMP) the proliferation of the pluripotent, spleen colony-forming, CFU-S, stem cells and other members of the rat hematopoietic cell populations, which need the microenvironment provided by docking on immature osteoblastic stromal cells (Byron, 1977; Gallien-Lartigue and Carrez, 1974; Perris et al., 1971). And as we shall learn in far more detail further on Calvi et al. (2003) and Weber et al. (2004) have much more recently confirmed that hPTH-(1-34), the potent adenylyl cyclase-stimulating forskolin, or a permanently switched-on mutant PTHR1 receptor, which does not need PTH to activate it, can enlarge a mouse's hemopoietic stem pool. It seems that the PTHR1 receptor's cyclic AMP signals cause the trabecular bone-lining cells to make large amounts of surface Jagged 1 which in turn binds to Notch receptors on adjacent hemopoietic stem cells attached by N-cadherin to the osteoblastic lining cells (Calvi et al., 2003; Ohishi et al., 2003; Schweisguth, 2004; Weber et al., 2004; Zhu and Emerson, 2004). The resulting Notch signals in the stem cells increase the pool of hemopoietic stem cells. When cells from the enlarged stem cell pool are activated and leave the osteoblastic niche, their macrophage progeny they can generate increased numbers of "rookie"osteoclast recruits, which carry on the stem cell tradition of making PTHR1 mRNA and receptors, but hold these in their cytoplasm and around their nuclei and maybe even in their nuclei without putting them out onto their surfaces as they mature (Faucheux et al., 2002; Kanatami et al., 1998; Langub et al., 2001; Watson et al., 2002). The PTH-induced burst of cyclic AMP formation also stimulates osteoclast formation by causing some immature marrow osteoblastic stromal cells with PTHR1 receptors to express and deploy on their surfaces the osteoclastogenesis-stimulating RANKL (Atkins et al., 2003; Fu et al., 2001; Galvin et al., 2001; Kanazawa et al., 2000; K. Lee et al., 1994; Stilgren et al., 2001). The PTH also induces cultured osteoblasts and possibly some immature osteoblastic stromal cells to make autocrine (autostimulating)/paracrine (neighbor-stimulating) IL-6 which causes the immature osteoblastic stromal cells to make RANKL (Fig. 20; Greenfield et al., 1999; Tamura et al., 1993).

But human osteoclasts appear *so far* to be different from rat osteoclasts—they actually have functional PTHR1 receptors on their surfaces! Dempster et al (2005) isolated human peripheral bood monocytes cultured them on bone slices or plastic culture dishes with human recombinant RANKL and recombinant human macrophage colony-stimulating factor (M-CSF) for 16-21 days. Mono-and multi-nucleated cells appeared which, like all osteoclasts, had calcitonin receptors and resorbed bone. But by 21 days both the mono-nucleated and multi-nucleated osteoclasts also expressed their *pthr1* gene and put PTHR1 receptor products out on their surfaces. Moreover, exposing the osteoclasts to 50 or 100 ng of rat PTH-(1-34)/ml of culture medium increased their ability to resorb bone slices 2- or 3-fold. If confirmed by others, this means that we must radically change our osteoblast-centered concept of PTH actions on osteoclasts and bone resorption in humans (Fig. 23). However, we (Cynthia M. Allen, Sue MacLean and myself) have been unable to confirm this. We did not find 15- to 21-day mature, TRAP-positive human osteoclasts derived from circulating preosteoclasts expressing their *pthr1* gene even using three PCR primers (including the one used by Dempster's group) with which we could readily demonstrate the *pthr1* gene's expression in engineered HKRKB7 porcine kidney cells with an average of 9 X 10^5 human PTHR1 receptors/cell.

iii b. The Target Cells

The basic requirement for a PTH to stimulate bone growth is that there must be enough PTHR1-bearing cells with which to start building up a crew of active PTHR1 receptor-bearing, osteoblasts. The response to a PTH will initially depend on the size of this receptored team. As stated above, in the bones of human children or rat pups the mature osteoblast teams will be large and located in osteoclast-free, bone-modeling clusters, but in an adult human it should be much smaller because most of the cells will belong to microcrack-filling BMUs. But there is also an extensive mobilizable reserve of PTH-responsive, BMU-independent retired osteoblasts in mature bones which have fewer, but still functional, PTHR1 receptors (Fig. 23). These are the lining cells whom we will also meet in section iii f, covering the walls of the osteons' central canals as well as the endosteal and trabecular surfaces and providing the niches on trabecular bone for hematopoietic stem cells (look ahead to Fig. 33). The PTH will increase the number of mature bone-making osteoblasts by increasing their generation, by extending their their lifespan by making them resistant to apoptosis, or summoning the retired lining cells to come out of retirement (Fig. 23). A mature osteoblast is at the end of a long line of cells stretching back to an ALP (alkaline phosphatase)-expressing "grandmother" cell in the bone marrow stroma or the periosteum and ultimately a mesenchymally-derived "great-great-great-grandmother" stem cell which had unlimited self-renewing potential and the options to found lines of adipocytes, chondrocytes, fibroblasts, or myoblasts (Aubin, 2000, 2001; Aubin and Triffitt, 2002). As is the case for all stem cells (Whitfield and Chakravarthy, 2001), these mesenchymally-derived stem cells normally proliferate only when needed to refill a pool of extensively proliferating transit-amplifying osteoprogenitors (i.e, transit amplifying cells) for growing bone in children and removing microcracks in adults (Fig. 23).

But let's look more closely at the PTH targets in the cortical bone of a mature skeleton. As we learned in the first chapter, a pack of continuously proliferating spindle-shaped immature transit-amplifying osteoprogenitor cells (i.e., juvenile osteoblasts) capable of expressing RANKL chases the BMU osteoclasts as the osteoclasts are pushing the PTHR1-bearing lining cells aside and tunnelling through the exposed bone, and getting rid of, a patch of microcracked bone. The osteoprogenitor cells put a layer of mature osteoprogenitors around the tunnel wall. These mature osteoprogenitors switch off and dismantle their cell-cycle machinery and become PTHR1-bearing preosteoblasts that mature into "smart", maximally receptor-bearing osteoblasts with primary cilia to measure, and respond to fluctuations in, the components and movements of the extracellular fluid (Tonna and Lampen, 1972; Whitfield,2003a; Xiao et al 2007) and start laying down the first of 7 or 8 pairs of lamellae around the tunnel wall to fill it in but save space for the blood vessel's channel which makes a new osteonal central canal. These nearly mature and mature osteoblasts plus the surrounding lining cells are the prime targets of injected, pill-delivered or inhaled PTH diffusing from the blood vessel in the center of the tunnel (Fig. 23). Presumably the several growth factors from the PTH-stimulated osteoblasts will "contaminate or dope" the new bone, diffuse down through the stack of cells and stimulate the proliferation of the receptorless outer immature osteoprogenitors and get into the blood to attract more progenitors to the construction site (Fig. 23). PTH-induced factors diffusing out of the stimulated, functioning, reactivated lining cells into the more extensive bone marrow and marrow sinusoids on the endosteal and trabecular surfaces should have greater overall impact because the products of the PTH-stimulated PTHR1-expressing osteoblastic cells and bone lining cells on endosteal and trabecular surfaces will immediately reach and stimulate immature osteoprogenitor cells in the adjacent marrow which have not yet started deploying PTHR1 receptors. The fact that endosteal surfaces but not periosteal surfaces have lining cells with PTHR1 receptors (though fewer than when they were maximally receptored osteoblasts) explains why PTH stimulates endosteal but not periosteal bone growth at least in in rats as does growth hormone (Andreassen and Oxlund, 2000).

iii c. Here We Meet Autocrine PTHrP and Emerging PTHR1s—
Proliferation Suppressors and Drivers of Normal Osteoblast Differentiation

And now let's look at the drivers of osteoblast differentiation. A menagerie of factors including hedgehogs (Indian and/or Sonic), Bapx 1, Msx-2 and signals from FGFR1 receptors promote a BMP-2 ⇒ BMP1B•II heterodimeric(2-component)receptor ⇒ phospho-Smad-1/5 second messenger-mediated surge of the type II Cbfa1/Runx-2 expression to close off the adipocyte option and start a clone of extensively proliferating "transit amplifying cells" with limited self-renewing abilitiy—the immature osteoprogenitors (Aubin, 2000, 2001; Aubin and Triffitt, 2002; Banerjee et al., 2001; Bianco et al., 2001; Bidwell et al., 2001; Chen et al., 1998; Chung et al., 2001; Dennis and Charbord, 2002; Ducy, 2000; Ju et al., 2000; Karsenty, 2000; Lian and Stein, 2003; Spinella-Jaegle et al., 2001) (Figs. 9 and 23). It is essential to know that for the BMPR-triggered surge of phosphoSmads to switch target genes on or off they must interact with Cbfa1/Runx-2 on the genes' promoters (Canalis et al., 2003).

A premature stimulation of the normally cycle-stopping mature osteoblast genes by the type II Cbfa1/Runx-2 is prevented and the progenitors kept proliferating by the "homeobox" protein, Msx2, which is expressed during the skeletal growth phase and binds to, and shuts down, type II Cbfa1/Runx-2 (Dodig et al., 1999; Lian et al., 1998; Ryoo et al., 1997; Shirakabe et al., 2001). Indeed MC3T3-E1 murine preosteoblasts throw their Cbfa1/Runx-2 into the proteasome garburettor barrel before they start replicating DNA and forced hyperexpression of the Cbfa1/Runx-2 protein reduces proliferation (Galindo et al., 2003). The level of Msx-2 might be determined by the expression of antisense Msx1-AS which downregulates the constitutively expressed Msx1 and maybe also Msx2 (Blin-Wakkach et al., 2001). Overexpression of Msx-2 would keep the osteoprogenitors proliferating to maintain an optimally sized precursor pool and prevent type II Cbfa1/Runx-2 from driving their maturation into post-proliferative preosteobolasts and osteoblasts whereas disabling the gene would reduce osteoprogenitor proliferation and cause premature maturation (Dodig et al., 1999). But at a critical point the signals from collagen-stimulated integrin mooring lines and other factors cause the appearance of another "homeobox protein", Dlx5, which downregulates Msx1 and pulls Msx2 off type II Cbfa1/Runx-2 which can then stimulate several key genes with the type 2 Cbfa1/Runx-2-binding, osteoblast-specifc element 2 (OSE2; the AACCACA DNA nucleotide sequence) in their promoter-signal boxes that include the osteoblastic-specific zinc finger protein transcription factor Osterix without which there can be no osteoblast maturation or bone formation (Blin-Wakkach et al., 2001; Ducy and Karsenty, 1995; Katagiri and Takahashi, 2002; Marijanovic et al., 2002; Nakashima et al., 2002; Zhang et al., 2002a). But Dlx5 also inhibits the expression of the osteocalcin gene by binding to a single "homeobox"-binding site in the gene's promoter –switch box (Ryoo et al., 1997). Why this seeming anti-differentiation action? As in all differentiation programs *the timing of a gene product's appearance is all-important.* So this is one of two strategies (we will meet the other one soon) to prevent type II Cbfa1/Runx-2 from prematurely stimulating the appearance of osteocalcin, which is expressed later on in the ostoblast maturation process to switch off and prevent excessive mineralzation. (Ducy et al., 1996).

At this point we must remember the surprising likelihood of signals from the osteoblastic cells' neuron-like NMDA glutamate receptors being involved in driving the pivotal expression of Cbfa1/Runx-2 (Hinoi et al., 2003). This hugely and surprisingly important role of glutamate receptors has been masked by the large amounts of glutamate that saturate the glutamate receptors of osteoblastic cells in the standard cell culture media (Hinoi et al., 2003, 2004). If we can somehow peel away the other functions of this multifunctional amino acid we can expect to see another part of the machinery that starts and drives osteoblast maturation.

A totally unexpected driver of an early transient prematuration stage of osteoblast differentiation has only recently been discovered (Fig. 23). This player is the 11-kDa FIS-1 (CGI-135), which is already famous for driving mitochondrial and peroxisome fission. This protein is best known for plugging its C-tail into outer mitochondrial membranes and peroxisome single membranes where it starts the the constriction of the organelle and organizes the components of a dynamin-like GTPase "pinchase" motor (Drp/Dnm1) that fissions these organelles in yeast and mammalian cells (Chan, 2006; Koch et al., 2005; Jofuku et al., 2005; Yoon and McNiven, 2001) (Fig. 24). But FIS-1 does a lot more than "just" drive mitochondrial and peroxisomal fissioning in preosteoblasts (Candeliere et al., 1999, 2002). It also operates briefly in the nucleus by complexing with, and thus taking out of action, INI1 (Snf5), a tumor-suppressing, cell cycle inhibitor which is one of the 10 or so subunits of the mega-dalton, ATP-fueled, SWI/SNF nucleosome-remodeling machines that initiate bone formation (Ae et al, 2002; Doan et al., 2004; Fujisawa et al., 2004; Imbalzano and Jones, 2005; Kalpana et al., 1994; Medjkane et al., 2004; Miller and Bushman, 1995; Muchardt and Yanniv, 1999, 2001; Pan et al., 2005; Peterson, 2002; Versteege et al., 2002; Vries et al., 2005; Young et al., 2005; Zhang et al., 2002) (Fig. 24). The importance of this transient INI1-suppressing nuclear operation of FIS1 in osteoblast differentiation is indicated by the fact that preventing its production in primary rat pup calvaria by injecting an antisense *fis1* gene-carrying adenovirus over the little animals' skull bones stops bone formation while overexpressing it by injecting a an adenovirus carrying a normal *fis1* gene greatly stimulates calvarial bone formation (Candeliere et al., 2002). It appears that for Fis1 to do its job in osteoblast differentiation it must appear in the

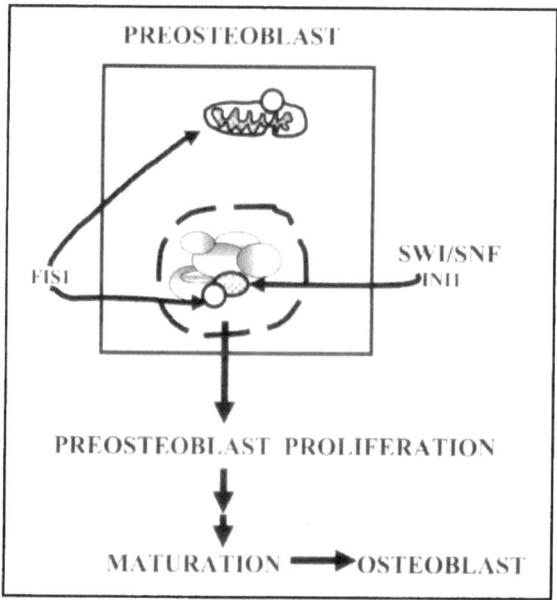

Figure 24. How the FIS1 protein might prevent nuclear INI1 from inhibiting immature osteoprogenitor proliferation when the cell is in a critical early stage of differentiation. In a non-cycling cell, FIS1 is plugged into the mitochondrial outer membrane as well as the peroxisomal single membrane where it promotes the organelles' fissioning. In the cell's nucleus INI1 is one of the several subunits of the mega-dalton, multi-component nucleosome remodeling SWI/SNF machine which somehow restrains the cell from starting the buildup to DNA replication and ultimately mitosis. But the appearance of a differentiation driver causes FIS1 to appear briefly in the preosteoblast's nucleus to derepresses the cell cycle machinery by binding to INI1 and putting the cell and its progeny along the transit-amplifying road to the expression of a panel of osteoblast-specific genes and terminal maturation.

preosteoblast nucleus when the cell is in an early, briefly responsive stage of preosteoblast development. Only when this time window is briefly open can Fis 1 release the proliferative block imposed by INI1and trigger what is likely a cascade of gene expressions that starts the transient amplifying proliferative progression toward final proliferatively shut down maturation (Fig. 23). When the window shuts, overexpressing Fis1can no longer stimulate bone formation nor can inhibiting Fis1 production by transfecting the cell with an antisense *fis1* gene stop bone formation (Candeliere et al., 2002). Obviously if we are ever to understand what pushes a preosteoblast along the road to PTH-responsiveness and functional maturity we must find out what happens during this brief FIS1-responsive/driven phase.

One of the type II Cbfa1/Runx-2's major targets is the PTHR1 receptor gene, specifically its P3 promoter (it has three such switch boxes) to which the protein binds as effectively as it does to the osteocalcin promoter-switch box (Karpieren et al., 2000). The now mature osteoprogenitors start expressing PTHrP and putting PTHR1 receptors out on their surfaces (Aubin, 2000,-2001; Aubin and Triffitt, 2002). Before reaching this advanced maturation stage primary human osteoblasts and murine MC3T3-E1 preosteoblasts do not respond to PTHs by making cyclic AMP (Arends et al., 2004; Cynthia M. Allen, Heather Douglas and I,unpublished observations; Schiller et al., 1999). The appearance of PTHrP and PTHR1 receptors coincides with the secretion of autocrine/paracrine PTHrP, the PTH-(1-34)-like N-terminus ("nose") of which both activates and stimulates the further expression of PTHR1 receptors. The signals from the PTHR1 receptors coincide with the shut down of the proliferative machinery and the start of the parade of mature osteoblast-specific genes leading to mineralization, matrix-embedded osteocytehood, post-osteoblast lining cell retiree or apoptotic death (Arends et al., 2004; Aubin, 1998, 2000, 2001; Aubin and Triffitt, 2002) (Figs. 9 and 23).

To become a mature osteoblast, a progenitor may use its PTHrP and PTHR1s to make the promoter-switch boxes of key osteoblast-specific genes accessible to transcription factors by bending the DNA to bring originally separated regions together to make functional switch boxes (Alvarez et al., 1998). This is done with "architectural" transcription factors, such as NmP4. They can bind to AT-rich regions of the minor groove of DNA and twist and/or bend the DNA to make functional gene promoters (switch boxes), for example by binding to an AT-rich region (between nucleotides 1971-1574 and -1555) which is upstream from the coding region of the collagen-I gene (Alvarez et al., 1998; Feister et al., 2000). Indeed PTH, and therefore autocrine PTHrP, causes NmP4 to get into the nucleus where it bends and twists the collagen-I gene's promoter-switch box into a functional switch for the transcribing machinery (Alvarez et al., 1998).

Since signals from activated PTHR1 receptors cause substantial cytoskeletal shifts as well as potentiate cells' responses to strain, they may use the same c-Src-triggered, Cas-Crk-mediated, nucleus-seeking mechanosomes that we talked about in Chapter 2 i that are assembled at the surface of strained osteocytes and then go to the nucleus to stimulate osteoblast-specific genes (Carvalho et al., 1994; Eagan et al., 1991; Goligorsky et al., 1986; Krempien et al., 1978; Lomri and Marie, 1990; Pavalko et al., 2003). In other words, PTHR1 signaling driven by autocrine PTHrP in later-stage preosteoblasts mimics strain deformation by sending architectural transcription factors such as NmP4 into the nucleus to bend and twist the promoters of osteoblast-specific genes into functional configurations that can receive and be activated by osteoblast-specific transcription factors.

A remarkable example of the ability of the PTHR1-fired cyclic AMP buck-shot to switch on some of the same genes they trigger in osteoblasts even in totally unrelated cells has been reported by Hefferan et al. (2002). They found that blood leukocytes from patients given 400 IU (activity units) of hPTH-(1-34)/day were expressing typical osteoblast genes such as those for alkaline phosphatase, BMPs-2,-4,-6, collagen I and IGF-I. This is a striking example of the leukocyte genome being turned on by PTH exactly like the peptide turns on the osteoblast's genome, but, of course,without being able to set up a functioning, osteoblast proteome in such an alien cell body. It also confirms my Group's now ancient, very pioneering, but equally very premature demonstrations of the ability of PTH to control and stimulate hemopoiesis and thymic lymphocyte generation (e.g., Perris et al., 1967, 1971; Whitfield, 2005b, 2006; Whitfield et al., 1973; 1998b).

Zuscik et al (2002) have shown that the expression of PTHrP's gene is negatively affected by the nuclear Ca^{2+} content. So the expression of the gene starting in preosteoblasts requires that the nuclear Ca^{2+} level be low enough. Unfortunately we don't know how the nuclear Ca^{2+} level changes during the the the progression from immature transit amplifying osteoprogenitor to proliferatively shut-down osteoblast. But maybe when something such as APRO (TIS21), which I will introduce further on, finally stops an osteoprogenitor proliferating, the Ca^{2+} transients which drive transit through the key stages of the cell cycle would stop, the nuclear Ca^{2+} level would drop and the expression of the PTHrP gene would be unleashed. But it also seems that the secretion of the emerging whole PTHrP molecule and its cleaved components would be driven by the very high, indeed dangerously high, Ca^{2+} concentration (e.g., 40 mM!) that hits the preosteoblasts and osteoblasts when they start working in the osteoclast tunnels and trenches. This Ca^{2+} would activate the osteoblastic cells' Ca^{2+}-sensing receptors which have their C-tails tied along with MAPK to filamin-A in signaling clusters in caveolar "message centers" (Awata et al., 2001; Hjälm et al., 2001). The signals from these clusters stimulate the PKCs/MEK1/2/ p38MAP kinases-mediated synthesis of PTHrP (MacLeod et al., 2003; Sanders et al., 2001; Tfelt-Hansen et al., 2002). Besides driving osteoblast differentiation, signals from PTHR1 receptors activated by the N-terminal part of secreted PTHrP would protect the cells from being killed by the large amount of phosphate that accompanies the Ca^{2+} (Adams et al., 2001; C.M.Allen et al., 2002; Meleti et al., 2000).

In Chapter 2 i we learned that strain turns on the PTHrP gene, possibly by opening TREK stretch-sensitice K^+ channels in osteoblastic cells (X. Chen et al., 2003). Therefore, the osteogenic PTHrP (Whitfield et al., 1997d) is very likely a principal mediator of loading-induced bone growth and maintenance. In other words straining osteoblastic cells causes them to make and secrete PTHrP much like pulling a skunk's tail triggers a spray of pungent perfume.

Indeed, the autocrine stimulation of preosteoblasts by signals from their own PTHrP-activated PTHR1s are major drivers of the cells toward mature osteoblasthood. Knocking out the PTHrP gene in mice impairs osteoblast development and bone formation (Miao et al., 2002). Schiller et al. (1999) have given an example of the coincidence of the emergence of PTHR1 receptors with proliferative shutdown, which was followed about 5 hours later by the initiation of bone nodule formation in cultures of MC3T3-E1 murine calvarial (skull bone) preosteoblasts. The cyclic AMP signals from the PTHR1 receptors activated by the emerging autocrine/paracrine PTHrP shut down the cell proliferation engine by turning off the Cyclin D1•CDK/4CDK6 kinases that start the chain of events leading up to DNA replication and along with this further increase the expression of PTHR1 (Datta et al., 2004; Lu et al., 2002). They push Cbfa1/Runx-2 into the cell nucleus (Fujita et al., 1999, 2001a) and into the arms of pRb, in this case a co-transactivator. After being pushed into the nucleus by the PTHR1 signals, Cbfa1/Runx 2 doesn't just wander aimlessly around looking for pRb and client genes. It has a nuclear matrix-targeting sequence (NMTS) in its C-tail (amino acids 397-434) that causes it to seek out specific nuclear matrix sites where it can form transcription complexes with suitably addressed osteoblast genes (Zaidi et al., 2001).The Cbfa1/Runx-2•pRb complex then promotes the effectiveness of the antiproliferative, cycle-blocking p21[CIP] and p27[KIP 1] proteins that have turned off the Cyclin D1•CDK4/CDK6 G1 build-up starter kinases and triggers the parade of gene expressions which ends with a mature fully mineralized matrix, loaded with various growth factors from the earlier stages (Datta et al., 2004; Lian et al., 1998; Stein et al., 1996; Thomas et al., 2001). The absence of the functional proliferation-restraining hypophosphorylated pRb that would otherwise work with Cbfa1/Runx-2 co-transactivator to open the mature post-proliferative osteoblast gene file explains the 500-times higher incidence of osteosarcoma associated with the disabled pRb in retinoblastoma patients (Gurney et al., 1995; Thomas et al., 2001).

This pivotal PTH/PTHrP- and Cbfa1/Runx-2-driven step in the progression of an osteoprogenitor cell such as MC3T3-E1 toward bone-making osteoblasthood includes just a brief but determining expression of the APRO1 (Antiproliferation1 or TIS21(mouse)/BTG2 (human) or PC3(rat)) gene (Matsuda et al., 2001; Raouf and Seth, 2002; Tirone, 2001) (Figs. 9, 23 and 25), the product of which blocks the G_1 build-up to chromosome replication (Matsuda et al., 2001; Guardavaccaro et al., 2000; Tirone, 2001). The highly labile APRO1 (TIS21) protein together with the Cbfa1/Runx-2•pRb complex's enhancement of the effectiveness of the p21^{CIP} and p27^{Kip} cell cycle blockers, prevents the transcription of the cyclin D1 gene and also directly interacts with, and inhibits, the CDK4•cyclin D1 protein kinase (Guardavaccaro et al., 2000; Matsuda et al., 2001; Tirone, 2001) (Fig. 25). Thus as mentioned above there is no CDK4•cyclin D1 to start the G_1 build up to chromosome replication by hyperphosphorylating and breaking the pRb lock on the cell cycle machinery. Unphosphorylated, pRB locks the cycle machinery by shutting down the activator of cell cycle genes—the E2F protein—and joins with Cbfa1/Runx-2 to drive post-proliferative osteoblast gene expressions (Harbour and Dean, 2000; Whitfield and Chakravarthy, 2001). If there is no APRO1, pRb can be hyperphosphorylated by CDK4•cyclin D1 which makes it let go of E2F which then collaborates with CDK4•cyclin D1's successor, the CDK2•cyclin E protein kinase, to turn on the genes for cell cycle engine proteins (Harbour and Dean, 2000; Matsuda et al., 2001; Tirone, 2001; Whitfield and Chakravarthy, 2001). However, APRO1 is doubly effective—it can also reduce the expression of cyclin E and with this the appearance of the CDK2•cyclin E protein kinase needed to directly stimulate the expression of the DNA-replicating proteins (Tirone, 2001).

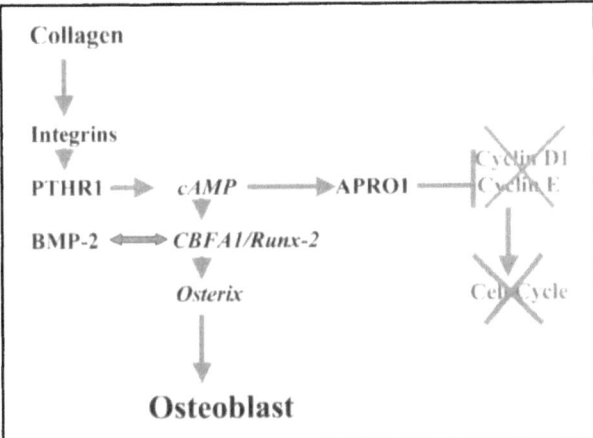

Figure 25. The pivotal moment in the initiation of osteoblast differentiation. The key event is the stimulation of the cyclic AMP/PKA system by signals from PTHR1 receptors activated by preosteoblasts' autocrine (autostimulating)/paracrine (neighbor-stimulating) PTHrP or exogenous PTH. The cyclic AMP system stops transit-amplifying proliferation by stimulating APRO1 which in turn prevents the activation of the major cell cycle protein kinases, cyclin D1•Cdk 4 and cyclin E•Cdk2. The cylic AMP signaling also turns on Cbfa1/Runx-2 expression which in turn sets up a self-reinforcing BMP2 ⇔ Cbfa1/Runx-2 cycle and stimulates the expression of the osteoblast-specific gene for osterix which is absolutely essential for osteoblast differentiation.

Parenthetically, this cycle-stopping expression of APRO1 (TIS21) is by no means limited to transit-amplifying preosteoblasts (Lim, 2006). Still in the context of bone it is also involved in osteoclast differentiation where it is induced by RANKL-induced RANK receptor signaling to stop osteoclast precursors proliferating as a prelude to terminal maturation (S.W. Lee et al., 2002). But it also seems to be part of a widespread mechanism for shifting a proliferating precursor cell's progeny into a post-proliferative stage of development. Thus, for example, when a mouse or rat neuroectodermal cell in the ventricular zone of the neural tube has reached the point of generating a post-proliferative neuron, it switches on its gene for the APRO1 (TIS21) protein (Iacopetti et al., 1994, 1999; Tirone, 2001). One of its daughters will shut her TIS21 gene off, but will still have some of mother's APRO1 (TIS21) protein and keep it during its outward migration and the initial stage of neuronal differentiation—the appearance of APRO1 (TIS21), the resulting lack of CDK4•cyclin D1 kinase and hence the failure of cycle-suppressing pRb's hyperphosphorylation and inactivation marks a neuron's birthday which over the subsequent decades the cell would celebrate as its "APRO1 (TIS21) day" (Iacopetti et al., 1994; Tirone, 2001). Knocking out the cycle-locking pRB gene prevents neuronal differentiation and keeps the neuroblasts proliferating which ultimately results in a massive apoptosis of the accumulating neuroblasts (Guardavaccaro et al., 2000; Lee et al., 1994). APRO1(TIS21) also associates with mCaf1, a component of the cell's mRNA deadenylase (Rouault et al., 1998; Tucker et al., 2001). This could mean that when the brief appearance of APRO1(TIS21) stops the cell cycling it ensures that the cycle block will persist by driving the deadenylation and deletion of the cycle-related mRNAs which clears the way for reprogramming the cell with mRNA messages from differentiation-related genes. By the time the APRO1 (TIS21) protein has gone, the cell, be it a neuron or osteoblast, is irreversibly committed to becoming a differentiated neuron or osteoblast.

APRO1(TIS21) is just one, albeit an important one, of a widely used set of genes, the "Maturation Gene Team" or the "M-Team" for short, the products of which turn off and lock down the cell proliferation-drivers and drive MC3T3-E1 cells into terminal maturation (Fig. 23). The quiescence-induction-associated Q (quiescin) 6 sulfhydryl oxidase protein also peaks and drops along with APRO1 (TIS21) (Raouf and Seth, 2002) (Fig. 23). Indeed Q6 expression is suppressed in tumor cells (Coppock et al., 1993, 2000). Another member of the team is HAX-1, which is part of the EDC (epithelial differentiation gene) complex (e.g., involucrin, loricrin, S100s, trichohyalin) on human chromosome 1q21-22 (Marenholz et al., 2001; Raouf and Seth, 2002; Whitfield and Chakravarthy, 2001) (Fig. 23). The Ca^{2+}-triggered expression of this EDC gene complex is at the heart of the terminal, apoptosis-like differentiation (i.e., diffpoptosis) of skin keratinocytes into keratin-loaded cadavers (Whitfield and Chakrvarthy, 2001).

But there has very recently emerged a major PTHR1-activated driver of osteoblastic maturation—Smad3-stimulated β-catenin (Tobimatsu et al., 2006) (Fig. 26). β-Catenin seems to be part of a coordinated maturation triggering mechanism involving the members of the M-Team. It appears that a burst of cyclic AMP-dependent protein kinase activity that stimulates APRO1(TIS21) expression also stimulates Smad3 expression which in turn stops the cytoplasmic turnover of β-catenin. β-Catenin's main job is to collaborate with E-cadherin to maintain the adherence of the cell to its neighbors and its cytoplasmic level is kept just high enough to do the job. Excessive β-catenin level is prevented by the protein being phosphorylated by GSK-3β (glycogen synthase-3β) which marks it for delivery to the cell's proteasome for shredding. The cyclic AMP-driven surge of Smad3 somehow stops this control and cytoplasmic β-catenin starts building up, enters the nucleus and forms a complex with the Tcf/Lef protein that stimulates key osteoblastic genes such as those for collagen I and alkaline phosphatase (Sowa et al., 2002, 2003; Tobimatsu et al., 2006) (Fig. 27).

In summary, the initiation of normal osteoblastic maturation probably goes like this. *Signals from collagen-stimulated integrins (Fig. 28) trigger the expression of PTHR1 receptor and*

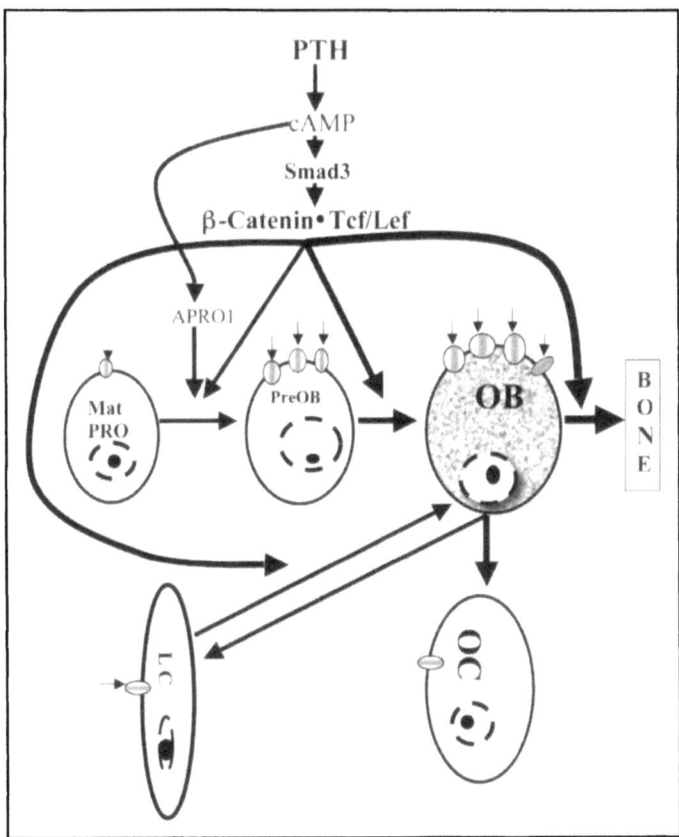

Figure 26. How PTH might act via cyclic AMP/PKA-stimulated APRO1 and Smad 3 and the Smad 3-stimulated β-catenin mechanism to drive the transit-amplifying cells' proliferative shut down and the maturation of preosteoblasts into functioning osteoblasts and the later stages of the osteoblast life cycle. LC, lining cell; Mat PRO, mature progenitor; OB, mature, terminally differentiated osteoblast; OC, osteocyte; PreOB, preosteoblast.

PTHrP to stimulate the emerging PTHR1 receptors to give cyclic AMP surges to upregulate Smad3 and start a cytoplasmic build-up and nuclear invasion of β-catenin (Tobimatsu et al., 2006) (Figs. 26 and 27). The cyclic AMP surge also triggers a brief appearance of the very labile APRO1(TIS21) (Fletcher et al., 1991; Tirone, 2001) which inhibits the expression of both cyclins D1 and E and the activity of any residual CDK4/6•cyclin D1 and CDK2•cyclin E kinases (Figs. 25-27). And they drive Cbfa1/Runx-2 into the nucleus (Figs. 29 and 31). The Ca²⁺ surge also caused by PTHR1 signaling may also stimulate the expression of HPX-1, which is another part of the differentiation machinery. The lack of CDK4/6•cyclin D1 prevents the phosphorylation of pRB, which, because of this won't let go of its icy grip on the replication genes' activator, E2F. But pRB can grab Cbfa1/ Runx-2 by its C-tail to form a complex that promotes the expression of Cbfa1/Runx-2's target osteoblast-specific genes. Meanwhile the briefly appearing APRO1(TIS21) also eliminates the mRNAs for the various parts of the cell cycle machinery. And the accumulated β-catenin surging into the reprogramming nucleus turns on key mature osteoblastic genes.

*   The take-home message from all of this is—By the time APRO1(TIS21) vanishes after its brief moment " on stage", the cell's key replication-driver genes are silent with their promoter-switch boxes*

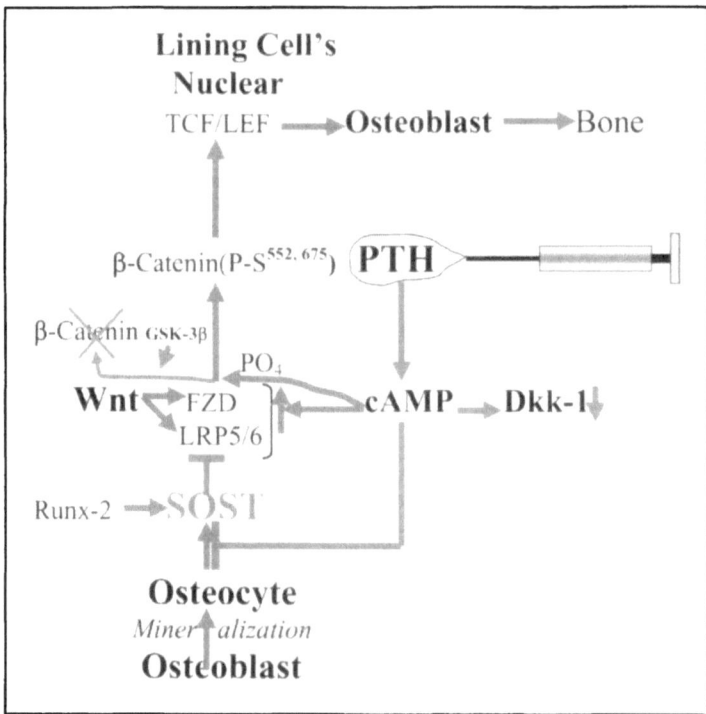

Figure 27. The Wnt mechanism via which PTH might drive the retooling of retired bone-lining cells into working osteoblasts. Briefly the PTH stimulates the mechanism by blocking the expression of the osteo-genesis-blocking SOST; increasing Wnt sensitivity by increasing the expression of the FZD/LRP5/6 Wnt receptor complex; downreulating Dkk-1 which would otherwise drive the endocytosis and destruction of LRP5/6; and preventing β-catenin turnover by phosphorylating its serine 552 and 675 aminoacids. See the text for further details

made unstimulable by having their CpG dinucleotide islands methylated and buried in chromatin condensed by having its histones deacetylated. But the cell emerges with a reconfigured nucleus with bone-making genes, such as osteocalcin, with their promoter-switch boxes now accessible for switch-ing on by having their histones acetylated, their chromatin decondensed, and their promoters' CpG islands demethylated (Villagra et al., 2002).

But (that ominous word again!) Fujita et al. (2002) have claimed that PTHR1 signaling does stimulate osteoblastic cell proliferation because they found that hPTH-(1-34) strongly stimulated cyclic AMP accumulation but only very slightly stimulated (about 50%) proliferation of ATDC5, MC3T3-E1 and B-Raf-transformed MG63 bone cells. The proliferogenic mechanism appeared to have to be triggered by the cyclic AMP binding to and activating Epac (an exchange protein directly activated by cyclic AMP), a cyclic AMP-GEF (Guanosine nucle-otide Exchange Factor) (de Rooij et al., 2000; Stork and Schmitt, 2002; Zwartkruis and Boss. 1999). Cyclic AMP•Epac stimulates the replacement of GDP with GTP on GDP•Rap1, a Ras-like GTPase (Zwartkruis and Bos, 1999). GTP•Rap1 stimulates B-Raf protein kinase that sets off the proliferation-stimulating MEK1/2 $\Rightarrow$ ERK1/2 cascade (Fujita et al., 2002; Pearson et al., 2001; Stork and Schmitt, 2002; Zwartkruis and Boss, 1999). And Tawfeek (2002) has since shown that a PTH such as [Gly[1],Arg [19]]hPTH-(1-28), which (the reader might recall from section i b) only stimulates adenylyl cyclase does indeed stimulate the MEK

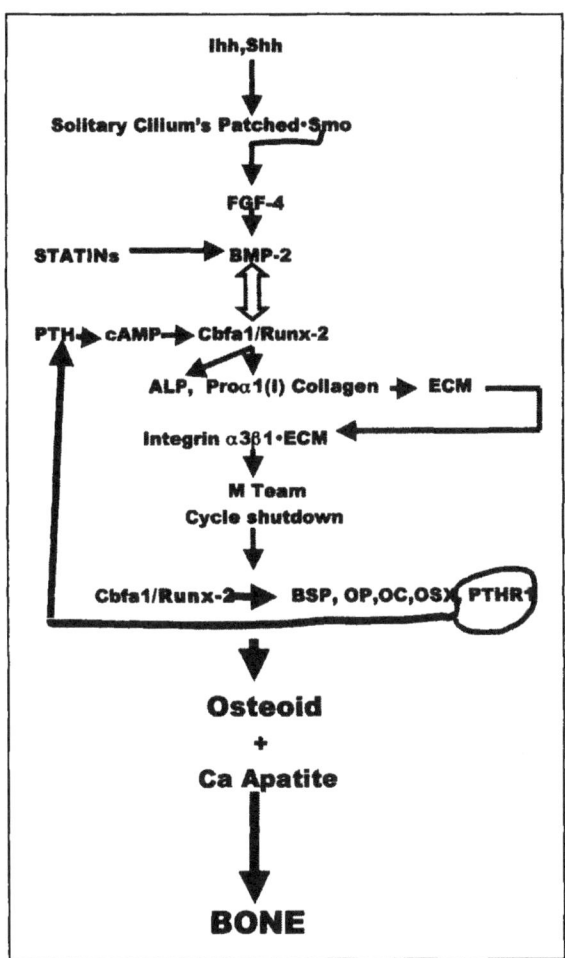

Figure 28. The points in the osteoblast maturation program where PTHs and maybe statins control the progression to mature bone-making osteoblast. This scheme includes the recent discovery that the hedgehogs' patched•smoothened receptor signal transducing complex may be located on the solitary cilium is discussed extensively in the text. Abbreviations: **ALP**, bone-specific alkaline phosphatase; **BSP**, bone sialoprotein; **Ca^{2+}•PO$_4$**, apatite-like calcium phosphate bone mineral; **ECM**, extracellular matrix; **FGF-4**, fibroblast growth factor-4; **Ihh**, Indian hedgehog; **OC**, osteocalcin; **OP**, osteopontin; **OSX**, osterix; **Shh**, sonic hedgehog.

⇒ ERK cascade which we would now expect, from a generation of cyclic AMP•Epac rather than a burst of PKA activity, to inhibit the cascade. On the other hand, Mahon and Segre (2002) have found that the PTHR1 receptor's C-tail connection to NHERF-2 is needed for hPTH-(1-34) to stimulate the MEK mechanism in PS120 fibroblasts by a pathway not involving cyclic AMP (i.e., it is not stimulated by the AC-stimulator forskolin) or PKC. Significant though this variously triggered proliferogenic MEK-ERK cascade seems to be for PTH's actions in vitro, it is probably irrelevant to how PTH, PTHrP and PTHR1 signaling actually stimulate normal osteoblast accumulation in a bone. But in the neoplastic or semi-neoplastic cells of various established bone cell lines used by Fujita et al. (2002) the normally incompatible

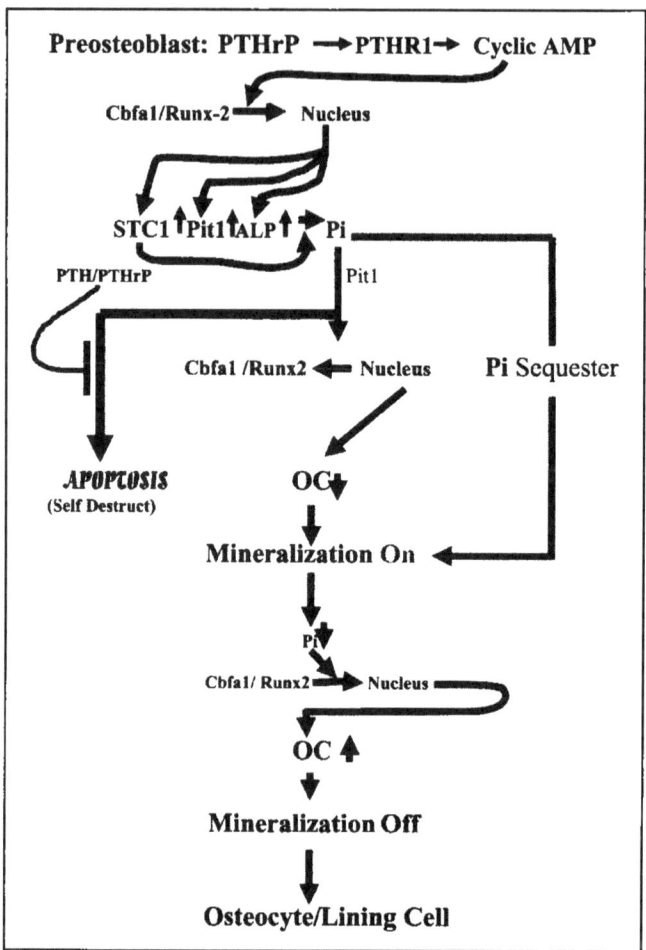

Figure 29. The control of matrix mineralization. Things begin with the expression of Cbfa1/Runx-2 and its shipment into the nucleus where it stimulates the expression of ALP (bone-specifc alkaline phosphatase), and Pit1 phosphate transporter and STC1 (stanniocalcin1). ALP is plugged into the cell membrane with its catalytic subunit (its businiess end) sticking out of the cell. ALP then clips phosphate off external compounds and STC1 and Pit1 load it into the cell. But the cell won't immediately start mineralizing the matrix because the nuclear Cbfa1/Runx-2 has stimulated the production of osteocalcin (OC) that inhibits mineralization. However, the accumulation of phosphate causes Cbfa1/Runx-2 to be kicked out of nucleus which stops OC and removes the block on mineralization. As phosphate is locked up in mineral, the cell phosphate content drops, Cbfa1/Runx-2 returns to the nucleus where it restores OC expression, stops mineralization. Now the cell has done its job and is ready to retire and become an osteocyte, bone lining cell or commit apoptic self destruction. Phosphate accumulation and exposure to the very high phosphate concentrations in osteoclast excavations can be deadly. But one of the many things PTHR1 signals can do is block phosphate-induced apoptosis.

proliferogenic and maturation mechanisms because of APRO1 (TIS21) and the M-team are more or less disconnected from the M-Team's control. Therefore, it is far more likely that the cyclic AMP fired from activated PTHR1 receptors summon APRO1(TIS21) to permanently shut down the cell cycle machinery and terminally differentiate normal bone cells.

iii d. PTHrP Signals Turn on Genes for Making Osteoblasts

iii d 1. PTHrP, the Hugely Complex Physiological Osteoblast Maturation Driver

The reader might recall from Section iii b that signals from PTHR1 can, like strain, load the nucleus with transcription factors such as β-catenin•Tcf/Lef and architectural transcription factors such as Cbfa1/ Runx2 and Nmp4 that bind target sites on chromosomes to produce genes with functionally reconfigured promoter-switch boxes accessible to being switched on by appropriate classical transcription factors. The accessibility of genes to these transcription factors driven into the nucleus by osteogenic short bursts of PTHR1 signaling is also likely to need the brief (6-24h) bursts of the enzyme UBP41 that could snip the ubiquitin off the nucleosomes' histone H2B which has until now enhanced gene-silencing chromatin coiling by promoting the methylation of certain lysines equivalent to the lysine 79 in the "non-tail" core of the yeast nucleosomes' H3 histones (Briggs et al., 2002; Latchman, 2004; Miles et al., 2002; Osley, 2004; Sun and Allis, 2002). (There is also evidence for deubiquitination of lysine 4 in the H3 tail being involved in turning off certain genes [Osley, 2004]). As a result, the chromatin-uncoiling by the PTHR1-stimulated UBP41 exposes the promoters of the genes to transcription factors for factors such as FGF-2, FGF receptors and IGF-I, and further stimulate the expression of the whole PTHrP chain and one or more of its chopped-out cytokine components (Figs. 23 and 30; Amizuka et al., 1996; Goltzman and White, 2000; Kartsogiannis et al. 1997; Walsh et al.,1997; Zhang et al.,1995), and TGF-βs (TGF-β1 and β2 if the PTH is hPTH-(1-34) that stimulates adenylyl cyclase and membrane-associated PKC activity, or only TGF-β2 if the PTH is hPTH-(1-31)NH$_2$ which only stimulates adenylyl cyclase in human osteoblasts [Wu and Kumar, 2000]). As you will see in section iii e, I call these PTHR1-induced factors "*missionaries*" because they carry the PTH/PTHrP message to multiply and generate osteoblasts to the transit-amplifying progenitors without PTHR1 receptors. Some of these factors will be locked up in the new matrix and will not be released until the osteoclasts of a future microcrack-repairing BMU dig them out without destroying all of them with the matrix-destroying proteases while digging a new tunnel or trench.

The production of mRNA for PTHrP-(1-139),-(1-141) and (1-173) by alternate splicing (i.e., cutting and pasting) of the PTHrP gene transcripts in primary human osteoblasts increases a dramatic 38-fold by 45 to 90 minutes after hPTH-(1-34) is added to the culture medium (Walsh et al., 1997). While we know almost nothing about the contributions of PTHrP and its autocrine/paracrine components to mature osteoblast activity and the osteogenic response, they are certainly important. PTHrP's PTH-like 1-34 N-terminal nose would bind to and activate the PTHR1 receptors, and through them adenylyl cyclase, as efficiently as a PTH (see Goltzman and White [2000], Lam et al.[1999] and Whitfield et al. [1998a] for reviews). Indeed because circulating PTH fragments with N-termini are virtually non-existant, autocrine (self-stimulating)/paracrine (neighbor-stimulating) PTHrP must be the osteoblasts' main adenylyl cyclase activator and maturation driver. But unlike PTH, the whole PTHrP molecule or a PTHrP fragment with the 85-107 NLS (nucleus-localizing sequence) or NTS (nuclear-targeting sequence) (Steggerda and Paschal, 2002) that is either made in the cell or rides into the cell from the surface on its PTHR1 receptor can function as an 'intrakine' or "intracrine" factor that using its NLS and NES (nuclear export sequence) signal sequences is shuttled back and forth between the cytoplasm and nucleus (Aarts et al., 1999,2001; Goltzman and White, 2000; Jans et al., 2003; Lam et al., 1999, 2000, 2002; Re, 2002a, 2002b; Re and Cook, 2005; Watson et al., 2000) (Fig. 21).

But let's take a closer look at this ancient string of cytokine "pearls". When its mRNA is translated into a protein with a so-called secretory signal sequence it is directed to the Golgi apparatus where it is put into the secretory machinery. But the mRNA can also be translated by the ribosomes from an alternative translation-start site into a protein without a secretory signal

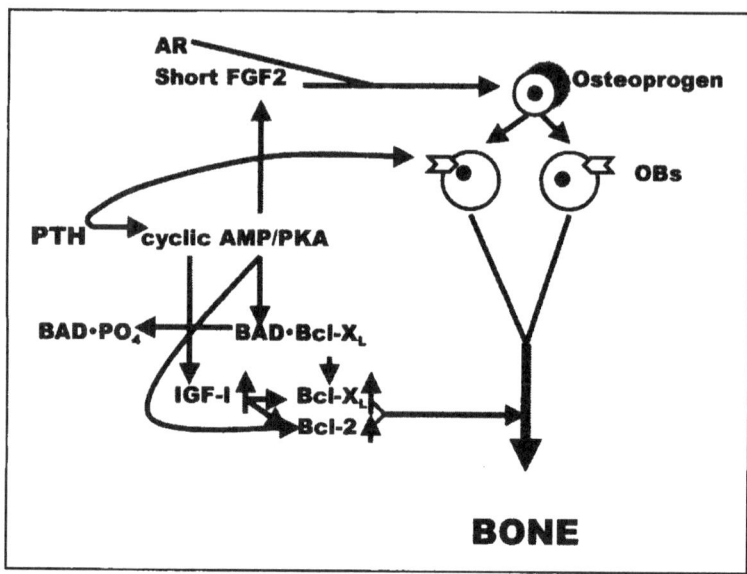

Figure 30. Two of the 3 ways by which a PTH stimulates bone formation. (1) It stimulates the expression and release of missionary growth factors such as amphiregulin (AR) and short FGF-2 which can stimulate osteoprogenitors (**Osteoprogen**) that have not yet begun to express PTHR1s. (2) It extends the working lives of osteoblasts (**OBs**) by releasing the anti-apopotosis Bcl-2 protein from the stranglehold of the pro-apopotsis BAD protein, trapping the phosphorylated BAD protein in the cytosol bound harmlessly to 14-3-3 protein, and stimulating the expression of IGF-I which stimulates the expression of the anti-apoptosis Bcl-2 and Bcl-X$_L$ proteins. The third way is to stimulate bone lining cells to come out of retirement and become working osteoblasts once again.

sequence (Jans et al., 2003). This protein is not fed into the secretory machinery. Instead it binds by its NLS sequence to the nuclear cargo carrier importin-β1 to form a complex that links to a microtubular trackway where it is pulled by a dynein motor to the mouth of a pore in the nuclear envelope and handed over to the pore channel transport machinery. The complex is then ratcheted along the pore channel by series of dockings to so-called nucleoporins ("nups") lining the pore channel and then dumped into the nucleus where it is separted from the importin carrier when the importin binds avidly to Ran•GTP and homes especially onto the nucleolus where it can *directly* affect nuclear functions such as the ribosome-making machinery. (Fried and Kutay, 2003; Jans et al., 2003; Lam et al., 2002).

It seems likely that this nuclear intrakine/intracrine stimulation would happen only in the mature, postmitotic preosteoblast or osteoblast because the cyclin-dependent protein kinases such as CDK4•cyclin D1 and CDK2•cyclin E that drive the cell cycle would disable the NLS by phosphorylating its Thr[85] (Goltzman and White, 2000; Lam et al., 1999, 2000). Once in the nucleus and nucleolus the peptide targets ribosome production and does its ancient job of coordinating the protein-synthesizing machinery with the signals from its activated surface receptors (Amling et al., 1997; Lam et al., 2000; Re, 2002a, 2002b). But it also seems likely that the nuclear PTHrP stimulates certain genes, among them the genes for the anti-apoptosis Bcl-2 and Bcl-X$_L$ proteins (Tovar-Sepulveda et al., 2002) that would prolong the cell's matrix-making activity by preventing it from self-destructing by apoptosis (Antonsson and Martinou, 2000; Fadeel et al., 1999). But in various rat tissues, murine MC3T3-E1 osteoblasts and rat ROS 17/2.8 ostesarcoma cells the endocytosed PTHR1 receptor itself is grabbed by its

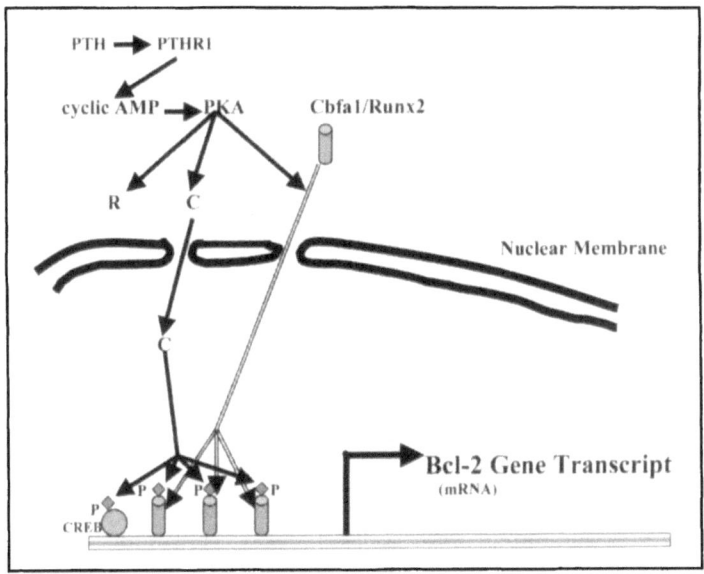

Figure 31. How a PTH could *directly* stimulate the expression of the apoptosis-preventing Bcl-2 protein. The cyclic AMP surge from the PTH-activated PTHR1 causes PKA's (cyclic AMP-dependent protein kinase) catalytic subunits (C) to separate from the enzyme's regulatory (R) subunits and travel into the nucleus. The cyclic AMP signal also causes Cbfa1/Runx-2 to move into the nucleus. The C subunits phosphorylate the **CREB** (cyclic AMP response element-binding) protein on Bcl-2 gene's promoter and the several Cbfa1/Runx-2 proteins bound to their specifc sites on the gene's promoter. The result of these nuclear migrations and phosphorylations is the activation of the *bcl-2* gene's promoter, Bcl-2 expression and the suppression of apoptotic self-destruction.

C-tail's NLS by importin-α_1 or importin-β nuclear transporters and loaded into the nucleo-plasm during G_0, G_1, S and telophase stages of the cell cycle when there is a nuclear membrane or a newly forming membrane (Pickard et al., 2006; Watson et al. 2000a, 2000b) (Fig. 21). Therefore, the PTHR1 receptor alone can get into the osteoblast nucleus to carry out some job such as gene activation as does the EGF receptor (Lin et al., 2001; Ni et al., 2001).

PTHrP is neither a hormone (i.e., it does not circulate throughout non-cancer-bearing mammalian bodies as it does in fish for example) nor a simple cytokine. It is really a string of different cytokine "pearls" with 8 endoproteolytic consensus sites marking their borders where prohormone thiol protease and prohormone convertases can cut them out to operate as individual cytokines, each with its own receptor and specific mission (Deftos et al., 2001; Hook et al., 2001). One of these cytokines is the phospholipase-C-activating PTHrP-(67-86), the role of which in bone is unknown (Orloff et al., 1996). But another of these is hPTHrP-(107-139) ("osteostatin"), with its highly conserved 107-111 (i.e., Thr(T)[107]-Arg(R)[108]-Ser(S)[109]-Ala(A)[110]-Trp(W)[111]) region, that by itself can stimulate the proliferation of cultured fetal rat osteoblasts (Cornish et al., 1999), bone nodule formation by cultured MG-63 human osteoblastic osteosarcoma cells (de Gortazar et al., 2004), and possibly bone formation in 7- to 8-months-old OVXed rats (Roufflet et al., 1994). More-over, concentrations of PTHrP-(107-139) and PTHrP-(107-111) as low as 10^{-15} and 10^{-13} M inhibit osteoclast generation and activity in vivo and in vitro (hence the name "osteostatin") (Fenton et al., 1991; Cornish et al.,1997; Zheng et al., 1994). hPTHrP-(107-139) operates against osteoclast generation in two ways—by lowering the expression of RANKL, the osteoclastogenesis-driver, while at the same time increasing the level of the anti-osteoclastogenic osteoprotegrin production in osteoblastic cells (de Gortazar et al., 2004). PTHrP-(107-139) is also angiogenic—it could collaborate with the angiogenic action of leptin pouring out of hypertrophic chondrocytes (Kume et al., 2002) by stimulating human osteoblasts to make the angiogenic VEGFs which would provide the blood vessels needed to feed the new bone being made in a trabecular/endocortical blister or cortical tunnel (Esbrit et al., 2000). And it might be the mysterious PTHrP-induced modulator of vascular invasion into cartilage that does not work through the PTHR1 receptor (Lanske et al., 1999). These fragments operate by stimulating PKCs on membrane scaffolds not by activating PTHR1 but by activating an apparently very high-affinity "TRSAW" (Thr(T)^{107}-Arg(R)^{108}-Ser(S)^{109}-ALA(A)^{110}-Trp(W)^{111}) receptor (Cuthbertson et al., 1999; Moonga and Dempster, 1995; Whitfield et al., 1996a, 1998a).

iii d 2. A PTHrP Region Reveals an Exciting Possible Role for Endothelin-1 in Bone Formation

But there is something else hiding in the PTHrP necklace which when it is cut out can *by itself* drive bone formation. PTHrP's 8-11 sequence (LHDK) is virtually identical to endothelin-1's residues 6-9 (Leu-Met-Asp-Lys) which means that PTHrP-(8-11) can stimulate the endothelin A receptor (Mohammad et al., 2003). And you may recall from Chapter 2 that endothelin-1 concentrations *as low as 10^{-10} (!)* can stimulate both proliferation and differentiation of the endothelin-A receptor-expressing osteoblastic cells in rat calvarial cell cultures (von Schroeder et al., 2003). Moreover stimulating the endothelin A receptor on mouse calvarial cells with either endothelin-1 or PTHrP-(8-11) strongly stimulates bone formation (Mohammad et al., 2003). Since endothelin-1 is made by vascular endothelial cells, this means, for example, that the blood vessels trailing behind the osteoclasts tunneling through bone can use it to stimulate osteoblasts to mature and start plastering new bone on the tunnel wall (Fig. 8). What an igenious way to link blood vessel extension to bone formation—let the endothelial cells following the osteoclasts produce a potent osteogenic factor!!!!

iii d 3. Some of the Many Genes Targeted by PTHs

Stromal cells are put on the osteoblast pathway when sonic hedgehog (Shh) activates its Patched• Smo signal complex probably on the cell's primary cilium (Corbit et al., 2005; Huangfu and Anderson, 2005) which stimulates them via FGF-4 to express BMP (bone morphogenic protein)-2 that in turn stimulates them to express type 2 Cbfa1/Runx-2 (Ducy, 2000; Karsenty, 2000a; Katagiri and Takahashi, 2002; Komori, 2000) (Fig. 28). As we shall see further on Cbfa1/Runx-2 is also involved in driving chondrocyte hypertrophy and the run-down to mineralization and coordinating this with the induction of osteoblasts in perichondrium/periosteum to make bone on calcified cartilage scaffolds in endochondral ossification (de Crombrugge et al., 2001; Shum and Nuckolls, 2001). Cbfa1/Runx-2 interacts with the R-Smads specifically from activated BMPR-1B receptors to switch on the Osterix gene, which has a Cbfa1/Runx2 element in its promoter switch box, then steers the cell irrevocably onto the osteoblast pathway (Katagiri and Takahashi, 2002; Lian et al., 2004; Nakashima et al., 2002; Yagi et al., 2003). Another hedgehog—Indian hedgehog (Ihh)—produced by pre-hypertrophic and hypertrophic chondrocytes together with BMP-2, BMP-6 and Cbfa1/Runx-2 stimulates differentiation of a distinct population of perichondrial cells into osteoblasts in the collar of the developing bone (Chung et al., 2001; Karsenty, 2001). The osteoblast-determining potency of the Cbfa1/Runx-2 gene product is illustrated by the fact that forcing *non*-osteoblastic, primary fibroblasts to make it causes them to switch on their osteoblastic genes (Ducy, 2000; Lian and Stein, 2003). And just overexpressing the Cbfa1/Runx-2 gene activity with an adenovirus construct is enough to stimulate bone formation in a neonatal rat metatarsal organ culture (Krishnan et al., 2003).

As expected from this, one of the osteogenically determining things an injected adenylyl cyclase-stimulating osteogenic PTH does is push progenitor cells along the osteogenic path by stimulating them to express more PTHR1 receptors (Lu et al., 2001; von Stechow et al., 2003), and via bursts of PTHR1-driven adenylyl cyclase activity turn up the volume of Cbfa1/Runx-2 gene expression and the loading of the cells' nuclei with the Cbfa1/Runx-2 transactivator (; Fujita et al., 2001a, 2001b; Krishnan et al., 2003; Moore et al., 2000; Selvamurugan et al., 2000) (Figs. 29 and 31). Cbfa1/Runx-2 has a NMTS (nuclear matrix-targeting sequence) domain in its C-tail (Zaidi et al., 2001) and a DNA-binding 'Runt' domain (Ducy, 2000; Ito, 1999). With its tail securely stuck to the matrix, Cbfa1/Runx-2 grabs the so-called OCE2 upstream sites of still silent osteoblast-specific genes such as the genes for ALP (alkaline phosphatase), BSP (bone sialoprotein), osteocalcin, osteopontin, OPG, proα1(I) and proα2(I) procollagen, PTHR1, and type I TGF-β receptor (Franceschi, 1999; Karperien et al., 2000; Kern et al., 2001; Komori, 2000) (Fig. 28). Thus, Cbfa1/Runx-2 is one of the DNA-benders and twisters that pins these key genes to sites on the nuclear matrix containing the RNA-polymerase II-gene-transcribing machine (Lian and Stein, 2003; Turner, 2001; Vilagra et al., 2002). These now primed genes with their configured promoters exposed by being pinned down by Cbfa1/Runx-2 stay quiet until pRb and other classical stage-specific transcription factors appear to form transcription-activating multicomponent enhanceosomes that can switch on the transcription printing press (Turner, 2001). And the expression of another gene, the gene encoding the zinc-finger-containing transactivator OSX (osterix), which, as you will remember, is a "must" for the subsequent maturation of the cell into an osteoblast (Katagiri and Takahashi, 2002; Nakashima et al., 2002; Yagi et al., 2003; Zhang et al., 2002a) (Fig. 23).

The "mature" or late osteoprogenitors emerging from the inter-trabecular pool of osteoprogenitors and just starting their spurt of PTHR1 receptor expression (one of the global markers of osteoblasts in all parts of the skeleton) and PTHR1-activating PTHrP (Fig. 9) can go either directly or from the red marrow's sinusoidal vasculature into trenches dug into trabecular surfaces and the nearby cortical endosteum (Bianco and Riminucci, 1998). Presumably much less bone will be made at fatty marrow sites as happens with FGF-2 treatment (Pun et al., 2001). However, they can reach cortical bone only from the marrow blood vessels after the cells in the trabeculae have had the first grab at them, which is one of the reasons why the adenylyl cyclase-stimulating PTHs are best at stimulating the growth of trabeculae-rich bones such as the vertebrae. Another reason for the PTHs' "trabeculophilia" besides the peptide first reaching the trabeculae after its injection or ingestion is that except for their PTHR1 receptors trabecular osteoblasts differ from cortical osteoblasts in their repertoire of expressed genes and their transcripts (i.e., their "transcriptomes") (Aubin, 2000, 2001; Aubin and Triffitt, 2002; Candeliere et al., 2001).

When the osteoprogenitor cells have arrived at the swept-up tunnel wall or trench and plastered enough collagen on it, the signals from autocrine/paracrine PTHrP-activated PTHR1 receptors, the expressions of which have been stimulated by collagen-bound/activated integrins, have led the cells' "M teams" to shut down the cycle-driving machinery and enabled Cbfa1/Runx-2 to up-regulate the differentiation-related genes (Fig. 28). But although they may look alike and make bone, the resulting post-mitotic mature osteoblasts do not have the same gene expression profile. The cells' gene expression profiles are specifically tailored to meet the different needs for working in different regions of the skeleton and even in different parts of a particular bone. Thus, for example, the relative expressions of bone sialoprotein, osteopontin, osteocalcin, and PTHrP at the mRNA and protein levels depend on in which part of a developing bone (e.g., endocranial or ectocranial surface of mouse calvaria) the cell is located as well as who its immediate neighbors are and consequently the microenvironment in which it finds itself. In another example, osteoblasts in rat calvaria express glutamate NMDA receptor/channels with subunits NR2A, NR2B and NR2D, but not NR2C while femoral osteoblasts express NR2C (Itzstein et al., 2001). But all osteoblasts, no matter where they are or who their neighbors are, express the genes for the PTHR1 receptor, type I collagen and alkaline phosphatase (Candeliere et al., 2001). The products of these three genes are absolutely needed—they are the *sine qua nons* for becoming a mature, post-proliferative, matrix-making and mineralizing osteoblast.

The PTHR1-expressing preosteoblasts and mature osteoblasts in modeling clusters making new bone and filling excavated microcracks in mature skeletons are the first or prime targets of injected PTH (K.Lee et al., 1994) (Fig. 9 and 23). But it must be remembered that the signals from the PTHR1 receptors cannot stimulate the proliferation of maturing preosteoblasts and mature osteoblasts that have been proliferatively disabled by their PTH/PTHrP-induced M-teams' APRO1(TIS21)s. By the time they have got rid of their cell cycle machinery and become working osteoblasts they have greatly (10-fold) increased the expression of PTHR1 receptors which probably enables the autocrine/paracrine N-terminal PTHrPs that they have made and secreted to give themselves and their neighbors the cyclic AMP signals needed to drive their further maturation and bone-building (Aubin, 1998, 2000, 2001; Aubin and Triffit, 2002).

Each daily PTH-triggered cyclic AMP transient releases active PKA catalytic subunits from inactive R•C holoenzymes (i.e., regulatory subunit•catalytic subunit) that surge from the cytoplasm into the target cell's nucleus through the nuclear pore complexes and turn on a set of early response genes with a CRE (cyclic AMP-responsive element) in their promoters and CREB•CBP/p300(cyclic AMP responsive element-binding protein•CREB-binding protein complexes) complexes on the CREs (Ptashne and Gann, 2002; Quinn, 2002) on their promoter-switch boxes (Fig. 31). However, after the nuclear catalytic subunit surge peaks around 15-30 minutes in cells such as MC3T3-E1 mouse preosteoblasts and ROS 17/2.8 osteoblastic cells, the catalytic subunits are no longer *personae gratae*. They are grabbed by the protein kinase A inhibitor PKIγ, which, like the various PKI isoforms, has a NES (nuclear export signal) domain, and are dumped out of the nucleus (X. Chen et al., 2002, 2003; Wen et al., 1995a, 1995b Wiley et al., 1999). This, of course, stops the immediate-early gene expressions that the catalytic subunits were stimulating (X. Chen et al., 2003). The evicted catalytic subunits reassemble inactive R•C holoenzymes in the cytoplasm. Together cyclic AMP transients may also feed back to stop the PTHR1 receptor attached to β-arrestin-2 from activating adenylyl cyclase by rapidly and briefly stimulating the expression of the gene for RGS-2, a GTPase activator, which converts receptor-generated adenylyl cyclase-activating GTP•Gsα to inactive GDP•Gsα (Pierroz et al., 2003; Thirunavukkarasu et al., 2002) (Fig. 21).

One of the earliest of the many hundreds of genes directly and indirectly booted by the cyclic AMP signals from PTHR1s are genes of the Fos family (*c-fos, fosB, fra*1, *fra*2) the products of which can associate with the products of members of the Jun family (*c-jun, junB, junD*) of genes to form AP-1(activator protein-1) transcription factor complexes that target an array of genes (X.D. Chen et al., 2004; Demiralp et al., 2001; K. Lee et al., 1994; McCauley et al., 1997, 2001; Onyia et al., 2001a; Tyson et al., 1999). The turning on of at least one of these genes, *c-fos*, is essential for PTHs to stimulate bone growth (at least in mice). Thus, knocking out the c-fos gene in mice also knocks out the ability of hPTH-(1-34) to stimulate femoral bone growth (Demiralp, et al., 2001).

The switching on of the *c-fos* gene has been used to mark the sequence of responses of the various bone cells in 4-week-old rats to subcutaneously injected full length hPTH-(1-84) (K.Lee et al., 1994). The *c-fos* gene expression in trabecular osteoblasts, and the few spindle-shaped stromal cells that made PTHR1 receptors, peaked between 15 and 30 minutes and stopped by 60 minutes. Most stromal cells did not make PTHR1 receptors nor could they and the osteoclast precursors express their *c-fos* genes until 60 minutes by which time the osteoblasts and the *few* PTHR1-expressing stromal cells had stopped. This delayed *c-fos* switch-on in the receptorless stromal cells is best explained by missionaries sent to the PTHR1-lacking cells by PTHR1-loaded osteoblasts. Only after the PTHR1-receptorless stromal cells had stopped expressing c-*fos* at 2 hours were the osteoclast precursors expressing the gene. The very late burst of c-*fos* expression in osteoclast precursors is likely due to RANKL flowing from the immature osteoblstic stromal cells which, unlike the densely PTHR1-receptored mature osteoblasts with their shut-down *rankl* genes, are the only ones that can make this osteoclastogenic factor (Atkins et al., 2003; Corral et al., 1998) (Figs. 23 and 37).

It is essential to understand that these brief PTHR1-induced pulses of cyclic AMP and both cytoplasmic and nuclear PKA catalytic subunits and the c-Fos•Jun transcription factor complexes they generate only *start* the bone-making machine. Daily PTH-induced cyclic AMP spikes can't *directly* make immature marrow osteoprogenitors or preosteoblasts and mature osteoblasts proliferate (Kostenuik et al., 1999; K. Lee et al., 1994; Y. Wang et al.2003; reviewed by Whitfield et al., 1998b). Indeed a striking example of this has been given by Y. Wang et al. (2003) who has reported that PTH-(1-34) stimulated the progression of preosteoblasts to bone nodule-making mature osteoblasts in primary calvarial osteoblast cultures *without stimulating proliferation*. However, PTH can stimulate the PTHR1-expressing osteoblasts that have been proliferatively switched off by the "Maturation Gene Team" of Figure 23 to make things that do stimulate the proliferation of the immature osteoprogenitors without PTHR1 receptors.

iii e. Speading PTH's Message to the PTHR1-Receptorless Transit-Amplifiers—The Missionary Factors That Suppress Cell Suicide and Stimulate Growth to Swell the Osteoblast Workforce

An osteogenic PTH works by both directly and indirectly producing a team of mature, mainly short-lived, proliferatively locked-down osteoblasts which themselves make and secrete various factors. These factors stimulate the osteoblast's own bone-making machinery as well as the proliferation of transit-amplifying immature osteoprogenitors that unlike the osteoblasts can go through a limited number of growth-division cycles but are too immature to have PTHR1s (Fig. 23). It is these factors such as amphiregulin (AR),apelin (APL), FGF-2, IGF-I, IGF-II growth factors and TGF-βs that take the wheel after the osteomotor is started by PTHR1 signals and then drive the expansion of the osteoblast team (Nishida et al., 1994; Whitfield et al., 1998b; Xie et al., 2006) (Fig. 23). But PTHrP from mature osteoblasts is actually another missionary-it can stimulate the proliferation and differentiation of immature progenitors without PTHR1 receptors by activating instead their endothelin A receptors (Fig.23).

The kick-starting PKA pulses also lengthen the working lifespans of mature osteoblasts by preventing them from committing the scheduled apoptotic suicide at the end of the job (Fig. 31). This increases the amount of bone they can make. Thus, as Gubrij et al.(2003) have shown, when osteoprogenitors can be made to constitutively overexpress an antiapoptotic protein such as Bcl-2 more of them survive to make more PTH-responsive, super-productive osteoblasts which, of course, make thicker trabeculae. The PTH/PTHR1-triggered PKA pulses' anti-apoptotic action results in part from phosphorylation and inactivation of the cells' apoptosis-initiating, appropriately named, BAD protein which makes BAD release the anti-apoptosis Bcl-X_L from its repressive grip and the phosphorylated BADis trapped in the cytosol bound to 14-3-3 protein and thus unable to cause cytochrome c to leak from the mitochondria and trigger apoptosis-executing caspase proteases such as caspase-3 (J.M. Adams, 2003; Bellido et al., 200; Chao and Korsmeyer, 1998; Stanislaus et al., 2000; Yin et al., 2003) (Figs. 30 and 31). Also they can be protected from suicidal thoughts and responses maybe by the stimulation of sphingosine kinase that produces the anti-apoptosis sphingosine phosphate, but most definitely by the stimulation of the expression of the very important IGF-I, a stimulator of the expression of the apoptosis-preventing Bcl-2 and Bcl-X_L family proteins (Calvi et al.,2001; Hill et al. 1997; Johanson and Rosen, 1997; Locklin et al., 2003; Machwate et al., 1998; Pfeilschrifter et al. 1995; Pugazhenthi et al., 1999; Tovar Sepulveda et al., 2002; Tumber et al. 2000; Virdee et al., 2000; Watson et al., 1995, 1999) (Fig. 30).

The Bcl-2 gene is also turned on directly by the PTH-triggered cyclic AMP pulses. And its Bcl-2 protein product seems to be one of the growing number of *"sine qua nons"* for PTH's anti-apopotsis action. Thus, preventing the translation of Bcl-2 gene transcripts with Bcl-2-specific, transcript-censoring siRNA (see the popular review of siRNAs' actions by Lau and Bartel, 2003) eliminates PTH's ability to prevent dexamethasone, etoposide (VP16), or the deprivation of substrate adhesion from triggering self destruction by apoptosis in OB-6 osteoblastic cells (Ali et al., 2003). How does PTH stimulate the all-important Bcl-2 gene? The gene has one cyclic AMP-targeted CREB and several Cbfa1/Runx-2 sites in its promoter and because of these, this anti-self-destruct gene is stimulated directly by the cyclic AMP boluses from PTHR1 (Plotkin et al., 2002). Thus, a cyclic AMP spike would phosphorylate and activate CREB and drive Cbfa1/Runx-2 into the nucleus and onto the Bcl-2 gene's promoter along with the activated phospho-CREB (Fujita et al., 2001a) (Fig. 31). The PKA catalytic subunits that also surge into the nucleus would phosphorylate and activate (i.e., increase the affinity of Cbfa1/Runx-2 for its Bcl-2 promoter sites) (Franceschi and Xiao, 2003; Selvamurugan et al., 2000). (Fig. 31).

The PTH-induced, apoptosis-supressing phosphorylations of BAD by PKA and the stimulation of cyclic AMP/CREB/Cbfa1/Runx-2-dependent Bcl-2 expression are transient, i.e., self-limiting (Bellido et al., 2001, 2002). One reason for this transience is the silencing and expulsion of CREB-activating PKA catalytic subunits from the nucleus by PKIγ. Another reason is that the PTHR1 signaling also causes Cbfa1/Runx-2 to be dumped into the proteasome shredder (Bellido et al., 2002). Thus, only well-separated short bursts of PTHR1 can maintain the anti-apoptotic action. It follows from this that turning off the cell's shredder with lactacystin or overexpressing and swamping the system with Cbfa1/Runx-2 should, and in fact does, prolong PTH's anti-apoptotic action (Bellido et al., 2002).

But Bcl-2 is only one of the 440 or so genes the products of which are involved in some 44 signaling cascades controlling cell cycle and apoptosis (X.D. Chen et al., 2004). Eight genes (6 upregulated *Jun*, MApKK3, AKT1, c-*fos*, p53, ADPRT; 2 downregulated *Fyn*, FADD) are the most often regulated by PTH and are directly linked to 144 of the 440 genes. Among these 144 genes are, besides Bcl-2, the genes for the anti-apoptosis factor survivin (Li, 2003) and integrin-linked kinase (X.D.Chen et al., 2004).

But in the case of PTH's anti-apoptosis action no less than 3 exceptions have been reported. The first exception has been found in murine C3H10T1/2 and MC3T3-E1 cells. As expected the hPTH-(1-34) triggered cyclic AMP buck-shot protected *preconfluent* cells from dexamethasone-induced apoptosis, but *the cyclic AMP triggered apoptosis in postconfluent* C3H10T1/2 and MC3T3-E1 cells (H.-L. Chen et al., 2002)! The second exception was found in the distal femurs of intact young Fisher 344 and Sprague-Dawley rats. Single daily injections of hPTH-(1-34) (Forteo™) into male rats that stimulate bone growth also stimulate a wave of osteoblast apoptosis in the proliferating zone just below the growth plate and osteocyte apoptosis in the osteocyte death zone that starts promptly during the first day, peaks around days 3-7 and then gradually drops off by 28 days (Stanislaus et al., 2000). And there was no detectable increase in Bcl-2 expression. However, there does seem to be an anti-apoptosis effect in other parts of the femur as indicated by a large drop in the activities of the apoptosis-driving caspases 2, 3 and 7 in the *whole metaphysis*. In fact these drops are so large that they can obscure any more limited regional increase in the caspase activities. In other words PTH *stimulates* apoptosis in certain zones of the rat femur, but it inhibits it elsewhere in the bone. The third exception has been found in 15 women with mild primary hyperparathyroidism (PHPT) (Zhou et al., 2004). In these patients there were large increases in apoptotic osteoblasts on cancellous, endocortical and intracortical surfaces along with increased bone formation. As stated by the authors *"…enhanced osteoblast number and/or activity seen with continuously elevated PTH in PHPT is associated with enhanced osteoblast attrition"*. .

The IGF-I secreted by the PTH-stimulated, Bcl-2 protected, longer-lived, working osteoblasts also increases the osteoblast population by serving as a missionary that delivers PTH's call to proliferate to nearby proliferating transit-amplifying osteoprogenitors on the walls of new osteonal tunnels under construction or to osteoprogenitors in the bone marrow or trabeculat trenches, none of which have PTHR1 receptors or are ready yet to make their own IGF-I (Calvi et al., 2001; Kostenuik et al., 1999). The importance of IGF-I for PTH's anabolic action is dramatically indicated by the fact that "knocking out" the IGF-I gene or the gene for its receptor in mice also knocks out the ability of hPTH-(1-34) to stimulate osteoprogenitor cell proliferation and differentiation in knock-out marrow stromal cell cultures and femoral bone growth in the knock-out animals as well as the fact that PTH doesn't stimulate the proliferation of osteoblasts isolated from IGF-I-lacking mice unless the cells are given exogenous IGF-I (Miyakoshi et al., 2001; Wang et al., 2004).

But neither PTH, nor the IGF-I it induces, can stimulate osteoprogenitor proliferation or the growth of unloaded bones, such as hindlimb tibias, because *the lack of loading disables the responsiveness of the osteoprogenitor cells to IGF-I* (Bikle et al., 1994; Kostenuik et al., 1999)! In fact if this missionary IGF-I is to carry the PTH message to proliferate and migrate to osteoprogenitor cells through the cells' IGF-1 receptors, the cells' $\alpha v\beta 3$ integrin signaling tethers must be being tugged if the IGF-I receptors are to be activatable (Sakata et al., 2004). It seems that the activated integrins drive the collection of separately signaling-incompetent IGF-I receptors into signaling-competent clusters at the cell surface (Miyamoto et al., 1996; Sakate et al., 2004; Zheng and Clemmons, 1998). In other words—no loading, no integrin signaling and so no signaling competent IGF-1 receptors! As if this is not enough, stopping the waving of the cells'solitary cilial flowmeters and silencing the matricrine squealing from tugged integrins reduce the accessibility and responsiveness of osteoblast-specific genes to Cbfa-1/Runx2 and hinder classical transcription factors from stimulating MAP kinases and pumping architectural transcription factors like NMP4 into the nucleus to reconfigure and prime these key genes for responsiveness (Franceschi and Xiao, 2003; Latchman, 2004; Parvalko et al., 2003; Whitfield, 2003a). Thus the ability of IGF-I to stimulate the Ras-ERK1/2-Akt mechanism is reduced and with it the expression of the genes for alkaline phosphatase, α-I collagen and osteocalcin.

The need for loading and the gene accessibility and responsiveness it creates is also seen on the cellular level, specifically in MC3T3-E1 mouse calvarial preosteoblasts, as a reduction by simulated low or micro-gravity of the ability of hPTH-(1-34) to stimulate the expression of the c-Fos and c-Jun transcription factors (Ontiveros et al., 2002). Unloading also stresses and thus increases the incidence of apoptotic suicide in the osteoprogenitor cells (Sakata et al., 2002). But osteoprogenitors are not the only cells affected. The reader may recall from Chapter 2 that the stress of canalicular fluid stasis drives osteocytes to suicide and to release lysophosphatidylcholine which attracts hungry osteoclasts to their corpses. This means that there is an overwhelming surge of osteoclastic vultures that eliminate PTH/IGF-I targets and override any possible PTH action.

However, once again comes the terrible "BUT" word. Sakai et al. (1999) have found that unloaded tibias in the paralyzed, neurectomized hindlimbs of ddY male mice can still respond to hPTH-(1-34) as effectively as the tibias in the sham-neurectomized contralateral limbs!! Here, of course we are dealing with a particular strain of male mice, instead of Kostenuik et al.'s female rats.

As expected A. Nakakjima et al (2002) have found that the cells in the PTH-stimulated calluses of rat femoral fractures express IGF-I with the peak expression occurring between 4 and 7 days. But they have concluded that IGF-I was not responsible for the stimulation of osteoprogenitor proliferation because the surge of DNA-replicating cells expressing PCNA, the DNA polymerase-δ's processive replication factor (Whitfield and Chakravarthy, 2001), did not coincide with the surge of IGF-I expression. Unfortunately they did not look at the expression of FGF-2, which, as we are about to see, could well have been the PTH-induced missionary mediator of osteoprogenitor proliferation.

IGF-I may also mediate the seemingly strong stimulation of DNA replication in the trabecular cells and chondrocytes of the rapidly developing bones of neonatal (2-days-old) mice injected with hPTH-(28-48) (Rihani-Bisharat et al., 1998). This PTH fragment binds to PTHR1, but it cannot stimulate adenylyl cyclase or bone growth in OVXed rats, but because it has the 28-32 PKC-stimulating domain it can still stimulate PKCs (Jouishomme et al., 1992; Strein, 1994) (Fig. 20). And it is these signals that trigger IGF-I expression in the receptor-bearing still proliferatively competent late progenitors and mature, non-proliferating preosteoblasts and osteoblasts.

But cyclic AMP-stimulated IGF-I might not be as important for stimulating bone growth in humans as it seems to be in rodents (Johanson and Rosen, 1997). Human osteoblasts make much more IGF-II than IGF-I (Mohan and Baylink, 1999). Moreover, PTHs do not stimulate human osteoblasts to make IGF-II. Indeed cyclic AMP actually seems to inhibit IGF-II expression. But the PTHs and cyclic AMP do trigger some production of IGF-I in these human cells (Mohan and Baylink, 1999). However, there is something very wrong here. Eriksen et al. (2004) have reported injecting 20 or 40 µg of rhPTH-(1-34)(Forteo™)/day into women stimulated bone growth and at least doubled IGF-II expression without affecting IGF-I expression. It would be interesting to know whether osteoblasts have IGF receptors on their primary cilia to detect IGFs in their surroundings.

IGF-I collaborates with the TGF-βs that are also stimulated by the PTHs (Wu and Kumar, 2000) to strongly stimulate the procollagen 1[I] gene (Vermes et al. 2001). Therefore, these autocrine (self-stimulating)/paracrine (neighbor-stimulating) factors together vigorously drive osteoblasts' matrix production.

The cyclic AMP surges also cause osteoblasts to make two very important things besides IGF-I—FGF-2s. One of these is the long (i.e., extended), 24-kDa, NLS (nuclear localization sequence)-bearing, isoform of FGF-2 (also known as bFGF [Okada-Ban et al., 2000; Ornitz and Itoh, 2001]) (Hurley et al., 1999, 2004; Liang et al., 1999; Montero et al., 2000; Power et al., 2002; Zhang et al., 2002b) (Fig. 32). But this long FGF-2 is no roving missionary. It is an intracrine factor that stays in the cell where it operates mainly in the nucleus of rapidly proliferating cells, such as immature osteoprogenitor cells (Olsnes et al., 2003; Re, 2002a, 2002b; Re and Cook, 2005) (Fig. 32).

How does the long FGF-2 get its NLS? The FGF-2 mRNA transcript has three CUG start codons upstream from the "canonical" or conventional AUG start codon (Vagner et al., 1995). Usually a ribosome binds to the transcript's 5α cap and "sniffs" along the transcript until it finds an AUG start codon and then settles down to translate the message. But upstream from the first start codon is a so-called IRES (internal ribosome entry sequence) where the transcript plugs directly into the ribosome which then starts translating the message into the long FGF-2 without looking any further for a AUG start codon. The important point here is that between the transcript's three start sites and the downstream conventional AUG there is a NLS before the secretion signal sequence (Jans et al., 2003; Stachowski et al., 2003). So the importins-α and -β, override the secretion signal sequence, grab the long FGF-2 by its NLS and drag it into the nucleus. There it directly or indirectly stimulates the expression of a set of genes which includes the genes for Cbfa1/Runx-2, collagen-I, IGF-I, osteocalcin, and PTHR1, as well as the ribosome genes and its own *fgf-2* gene, products of which can also inhibit apoptosis (Bouche et al., 1987; Debiais et al., 2002; Olsnes et al., 2003; Re, 2002a, 2002b) (Fig. 32).

The other FGF-2 is the short (18-kDA) FGF-2, which stays mainly in the producer cell's cytoplasm and, like PTHrP, is exported as an intracrine/autocrine/paracrine factor that can activate both its own cell's and neighboring cells' receptors (Olsnes et al., 2003; Re and

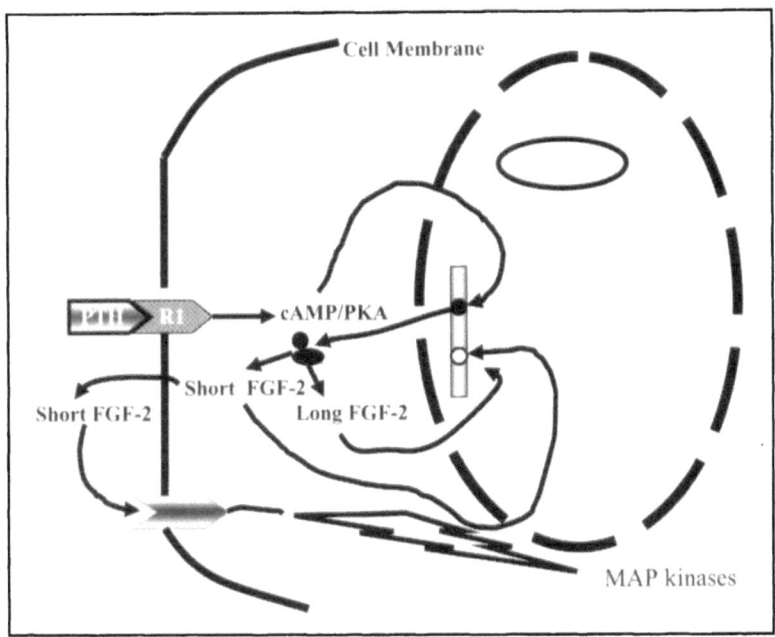

Figure 32. FGF-2 is one of the major missionary growth factors sent out from a PTH-stimulated osteo-blast to stimulate the transit-amplifying precursors that do not yet have PTHR1 receptors (see also Figs. 9 and 23).Here we see how an adenylyl cyclase-stimulating, cyclic AMP-producing PTH stimulates the expression of this factor in osteoblasts with PTHR1 receptors and how it stimulates the proliferation of transit-amplifying immature osteoblasts without PTHR1s but with FGF receptors. The cyclic AMP/ PKA signal from PTHR1 stimulates the expression of short and/or long forms of FGF-2. The long form is translated but the protein stays in the cell, and because it has a NLS (nuclear localization sequence) it moves into the nucleus where it stays and stimulates target genes. The short form is released from the cell and travels to cells with FGF receptors. There it activates the receptor that sends signals such as a burst of MAP protein kinase activity. But the short FGF-2 can also get into the target cell either by itself or with its receptor and use its NLS to get itself and its receptor into the nucleus to stimulate genes such as those for Cbfa1/Runx-2, collagen I, IFG-I, osteocalcin, and PTHR1.

Cook, 2005). The short FGF-2 is translated from the transcript's conventional downstream AUG start codon and so does not have the upstream NLS and therefore cannot be kid-napped by the importins and dragged into the nucleus. But it can still get into the nucleus because it is small enough to simply diffuse through the nuclear pores (Stachowski et al., 2003). This is the missionary FGF-2 which when secreted can activate FGF receptors on its producer cell as well as on immature (i.e., PTHR1-lacking) osteoblastic progenitors or it can ride into the cell and into the nucleus on its receptor (Fig. 32). The key point here is that the signals from the activated FGF protein kinase receptors drive the differentiation of osteoprogenitors, not necessarily by increasing the expression and production of Cbfa1/ Runx-2, but by stimulating the phosphorylation and therefore the activity of existing Cbfa1/ Runx-2 by stimulating ERK 1/2 protein kinases (Franceschi and Xiao, 2003).

The likelihood of FGF-2 being one of the prime mediators, if not *the* prime mediator, of PTH's bone-forming action is indicated by its osteogenicity in rats (e.g., Yao et al., 2004) and hPTH-(1-34)'s failure to stimulate bone formation in mice which have had their *fgf*-2 gene knocked out (Hurley et al., 2002). FGF-2 also stimulates cell migration, which means that it can lure osteoprogenitors to clusters of active FGF-2-producing osteoblasts in osteonal tunnels for example. It should be also be noted that while PTH works largely in OVXed rats by increasing trabecular thickness, injecting FGF-2 *by itself* into such animals can also increase trabecular number and connectivity probably by affecting a broader group of osteoblastic cells (Lane et al., 2003).

Therefore, PTH-induced surges of FGF-2 and IGF-I drive osteoblast accumulation and bone formation in part by stimulating the proliferation of juvenile osteoprogenitors and their ultimate differentiation into PTHR1-receptor-bearing osteoblasts with extended working lives and all thoughts of committing apoptotic suicide suppressed (Fig. 23). As would be expected if FGF-2 mediates PTH action, pretreating OVXed rats for 2 weeks with daily intravenous boluses (250 µg/kg) of FGF-2 before 8-weeks of daily subcutaneous injections (80 µg/kg) of hPTH-(1-34) and (of course a further round of FGF-2 boluses, this time from the PTH-treated cells) increases trabecular thickness and connectivity more than PTH injections alone (Iwaniec et al., 2002). This is probably due to the FGF-2 pretreatment increasing the pool of PTH-responsive osteoblastic cells.

PTHR1-induced bursts of PKA activity in UMR-106-01 rat osteosarcoma cells also rapidly (e.g., within 1 hour) and massively (e.g., by 12 times) increase the expression of the gene encoding amphiregulin (AR), a member of the EGF family without affecting the other family members (EGF itself, betacellulin, epiregulin, HB-EGF, TGF-α) of the EGF and ErbB2 receptor-activating family (Qin et al., 2003) (Fig. 23 and 30). Clearly such a potent proliferogen could collaborate with the other missionaries such as short FGF-2 to stimulate osteoprogenitor proliferation (Fig. 23 and 30).

Obviously bone formation would also be enhanced by blocking osteoclast generation, As we have seen mature osteoblasts have their RANKL genes shut off and instead make the anti-osteoclastogenic osteoprotegerin (Fig. 23 and 37). But these amazing cells, when stimulated by PTH, give osteoclastogenesis a double whammy by also making the anti-osteoclast IL-18, the gene for which is stimulated by cyclic AMP/protein kinase A signals from the activated PTHR1 receptors (Raggatt et al., 2003).

But RANKL has unexpectedly, indeed astonishingly, joined the alarmingly growing mob of factors (Fig. 19) all clamoring for recognition as principal directors of the PTHs' osteogenetic action. It has been rather astonishingly suggested that RANKL, or more precisely, oligomerized (aggregated, clustered) RANKL mediates PTH's cyclic AMP-driven anabolic action on primary mouse osteoblastic cells. Why? Well, a cluster (oligomer) of RANKLs fused to the enzyme GST (glutathione-S-transferase)as a platform stimulates bone formation when injected into mice (Faccio et al., 2003). This cluster of RANKLs bound to RANK receptors stays on the surface of murine calvarial osteoblasts much longer than single molecules of RANKL bound to RANKreceptors (Faccio et al., 2003)!!! The oligomerized RANKL stays there long enough to drive RANK receptor signaling that is sufficiently prolonged to stimulate collagen production via the NF-κB/ERK1/2 mechanism which according to Franceschi and Xiao (2003) would, among other things, phosphorylate and activate Cbfa1/Runx2. The mediation of PTH's anabolic action in mice by oligomeric RANKL•RANK complexes is also strongly indicated by the ability of osteoprotegerin to prevent PTH action (Faccio et al., 2003). But the question is what cells are stimulated by a PTH to make RANKL oligomers? Since mature osteoblasts' RANKL genes are locked shut by having CpG islands in their promoter-switch boxes methylated, the RANKL-making targets of PTH would have to be immature osteoblastic cells. At any rate, both mouse osteoblasts and human osteoblasts (Reinholz et al., 2002) appear to have the RANK receptors needed to respond to RANKL and thus counterbalance osteoclast activity.

There may be another source of missionary growth factors with which PTH boluses could drive osteoblast accumulation—none other than developing osteoclasts! Kubota et al. (2002) have found that RANKL-stimulated developing osteoclasts make PDGF (platelet-derived growth factor)-BB, which can drive FIS-1 into the nucleus (G. Candeliere unpublished observation), knock the anti-proliferative INI1 out of action and prevent MC3T3-E1 semi-transformed, transit-amplifying preosteoblasts from maturing by keeping them in the cycling mode. Therefore, the PDGF-BB from osteoclasts that have been induced by RANKL to develop from PTH-stimulated immature osteoblastic stromal cells could stimulate the proliferation of nearby transit-amplifying osteoprogenitors. But this interaction would not last very long because the stimulated osteoprogenitors would increasingly make and release OPG and turn off their RANKL genes and RANKL production which would combine to reduce osteoclast development and with it PDGF-BB production (Atkins et al., 2003; Kubota et al., 2002).

As if all of this is not enough, the reader may recall that the signals from PTH- or PTHrP-activated PTHR1 receptors probably stimulate the expression of leptin, which we met in Chapter 3 i (Torday et al., 2002) (Fig. 11). This autocrine and paracrine cytokine does a lot of things to promote bone formation when operating locally and away from the central nervous system where it could stimulate anti-osteoblast adrenergic activity (Fig. 11). It opposes osteoclast generation by stimulating osteoblasts to make the anti-osteoclastogenic OPG and reduce the pro-osteoclastogenic RANKL production by immature osteoblastic cells, and it prevents transit-amplifying osteoprogenitors from choosing the adipocyte option, stimulates osteoblast maturation, stimulates the proliferation and migration of endothelin-1-BMP-2-and BMP-7(OP-1)-secreting/osteoprogenitor-stimulating vascular endothelial cells to the bone construction site, and finally promotes matrix mineralization.

Finally, another missionary has just been identified.—the angiotensin II-like APL (apelin) and its GPCR/7 TM receptor APJ. Like leptin it is secreted by adipocytes and its secretion is stimulated by insulin (Lee et al., 2006) (Fig. 33). It is also found in regions of the cerebellum, hypothalamus, vascular endothelium, heart, lung and kidney and controls an impressively wide variety of things such as fluid homeostasis, heart musle contraction and blood pressure (Lee et al.,2006). We now know that human osteoblasts make and secrete APL and decorate their surfaces with its APJ receptor (Xie et al., 2006). It seems that APL from osteoblasts stimulates osteoprogenitors but does not affect differentiation (Xie et al., 2006) (Fig. 23).

iii f. The Important Contribution of the Extensive Lining Cell Pool to the PTHs' Osteogenicity

So far we have focused on PTHR1-bearing osteoblasts in the short-lived microcrack-repairing BMUs. But there is a much larger permanent population of mobilizable bone cells that can respond to a PTH. These other cells are the bones' osteogenic National Guard—their Militia—the extensive sheets of bone-lining cells which are retired post-osteoblasts. Indeed, the osteogenic response to PTH in rats may be due mainly to the temporary return of these thin, flat, ex-osteoblasts to plump, bone-making osteoblasts (Dobnig and Turner, 1995; Leaffer et al., 1995; Whitfield et al., 1998a) (Figs. 9 and 23). This reversion is due to the direct stimulation of the osteocytes and lining cells by PTH because although they have reduced their PTHR1 receptor expression and other traces of their past bone-making osteoblast lives, they still keep some of these receptors on their surfaces (at least in humans) (Langub et al., 2001). In other words PTH directly boots the lining cells into active osteogenicity. But to do this the peptide must help unlock the lining cells' osteogenic tool boxes by preventing connected osteocytes from making SOST (sclerostin) which has been keeping the lining cells flat and quiet (Grey and Reid, 2005; Kulkarni et al., 2005; Kusu et al., 2003; Poole and Reeve, 2005; Poole et al., 2005; Semenov et al., 2005; van Bezoojen et al., 2004, 2005; Winkler et al., 2003) (Fig. 20).

Presumably the reverting cells would raise their PTHR1 receptor display to the level in mature osteoblasts and once again start making and pumping out missionary growth factors that would expand the pool of osteoprogenitor cells, for example, in the adjacent bone marrow (Fig. 23). While most of the osteoblasts in a BMU must die when they have done their job, things are very different when the PTH injections stop. The osteocytes simply resume making SOST. The ex-lining cells relax, stop making bone, flatten out again go back to their strain-monitoring job in the osteocyte-lining cell osteointernet with their glutamate receptors poised to respond to glutamate from strained osteocytes (Dobnig and Turner, 1995; Leaffer et al., 1995) (Figs. 2 and 23). This PTH-driven recruitment of osteoblasts from the lining cell pool is likely why hPTH-(1-34) stimulates the appearance of functioning osteoblasts on *quiescent* trabecular surfaces in human bones before the appearance of osteoclasts to "prepare" the surfaces—the basis of cyclic PTH therapy discussed in the previous chapter (Hodsman and Steer, 1993).

It is very important for the reader to understand that this direct summoning by PTH of lining cells to active, osteoblastic duty flatly contradicts one of the widely held beliefs of the Bone World that osteogenesis requires a preliminary preparation of the bone surface by osteoclasts. As discussed in Chapter 2 using the "Highway Repair Analogy", the seemingly very tight osteoclast-osteoblast linkage is based on the BMUs that repair microcracks. Microcrack repair requires that osteoclasts first dig out the damaged patch and only then are osteoblasts generated in response to signals and factors from the excavation to fill the excavation with new bone. But this dependence on osteoclasts does not apply when the PTHR1 buttons on dormant post-osteoblasts such as the lining cells are pushed by a PTH (Fig. 23).

How does an osteogenic adenylyl cyclase-activating PTH turn on lining cells? There are reasons to think that expression of a lipid-modified glycoprotein called Wnt (from *wingless* the fruit fly homolog and *int* the equivalent human gene) might be the mediator (Fig. 27). A surging Wnt would cause β-catenin to accumulate in the cytoplasm and get into the nucleus to form a complex with Tcf/Lef (T-cell factor/Lymphoid Enhancer Factor) that stimulates the expression of genes, the products of which stimulate osteoblast activity (Kulkarni et al., 2005; Nusse, 2005). Wnt works by binding to its receptor called frizzled (FZD) to produce a complex with the LRP-(low density lipoprotein receptor-related protein)5/6 (He et al., 2004; Pinson et al., 2000; Westendorf et al., 2004). Before this, any excess β-catenin protein not needed to link the cell to its neighbors has been delivered to an APC (adenomatous polyposis coli) platform containing GSK-3β (glycogen synthase kinase-3β) and the axin protein where it is phosphorylated by GSK-3β and thus marked for ubiquitinylation and delivery to the cell's proteaseome shredder (reviewed in Westendorf et al., 2004; Whitfield and Chakravarthy, 2001). In other words, before the arrival of Wnt the cell has been keeping the cytoplasmic β-catenin level low enough to do its job at the cell surface helping E-cadherin keep the cell adhereing to its neighbors without "moonlighting" as a nuclear transcription factor. However if Wnt should appear, the Wnt•FZD•LRP receptor complex activates Dsh(disheveled)•GBP(GSK-binding protein) complex which grabs GSK-3β and breaks up APC•axin•. GSK-3β•β-catenin complex in the process. Then the scaffold-promoter axin is taken out of action by binding to the LRP receptor component. Now β-catenin can no longer be bound to the scaffold where it can be phosphorylated by GSK 3 and sent to the shredder. Now a free-agent, β-Catenin starts accumulating in the cytoplasm and invades the nucleus where it associates with Tcf/Lef and stimulates osteogenesis-driving genes (Nusse. 2005).

A PTH such as hPTH-(1-38) can indeed potentiate this Wnt-driven mechanism in rat bone and cultured UMR 106 rat osteosarcoma cells by stimulating the expression of the osteoblastic FZD and LRP-6 genes, i.e., to make more receptor complexes, and also to make these complexes more effective by downregulating the gene for Dkk-1 (Dickkopf-1) which would otherwise collaborate with Kremen-1 to cause the endocytosis and lysosomal destruction of LRP-5/6 (Kulkarni et al., 2005; Nusse, 2001,2005; Westendorf et al., 2004) (Fig. 27).

But do osteogenic PTHs really have to stimulate Wnt expression to trigger the β-catenin mechanism? Not at all!! They can bypass Wnt entirely. In fact there are at least two complementary ways by which the PTHs can bypass Wnt and stimulate the β-catenin-dependent osteogenicity. First, Tobimatsu et al (2006), using MC3T3-murine preosteoblasts, have reported that hPTH-(1-34) activates the β-catenin-elevating Smad3 signal transducer within 1 hour (Fig. 26). Second, bursts of cyclic AMP-dependent protein kinase activity such as those from the PTHR1 receptor-activated adenylyl cyclase prevents the surging β-catenin from from being ubiquitinylated and shredded by phosphorylating its $Ser^{552}$and Ser^{675} residues (Hino et al., 2005; Taurin et al., 2006) (Fig. 27).

No matter how the PTHs turn on the β-catenin mechanism, nothing will happen if the osteocyte-produced SOST brakes are still on (Keller and Kneissel, 2005; Poole and Reeve, 2005; Poole et al., 2005; Sevetson et al., 2004; Sutherland et al., 2004; van Bezooijen et al., 2004, 2005) (Figs. 9, 23 and 27)! But the PTHs also look after this problem by using the cyclicAMP and cyclic AMP-dependent protein kinase from the osteocytes' PTHR1s to shut down the osteo-cytes' Cbfa1/Runx-2-driven SOST genes (Bellido et al., 2005; Keller and Kneissel, 2005; Sevetson et al., 2004) (Fig. 27). Another connection between SOST and the osteogenesis-promoting Wnt/FZD/ β-catenin mechanism is SOST's ability to bind to the LRPs and block the Wnt/FZD/β-catenin mechanism (Li, X. et al., 2005). But of course this may be, indeed probably is, irrelevant if the PTHs do not trigger the β-catenin-Lcf/Lef mechanism via Wnt.

iii g. Matrix Mineralization—The Grand Finale

The main foundation of bone is its collagen I matrix. Matrix production starts with the production of collagen α1 and α2 propeptides. The propeptides are hydroxylated by the oxygen-requiring procollagen prolyl 4 hydroxylase to form stable triple helices. Then certain lysines are hydroxylated in preparation for secretion (Uzawa et al., 1999). The propeptides are then snipped off and the triple helices are rendered insoluble and assemble spontaneously into still-unlinked fibrils. Finally lysl oxidase arrives on the scene to produce the covalent cross links that tie the fibrils together. Now the new matrix is ready to be made into bone by mineralization.

Inorganic phosphate and the mature osteoblast's Na^+gradient-driven, type III Pit1phosphate transporter that ships it into the cell are key players in bone formation (Caverzasio and Bonjour, 1996; Nielsen et al., 2001; Selz et al., 1989; Takeda et al., 1999. Moreover, we are now learning that inorganic phosphate is a major signaler, an activator of certain gene promoters (Conrads et al., 2005; Beck, 2003; Beck and Knecht, 2003; Beck et al., 2003). The PTHrP-triggered signals from the emerging PTHR1 receptors on the young matrix-making cells drive Cbfa1/ Runx-2 into the nucleus to stimulate the *alp* (alkaline phosphatase) gene, the product of which is a Ca^{2+}-binding enzyme that is plugged into the cell membrane with its catalytic subunit (i.e., its "business end") sticking out into the extra-cellular space (Balcerzak et al., 2003; Beck and Knecht, 2003; Beck, 2003; Beck et al., 1998, 2000; 2003; Mornet et al., 2001). But the cell won't immediately start mineralizing matrix because the Cbfa1/Runx-2 pushed into the nucleus by PTHR1 signaling (Fujita et al., 2001a,b) stimulates the expression of the mineralization-blocking osteocalcin (Ducy et al., 1996) (Fig. 28). The ALP then hydrolyzes pyrophosphate released from nucleotide triphosphates and snips phosphate off of various external substrates such as the β-glycerophosphate that is added to culture media for bone nodule formation and in the Real World snips phosphate out of bone matrix components (Balcerzak et al., 2003). As the maturing osteoblast nears the mineralizaton threshold, it upregulates the expression of SCT1 (stanniocalcin 1) and the gene for the Pit1 isoform of type III Na-P_i transpoprter to carry phosphate into the cell (Beck et al., 2003; Yoshiko et al., 2003) (Fig. 29). The SCT1-driven phosphate uptake stimu-lates the expression of a group of genes whose products drive the transition from matrix-making to mineralization (Beck and Knecht, 2003; Beck et al., 2003). First, the incoming phosphate downregulates the expression of the major matrix components, collagen-1α1 and collagen-1α2, because the emphasis must now shift to mineralization—enough collagen has been made (Beck and Knecht, 2003; Beck et al., 2003). The incoming phosphate also stimulates the gene for Nrf2, the transcription factor that homes on the antioxidant response element (ARE) and the battery of genes that have it in their promoter-switch boxes (Nguyen et al., 2003). Presumably the Nrf2/ARE-driven genes and their products will help the cell survive the formidable challenges of a lot of phosphate and a blebbing membrane (Beck et al., 2003). Of course the phosphate also stimulates the expression of calcyclin and annexin V which, as we shall see in the next paragraph, are needed for mineral making (Beck et al., 2003). And the phosphate also stimulates the gene for cyclin D1 which has some role in differentiating osteoblasts besides its more familiar role of triggering the build-up to DNA replication in proliferatively competent cells (which the mature osteoblast most certainly is *not*). However, when the Pit-1-pumped phosphate reaches a critical level it stimulates the nuclear export machinery, which empties the nucleus of its Cbfa1/Runx-2 (Fujita et al., 2001a,b) (Fig. 29). This dumping of Cbfa1/ Runx2 out of the nucleus shuts down osteocalcin expression (Fujita et al., 2001b), which lifts the block to mineralization (Ducy et al., 1996; Fujita et al., 2001a) (Fig. 29). (Of course the block will have to be reimposed later on to avoid runaway mineralization.) Again one must remember that timing is everything.

To mineralize the new matrix, the osteoblast loads Pit transporters, alkaline phosphatase and annexin V(5) Ca^{2+} channels into pre-vesicle patches in their cell membrane that will be budded off to make vesicles that are detached calcium-apatite-making nano-machines (Balcerzak et al., 2003). To start out, the large amount of CaR-activating Ca^{2+} released by the osteoclasts is sucked up and the osteoblasts' CaRs are silenced when the cells combine the Ca^{2+} with phosphate to mineralize the osteoid matrix they have just finished making. The Cbfa1/Runx-2-stimulated alkaline phosphatase and Pit1 transporter (the gene for which has been stimulated by incoming phosphate) are now plugged into the pre-vesicle patches (Beck, 2003). But it seems that the transporter's operation must be protected by the cell's PHEX endopeptidase which would destroy any transport-inhibiting FGF-23 that the cell might be making or be exposed to (Bowe et al., 2001; Shih et al., 2002). The new osteoid is mineralized when alkaline phosphatase-generated, Pit1 transporter-delivered phosphate and Ca^{2+} flowing through the channels formed in the vesicle walls by AnxV(5), the gene for which has also been stimulated by incoming phosphate, combines to form needles of impure hydroxyapatite ($Ca_{10}[PO_4]_6[OH]_2$) in the matrix-bound vesicles (Bandorowicz-Pikula et al., 2001; Beck, 2003; Caverzasio and Bonjour, 1996; Hoshi and Ozawa, 2000; Selz et al., 1989). These vesicles armed with alkaline phosphatase, the Pit1 Na-P$_i$ transpoprter, and annexin V(5) channels in their walls are now budded from the surfaces of the osteoblasts and when they have loaded themselves to bursting with calcium apatitie needles, they dump the needles onto the osteoid (Caverzasio and Bonjour, 1996; Nielsen et al., 2001). Since PTH-induced cyclic AMP transients *stimulate* type III transporters, well-separated boluses of adenylyl cyclase-stimulating PTHs promote mineralization by selectively increasing the transporter's V_{max} (its top carrying speed), which persists even after the vesicles bud from the osteoblast's surface (Caverzasio and Bonjour, 1996).

At this point we might ask what happens to the PTHR1 receptors when the mineralizing osteoblast's surface is budding off vesicles. Are they budded off along with the vesicles? We don't know. Certainly it appears that the post-mineralization osteocytes have downregulated their PTHR1 receptor expression and maybe this is how they do it.

If an osteoblast is lucky enough to survive surging Ca^{2+} and phosphate and a roiling membrane to become an osteocyte, the incoming phosphate will have triggered a second burst of osteopontin gene expression and osteopontin synthesis to glue the cell and its processes to the walls of its cubicle and its dendrites to the walls of their canaliculi (Beck and Knecht, 2003; Beck et al., 1998, 2000, 2003; Denhardt and Noda, 1998; Denhardt et al., 2001; McKee and Nanci, 1996). The emerging osteopontin also prevents the incoming phosphate from inducing apoptosis (Koyama et al., 2003).

The incoming phosphate stimulates osteopontin expression by somehow causing the biphasic activation (by phosphorylation) specifically of ERK 1/2 (Beck and Knecht, 2003). In other words, there is an early surge of phosphorylated ERK 1/2 between 15 and 45 minutes after adding 10mM phosphate to the MC3T3-E1 cells' culture medium and a second surge beginning at 8 hours and building up at least until 32 hours. Other members of the MAP kinase family such as p38 are not involved, but a separate stimulation of PKCs activity also seems to be needed. As usual—complexity, complexity!!

iii h. PTHs and Mechanosensitive Ca^{2+} Channels

Again we turn to to the seemingly intimate relation between PTH and strain respon-siveness. PTH might also sensitize the strain-responsive osteocytes, (which still express PTHR1 receptors, though at much lower levels than when they were osteoblasts (Aubin, 2000; Langub et al., 2001) to strain (Duncan et al., 1992). Thus, adenylyl cyclase-stimulating PTHs phosphorylate the mechanosensitive Ca^{2+} channels (MSCCs) of MC3T3-E1 mouse preosteoblasts and UMR 106.01 rat osteoblasts by cyclic AMP-stimulated PKA kinase (Duncan and Misler, 1989; Duncan et al., 1992; Ryder and Duncan, 2001). This causes the MSCCs to open wider (increase their conductance) and stay open longer when they are pulled and stretched by shear stress. However, at high levels of shearing both the adenylyl cyclase and PLC pathways may be required to increase the MSCCs'responsiveness (Ryder and Duncan, 2001). A need for PTH to maintain MSCC responsiveness to strain could be why thyroparathyroidectomy has been reported to eliminate the responsiveness of rat bones to mechanical stress (Chow et al., 1998). This shearing also stimulates the cells to pump out ATP and UTP, which by stimulating their P2 nucleotide receptors, sensitize the cells to signals from PTHR1 receptors (Bowler et al., 2001). Since these nucleotides are very short-lived, they accumulate at high levels only around the producing cells—they are autocrine (self-stimulating) but not paracrine (neighbor-stimulating) (Bowler et al., 2001). Thus they selectively focus and magnify PTH's action on the strained cells and their MSCCs. A practical use of this sensitization of MSCCs by PTH would be to use the peptide along with exercise therapy programs to increase bone formation upon return to macro-gravity or remobilization after prolonged inactivity (Chow et al., 1998; Ma et al., 1999).

iii i. PTHs and Intraosseus Blood Circulation

Obviously the amount of blood flowing through a bone is a very important factor con-trolling its growth. And PTHs can considerably affect this flow. First, although vascular muscle cells generally have PTH receptors (Whitfield et al., 1997a), bone vascular endothelial cells, unlike other endothelial cells, also have PTH receptors on their surfaces (Brandi and Collin-Osdoby, 2006; Streeten and Bondi, 1990). Then, it is likely that boluses of AC-stimulating PTHs and N-terminal PTHrP fragments can cause vasodilation and pulses of blood through the bones (Laroche, 2001; Whitfield et al., 1997e). And such pulses should contribute to the stimulation of bone formation because according to Reeve et al. (1988) the bone-layering rate of osteoblasts is directly proportional to the skeletal blood flow.

iii j. PTHs, Hematopoiesis and the Treatment of Chemotherapeutic Marrow Damage

A life-threatening consequence of chemo- and radiotherapies is the destruction of he-matopoietic stem cells, the depletion or even "emptying" of a cancer patient's bone marrow. Thus, a drug that builds hematopoietic stem cell niches and promotes the engraftment of stem cells harvested from peripheral circulating blood would be a valuable addition to a cancer therapist's tool box. During the last 49 years evidence has been accumulating in fits and starts for the parathyroid hormone (PTH) being just such a marrow-stimulating tool.

The story began in 1958 when Rixon et al (1958) reported that the only commercial source of PTH at that time, Eli Lilly's PTE (bovine parathyroid extract), surprisingly and significantly increased the survival of X-irradiated rats. The reader might recall that this was the same PTE that H. Selye (1932) had used 26 years earlier for the first demonstration of PTH's potent ability to stimulate osteogenesis which laid the foundation of today's most effective treatment for established osteoporosis. Three years later, Rixon and Whitfield (1961), reported the results of a further 59 experiments using 1296, 300-g male hooded rats. They found that only one dorso-thoracic subcutaneous injection of PTE containing 50-200 USP units of PTH activity at any time between 18 hours before and 3 hours after irradiation with 7.0-8.5 Gy (700-850 rads) of 2 MeV X-rays significantly increased the number of animals surviving 30 days later. For example, 33% of vehicle-treated rats irradiated with 8.0 Gy were still living, but in strikingly poor condition, 30 days later while 73% of rats that had received PTE with 200 USP PTH units 5 minutes after irradiation were alive and in excellent condition by 30 days (p <0.0001). But injecting as much as 200 units of PTH activity at 5 hours after irradiation could no longer affect the 30-day survival.

The fraction of rats that were still living 30 days after receiving an X-ray dose between 7.0 and 8.5 Gy would have depended on the fraction of their hematopoietic stem cells and the stem cells' transit-amplifying progeny that survived the irradiation and repopulated the bone marrow above the threshold for survival. Then Perris et al (1967) showed that injecting 200 USP units of the PTE immediately after irradiation of rats with 3 Gy of γ-radiation from ^{60}Co did significantly increase the flow of femoral bone marrow cells into mitosis starting as soon as 4 hours later. Therefore, PTE's survival-enhancing action in the irradiated rats was due to it somehow stimulating a timely supra-survival threshold repopulation of their radiation-depleted bone marrows. But was it the PTH activity in the PTE extract that saved the irradiated rats in 1958 and 1961?

There was no pure bovine PTH in the early 1960s with which to answer this question. The now universally used osteogenic hPTH-(1-34)OH (teriparatide; Lilly's Forteo™) did not appear until the early 1970s (Potts et al., 1971; Tregear et al., 1974). So we had to use the animals' own PTH. We did this by injecting the Ca^{2+}-chelating EDTA or the Ca^{2+}-binding Na-caseinate, either of which would lower the circulating free (ionic) Ca^{2+} concentration. This circulating Ca^{2+} drop would silence the parathyroid cells' Ca^{2+}-monitoring CaRs (Ca^{2+}-sensing receptors (Mithal and Brown, 2003; Nemeth et al., 2001) and cause the cells to fire a bolus of PTH into the circulation *within seconds* to restore the normal circulating Ca^{2+} concentration and the signaling from the CaRs.

In 1967 Perris and Whitfield reported that intraperitoneally injecting 125-g male Sprague-Dawley rats with EDTA almost halved the circulating free Ca^{2+} concentration within 10 minutes and significantly increased the flow of femoral bone marrow cells into mitosis by 4 hours. Then Rixon and Whitfield (1969) showed that injecting enough Na-caseinate to halve the circulating free Ca^{2+} concentration in normal female CF_1 mice doubled the proliferation of their femoral bone marrow cells and tripled the proliferative activity in the femoral bone marrows of mice irradiated with 6.0 Gy (600 rads) of 300 kVp X-rays. Moreover the Na-caseinate significantly increased the 30-day survival of the irradiated mice from 45% to 80 %. Na-caseinate also stimulated femoral bone marrow cell proliferation in normal, but *not* TPTXed (thyroparathyroidectomized), male Sprague-Dawley rats, which indicated the mediation of the proliferative response by endogenous PTH (Rixon and Whitfield, 1969).

But TPTX removes calcitonin and thyroid hormone. So Rixon and Whitfield (1972) showed that removing just the parathyroid glands (PTX) or the whole thyroid-parathyroid complex (TPTX) from male Sprague-Dawley rats greatly reduced the mitotic activity in the femoral bone marrow. This drop was followed within 8 days by a 40% drop in the marrow's total nucleated cell population which included a dramatic (ca. 70%) reduction of the erythroid and lymphoid subpopulations. As expected, the dramatic erythroid hypoplasia in the bone marrows of PTX and TPTX rats was accompanied by a 68 % drop in the reticulocyte fraction of the non-nucleated marrow cell population, an equivalent reduction of ^{59}Fe incorporation into peripheral erythrocytes and an increase in the post-hemorrhage hematocrit restoration time from the normal 5 days to 9 days [Perris and Whitfield, 1971; Perris et al., 1971). Erythropoiesis in these parathyroidectomized rats and the speed of hematocrit restoration after hemorrhage were restored to the normal values by daily subcutaneous injections of purified native bovine PTH purchased from the now long-gone Wilson Laboratories in Chicago. It should also be added in passing that PTX reduced, and the purified bovine PTH restored, the primary immune response of rats to injection of sheep erythrocytes (Swierenga et al., 1976). Thus, it was the PTH activity in the Lilly PTE that had saved Rixon et al's heavily X-irradiated rats and that PTH somehow controls hematopoiesis in mice and rats. But did the hormone act directly on hematopoietic progenitors or did it work only indirectly by raising the blood Ca^{2+} concentration?

It was Gallien-Lartigue and Carrez (1974) who partly answered this question. To do this they used Till and McCulloch (1961)'s spleen colony assay in which mouse femoral bone marrow is removed, its cells are suspended in a suitable medium and an appropriate number of them are then injected into lethally irradiated mice. Ten days later the numbers of colonies with variously differentiated cells that were formed in the irradiated animals' spleens by pluripotent hematopoietic stem cells and early-stage precursor cells in the injected marrow suspension are counted (Gallien-Lartigue and Carrez, 1974; Stem Cells, 2001; Till and McCulloch, 1961). Gallien-Lartigue and Carrez also used the "thymidine suicide" technique (Becker et al., 1965; Byron, 1977; Schofield, 1978, 1979) to find out whether the pure native bovine PTH they obtained from Calbiochem could directly stimulate CFU-S cell proliferation. They incubated the suspended donors' femoral bone marrow cells for 30 minutes in medium containing a relatively high radioactivity (e.g., 200 µCi [7.4 M Bq]/ml) from high-specific activity ^{3}H-thymidine and then determined the number of colonies in the spleens of the recipient irradiated animals. If the suspended CFU-S cells were stimulated to make DNA by the PTH, the number of spleen colonies in the recipient animals' spleens would have been lower than the number produced in the recipient spleens by injecting untreated cells because the DNA-synthesizing cells would have incorporated ^{3}H-thymidine from the medium into their DNA and been killed by the β-particles from the ^{3}H.

PTH did indeed stimulate CFU-S cells in the donor bone narrow to suicidally start replicating DNA, and it did so within just a couple of hours as indicated by the significantly fewer colonies that appeared in the spleens of the lethally irradiated mice injected with PTH-treated donor marrow cell suspensions instead of untreated marrow cell suspensions (Gallien-Lartigue and Carrez, 1974). In fact these colony-forming cells responded to the PTH as fast as the flow of bone marrow cells into mitosis in PTH-treated rats. Thus, PTH could at least stimulate the initiation of DNA replication by a population of cycling CFU-S cells that had exited the primitive quiescent, G_0, niche-bound LTR ("long-term repopulating")-HSC cell state and paused in a late G_1 (i.e., prereplicative) state(Gallien-Lartigue and Carrez, 1974) (Fig. 33). Therefore, these CFU-S progenitor cells had PTH receptors and could promptly start replicating DNA, enter mitosis and presumably start generating transit-amplifying and ultimately diverse terminally differentiated progeny in response to signals from these receptors (Fig. 33).

What are these PTH-responsive CFU-S cells? They are pluripotent, but they have only a limited self-renewing potential (Schofield, 1978). This is not "stemness". True stem cells nestle in their niches holding onto their proliferatively quiescent G_0 state in order not to lose their self-renewing and marrow-repopulating potential with unnecessary cycling (Fig. 33). When a niche-bound stem cell does receive a signal to initiate and complete a growth-division cycle, one daughter cell stays in the niche and lapses back into a G_0 state while the other leaves home. If it cannot find another niche, it will keep its cell cycle engine running as a CFU-S cell, the rapidly proliferating, transient amplifying progeny of which lose a fraction of stemness with each round of replication. After a limited number of rounds of replication the CFU-S progeny move along trails signposted on the marrow stroma with appropriate chemokines, cytokines and adhesives to find vascular endothelial cell niches where they terminally mature and from which they enter the blood (Kopp et al., 2005) (Fig. 33).

Since injecting CFU-S cells can save lethally irradiated mice (Nakorn et al., 2002; Schofield, 1979), increasing the size of the quiescent LTR-HSC pool, increasing the proliferation and preventing the apoptosis of their CFU-S transit-amplifying progeny were ways by which the PTH activity in the Lilly PTE could have saved Rixon et al's irradiated rats. Obviously by 1974 it had become very important to locate the LTR-HSC niches in the bones and find out how PTH might control their maintenance and numbers in these niches. But 30 years elapsed before we began finding out.

For the last 30 or so years it was suspected that the LTR-HSC cells' bone marrow "niches" were on the inner surfaces of bones instead of in the more central regions of the marrow (Kopp et al., 2005; Lord, 1990; Lord et al., 1975; Mason et al., 1989; Schofield, 1978, 1979; Visnjic et al., 2004). The bone marrow is a good place for a niche because it is armoured with cortical bone and relatively anoxic which means low levels of dangerously mutagenic reactive oxygen species (Lennon et al., 2001). Thus, during embryonic development, the first hematopoietic colonies appear next to the endosteum and that is where hematopoietic foci first reappear in irradiated dogs (Fliedner et al., 2002). And Visnjic et al. (2004) vividly demonstrated the coupling of bone marrow hematopoiesis to osteoblasts using transgenic mice with the ganciclovir-inhibitable herpes virus thymidine kinase (HVTK) gene under the control of a fragment of the collagen gene's Col $\alpha 1$ type 1 promoter (Col2.3δ TK). In these animals only the collagen $\alpha 1$-expressing osteoblasts express the Col $\alpha 1$ type 1 promoter-driven HVTK during their maturation and are thus selectively slaughtered by ganciclovir. The selective ganciclovir-induced massacre of maturing osteoblasts caused a progressive loss of bone, a loss of bone marrow cellularity and early hematopoietic progenitors, and shifting of hematopoiesis from the adult bone sites to the former fetal sites in liver and spleen. But when the osteoblast-killing ganciclovir treatment was stopped, hematopoiesis returned to the bones as osteoblasts reappeared along with pockets of hematopoiesis at sites of new bone formation by the returning osteoblasts.

We now know that the mouse bone marrow LTR-HSC cells' niches are on the trabecular bone surfaces covered with retired osteoblasts—the SOST-restrained, but PTH-responsive, bone-lining cells discussed above (Yin and Li, 2006) (Fig. 33). The LTR-HSC cells are also equipped with CaRs, the signals from which are essential for attaching the cells to the lining cells' collagen I and their docking in the niche (Adams et al., 2006). It appears that the LTR-HSC cells are attracted to remodeling sites where osteoclasts have been releasing large amounts of Ca^{2+} that serve as chemokines which attract the cells by the intesity of signaling from their CaRs (Olszak et al., 2000). Presumably the CaR signals stimulate the movement of the cells as well as inducing them to express collagen-binding receptors on their surfaces.

Once they haved lodged in the the niche, the LTR-HSC cells are connected to the bone-lining cells by a dense forest of receptor•ligand complexes, signaling wires, matrix-anchored cables and matrix components such as collagen and osteopontin (Arai et al., 2004,2005; Haylock and Nilsson, 2005; Moore and Lemischka, 2004, 2006; Naveiras and Daley, 2006; Stier et al., 2005; Yin and Li, 2006; Zhang et al., 2003; Zhu and Emerson, 2004), only a few of which could be included in Figure 33. The osteoblastic lining cells make Ang-1 (Angiopoietin-1) which in turn stimulates the HSCs to make Ang-1's Tie-2 tyrosine kinase receptor which "ties" the HSCs to the trabecular bone-lining cells [Arai et al., 2004, 2005; Moore and Lemischka, 2004) (Fig. 33). The signals sent into the LTR-HSC cells from the Ang-1•Tie-2 complex do an extremely important thing for the maintenance of LTR-"stemness". As with epidermal basal keratinocyte stem cells (Whitfield, 2004), the HSCs' self-renewing LTR-"stemness" and the associated proliferative quiescence in a G_0 state depend in part on matrix adhesion-stimulated signaling from $\beta 1$-integrins which are lost when

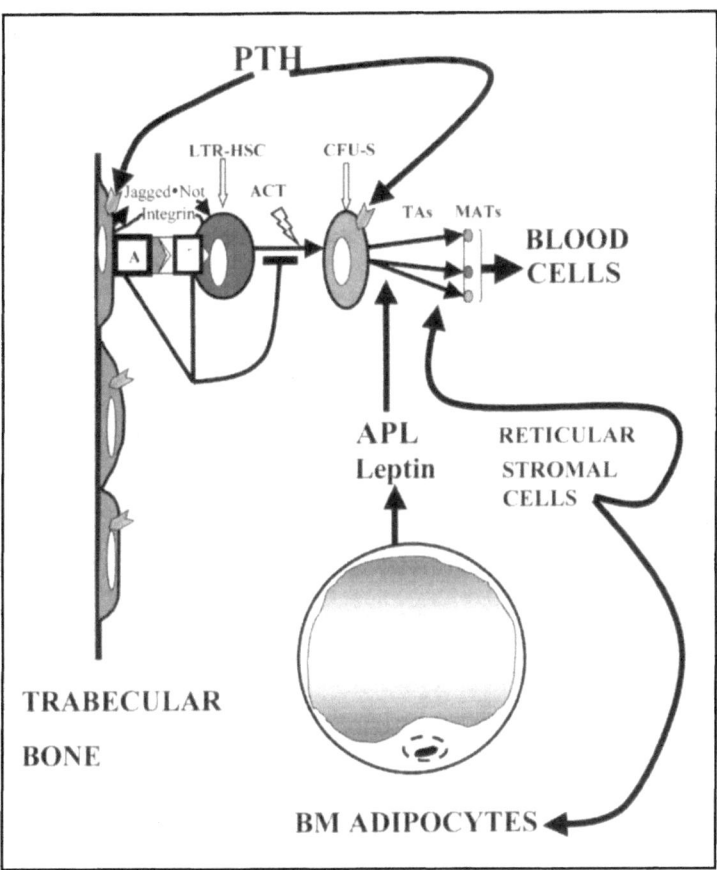

Figure 33. Post-osteoblast bone–lining cells have PTHR1 receptors and they form the anchorages (niches) on trabecular bone (Figs. 1 and 16A,B) for quiescent, long-term repopulating (LTR) hematopoietic stem cells (HSCs). Like boats tied to a dock, the HSCs are tied to trabecular bone-lining cells by a forest of connected restraining signaling "wires" among which are Ang-1(A)-Tie-2(T), Jagged 1-Notch 1 (Not) and Integrins that keep the HSCs fron losing their "stemness" by proliferating and generating the various transit-amplifying (TA)precursors of the various kinds of blood cell. When activated (ACT) by precursor pools depletion, the HSCs unplug themselves from their osteoblastic niche, and change into CFU-S cells which then generate the transit amplifying (TA) precursors. The differentiation of these precursors into various mature (MAT) blood cells is promoted by leptin and maybe apelin (APL) from the bone marrow's (BM) adipocytes. Adenylyl cyclase-stimulating PTHs cause the bone-lining dock cells to make the Jagged 1 which by binding to the HSC's Notch 1(Not) keeps the HSCs proliferatively quiet and expands the HSC pool and with this an increased potential for restoring bone marrow when it is depleted by chemotherapeutic drugs or ionizing radiation.

the LTR-HSCs start cycling and disconnecting their various tethers to the lining cells. Repeated cycling would deplete the pool of LTR-HSC cells which would irreversibly lose their abilites to renter the G_0 quiescent state and to self-renew when starting to cycle (Glimm et al.,2000). So the signals from Ang-1•Tie-2 complexes prevent the HSCs from cycling, keep them sticking to the bone surface, keep their β1-integrin signaling from fading, keep them resistant to apoptosis without losing the ability to be activated and generate rapidly proliferating transit-amplifying progenitors when needed (Arai et al., 2004, 2005) (Fig. 31).

Although the signals from lining cell-Ang-1•HSC Tie-2 complexes and β1-integrins are major keys to the maintenance of the LTR-HSC pool, there are several other important players in LTR-HSC control. There is the HSCs' Wnt-activated Frizzled /LRP (low density lipoprotein receptor-related protein)5/6 receptor mechanism (Brandon et al., 2000). As we learned above when discussing Wnt signaling and the SOST-suppressor in lining cells the signals triggered by the Wnt glycoprotein stop cytoplasmic β-catenin from being phosphorylated, ubiquitinylated, and dumped into the proteasome for proteolysis (Westendorf et al., 2004). This enables β-catenin to build up and enter the nucleus where it combines with LEF/TCF protein to form a transcription factor that affects the expression of several genes, the products of which contribute to the maintenance of the LTR-HSC pool by promoting LTR-HSC self-renewal rather than LTR-HSC-depleting conversion to transit-amplifying progenitors and terminal differentiation (Brandon et al., 2000; Dazzi et al., 2005; Reya et al., 2003; Staal et al., 2005; Van Den Berg et al., 1998; Willert et al., 2003). The accumulating cytoplasmic β-catenin also contributes to the mooring of HSCs to the lining cells by binding to the bone-lining cells' N-cadherins [34]. But, as we shall see further on, the niche has yet another important pair of denizens—Notch1 and its activator Jagged1 which collaborate with the Wnt receptor-driven mechanism in maintaining the LTR-HSC cells' self-renewing "stemness" (Duncan et al., 2005).

But how might a PTH drive such a complex system? The native PTH and its 31-34-amino acid N-terminal fragments strongly stimulate the reversion of lining cells to functioning osteoblasts that can make trabecular bone and thus increase the LTR-HSC niche space provided, of course, that they do not excessively stimulate trabecular bone growth and occlude marrow space (e.g., Fig 33). Since the size of the LTR-HSC pool depends on the available trabecular bone niche space (Zhang et al., 2003; Zhu and Emerson, 2004), the osteogenic PTHs should increase the LTR-HSC pool size and with this the effectiveness of the response to bone marrow injury. Since proliferatively quiescent (i.e., G_0-phase), Ang-1•Tie-2-restrained LTR-HSCs' have been shown to resist apoptogenesis, and since intravenously injecting Ang-1protein or *ang-1* gene-bearing adenovirus into mice increases the total survival of 5-flurouracil-treated mice from 0 to 40% and to prolong the time required for x-irradiated mice to die by about 30% (Arai et al., 2004), a PTH-induced increase of the apoptosis-resistant LTR-HSC pool and thus the mobilizable reserve of transit-amplifying progenitors should also increase the resistance of an animal or human to myelosuppressive drugs and ionizing radiation. The PTHs might also promote the attachment of LTR-HSC cells to the expanding niches by stimulating the bone-lining cells to make more N-cadherin for binding to the LTH-HSC cells' β-catenin (Marie, 2002; Suva et al., 1994).

Then there is the Jagged-Notch 1 couple (Fig. 33). Weber et al. (2006) have shown that the adenylyl cyclase plus PKCs-stimulating hPTH-(1-34) (Forteo™), the adenylyl cyclase-stimulating hPTH-(1-31) or forskolin, but *not* the PKCs-stimulating TPA or hPTH-(13-34), induce osteoblasts to make Jagged 1. Calvi et al. (2003) have shown that injecting hPTH-(1-34) into mice, and therefore triggering adenylyl cyclase activity, or producing mice with osteoblasts expressing permanently switched-on mutant PTHR1 receptors (which do not need a PTH to activate them) enlarges the LTR-HSC pool. And it does this by causing the bone-lining cells to make large amounts of Jagged 1 which binds to and activates the cleavage of the attached LTR-HSC cells' Notch 1 receptors by ADAM metalloproteinases and γ-secretases. The released intracellular ICN portions of Notch 1receptors then move from the cell surface into nucleus. There they combine with the DNA-binding CSL/RBP-J protein to form the ICN• CSL/RBP-J gene transactivator (Suzuki and Chiba, 2005). It seems that PTH-enhanced Jagged 1•Notch 1 signaling will collaborate with the Ang-1•Tie-2 signals and the Wnt cascade to give LTR-HSCs a complete stemness maintenance package by stimulating β1 integrin and

N-cadherin expression, cell cycle suppression and nuclear β-catenin•Lef/Tcf and ICN•CSL/ RBP-J transactivators to suppress the expression of transit-amplifying progenitors' gene expressions (Calvi et al., 2003; Duncan et al., 2005; Suzuki and Chiba, 2005; van Es et al., 2005).

But the LTR-HSC pool size and hematopoiesis in mice can be oppositely and dramatically affected by continuously stimulating osteoblastic cells by arming them with a mutant, constitutively (i.e., ligand-independent) activated PTHR1 receptor, the expression of the gene for which is limited to osteoblasts by attaching it to a collagen I A1 promoter-switch box (Calvi et al., 2001; Kuznetsov et al., 2004). The demand for osteoblast precursors caused by the maturation-driving signaling from these PTHR1 receptors on collagen-expressing preosteoblasts and osteoblasts diverts the flow of the progeny of multipotent mesenchymal skeletal stem cells (Fig. 23) from the hematopoiesis-supporting, leptin-making adipocytes (Bennett et al., 1996; Corre et al., 2004; Fantuzzi and Faggioni, 2000; Fietta, 2005; Gainsford and Alexander, 1999; Gimble et al., 2006; Laharrague et al., 1998, 2000) and other stromal cells in the inter-trabecular marrow cavities. As expected from the continuous cyclic AMP-dependent Jagged-Notch 1 signaling and the massive production of trabecular bone with lining cell-covered HSC niche space causes the LTR-HSC pool to increase. But, hematopoiesis from the transit-amplifying progeny of the CFU-Ss is hobbled because of the lack of differentiation-supporting stromal cells and marrow cavities to accommodate them. In other words, the LTR-HSC/CFU-S pools have a much reduced demand for progenitor production. Eventually, the drain on the skeletal stem cell population reaches a point where the overproduction of osteoblastic cells fades, other stromal cells appear and marrow cavities appear and hemopoiesis starts or increases.

Finally, as found by Gallien-Lartigue and Carrez (1974), CFU-S cells have PTH receptors and the adenylyl cyclase-stimulating native PTH can directly stimulate a late-G_1 population of CFU-S cells to start cycling (Fig. 33). Moreover, another adenylyl cyclase stimulator, the β-adrenergic agonist isoproterenol and dibutyryl-cyclic AMP itself also stimulate mitotic activity in rat bone marrow and the proliferation of murine CFU-S cells (Byron, 1977; Rixon and Whitfield, 1970).

It has likely become obvious to the reader that the PTHs could be very valuable tools to put in the cancer therapist's tool box. They should be able to promote the repopulation by HSCs of cancer patients' bone marrows that have been emptied by ionizing radiation or cytotoxic chemotherapeutic drugs. If so, they would enhance blood-harvested HSC engraftment and bone marrow repopulation in at least three ways: first by making more trabecular bone to accommodate more LTR-HSCs; second, by expanding transplanted LTR-HSC pools after the injected HSCs have set up their niches with bone-lining cells; and third by stimulating transit-amplifying progenitor proliferation until at least an adequately functional bone marrow has been established and steady state feedback controls have been established. A tantalizing glimpse of the PTHs' potential for enhancing transplant engraftment and repopulating bone marrow that validates Rixon's and my ancient discovery has been provided by Calvi et al. (2003). They lethally x-irradiated mice and then injected just enough bone marrow cells from normal donor animals to allow 27% of the irradiated animals to survive for at least 28 days. But when the irradiated animals also received daily injections of hPTH-(1-34) (Forteo™), *all* of them survived with larger loads of transplanted marrow cells in their hind limbs.

Of course there is a lot more to learn about how the PTHs interact with the complex tangle of signal wires connecting HSC cells to the bone-lining cells in their niches. We must also find out whether the several PTH peptides either on the market or nearing the market, which are known to be clinically safe at an osteogenic dose for osteoporotic humans, can also promote the engraftment of peripherally harvested HSCs and the subsequent bone marrow repopulation in rats, monkeys and ultimately cancer patients as effectively as at least one of them, Forteo™, appears to do in mice.

iii k. But Are We Building Cyclic AMP Sand Castles?

So far the *assumption* has been that because only PTH peptides with intact N-terminals can stimulate bone formation in rats, cyclic AMP must be the osteogenesis starter. We don't know with what concentrations the boluses hit their targets in the bones. Therefore what do we really know about the signals the PTH peptides send into their targets in the bones? For example when 25 μg of rhPTH-(1-34) (Forteo™) is injected subcutaneously into osteoporotic women the transient peak circulating concentration 15-45 minutes later is 0.09 nM (Lindsay et al 1993).

Would this concentration stimulate adenylyl cyclase in osteoblastic cells in a human or a rat? The short answer is—we don't know. In cultured rat ROS osteoblast-like cells and rat kidney proximal convoluted tubule cells, concentrations of 1 nM or less would not stimulate adenylyl cyclase, but they would stimulate PKCs such as PKC-δ. This problem has been been around for a long time, but only now has it been brought to the fore by Buckley et al (2006). They have reported that concentrations of 0.01-0.1 ng/ml of medium (i.e., 0.0025-0.025 nM) stimulate human SAOS osteoblastic cells and primary bone osteoblasts to express and make collagen and OPG and decrease RANKL expression without stimulating cyclic AMP formation or *c-fos* expression. In other words, cyclic AMP and c-Fos were not needed to increase osteogenic genes and oppose osteoclastogenesis. By contrast stimulating cyclic AMP formation and *c-fos* with higher PTH concentrations (10-100 ng/ml) do exactly the opposite. Therefore, according to these experimenters cyclic AMP and *c-fos* are anti-osteogenic in primary human osteoblasts.

But if this is the case, why don't N-terminally truncated PTHs such as hPTH-(3-34) or hPTH-(13-34) stimulate bone formation in rats (Rixon et al., 1994)? It might be because a PTH needs its N-terminus to reach its targets after injection (Whitfield et al., 1998a). Of course the N-termus would also be needed to get the cyclic AMP for blocking apoptosis and extending osteoblasts' working lives. It might also be that both cyclic AMP and PKC-δ are needed to start the complete osteogenic mechanism in the animal but not in culture. Obviously Buckely et al.'s work must be repeated and, if repeatable, extended.

iv. PTHs Speak to Bones with Two-Pronged Forked Tongues— The Other Prong Says Remove It!

While the osteogenic action of single daily injections of an adenylyl cyclase-stimulating PTH predominates over their osteoclast-stimulating, cortex-porating, bone-demolishing action, the balance shifts to net resorption if the PTH is continuously infused for longer than 1 hour or injected too frequently (Dobnig and Turner, 1997). And it appears that although intermittent injections of hPTH-(1-34) have been found to at least transiently increase cortical porosity in humans, monkeys and rabbits, it has been found in mice and rats that anabolic intermittent injections of medium and high doses of the peptide actually *decrease* osteoclast surface in OVX rats (Lane et al., 1996) and suppress the final steps of osteoclast differentiation in male ddY mice (Sakai et al., 1999). Why are intermittent and continous treatments with a PTH so strikingly different?

Watson et al. (1999) found part of the answer in rats. As expected, intermittent injections of the full-size hPTH-(1-84) caused a massive accumulation of IGF-I-making osteoblasts (Watson et al., 1995, 1999). As we have just seen seen one of PTH's major, cyclic AMP-mediated osteogenic actions is the inhibition of SOST expression. Indeed according to Bellido et al. (2005) continuously elevating PTH in mice causes SOST expression to crash (Fig. 34). This would of course release the brakes on osteogenesis not resorption. And in Watson et al.'s experiments, continuously infusing the PTH did indeed cause the "unbraked" osteoblasts to accumulate. But they did not make more bone. They made less autocrine/paracrine IGF-I and

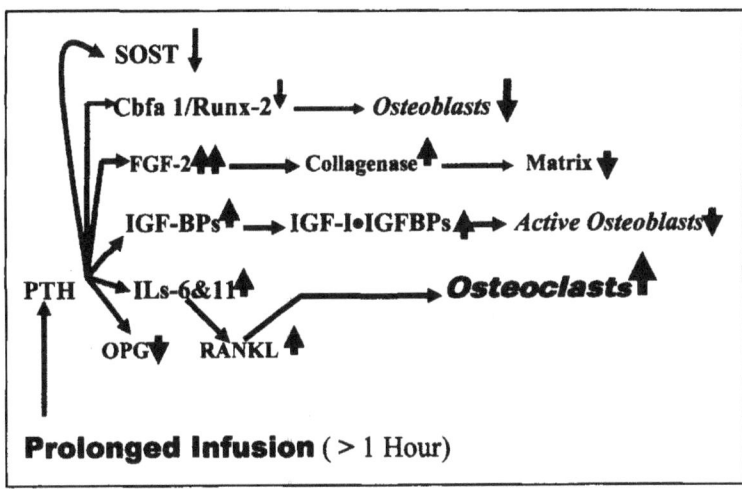

Figure 34. The various things that happen to stimulate osteoclast activity and bone resorption rather than bone growth when a PTH is continuously infused instead of being injected once a day, in an inhaled puff, or swallowed once a day in a pill.

started making IGF-binding proteins, IGFBP-3, IGFBP-4 and IGFBP-5 (Fig. 34). The IGF-I-trapping IGFBPs, particularly IGFBP-4 (Mohan et al., 1999), would have prevented the smaller amounts of IGF-I from stimulating bone formation. But we must be careful here. IGFBP-4 is rather special. It does bind IGF-I, but the cell can then chop the protein up with an IGFBP-4-specific protease which releases the IGF-I which can then stimulate bone formation (Miyakoshi et al., 2001).

A common experience with various cyclicAMP-PKA-initiated processes is that while short bursts of cyclic AMP-PKA activity stimulate them, prolonged cyclic AMP-PKA activity inhibits them. It would be expected that this would also apply to PTH-stimulated osteogenesis because it too is triggered by a burst of cyclic AMP-PKA activity. This relation is illustrated in Figure 35 where a short burst of cyclic AMP-PKA activity stimulates the ignition phase of osteogenesis which in turn causes the expression of IGI-I which takes over the steering wheel and guides the maturation stage. But if the cyclic AMP-PKA activity is prolonged beyond a critical time, IGF-I still appears though reduced, but key reactions are inhibited and others are stimulated, the result of which is bone resorption (Fig. 35). It is also likely that when the PTH boluses are too closely spaced as in Figure 35 something much more dangerous than reducing IGF-I expression happens because adenylyl cyclase-stimulating PTHs increase the top speed (V_{max}) of the osteoblasts' Pit1 phosphate transporter (Caverzasio and Bonjour, 1996; Nielsen et al., 2001; Selz et al., 1989). This transporter normally provides the inflow of phosphate needed to drive the expression of osteopontin and the accumulation of mineral in the matrix vesicles budded off the osteoblasts (Beck et al., 2000; Caverzasio and Bonjour, 1996; McKee and Nanci, 1996; Selz et al., 1989). Thus, while well-spaced cyclic AMP/PKA pulses would promote osteoid mineralization by causing phosphate to surge into the matrix vesicles continuous infusion or more closely spaced boluses of a PTH might raise the flow of phosphate into the cell to an apoptosis-triggering level (Allen et al., 2002; Mansfield et al., 2001; Meleti et al., 2000).

This model is supported by Locklin et al. (2003) who reported that intermittent exposure of cultured primary mouse marrow cells to hPTH-(1-34) caused an IGF-I-dependent stimulation of osteoblast differentiation markers while continuous exposure could not stimulate these

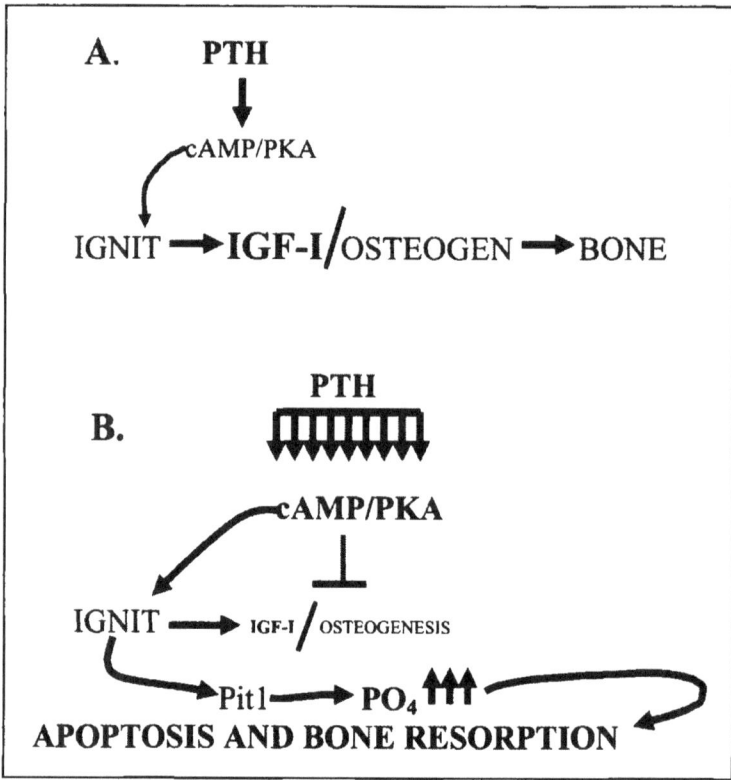

Figure 35. A. Once-daily bursts of adenylyl cyclase activity from PTHR1 receptors turn the ignition switch (**IGNIT**) of the osteogenic mechanism which in turn stimulates the expression of IGF-I. The resulting surge of IGF-I drives osteogenesis as described in the text. B. However, many closely spaced bursts of cyclic AMP production are counterproductive. They still stimulate, though less effectively, IGF-I expression and osteogenesis and instead override these by stimulating bone resorption by doing such things, besides those in Figure 34, as causing the Pit1 phosphate transporter to overload the cells with apoptogenic levels of phosphate.

markers and reduced IGF-I mRNA production. Added to the IGF-I blockade is the switching on of the accumulating osteoblasts' matrix-destroying collagenase-3 by overproduction of FGF-2 (Fig. 32; Hurley et al., 1995, 1999). PTH itself would also have stimulated collagenase-3 expression by stimulating Cbfa1/Runx-2 expression (Locklin et al., 2003; Selvamurugan et al., 2000), but as noted in the next paragraph this might not happen with excessive exposure to PTH. This combination of IGF-I blockage and a surging, matrix-destroying collagenase should be enough to tip the balance over to bone demolition.

Intermittent exposure of mouse marrow to adenylyl cyclase-stimulating PTHs stimulates Cbfa1/Runx-2 that drives osteoblast differentiation and bone building (Locklin et al., 2003). Single daily boluses of adenylyl cyclase-stimulating PTHs actually trigger the loading of nuclei with Cbfa1/Runx-2 and the expression of about 158 genes in rat femoral metaphyseal bone cells (Fujita et al., 2001a,b; Moore et al., 2000; Onyia et al., 2001) (Fig. 31). Among these genes is the one encoding a still functionally mysterious, 145-aminoacid cytoplasmic protein—PAIGB—in the tibial cells of Sprague-Dawley rats (Robinson et al., 2003). These once-daily boluses also promptly trigger only brief (6-24 h) bursts of ubiquitin-specific protease UBP41, expression of which could save key gene transactivators by removing any attached ubiquitin chains that would mark them for the proteasome shredder (Chung and Baek, 1999; Huang et al., 1995; Miles et al., 2002). However, as I said in section iiid of this chapter, it seems likely that these brief bursts of UBP41 activity could promote the reconfiguration of chromatin that makes genes more accessible to appropriate transcription factors (Latchman, 2004). Ubiquitination of histone H2B facilitates the methylation on lysine residues 4 and 79 of histone H3 which promotes a more tightly packed chromatin structure and gene silencing (Latchman, 2004). Therefore, removing these ubiquitins with PTH-triggered, cyclic AMP-driven bursts of UBP41 could result in the demethylation of H3 histones and the unlocking of dozens of genes such as the osteogenically key genes that are responsive to the Cbfa1/Runx2 that is at the same time being loaded into the nucleus by the PTH boluses.

But continuously infusing these PTHs into rats selectively switches on a different and very much larger, set of 759 genes in the metaphyseal cells, which, for example, no longer includes the gene for the PAIGB protein, but instead includes genes for bone matrix-shredding proteases (Onyia et al., 2001a, 2001b; Robinson et al., 2003). Continuous PTH exposure also locks on UBP41expression, (Miles et al., 2002; Murray et al., 1998), which, by increasing the free ubiquitin pool along with a cyclic AMP-induced stimulation of the proteasomal endopeptidases would shift the cell's proteome (its protein complement) from one for bone making to one for bone destruction. This prolonged activity of UBP41 and increased proteasomal enzymes drives a PKA-stimulated ubiquitination and destruction of Cbfa1/Runx-2 with a consequent shutdown of Cbfa1/Runx-2-dependent osteoblast-specific genes such as the genes for Bcl-2 and OPG (Bellido et al., 2002; Tintut et al., 1999). The upshot of these changing gene expressions during continuous PTH infusion are drops in the expression of the anti-osteoclastogenesis OPG and osteoblasts' bone-making genes such as those for BSP, collagen I and osteocalcin as well as a reciprocal increase in RANKL and the products of other genes that drive osteoclast differentiation and activation and bone resorption (Ma et al., 2001) (Fig. 32).

How could continuous PTH infusion turn on or off so many more genes than single daily boluses of the peptide? Evidently, continuous infusion activates many more transcription factors. According to Chauvin et al. (2001), spacing the boluses 24 hours apart would enable an osteoblast's PTHR1 receptors to fully recycle once and enable the cell to present a fully receptored surface to the next PTH bolus. Each bolus would cause a brief burst of cyclic AMP/PKA signaling from the full complement of PTHR1s on the various target cells (Fig. 21). But a continuously infused PTH might cause a continuous, β-arrestin-2-mediated endocytic cycling of PTHR1s with reduced steady-state level of receptors and very low-level adenylyl cyclase/ cyclic AMP signaling (Locklin et al., 2003) but with a steady, β-arrestin-2-mediated pumping of MKK4-activated JNK kinase (and maybe PTHR1 receptor itself) into the nucleus (Fig. 21). Such a sustained injection of a transcription factor(s) activator such as JNK into the nucleus could massively stimulate the expression of a much larger set of genes and their resorption-driving products.

The ultimate arbiter of whether a PTH builds or destroys bone is the RANKL/OPG expression ratio it establishes. Halladay et al (2002) have used UMR106 rat osteoblast-like cells transfected with an *opg* gene promoter-switchbox construct pOPG5.9βgal (-5917 to +19) to see the effects of short and long exposures to hPTH-(1-38) and a PTH of the Ostabolin Family, hPTH-(1-31) (Ostabolin™), on the activity of the *opg* gene's promoter as indicated by the extent of stimulation of a β-galactosidase "reporter gene" attached to the promoter construct. Exposing the UMR106 cells to the PTHs or to cyclic AMP-elevators such as the potent adenylyl cyclase-stimulating forskolin or to a cyclic AMP turnover inhibitor, the cyclic nucleotide phosphodiesterase-inhibiting, IBMX, or to cyclic AMP itself for 8 hours nearly doubled the *opg* promoter-switchbox activity with the hPTH-(1-31) being slightly more effective than hPTH-(1-38). However, exposure to either PTH or one of the cyclic AMP elevators for 24 or 48 hours switched the promoter off. (Unfortunately they did not look at the effects of the peptides at shorter intervals such as the physiologically relevant 1 or 2 hours.) On the other hand, a PKC stimulator such as TPA (12-O tetradecanoyl phorbol-13 acetate) (at 10^{-8} or 10^{-7} M) reduced the *opg* gene promoter activity in cells exposed to the drug for 8 hours but strongly stimulated the *opg* promoter in cells exposed for 24 or 48 hours. Thus, the stimulatory effect of a short exposure to an adenylyl cyclase-/PKC-stimulating PTH (don't forget that this ratio depends on the cell's NHERF-2 content) on *opg* gene activity would be reduced by PKC stimulation while the inhibitiory effect of a prolonged exposure would be blunted by PKC stimulation. This intervention of PKC into the process could at least partly explain why intermittent injections of members of the Ostabolin™ family (hPTH-(1-31NH$_2$ [Ostabolin™] and [Leu^{27}]Glu^{22}-Lys^{26}hPTH-(1-31)NH$_2$ [Ostabolin-C™]), which tend not to stimulate PKC, are poor stimulators of osteoclast generation and do not cause hypercalcemia in various animal models and humans even at very high doses (Fraher et al., 1999; Jolette et al., 2003, 2005; Mohan et al., 2000; Whitfield et al., 1998c).

Added to direct *opg* gene suppression by prolonged cyclic AMP pulsing is the ability of a prolonged PTH exposure to prevent osteoblastic cells from making OPG by stimulating the proteasome that destroys Cbfa1/Runx-2 which normally keeps the *opg* gene working by binding to the opg promoter's 12 so-called OSE2 sites (Bellido et al., 2002; Fu et al., 2001; Galvin et al., 2001; Hofbauer et al., 1999, 2000; Kanazawa et al. 2000; Miles et al., 2002; Murray et al., 1998; Onyia et al., 2000; Stilgren et al. 2001; Thirunavukkarasu et al., 2000; Tintut et al., 1999).

But what about the other player, the pro-osteoclast RANKL? Within an hour of a single injection of the adenylyl cyclase/PKC-stimulating hPTH-(1-38) into rats, RANKL mRNA production surges upwards and OPG mRNA reciprocally crashes in the distal femoral metaphysis and diaphysis (Fig. 36). But OPG expression soon *rebounds* as its gene's promoter-switch box is turned on and *rankl* gene expression is reciprocally turned off by 3 hours which means that the OPG/RANKL ratio must be high by this time (Ma et al., 2001; Onyia et al., 2000). According to the observations of Locklin et al., (2003) on mouse marrow cells if the PTH treatment is continuous *rankl* gene expression should rise while the *opg* gene's promoter and thus its expression should be inhibited in the rat. Indeed, as expected, continuous infusion of hPTH-(1-38) into rats causes a sustained rise in *rankl* expression, a sustained drop in OPG expression, and increases in serum Ca^{2+} concentration and osteoclast generation (Ma et al., 2001) (Fig. 34). Locklin et al. (2003) have also found that although a continuous exposure to the adenylyl cyclase-only stimulating hPTH-(1-31) (Ostabolin™) is a 2 to 3 times more effective stimulator of *rankl gene expression* than continuous exposure to hPTH-(1-34) in their mouse marrow cultures, the two peptides only equally increased osteoclast generation. This indicates a huge difference between the actions of thesePTHs on message processing. And it could mean that hPTH-(1-31) signaling lacks something needed to get the additional number of osteoclast-stimulating *rankl* gene transcripts translated into active RANKL. But as we will see further on we are being confused by data from "pure" and mixed-stage cultures and the whole animal.

At this point you might ask why the expressions of RANKL and OPG are reciprocal *in the same cell?* The answer might lie in the phosphorylation of the CREB protein's Ser^{133} by PKA activated by the cyclic AMP signals coming from PTH-activated PTHR1 (Fu et al., 2002). The PKA-phosphorylated/activated CREB plus a CBP/p300 (Ptashne and Gann, 2002; Quinn, 2002) on the *rankl* gene promoter's CRE turns on the *rankl* gene (Fig. 36). At the same time CREB•CBP/p300 on the c-*fos* gene's promoter stimulates this gene's expression (Tyson et al., 1999). The c-Fos product turns off the *opg* gene via the AP-1 complexes (Latchman, 2004) it forms with c-Jun on the *opg* gene's promoter (Fu et al., 2002) (Fig. 34). (Interestingly, the

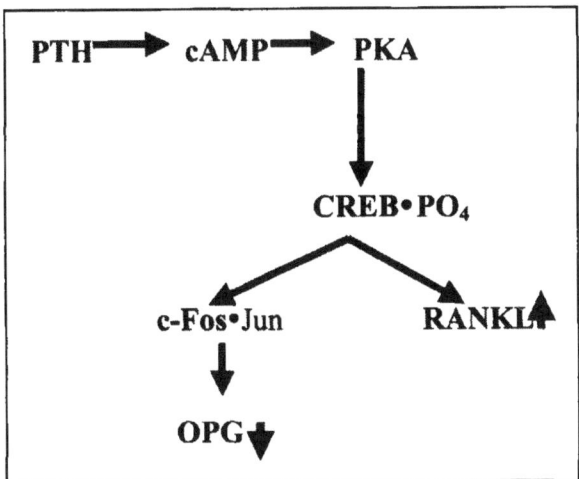

Figure 36. The reciprocal control of osteoprotegerin (OPG) and RANKL expression by a cyclic AMP/PKA signal from PTH-activated PTHR1 receptor. This mechanism would operate in a single cell, but the striking inability of PTHs belonging to the Ostabolin™ family to stimulate osteoclast activity and bone resorption is not due to this kind of mechanism but to the abilities of juvenile osteoblastic cells to make RANKL but not OPG and the abilities of mature osteoblasts to make OPG but not RANKL as discussed in the text and illustrated in Figure 37.

hPTH-(1-34)-induced increase in AP-1 activity and suppression of *opg* gene expression in osteoblasts could be enhanced by signals from serotonin receptors activated by serotonin coming from strain-stimulated osteocytes [Bliziotes et al., 2001]). This mechanism explains the prompt stimulation of *rankl* and suppression of *opg* by the brief bolus of hPTH-(1-38) in the rats of Ma et al. (2001) But why did *opg* expression soar and *rankl* crash during the next hour or so? Obviously the inhibitory AP-1 complexes on the *opg* promoter must rapidly turnover leaving a still activated *opg* promoter. According to this sort of scenario, a continuing supply of AP1 and suppression of *opg*, activation of *rankl* and the resulting increased osteoclast generation would require a sustained barrage of PTH boluses. But there is a problem here. This reciprocity mechanism at the single cell level is irrelevant to the whole bone or primary mixed bone cell cultures because the transit-amplifying immature osteoblastic cells express their *rankl* gene, but then turn the gene off and switch the *opg* gene on when they become functioning osteoblasts (Atkins et al., 2002; Corral et al., 1998) (Figs. 23 and 37).

But in mice, the responses of OPG and RANKL expressions to PTH injections appear to depend on whether the animal is OVXed or not (Lu et al., 2001). Lu et al. have found that OVX alone caused RANK expression in the tibias of C57BL/36 mice to double by 4 weeks and then drop back to only one-third of the starting level during the next 7 weeks and the OPG expression dropped to only 20% of the starting level by 11 weeks. Injecting hPTH-(1-34) into intact mice once daily for 5 days a week caused RANKL expression to increase 4 times and PTHR1 expression to triple. But, OVX inverted the responses of RANKL and OPG to PTH. The same PTH injections into OVXed mice reduced RANKL mRNA production to 37% of the starting level and a 2- to 3-fold increase in the level of OPG mRNA and PTHR1 expression probably because the PTH-stimulated the estrogen-lacking RANKL-competent immature cells to mature into RANKL-incompetent, OPG-making osteoblasts without being replaced by new RANKL-making immature cells. The estrogen lack could indeed have caused a shortage of immature osteoblastic stromal cells by diverting the flow of bi-potential precursor cells from the osteoblastic to adipocyte

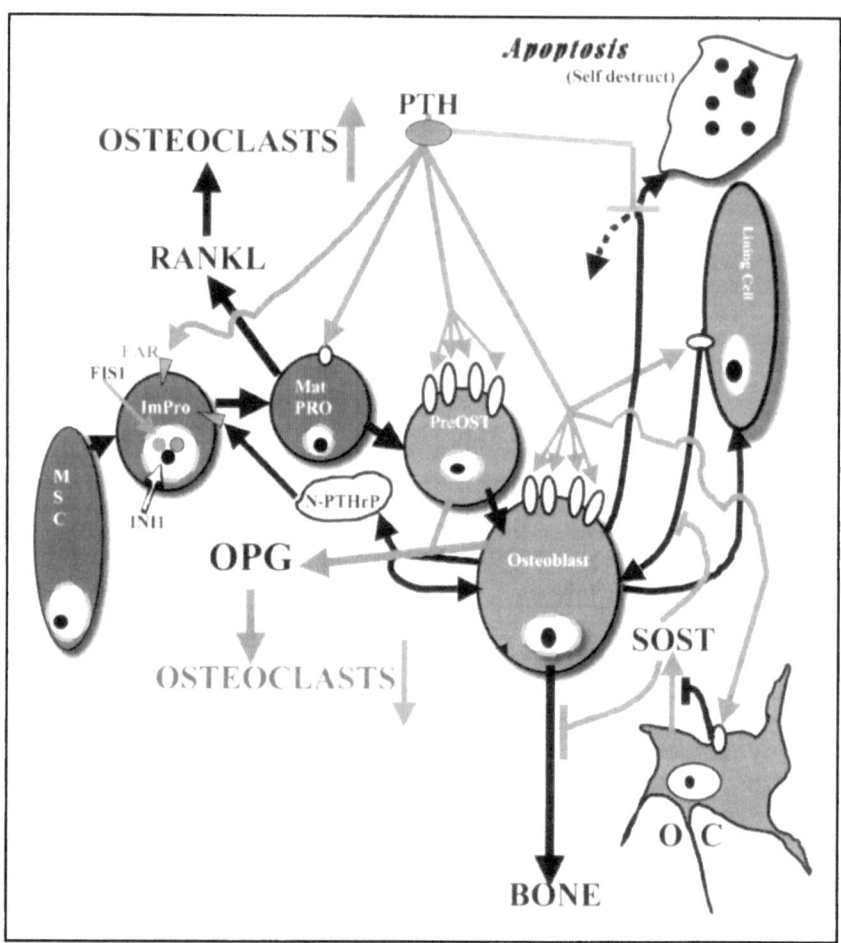

Figure 37. When a PTH is injected and peaks in the blood only for 15 miniutes or so once each day, the late preosteoblasts and mature osteoblasts with the most PTHR1 receptors (the oval buttons) on their surfaces will be the most likely responders. However when the PTH is infused or injected at close intervals there will be enough time to give the receptor-poor mature progenitors at the end of the transit-amplifying stage a greater chance to respond. Since the receptor-dense osteoblasts cannot express the RANKL osteo-clast-promoter (they have considerably downregulated its gene) but they can express the anti-osteoclast OPG (osteoprotegerin), the response to intermittent PTH injection would favor a high OPG/RANK ratio and osteogenesis. By contrast continuous infusion would reduce the the ratio and tip the balance toward resorption. Therefore, the actual response to a PTH at any given moment will be a function of the fraction of receptor-dense OPG-expressing mature or nearly mature osteoblasts in the bone cell populations. Of course, as noted in the text, the resorption or no resorption response will also depend on things such as the ability of the peptide to stimulate protein kinase Cs. The abbreviations are the same as in Figure 23.

pathway (Okazaki et al., 2002). This suggests that PTHs might be less able to stimulate bone growth in ovary-intact than in OVXed rats. And this is indeed the case. The abilities of hPTH-(1-34) and the two members of the Ostabolin Family of PTHs to stimulate femoral trabecular growth in ovary-intact rats are only about half of their abilites to stimulate trabecular growth in OVXed rats (unpublished observations in my laboratory). It is obviously not a little important to know whether this inversion in mice can be repeated in mice and extended to postmenopausal women.

i v a. The Ostabolins—Feisty Little PTHs with a Seeming Reluctance to Stimulate Bone Resorption

The ability of PTHs to stimulate resorption has bedeviled the clinical application of these peptides since the 1930s. So the dream of all "PTH buffs" has been to make a PTH that could stimulate bone growth, but not bone resorption. Higher doses of such a PTH could be safely used to maximally accelerate fracture healing and get a bigger anabolic bang for the osteoporotics' buck. It appears that two such PTHs have at last been made. These PTHs belong to the Ostabolin™ family—hPTH-(1-31)NH$_2$ (Ostabolin™) and [Leu^{27}]*cyclo*(Glu^{22}-Lys^{26})hPTH-(1-31)NH$_2$ (Ostabolin-C™; i.e., cyclized Ostabolin™). The potently anabolic Ostabolin ™ has been shown to be both orally administrable and a very poor stimulator of resorption in humans and mice (Fraher et al., 1999; Mehta et al., 2000; Mohan et al., 2000). And Jolette et al. (2003, 2005) have shown that the potent stimulation of bone formation in cynomologous monkeys even by daily injections of such a formidable dose of Ostabolin-C™ as 80 μg/kg of body weight for 7 days *decreased* the indices of bone resorption!

The possibility of Ostabolin-C™ decreasing bone resorption while potently stimulating bone growth in monkeys seems at first sight questionable—perhaps Jolette et al. (2003, 2005) have made a mistake. However, it may "simply" be that intermittent injections of Ostabolin-C™ are better at suppressing osteoclast differentiation than Forteo™ (Lane et al., 1996; Sakai et al., 1999). We have found that injecting 2 nmoles of either Forteo™ or Ostabolin C™/100g of body weight of into young sexually mature Sprague-Dawley rats during the first 2 weeks after OVX significantly and equally drops the number of osteoclasts from that in the sham-operated rats, but Ostabolin C™ is significantly more effective than Forteo™ in dropping osteoclasts as expressed as numbers of osteoclast nuclei. Also during the first 8 weeks after OVX, the metaphyseal trabecular number in distal femurs of Sprague-Dawley rats fell 52% from the normal (sham-operated) 4.6 ± 0.1/mm to 2.2 ± 0.1(*p* <0.01) When 0.6 nmole of hPTH-(1-34) (Forteo™)/ 100 g of body weight was injected once daily into the OVXed rats starting 2 weeks after the operation and ending 6 weeks later, the mean number of metaphyseal trabeculae had dropped from the normal (sham-operated) value of 4.6 ± 0.1 to 2.0 ± 0.1/mm (*p*<0.01) which was not significantly different from the 2.2 ± 0.1 in the vehicle-injected OVXed rats. However, when the same dose of Ostabolin-C™was injected instead of hPTH-(1-34) (Forteo™) the trabecular number dropped by only 28% to 3.3 ± 0.1/ mm which was 1.7-fold higher than the 2.0 ± 0.1/mm (*p* <0.01)in the hPTH-(1-34) (Forteo™)-treated rats. If the PTH injections were delayed for 9 weeks after OVX when the maximum osteoclast activity had passed, there was no difference between the trabecular numbers in the animals treated daily with either peptide between 9 and 15 weeks. In both kinds of experiment the two peptides *equally and greatly increased* the mean trabecular thickness to a value about 1.7 times higher than in the femurs of the sham-operated animal. Thus it seems from these numbers that Ostabolin-C™ and hPTH-(1-34) (Forteo™) are equally osteogenic, but Ostabolin-C™is the better osteoclast suppressor at least in rats and monkeys.

Why would these Ostabolin™-family peptides be such poor stimulators of bone repsortion? On the basis of the available information from *cultured cells* it simply makes no sense! To try to answer this question we must look more closely at what we have learned in the past few paragraphs. RANKL expression is unquestionably stimulated by cyclic AMP/PKA. And all of the osteogenic peptides, including the Ostabolin™ family peptides, activate PTHR1 which stimulates adenylyl cyclase which stimulates RANKL. So why would Ostabolin™ family members be poorer osteoclast stimulators than hPTH-(1-34) (Forteo™). Why would the potent adenylyl cyclase-stimulator and potently anabolic [Leu^{27}]*cyclo*(Glu^{22}-Lys^{26})hPTH-(1-31)NH$_2$ (Ostabolin-C™) actually *reduce* osteoclast activity in OVXed rats as wells as monkeys?

The reason may stem from two things. First, the density of PTHR1 receptors on an osteoblastic cell depends on its "maturity" (Aubin, 1998, 2000, 2001; Aubin and Triffitt, 2002; K. Lee et al., 1994; McCauley et al., 1996; Schiller et al., 1999). Second, immature progenitors and mature osteoblasts differently express RANKL and OPG (Atkins et al., 2003). RANKL is made by immature marrow osteoblastic cells with few emerging PTHR1 receptors while the mature osteoblasts with maximum PTHR1 receptor densities do not make RANKL but can make OPG (Figs. 9, 23 and 37). As I said back in Chapter 2ii, this maturity-dependent expression of RANKL and OPG is a smart move on Nature's part because the two populations live in different places with different neighbors and jobs to do in the crowded spaces of an osteonal tunnel or a trabecular trench. This means that a PTH will stimulate RANKL expression in immature osteoblasts and OPG but *not* RANKL in mature osteoblasts and the relative levels of expression will depend on the population's maturity/receptor density profile. Thus, PTH-stimulated immature, RANKL-competent immature osteoblastic cells following osteoclasts in a new osteonal tunnel or in a trabecular trench would help the osteoclasts, but PTH-stimulated, mature, working osteoblasts with their RANKL genes shut down (Kitazawa et al., 1999)would stop osteoclast generation with OPG to protect the bone they are making.

Now I come to the point. The Ostabolin™ family peptides have *lower* affinities for PTHR1 than hPTH-(1-34) (Forteo™) (Whitfield et al., 2000c). This means that during their brief appearance in a single short-lived injected bolus the Ostabolin™ family peptides would be less able than hPTH-(1-34) (Forteo™) to grab onto enough of the fewer PTHR1s to stimulate the immature RANKL makers. But they would have enough time and not be significantly less effective than hPTH-(1-34) (Forteo™) in grabbing enough PTHR1s to stimulate the OPG-making preosteoblasts and mature osteoblasts with many more PTHR1 receptors on their surfaces. However, the advantage should vanish when the peptides are given much more time to grab PTHR1s during continuous infusion. This should be the case for Ostabolin-C™ which has a higher affinity for PTHR1 than Ostabolin™ but still a lower affinity for PTHR1 than hPTH-(1-34)NH_2 (Whitfield et al., 2000c). Indeed that is exactly what happens. When Ostabolin™ is continuously infused (at a dose of 4 nmoles/kg per day) into a Sprague-Dawley rat the blood Ca^{2+} concentration rises at about the same rate as in rats receiving hPTH-(1-34)NH_2 but then stops rising during the next 5 days while the blood Ca^{2+} concentration continues rising in the animals receiving hPTH-(1-34)NH_2. However, in striking contrast to its total inability to raise the blood Ca^{2+} level when injected intermittently (Jolette et al., 2003, 2005), Ostabolin-C™ infusion causes the blood Ca^{2+} concentration to rise as steadily and at least as high as, or even higher than, in the rats receiving hPTH-(1-34)NH_2.

This explanation could also account for the strange observation of Fujita et al. (1999) that a *once-a week* injection of 60 μg of hPTH-(1-34)OH (Forteo)™ significantly increased spinal BMD but at the same time significantly *reduced* the level of collagen breakdown bits in osteoporotic patients. In this case a weekly brief bolus of PTH could only target the currently available mature, receptor-rich, OPG-making osteosteoblasts with their shut-down *rankl* genes, but it would be too ephemermal to significantly stimulate the immature RANKL-makers.

Another reason for the reluctance of the intermittently injected Ostabolins to stimulate bone resorption is their lack of the C-terminal, PLC-independent, PKC-stimulating 28-32 region—they lack the critical residue 32 (Jouishomme et al., 1994). Because of this they cannot stimulate PKCs in ROS 17/2 cells and human osteoblasts (Wu and Kumar, 2000). And this could be one reason why they cannot stimulate bone resorption. Thus, Sprague et al. (1996) have shown that preventing PKCs' activating translocation to the cell membrane completely prevents PTH-(1-34)'s potent ability to stimulate Ca^{2+} release from neonatal rat calvariae while inhibiting adenylyl cyclase only partially reduces the efflux.

v. How an Odd Couple—PTHrP and an Indian Hedgehog Build Bones

An odd couple—PTHrP and a hedgehog, specifically an Indian hedgehog—controls the growth of a long bone such as the femur. Indian hedgehog is one of a family of several proteins that are called hedgehogs, because disabling the single hedgehog gene in the fruit fly *Drosophila* causes the fly's larvae to have a pattern of cuticular denticles that make it look like a prickly hedgehog (Bier, 2000).

The femur starts with the condensation of mesenchymal cells. Before condensing, the mesenchymal cells in the bone field are separated by matrix and not interested in each other. But then something happens! The dispersed cells start expressing N-cadherin, a so-called homophilic cell-cell adhesion protein (Davies, 2005). As they start sticking together, their N-cadherin expression escalates and soon a bone-shaped clump of cells appears. The cells in the clump now start expressing the master cartilage-inducing transcription factor known as Sox 9 (de Crombrugghe et al., 2001; Shum and Nuckolls, 2001). Sox 9 probably stimulates the expression of the genes for the cell surface components needed for chondrocytic condensation (de Crombrugghe et al., 2001). Then Sox 9 causes the cells to express L-Sox 5 and Sox 6 that are "architectural" transcription proteins that reconfigure nuclear chromatin structures to enable Sox 9 and other classical transcription factors to turn on chondrocye-specific genes (de Crombrugghe et al., 2001; Shum and Nuckolls, 2001). The condensed mesenchymal cells differentiate into chondrocytes that lay out a cartilage model of the future bone and perichondrial cells that surround and enclose the emerging model. The cells at the center of the model stop proliferating, stop expressing Sox 9 and start maturing into terminal hypertrophic (swollen) cells, which calcify the matrix to enable it to be resorbed by osteoclasts and replaced by bone. Meanwhile in the center, the model's perichondrium is invaded by capillaries with osteogenic endothelin-1-secreting endothelial cells and changes from a perichondrium into a periosteum. In other words, the perichondrial layer spawns osteoblasts that deposit bone on the inner surface of the periosteum to make a hollow cylinder—the bone collar. Meanwhile the hypertrophic cells' matrix calcification reduces their already low oxygen and nutrient supplies, which causes them to trigger the suicidal apoptosis. The calcified matrix is removed by "chondroclasts" leaving only trabeculae-like calcified cores or spicules in a primary marrow space. Vascular and perivascular cells from the periosteum then invade this marrow space. The blood vessels produce the marrow sinusoids and the perivascular invaders generate the bone marrow stromal cells, osteoprogenitors and the first osteoblasts, which layer bone onto the calcified cartilage cores to make the first trabeculae. The "intramembranous" making of the bone collar directly from mesenchymal cells, and the endochondral bone formation (i.e., the "lost-wax"-like replacement of cartilage by bone) in the center of the cartilage model, together make up the so-called *primary ossification center* (Kerr, 1999).

As the bone lengthens, the cartilage layers attached to the inner surfaces of the bony epiphyses at both ends of the growing bone are reduced to thin growth plates (also called *physes*). Now the bone has proximal (near the body) and distal (distant from the body) growth plates, which contain chondrocyte stem cells and are the engines that drive the lengthening of the bone. The round moderately sized chondrocytes at the top of a growth plate nearest the epiphysis, the *reserve* or *resting zone*, are the stem or chondroprogenitor cells (R.B. Martin et al., 1998). They are fed with nutrients *diffusing* down through the cartilage matrix from the blood vessels in the overlying epiphyseal bone. Thus, the core of the growth plate and the columns of cells extending down from it are hypoxic, so to do their jobs they have to turn to HIF-1α and HIF-1β (which we met in Chapter 1) and the large number of oxygen-regulated genes it controls to stimulate the genes for the the glucose transporter 1 and the enzymes of the anerobic glycolysis machinery to replace the mitochondrial oxygen-driven ATP production (Davies, 2005; Marx, 2004; Schipani et al., 2001; Wenger, 2002). They must stop trashing HIF-1α.

When the oxygen level in the core drops below a critical level, the cells' oxygen-driven protein hydroxylases stop working which means that prolines in their HIF-1s' oxygen-dependent destruction domains are no longer hydroxylated and the peptides are no longer polyubiquitinated and dumped into the cells' protein-shredding proteasome garburetor barrels (Schipani et al., 2001; Wenger, 2002). The accumulating HIF-1α moves into the nucleus together with HIF-1β and stimulates the transcription of the genes for the components of the glycolytic machinery.

When a stem cell gets the first of three differentiation signals that starts its proliferative engine and divide, one of its two daughter cells stays home to keep up the stem cell tradition while the other daughter leaves home to join the parallel columns or stacks of flat or disk-like transit-amplifying cells in the *proliferative zone* where they express the genes for aggrecan core protein, link protein and type II collagen and thus lay down the gel-like permeable matrix (e.g., Q. Chen, 2004; Q. Chen et al., 1995). Then when they reach end of the proliferative column they generate signal 2 to shut down the cell cycle machinery and stop expressing the genes for aggrecan core protein, link protein and type II collagen and start expressing the gene for CMP (cartilage matrix protein or matrilins) the specific marker for the *maturation* or *pre-hypertrophic zone* (Deák et al., 1999; Q. Chen, 2004; Q. Chen et al., 1995). Then they plunge into the deadly ionic turbulence of the *hypertophic zone* (R.B. Martin et al., 1998) where they switch off the CMP gene and start their apoptosis-terminated hypertrophic maturation program in which they increase their volume (i.e., hypertrophy) 5-to 10-fold, resorbing adjacent matrtix in the process and increasing their expression of alkaline phosphatase, annexin V(5), BMP-6, PTHR1 receptors, type X collagen and, at this point, the very important Indian hedgehog (e.g., Balcerzak et al., 2003; Q. Chen, 2004; Q. Chen et al., 1995; Zuscik et al., 2002). The purpose of all of this activity driven in the osteoblastic direction by Cbfa1/Runx2 is to lay down a type X cartilagenous matrix primed for calcification which is necessary for osteoclasts to attach to the surface, remove it and thus enable its ultimate relacement by bone. However, it is important here to know that there are permanent cartilage bodies such as articular cartilage which do not express Cbfa1/Runx2, do not plunge into a suicidal, Cbfa1/Runx 2-driven ossification, but have chondrocytes that when appropriately stimulated can stop making collagen X, restore their proliferative machinery and produce progeny that can go though the maturation and hypertrophic stages of differentiation (Q. Chen et al., 1995; Shum and Nuckolls, 2001).

Finally the cells start depositing Ca^{2+}-hydroxyapatite crystals on the type X collagen matrix between the cell columns. The calcification of the cartilage drives the hypertrophic cells to apoptotic self destruction by impeding the diffusion of food supplies and disposal of wastes (there is no lacunocanalicular network in cartilage to efficiently distribute nutrients and dispose of waste via the marrow vasculature as is needed because of the rigid impermeable matrix in cortical bone). The dying (apoptosing) cells release proteases, which shred the cartilaginous matrix in the distal segments of the cell columns, but leave the calcified matrix between the columns intact. The resulting proximal calcified "stalactites" and distal calcified "stalagmites" are targeted by osteoclasts and osteoblasts and become the foundations of the primary proximal and distal metaphyseal trabeculae. The invasion of these osteoclasts and osteoblasts is enabled by the proliferation of blood vessels with endothelial cells firing off osteoprogenitor cell stimulants and attractants such as BMP-2, endothelin-1, and OP-1 driven by leptin-induced proteases and the HIF-1α-induced expression of the chondrocytes' vascular endothelial growth factor (VEGF) genes (Bouletreau et al., 2002; Carano and Fivaroff, 2003; Mayr-Wohlfart et al., 2002; Shum and Nuckolls, 2001; Wenger, 2002). Along with the osteoclast and osteoblast precursors come the hematopoietic stem cells from the liver attracted by signals from their CaRs (Ca^{2+}-sensing receptors) to their new Ca^{2+}-rich, collagen I-lined niches on the forming trabeculae (Adams et al., 2006; Olszak et al., 2000).

But what do PTHrPand its Hedgehog companion do in this process? The Sox 9-driven cells of the embryonic bone's starting cartilage model make Indian hedgehog (Ihh), but Ihh production is eventually confined to the pre-hypertrophic stack cells (Fig. 38). Ihh makes its way from the producer cells (Bitgood and McMahon, 1995; de Crombrugghe et al., 2001; Minina et al., 2001; Karplis, 2001; T. Kobayashi et al., 2002; Shum and Nuckolls, 2001;Vortkamp et al., 1996; Vortkamp et al., 1998) to the resting peri-articular (around the joint) cells which have receptors called "Patched or Ptch for short" which grab and suppress, other receptors called "Smoothened" or just "Smo" to their friends (summarized for example, in Bijlsma et al., 2004; Whitfield and Chakravarthy, 2001). When Ihh binds to the peri-articular cells' Ptch receptors they let go of Smo, which then emits signals that send the transcription factor Gli into the nucleus to stimulate the expression of TGF-β2 which in turn stimulates the expression of the PTHrP gene (Alvarez et al., 2002; Karaplis, 2001; Stark, 2002; Whitfield and Chakravarthy, 2001) (Fig. 38). This Ihh ⇒ TGF-β ⇒ PTHrP sequence also stimulates the peri-articular cells to start proliferating by expressing cyclin D1, the first of the parade of stage-specific cell cycle-driving cell cycle-engine kinases (Beier et al., 2001; Whitfield and Chakravarthy, 2001), and join the parallel columns of dividing chondrocytes (Kobayashi et al., 2002). And Ihh stimulates the appearance of BMP-2 which in turn stimulates the expression of Cbfa1/Runx-2 in mesenchymally derived perichondrial cells, which commits the cells to the osteoblast lineage and formation of the periosteum (Canalis et al., 2003; de Crombrugghe et al., 2001; Shum and Nuckolls, 2001).

But remember what Zuscik et al. (2002) said about Ca^{2+} and PTHRP gene expressionn. The intracellular Ca^{2+} concentration must be low enough to enable PTHrP expresson. This Ca^{2+} sensitivity resides in the -1498 to the -863 region of the PTHrP gene's promoter (Zuscik et al., 2002). This part of the PTHrP gene's promoter has consensus sites for eleven transactivators (Zuscik et al., 2002). One of these is YY1, a possible silencer the Ca^{2+}-driven binding of which to the PTHrP gene promoter interferes with a positive factor (Roy et al., 1996; Zuscik et al., 2002). Alternatively one of these eleven may be a transcription promoter, the binding of which to its target site on the PTHrP gene promoter is reduced by nuclear Ca^{2+}just as nuclear Ca^{2+} reduces the binding of the transcription enhancer, CBF/NF-Y, to its target site on the grp78/BiP gene (Roy et al., 1996). In either case, the Ca^{2+} concentration in the peri-articular region must be kept low enough or the cells' Ca^{2+}-sequestering pumps must work hard enough to keep the nuclear Ca^{2+} level low enough allow PTHrP to be made in response to Ihh stimulation. The PTHrP from the peri-articular cells then percolates along the stacks of proliferating cells, which deploy increasing numbers of PTHR1 receptors on their surfaces as they approach the deadly seething hypertrophic zone (Amizuka et al., 1994; Karaplis, 001; Lee et al, 1995; Weisser et al., 2002).

The cyclic AMP signals from the growing numbers of PTHR1 receptors and probably PTHrP's intracrine nuclear action (Re and Cook, 2005) keep the cells from switching off their proliferative machinery and tearing open the "HT (hypertrophic)" gene envelope, which contains the genes for alkaline phosphatase, annexin V, BMP-6, collagen X, PTHR1 and, of course, Ihh and eventually killing themselves (e.g., Zerega et al., 1999) (Fig. 38). The PTHR1 signals do this via PKA, which by phosphorylating two sites on the Sox 9 protein increases Sox 9's DNA-binding ability and its ability to prevent the transition from proliferation to hypertrophy (de Crombrugghe et al., 200; Shum and Nuckolls, 2001) (Fig. 38). This prevention is ultimately due to the suppression of the expression and production of p57^{Kip2}, an inhibitor of the cyclin-dependent protein kinases that drive the build-up to DNA replication (MacLean et al., 2004; Whitfield and Chakravarthy, 2001). As we have learned above, PTHR1-induced bursts of cyclic AMP-dependent PKA activity should prevent the proliferating chondrocytes from triggering apoptosis by phosphorylating and inactivating their pro-apoptotic BAD and Bax proteins and stimulating them to make the anti-apoptosis Bcl-2 protein (Amling et al., 1997; Bellido et al., 2001, 2002). Indeed, PTHrP does stimulate Bcl-2 expression in growth plate chondrocytes, and knocking out the Bcl-2 gene causes premature maturation and apoptosis of

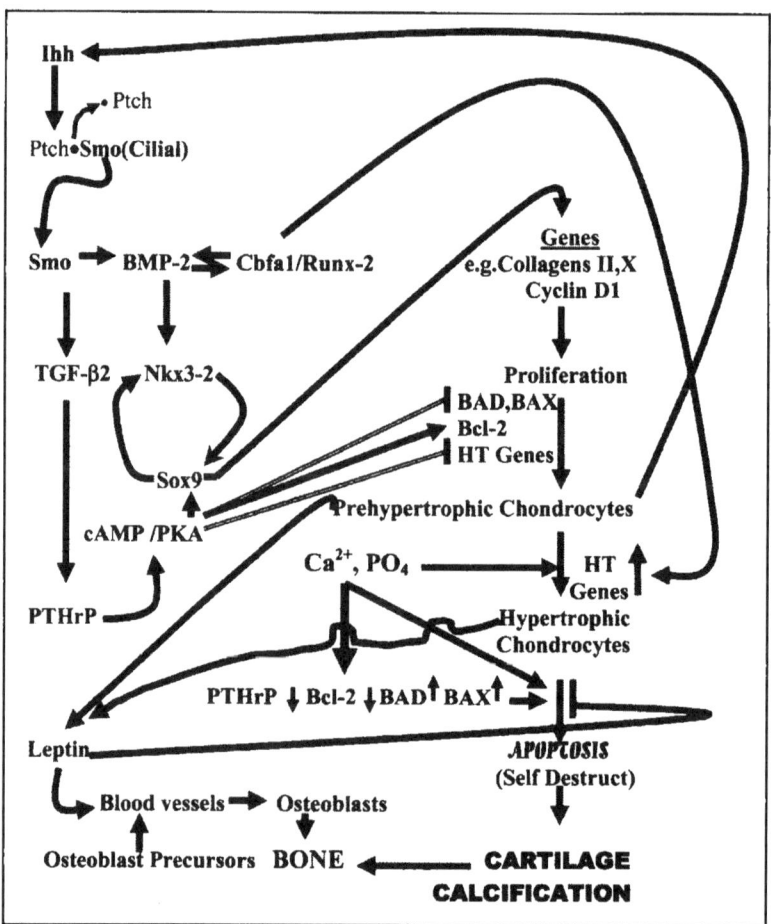

Figure 38. The roles of PTHrP and its friend, Ihh (Indian hedgehog), to steer the voyage of chondroblasts from the the the stem cell niche to hypertrophic self destruction and cartilage calcification. **HT Genes** refer to the set of genes that are turned on in the swelling, hypertrophic cell; **Ptch**, the inhibitory Patched; and **Smo**, the stimulatory Smoothened components of the hedgehog receptor. See the text for details of this complex story

the chondrocytes and shortened long bones in mice (Amling et al., 1997). The PTHrP might also hold off maturation by using its NTS to get into the nucleus and localize to the fibrillar component of the nucleolus where it can inhibit ribosome production (Aarts et al., 2001).

Leptin also has an important role in cartilage function. As we learned in Chapter 3i large amounts of leptin are made specifically by a subpopulation of hypertrophic chondrocytes adjacent to invading capillaries but not by hypertrophs distant from blood vessels (Kume et al., 2002). It appears that these hypertrophic chondrocytes use their leptin to get BMP-2-and BMP-7 (OP-1) secreting vascular endothelial cells moving and proliferating and make proteases to tunnel through the matrix to enable the stimulation and delivery of osteoprogenitor cells to the new bone sites (Kume et al., 2002). Human articular chondrocytes express the complete functional large isoform leptin receptor LepRb and leptin stimulates the proliferation and extracellular matrix production of these cells (Figenschau et al., 2001). And R. Nakajima et al. (2003) have found that leptin also stimulates the proliferation and differentiation of rabbit growth plate chondrocytes. Kishida et al. (2005) also found that leptin is one of the controllers of chondroctye maturation and apoptosis in the mouse bones. But they have found that in mice it is the prehypertrophic (maturation-zone) chondrocytes, but not hypertrophs, that make leptin and the swollen hypertrophic chondrocytes have the the Lep Rb receptors to respond to it. However, while Kume et al (2002) would agree that most hypertrophs do not make leptin, the blood vessel-associated ones make a lot of it. In other words, it seems that leptin from the prehypertrophs can influence most of the the hypertrophs who have the means to respond. And this leptin restrains the hypertrophs from committing apoptotic suicide. When they turned to ATDC5 chondrocytes that express LepRb receptors, they found that a normal serum concentration of leptin (i.e., 1-10 ng/ml of medium), *acted like PTHrP* to stop matrix calcification and delay or prevent apoptosis (Fig. 38). The importance of the chrondrocyte leptin is indicated by the growth plates in leptinless *ob/ob* mice being nearly twice as fragile as the growth plates in normal mice (Kishida et al., 2005). Now Gat-Labonski et al (2006) have shown that leptin does in fact stimulate murine tibial hypertrophic chondrocytes to make PTHrP which retards the progression to hypertrophism and apoptosis.

Leptin also appears to be the missing link between obesity and osteoarthritis! At first sight this appears to be nonsense because surely the obesity-driven damage in knees for example would simply be due to excessive mechanical strain driving the collapse of cartilage and the growth of osteophytic spikes. But what about osteoarthritic finger joints of obese people which are not strained like their knees! This indicates that there is something made by fat cells that circulates throughout the body and hits all joints regardless of mechanical strain, although it would be expected to amplify the effects of strain. That circulating joint wrecker is most likely leptin. Indeed in humans leptin is found in the synovial fluid from osteoarthrtitc patients and there is a marked expression of leptin in osteoarthritic cartilage and osteophytic bone out-growths, but not in normal cartilage (Dumond et al., 2003; Loeser, 2003; Teichtahl et al., 2005; Terlain et al., 2005). In fact, the fraction of chondrocytes making leptin rises with the severity of joint damage (Teichtahl et al., 2005; Terlain et al., 2005). And injection of leptin into the tibial-femoral (knee) joints of rats stimulates the leptin-receptor-expressing chondrocytes to make IGF-I, TGF-β1 and cartilage proteoglycans (Dumond et al., 2003; Terlain et al., 2005). So the escalating leptin expression probably does its damage by stimulating chondro-cyte activities and with this the growth of joint-destroying osteophytes (Dumond et al., 2003; Teichtahl et al., 2005; Terlain et al., 2005). Thus, because of the aging and fattening of North Americans the osteophytic influence of leptin on joints will likely become very important in the near future.

Now back to the growing bone! The younger proliferating cells push the older chondrocytes into an increasing external Ca^{2+} concentration and a strongly increasing, maturation-driving inorganic phosphate concentration on the approaches to the hypertrophic calcification zone (Wuthier, 1993).

The Ca^{2+} turns on the cell's Ca^{2+} receptors (the CaRs; W.Chang et al., 2002, 2003; Chattopadhyay and Brown, 2003; Wang et al., 2001; Yamaguchi, 2003), the signals from which raise the $[Ca^{2+}{}_i]$ enough to "activate" the Ca^{2+}-sensitive silencer seeking the -1498 to the -863 region of the PTHrP gene's promoter. Once the signals from the Ca^{2+} receptors become loud enough, the silencer shuts off PTHrP expression and signaling from the high density of PTHR1 receptors on the mature hypertrophic cells' surfaces fades away (W. Chang et al., 2002; Iannotti et al., 1994; Weisser et al., 2002; Wu et al., 1995; Zuscik et al., 2002). Turning off PTHR1 signaling switches off the proliferation engine and Sox 9 expression which lifts the restraints on "HT" genes which can be stimulated by, among other things, Ihh-stimulated Cbfa1/Runx-2 (de Crombrugghe et al., 2001) (Fig. 38). And perhaps most importantly the silencing of the PTHR1 receptors by the PTHrP crash removes the anti-apoptosis shield. In other words when the external Ca^{2+} concentration reaches a critical level, the progression of the cells to the hypertrophic stage and their Cbfa1/Runx2-directed osteogenic doom is triggered.

As in the case of osteoblasts on the threshold of mineralization, the key driver of chondrocyte hypertrophic expansion and mineralization is probably the upregulation of STC1 (stanniocalcin1) and the Pit1 Na-phosphate co-transporter (Wu et al., 2006; Yoshiko et al., 2003). As you learned above when the inorganic phosphate reaches a critical level in cells such as osteoblasts and chondrocytes, it turns on the caspase 9-mediated, mitochondrial-driven apoptotic mechanism (Meleti et al., 2000; Sabbagh et al., 2005), unless stopped by PTH (Allen et al., 2002). The shutdown of PTHrP expression and PTHR1 signaling would thus enable the inorganic phosphate flowing into the hypertrophying cells through the STC 1-stimulated Pit1 co-transporters to do its ultimately deadly job in collaboration with Ca^{2+} (Magne et al., 2003). The phosphate stimulates key genes such as the type II collagen gene and together with Ca^{2+} enables apoptosis by reducing the anti-apoptosis Bcl-2 level and increases the pro-apoptosis Bax level (Coe et al., 1992; Magne et al., 2003).

The now gasping, hypoxic, hungry HT cells in the increasingly impermeable calcifying matrix, no longer protected by PTHR1 signals and with no longer adequate HIF-1-driven glycolytic machinery and consequently dwindling glycogen reserves, rapidly emptying ATP fuel tanks, and spluttering ion pumps, are now plunging toward their self-inflicted apoptotic doom in the hypoxia of the lower hypertrophic zone as their internal phosphate level rises menacingly and they fill their endoplasmic reticular storage vesicles with in-rushing Ca^{2+} and, in a last futile attempt to avoid a Ca^{2+} catastrophe, dump it into their leaking mitochondria (Brighton et al., 1976, 1978; Iannotti et al., 1994; Magne et al., 2003; Mansfield et al., 1999; Rajpurohit et al., 1999; L.N.Y Wu et al., 1995). The pH of the late hypertrophic and calcifying chondrocytes reaches its minimum which by protonating the annexins V(A5)s makes them more hydrophobic which in turn promotes their insertion into the cell membrane (Balcerzak et al., 2003). As the now mineralizing hypertrophs start apoptosing (mineralizing osteoblasts do not apoptose!), their membranes sprout blebs with annexin V (A5) plug-ins. These blebs are released as 100-nm vesicles, loaded with alkaline phosphatase, Ca^{2+}-binding proteins, collagens II and X and proteases, with their membranes studded with channels formed by annexin V(5) through which Ca^{2+} flows and the type III Na-phosphate co-transporters through which phosphate flows to form hydroxyapatite crystals on the inner surface of the vesicle membranes (Balcerzak et al., 2003). When the vesicles are loaded, their membranes are ruptured by the proteases and/or phospholipases, the hydoxyapatite needles are dumped into the extracellular fluid where they trigger a self-sustaining mineral nucleation on the matrix using the ambient normal levels of Ca^{2+} and phosphate (H.C. Anderson, 1995, 2003; Balcerzak et al., 2003; Bandorowicz-Pikula et al., 2001; Boskey, 1992; Hoshi and Ozawa, 2000; Kirsch et al., 1997, 2000; Plate et al., 1996). The cartilage calcifiers also now express collagen I which bridges the transition from a mineralized type II/IX/XI collagen cartilage matrix to a type I collagen-rich bone matrix made by the progeny of the Ihh/Cbfa1/Runx2-stimulated perichondrial/periosteal cells (Kirsch et al., 1997).

As these hypoxic chondrocytes have been pushed downward something else may happen that contributes to their injury and death. Their HIF-1αs have been busily stimulating the production of VEGF which has been stimulating the extension of blood vessels into the bone to deliver osteoclast and osteoblast progenitors to respectively clear away the cartilage and replace it with bone. With these vessels comes oxygen which instead of being a welcome relief for the hypoxic chondrocytes delivers the coup de grace because by causing a wave of acetyl CoA from the accumulated lactate to surge into the reawakened mitochondria. This causes the mitochondria to spray deadly ROS (reactive oxygen species) into the cytoplasm because the cytochrome oxidases are overwhelmed and cannot put all of the flood of electrons safely into water.

Interestingly, elasmobranchs (sharks and rays) appear to use part of the same PTHrP-dependent mechanism to make their permanent cartilagenous skeleton (Trivett et al., 2002). But the final stages are missing. There is no cartilage calcification, destruction and replacement by bone. Trivett et al. (2002) have suggested that ancestral elasmobranchs actually had the full endochondral bone-making process but lost it along the way. Indeed there is still some calcification of these beasts' outer notochordal sheath.

Why don't cortical osteocytes strangle themselves like chondrocytes when buried in calcified matrix? The answer is really quite simple—their elegant network of canals, the lacunocanalicular osteointernet. The buried osteocytes are continuously supplied with food and oxygen from the blood flowing through the vessels of the cortical osteonal canals and their waste is carried out and dumped into the blood through their lacunocanalicular network that is connected to blood vessels and nerves while chondrocytes are solitary, anti-social creatures who pay the price of their splendid isolation by having limited access to marrow blood vessels which forces them to depend entirely on diffusion through the cartilage matrix to get most of their supplies (Archer and Francis-West, 2003). While diffusion from the bone vascular "railheads" at the growth plate is at first quite adequate in the proliferative zone because the matrix is largely water and the cells can operate under the hypoxia prevailing in the core regions conditions by upreulating their HIF-1 (R.B. Martin et al., 1998; Schipani et al., 2001) it becomes lethally inadequate when the matrix is mineralized in the hyertrophic zone.

Disabling the genes for Ihh, PTHrP or PTHR1 releases the brakes on the proliferative "off-switch" and the cells prematurely rip open the "HT" envelope, commit apoptosis, and the bone under construction is abnormally short and its owner is a dwarf (Amizuka et al., 1994; Karaplis, 2001; Karaplis et al., 1994; Lanske et al., 1996; St-Jacques et al., 1999). However, if the Ihh gene is still working, accelerating maturation, and therefore the production of Ihh-making pre-hypertrophic cells by disabling the PTHrP or PTHR1 genes, increases Ihh production by the pre-hypertrophic cells. The increasing Ihh tries to make up for the chondrocyte shortage caused by the premature maturation by increasing the flow of peri-articular cells into the dividing columns (T. Kobayashi et al., 2002). On the other hand, uncontrolled, persistent PTHR1 signaling delays the proliferative switch-off and maturation which would tend to reduce the appearance of Ihh-making pre-hypertrophic cells and thus slow up the flow of cells into the dividing columns (Schipani et al., 1995, 1997; Weir et al., 1996). Excessive Ihh production also inhibits the initiation of maturation by swiftly turning up the peri-articular cells' PTHrP production which in turn would tend to feed back and reduce the appearance of Ihh-making pre-hypertrophic cells (Kobayashi et al., 2002; Minina et al., 2001; Vortkamp et al., 1996). Finally, Ihh-stimulated chondrocyte proliferation together with persistently firing, maturation-blocking PTHR1 receptors could be a deadly combination—it could cause tumors. And indeed it does! Hopyan et al. (2002) have found an abnormally signaling PTHR1 receptor in the cells of benign cartilage tumors from a human suffering from endochondromatosis (i.e., Ollier and Maffucci diseases).

Before leaving the cartilage and its chondrocytes one might ask whether the chondrocytes in a non-hypertrophying, non-ossifying *permanent* cartilage tissue such as articular (joint) cartilage, as opposed to the ephemeral cartilage model in a growing femur, are equipped like osteocytes to measure and appropriately respond to strain. Articular chondrocytes, unlike osteocytes, do not form intercommunicating fluid-filled canalicular networks filled with gap-junctionally connected cells with direct access to blood vessels. They rely on diffusion of nutrients through the matrix. Therefore, they must function with very low oxygen levels. This means that they must use anaerobic glycolysis instead of aerobic respiration to make their ATP fuel and thus do not have the large mitochondrial complements of osteoblasts and osteoclasts (Archer and Francis-West, 2003). Nevertheless they obviously do their job. To do this they express the stress-reducing, function-maintaining hypoxia-inducible factor compnents (HIF-1α

and HIF-1β) and use the HIF-1α•HIF-1β pair and the AP-1 transcription factor complex to drive various gene expressions such as the genes for TGF-βs and for the components of the glycolytic mechanism (Archer and Francis-West, 2003; Bilton and Booker, 2003; L.E. Huang and Bunn, 2003; Lando et al., 2003). But unlike growth plate chondrocytes, joint chondrocytes do not normally express annexin As and mineralize their matrix. But annexin As including atypical ones such as annexin VIII (A8) are expressed and mineralization develops with age and osteoarthritis.

How does an articular chondrocye feel the movement of its joint? Again we find the solitary cilium protruding from a chondrocyte with pericellular CMP (cartilage matrix protein or matrilins) linked to tensile force-concentratinmg collagen fibrils and aggrecan (Balcerzak et al., 2003; Q. Chen, 2004; C.G. Jensen et al., 2004). The cell responds with a Ca^{2+} signal when the cilium is tugged by the collagen fibrils attached to it when the cartilage matrix is alternately compressed and stretched by joint bending and twisting (Archer and Francis-West, 2003; Gillespie, 1999; C.G. Jensen et al., 2004; Meier-Vismara et al., 1979; Poole et al., 1997, 2001; Whitfield, 2003a; Wilsman, 1976; Wilsman and Fletcher, 1978) (Fig. 39).The response includes the stimulation of Ihh gene expression and the Ihh-driven upregulation of BMP 2/4 and the stimulation of chondrocyte proliferation, maintenance and cartilage repair if needed (Q. Chen, 2004). But there is a huge difference between an articular chondrocyte and an osteocyte. The signal from a collagen-pulled cilium of an anti-social chondrocyte embedded in its elastic collagen-aggrecan gel would not spread to its neighbors as would a signal triggered by a cilium bending in the canalicular fluid of an extensively connected, networking osteocyte "syncytium" (Whitfield, 2003a). Of course we really don't know what processes the signal triggers, but C.G.Jensen et al (2004) have suggested that the signal might stimulate the secretion of matrix components.

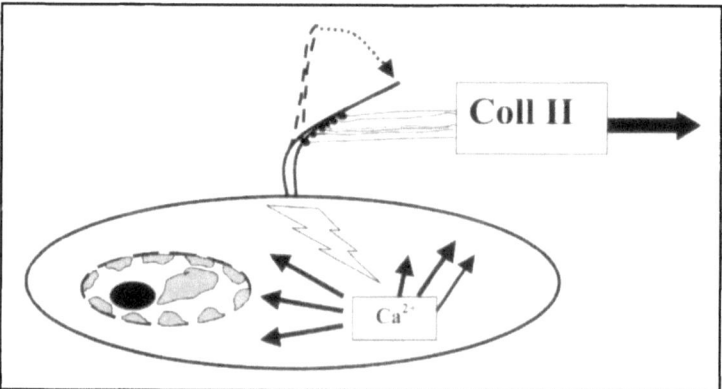

Figure 39. One of the two ways a chondrocyte in the permanent articulate cartilage can detect and respond to movements of the joint in which it resides. According to Jensen et al. (2004) the chondrocyte's solitary signal cilium is attached to collagen fibers. It follows that variously deforming the matrix will tug and bend the cilium back and forth. By analogy with the kidney tubule cell's cilium (Nauli and Zhou, 2004; Praetorius and Spring, 2001,2005), this tweaking of the cilium should send pulses of Ca^{2+} into the cell through the waving cilium's stretch-opened ion channels. These Ca^{2+} pulses trigger a burst of events such as gene activations leading to the production of matrix components and with them matrix maintenance and repair. This cilium signaling is supplemented by the matricrine signaling from the tugging of surface integrins linked to the collagen fibers in the extracelluar matrix for example by CMP (cartilage matrix protein or matrilins) (Q.Chen, 2004). The cilium in the cartilage cells of a growing bone may also bear the Ptch/Smo receptor complexes for hedgehogs that drive the initial stages of endochondral bone formation (Corbit et al., 2005; Huangfu and Anderson, 2005; Lum and Beachy, 2004).

vi. A Brief PTHrP-Sponsored Side-Trip to the Skin

Hopefully you will not mind taking a short side-trip hosted by PTHrP to look at it doing another of its many jobs controlling tissue development—in this case the skin epidermis. (We must not forget that some of our ancient predecessors had boney armorplates in their skin although these were the products of neural crest epithelial cells [Smith and Hall, 1990].) Here too we see the same pattern as in growing bone—stem cells giving rise to rapidly proliferating transit amplifying progeny which stop proliferating, mature and die in a curious hybrid process we have called *diffpoptosis* (differentiation and apoptosis) (Whitfield and Chakravarthy, 2001). The cycling transit amplifying basal keratinocytes start making and deploying PTH/PTHrP receptors that respond differently from the PTHR1 receptors of chondrocyte and osteoblast—normal keratinocytes may (Errazahi et al., 2003) or may not (Orloff et al., 1995; Whitfield and Chakravarthy, 2001) express their gene for the conventional PTHR1. At any rate the signals from this receptor, whatever it may be, do not include adenylyl cyclase activation, although the cells can respond with a large burst of adenylyl cyclase activity to the β-adrenergic receptor-activating isoproterenol or to PTH-(1-34) when they have been artificially engineered to express PTHR1 (Orloff et al., 1995; Whitfield et al., 1992). However, the keratinocytes do not make the PTHrP to stimulate these unconventional PTH receptors until they lift off the basal lamina otherwise the PTHrP would interfere with their proliferation. Indeed Menon et al. (1991) have shown that the proliferating basal cells are the only keratinocytes with substantial amounts of internal Ca^{2+} which according to the findings of Zuscik et al. (2002) should prevent PTHrP expression. These cells would be generating nuclear Ca^{2+} transients to start replicating their chromosomes and then to enter and complete mitosis (Whitfield and Chakravarthy, 2001). Then when the cells lift off the basal lamina they stop proliferating, nuclear Ca^{2+} drops and PTHrP is expressed (Juhlin et al., 1992; Menton et al., 1991; Whitfield et al., 2001). The amount of PTHrP coming down from the suprabasal cells tells the basal cells how many suprabasal cells are there and accordingly controls the rate of proliferation—a lot of cells in the stack means a lot of PTHrP coming down and a reduced production of basal cells while a depleted cell stack means less PTHrP coming down and increased basal cell production. Also the PTHrP trickling down reaches the underlying subdermal fibroblasts and stimulates them to make KGF (keratinocyte growth factor) which would, of course, drive the proliferation of transit-amplifying keratinocytes (Blomme et al., 1999). However, the PTHrP expression in the lifted cells does not last. The external Ca^{2+} concentration triples or more when the cells rise through the granular layer and into the lower corneum and, as happens at the threshold of the hypertrophic cartilage zone, this is enough to shut off PTHrP expression and trigger diffpoptosis which converts granular keratinocytes into cornified corpses (Whitfield and Chakravarthy, 2001).

Before going any further we must quickly look at what causes some basal stem cell daughters to lose their "stemness" and start proliferating and ultimately differentiating. When these daughters reach a certain point on the edge of the stem cell niche they stop expressing Jagged or Delta molecules, but express a receptor for these molecules known as Notch. This means that the Jagged or Delta on their stem cell sisters will bind to Notch, the signals from which causes the borderland daughters to become transit amplifying cells each of which produces a limited number of terminally differentiating offspring.

The Delta/Jagged-activated Notch receptors are cut by γ-secretase and the cytoplasmic fragment cut loose, ICN, travels to the nucleus where, as we learned above for hematopoietic control, it mates with CSL/RBP-J protein to form the ICN•CSL/RBP-J gene transactivator which eventually stimulates the expression of the cycle-stopping p21$^{WAF1/Cip1}$ via the so-called RPB- binding site in the gene's promoter switch box and the expression of terminal differentiation markers such as involucrin via the Notch ankyrin domain (Lefort and Dotto, 2004). Without Notch signaling and the switching off of the cycle-driving engine there could be hyperproliferation, deregulated expression of differentiation markers, and invasion of once-forbidden suprabasal non-proliferative zones by cycling keratinocytes (Lefort and Dotto, 2004; Rangarajan et al., 2001).

If the PTHrP from the suprabasal cells should also reach the dermal fibroblasts it would activate their normal adenylyl cyclase-coupled PTHR1 receptors. The burst of cyclic AMP production could cause the fibroblasts to make Jagged-1 as happens in hPTH-(1-34)-treated bone marrow osteoblasts and cultured rat UMR106 osteoblasts (Calvi et al., 2004; Weber et al.,2006). This Jagged-1 (the skin's most expressed Notch ligand) might be the same as the soluble Jagged1 made by neonatal keratinocytes (Aho, 2004) which would stimulate the Notch receptors on the overlying keratinocytes and drive their differentiation (Aho, 2004; Lowell et al., 2000; Nickoloff et al., 2002; Nicolas et al., 2003; Okuyama et al., 2004; Rangarajan et al., 2001; Thélu et al., 2002). This could be another way the amount of suprabasal PTHrPcould control keratinocyte production. And it would also mean that applying hPTH-(1-34) to the skin could reduce keratinocyte proliferation and drive differentiation.

It follows that a failure of keratinocytes to express PTHrP would cause a massively dysregulated cell production like that in psoriasis (Juhlin et al., 1992). The beefy red psoriatic patch is produced by the hyperproliferation of keratinocytes which causes a pile up of cornified squames with an impaired ability to slough off (desquamate). It appears that the reason for the psoriatic hyperproliferation is underexpression of the Jagged-Notch-Fringe signaling system which is upregulated in the basal and differentiating layers of the normal epidermis and maintains the normal relation between proliferation and differentiation (Thélu et al., 2002). The hyperproliferation of the basal keratinocytes and the suppression of the Delta-Notch control mechanism is somehow triggered locally by bacterially or otherwise activated Langerhans dendritic cells and driven for example by IL-8 from neutrophils streaming out of local blood vessels (Nickoloff et al., 2000). The keratinocytes in turn secrete autocrine (autostimulating)/paracrine (neighbor-stimulating) NGF that prevents them from self destructing while stimulating the T-cells, which have been directed to the battle zone by the activated Langerhans cells coming to the lymph nodes from the skin, and recruiting other factor-spewing inflammatory cells (Lebwohl, 2004; Nickoloff et al., 2000). And the keratinocytes also emit NO that causes the local dermal blood vessels to dilate which gives the patch its swollen redness and enhances the movement of activated Langerhans cells to the nodes, T-cells and neutrophils to and from the battle zone (Nickoloff et al., 2000). This torrent of pro-inflammatory cytokines and other factors results in the basal layer having a proliferation-favoring, lower than normal, level of *extracellular* Ca^{2+} and the suprabasal cells having an abnormally high extracellular Ca^{2+} level which could be why suprabasal psoriatic keratinocytes do not make PTHrP, the proliferative terminator (Holick et al., 1994; Menon and Elias, 1991; Zuscik et al., 2002).

Is this lack of PTHrP the reason for the excessive proliferation? Indeed it seems to be! Topically applying an adenylyl cyclase-stimulating PTH such as hPTH-(1-34) or one of the newer smaller Ostabolin™ Family PTHs, but *not* hPTH-(7-34) which cannot stimulate adenylyl cyclase, stops the wild cell production and eliminates psoriatic plaques (Holick et al., 2003). How? Keratinocytes don't have a PTH/PTHrP receptor that activates adenylyl cyclase! The answer likely lies with the dermal fibroblasts which do have the adenylyl cyclase-coupled PTHR1(Blomme et al., 1999) and thus may share with the closely related PTHR1-bearing, trabecular bone-lining cells of the hematopoietic stem cell niches, the ability to make large amounts of Jagged-1 when treated with hPTH-(1-34) (Calvi et al., 2004; Weber et al., 2006). Such Jagged-1 could restore order to the disturbed keratinocytes by reviving their Notch signaling-driven differentiaion mechanism (Lowell et al., 2000; Nickoloff et al., 2002; Nicolas et al., 2003; Okuyama et al., 2004; Rangarajan et al., 2001; Thélu et al., 2002; Whitfield, 2004a).

vii. PTHs and Rheumatoid Arthritis

In rheumatoid arthritis, a systemic autoimmune disease, immune-response cells such as macrophages, polymorphonuclear leukocytes and T-lymphocytes invade the joints and start pouring out cytokines that produce an inflamed synovium, the membrane of the sac that encloses the fluid that lubricates the joints. In mice and rats this cytokine storm can be started by injecting type II collagen intradermally along with Freund's adjuvant to induce an autoimmune reaction to the animal's own cartilage (Mori et al., 2002). Cytokines such as IL-1 and interferon-γ from this autoimmune response cause the formation of clusters of tall endothelial cells in the postcapillary venules of the synovial membrane that serve as the ports of entry of lymphocytes that form perivascular lymphoid aggregates or cuffs around the blood vessels like those of tonsil-like lymphoid tissues (Iguchi and Ziff, 1986; Ziff, 1993). These lymphocytes plus macrophages and plasma cells swarming through the layer of flat endothelial cells of the post-capillary venules release a swarm of cytokines which includes the mighty TNF-α, that pushes the synovial fibroblasts to proliferate and make things such as PTHrP and RANKL which stimulates osteoclast generation and with it bone destruction (Funk, 2001; Funk et al., 2002; Goldring, 2002; Goldring and Gravallese, 2000, 2002; Gravallese and Goldring, 2000; Iguchi and Ziff, 1986; Mori et al., 2002). The surface of the underlying bone is primed for an osteoclast onslaught by proteases from both the invading leukocytes and the proliferating synovial fibroblasts. It becomes the site for the generation of osteoclasts from invading precursors that is driven initially by RANKL from the invading T-lymphocytes but later mostly from the cytokine-crazed, proliferating synovial fibroblasts (Goldring, 2002; Goldring and Gravallese, 2000, 2002; Gravallese and Goldring, 2000; Mori et al., 2002).

The synovium now relentlessly thickens into a predatory, giant ameba-like, bone-eating blanket or *pannus* (Latin for cover), a tumor-like blanket of inflammatory factor-secreting granulation tissue made up of polymorphonuclear leucocytes and lymphocyte invaders and the proliferating fibroblasts which vigorously make PTHrP that, like TNF-α, stimulates RANKL expression by the underlying osteoblastic cells (Funk et al., 2002; Goldring, 2002; Goldring and Gravallese, 2000, 2002; Gravallese and Goldring, 2000; Mori et al., 2002).

It must clearly be understood that inflammation and its associated cytokines do not cause the osteoclast attack and bone loss—something special is needed. This special something is the RANKL from T-cells and synovial fibroblasts as well as bone marrow stromal cells and osteoblasts (Bolon et al., 2002a, 2002b). Therefore, mice without functional RANKL genes can still develop severely inflamed joints in an autoimmune response response, but without RANKL no osteoclasts arrive on the scene to dig holes in the bones (Goldring, 2002; Goldring and Gravallese, 2002). However, the IL-1 cytokine from the immunocytes does stimulate osteoclast generation by its sharing with RANKL the ability to stimulate NF-κB in osteoclast precursors (Xing et al., 2003).

As the pannus grows, it, like a tumor, must make angiogenic factors such as VEGF to make new blood vessels to get its oxygen, nutrients and along with these more blood-borne cytokine-producing lymphocytes, macrophages, neutrophils and osteoclast precursors to help drive it forward (Funk et al., 2002). The endothelial cells of the pannus's blood vessels make PTHrP, but they don't have the PTHR1 receptors to respond to it, although they can ship some of it into their own nuclei instead of outside the cell to do stimulate genes the products of which can prevent them from committing apoptotic suicide (Funk et al., 2002; Lam et al., 2002). The PTHrP from the endothelial cells hits the underlying vascular smooth mucle cells, which do have PTHR1 receptors (Funk et al., 2002). The cyclic AMP signals from the PTHR1 receptors cause the smooth muscle cells to relax and the blood vessels to dilate and thus carry more blood into the advancing pannus.

The steady pounding from the rising tide of RANKL-stimulating PTHrP from the pannus cells crawling over the joint causes the excessive osteoclast generation and digging to spread from the superficial subchondral bone to endosteal bone which can eventually spead systemically to osteoporosis(Fu et al., 2002; Funk, 2001; Funk et al., 2002).

Obviously the bone erosion by osteoclasts in a rheumatoid arthritic joint can be stopped for example with the very long-acting OPG, the RANKL decoy ligand, that prevents mature osteoclasts from setting up their sealing ring and resorbing bone by directly binding to a 140-kDa protein on their surfaces and takes RANKL out of action which prevents osteoclast generation by depriving osteoclast precursors of signals from their RANK receptors, (Bolon et al., 2002a, 2002b; Goldring, 2002; Goldring and Gravallese, 2002; Hakeda et al., 1998). Alternatively osteoclasts can be disabled and killed by one of the bisphosphonate gang or calcitonin. It seems likely that a short treatment with single daily shots of one of the osteogenic PTHs, particularly hPTH-(1-31)NH$_2$ (Ostabolin™) or [Leu^{27}]cyclo(Glu^{22}-Lys^{26})hPTH-(1-31)NH$_2$ (Ostabolin-C™), that are very poor stimulators of osteoclasts (Fraher et al., 1999; Jolette et al., 2003,2005; Mohan et al., 2000; Whitfield, 2003b) should be able to do something that OPG or the antiresorptives cannot do—re-grow the bone in the osteoclast-gnawed joints and as well as restore bone lost throughout the skeleton because of the spread of osteoclastogenic factors from the damaged joints, a lack of activity due to severe joint pain, and the use of osteoporosis-inducing glucocorticoids (Lane et al., 2000; Whitfield, 2003b). Indeed the very promising first indications of this have been provided by Fukata et al. (2003)'s report that intermittent hPTH-(1-34) injections (3 times per week for 4 or 6 weeks) can prevent or stop the drop in cortical and trabecular BMD, increase trabecular thickness and increase the trabecular mineral apposition rate, i.e., increase trabecular bone formation, in the tibias of female Sprague-Dawley rats with collagen-induced arthritis.

viii. PTHs and Cancer—A Concern with All Growth Stimulators

Anything that directly or indirectly stimulates proliferation may also drive tumor formation. The level of the risk depends on the intrinsic genomic instability of the target cell population, the mutagenicity of the agent and the number of cycles it forces its target cells to pass through (i.e, the duration of exposure) as well as the levels of other mutagens in its surroundings (Gibbs, 2003; Greaves, 2000). The question immediately arises as to whether osteogenic PTHs are also carcinogenic. Of course no osteogenic, PTHR1-activating N-terminal PTH is going to switch on the cell's cycle engines to get an M-team shut-down osteoblast, osteocyte or lining cell to start proliferating. But it will switch on the expression and secretion of missionary growth factors that can stimulate the proliferation of the transit-amplifying progenitor cells which are too immature to have PTHR1 receptors (Fig. 23). This means that the cells of any osteosarcoma that might appear should not have PTHR1 receptors. The PTHs are not genotoxic (mutagenic), but if there should be a carcinogenesis-initiating mutation in one of the osteoprogenitors stimulated to proliferate by the missionary short FGF-2, IGF-I or IGF-II from the PTH-stimulated post-proliferative osteoblasts, there would be no immediate danger if the mutant osteoprogenitor or its progeny could still eventually mature into irreversibly post-mitotic preosteoblasts and beyond after only a few additional cycles. But things could become dangerous if the the PTH-driven bombardment with growth factors persists and mutant progenitor cells appear which have disabled cycle suppressor genes or permanently activated genes for cell cycle drivers or have acquired the ability to constitutively make their own autocrine growth factors. These mutant cells keep proliferating and further accumulating mutations that produce dangerous tumor-forming mutant progeny that are immune from being induced to commit apoptotic suicide probably because of the disabling or loss of p53, the cell's mutation censor that would normally force it to apoptotically self-destruct (Whitfield and Chakravarthy, 2002) (Fig. 40). Then in one of these rogue clones a cell may undergo an additional, deadly mutation which enables it to found a clone of widely metastasizing osteosarcoma cells which successfully escape from the constraints of multicellularity (Fig. 40).

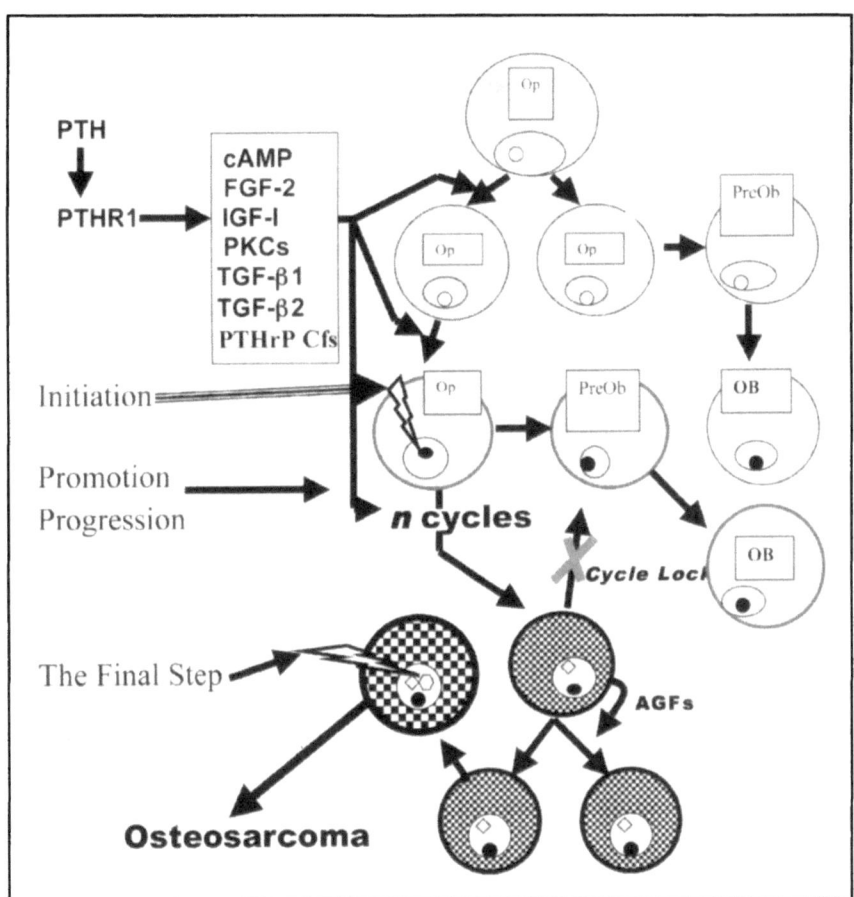

Figure 40. How a lifelong barrage of PTH boluses can induce osteosarcoma. Basically the signals from PTHR1 receptors and the release of the several missionary growth factors listed in the panel (note that **PTHrP Cfs** refers to cleavage products of the PTHrP chain of factors) from preosteoblasts and mature osteoblasts drive the proliferation of immature osteoprogenitors. Over time mutations happen in the continuously proliferating cells and accumulate (as symbolized by the progression of nuclear circles) up to the point where cells appear that continuously make their own autocrine growth factors which lock them into the cycling mode (**cycle lock**) and prevent them from shutting down their cell cycle machinery and mature. At some point in their carcinogenic promotion they also lose their ability to check for, and eliminate, mutations or self destruct if a mutation can't be eliminated. This loss causes the mutation load in previously intercepted and discarded, but now suvivning, cells to freely generate metastatic clones and cancer. Abbreviations: **AGF**, autocrine (autostimulating) growth factors; **OB**, mature osteoblasts **Op**, juvenile osteoprogenitor; **PreOb**, preosteoblast.

This appears to be what actually happens in a rat subjected to a life-long treament with rhPTH-(1-34) (Forteo™). At the July 27, 2001 meeting of the FDA Endocrinologic and Metabolic Drugs Advisory Committee it was revealed that a lifelong 2-year treatment with Lilly's rhPTH-(1-34) (Forteo™) is a potent osteosarcoma inducer in both female and male Fisher rats (Center for Drug Evaluation and Research, 2001; Vahle et al., 2002, 2004). Treating the rats with the peptide at doses of 5, 30, or 75 µ/kg of body weight respectively caused the appearance of osteosarcomas in 3, 21, and 31 of 650 male rats and 4, 12, and 23 of 60 female rats. And as predicted above, none of the cells in the osteosarcomas had PTHR1 receptors as did the surrounding normal bone cells. This indicated that these tumors arose from the PTHR1 receptorless progenitors that were driven by daily pulses of missionary growth factors from PTH-stimulated PTHR1-bearing mature osteoblastic cells.

At first glance, this seems to mean that PTH treatment for osteoporosis is dangerous. But it most likely isn't! But this kind of demonstration is ludicrously unrealistic. The daily dose of the fragment given to the rats (75 μg/0.4-0.6 kg of body weight) was about 400 times the 20 μg/40-60 kg of body weight recommended for treating osteoporotic humans (Center for Drug Evaluation and Research, 2001). Although extensively proliferating rat and human osteoprogenitors obey the same rules, they are probably more genomically unstable in an in-bred rat than in a "wild-type" human as indicated by the very high basal osteosarcoma inci-dence of 2000-4000 x10^{-6} in the Fisher rats compared to the 1000-fold lower (3-5 x 10^{-6}) incidence in humans (Gurney et al., 1995). Moreover, the short-lived, small-boned rats keep growing throughout their lives (though at a decelerating rate and with a decreasing osteogenic responsiveness to PTH [Walker, 1971]) and therefore always have a substantial pool of imma-ture osteoprogenitors and PTH-responsive mature osteoblasts and lining cells for continuous bone-making. But because the long-lived large-boned humans close their bones' growth plates and stop growing, the size of their osteoprogenitor pools are limited to providing osteoblasts for the microcrack-repairing BMUs. The PTH-stimulable bone-lining cells are not dangerous because they can only become proliferaively incompetent functioning osteoblasts. Indeed the osteosarcoma incidence in humans peaks in growing children around 13-15 years of age and then virtually vanishes after growth stops (Gurney et al., 1995) while a rat, especially a male rat, stays equivalent to a human child throughout its life. Neither the humans nor 80 cynomolgus monkeys treated in the trials with 8 times the human dose of Forteo™ for 18 months devel-oped osteosarcoma (Center for Drug Evaluation and Research, 2001). Of course if osteosarco-mas are as rare in monkeys as in humans, the Lilly Forteo™ team would have needed a huge number of monkeys and a decade or more to detect an increased osteosarcoma incidence.

Vahle et al (2002,2004) have experimentally addressed the contributions of the dose and duration of Forteo™ treatment to its oncogenicity and whether there is a dose that is osteo-genic, but not oncogenic. As expected when they gave Fischer rats 344 rats 30 μg of peptide/kg of body weight for 20 or 24 months 27.5% developed osteosarcoma, but when rats were given the same daily dose for only 6 months between the ages of 6 and 8 months (i.e., after the maximum growth phase) only 3% developed the cancer which was the same as in the vehicle-treated control animals. But when the animals were given daily injections of 5 μg of the peptide/kg of body weight starting when they were 6 months of age and continued for either 6 or 20 months they developed no osteosarcomas, but had substantially increased bone masses. Therefore, there was a "no-effect" dose for oncogenicity that still caused a considerable rise in bone mass (Vahle et al., 2004). In other words they had established once again the "2 great Ds" (dose and duration of treatment) of oncogenesis, but this time for a PTH.

We were falsely reassured by the fact that none of 12,000 chronically hyperparathyroid Swedes has developed osteosarcoma (Center for Drug Evaluation and Research, 2001). But they didn't look hard enough—as of the first half of 2003 there were reports of four women, two of whom are Swedes, with parathyroid adenomas and high blood concentrations of PTH. Wiig et al., (1971) reported two cases of hyperparathyroid Swedish women, a 46-year-old with tibial dedifferentiated chondrosarcoma and a 34-year-old with a high-grade mandibular osteosarcoma. Smith et al (1997) described the case of a 49-year-old woman with a surgically removed parathyroid adenoma and multiple malignant bone tumors. The latest is a recent report by Betancourt et al. (2003) of a 69 year-old black woman who had a parathyroid adenoma that had raised the circulating PTH concentration above 1000 pg/ml (a normal concentration would have been 10-65 pg/ml) and along with this she had developed a high-grade fibroblastic osteosarcoma. These osteosarcomatous women seem at first sight to be very significant because they were all much older than than the usual osteosarcoma patient who is 15 or less years old. However, a very recent study by Cinamon et al.(2006) have scanned the records of patients with primary hyperpathyroidism to try to find any indications of osteosarcoma. None of 582 patients with primary hyperparathyroidism had osteosarcoma. On the other hand, none of 126 patients with osteosarcoma had primary hyperparathyroidism or its biochemical markers.

Unfortunately this lack of a connection between primary hyperparathyroidism and primary osteosarcoma might not be comforting. Certainly,these hyperparathyroid people secrete a lot of hPTH-(1-84). But the N-terminal segments of the secreted native PTH molecules are promptly and selectively destroyed (Bringhurst, 2002; Bringhurst et al., 1988), as I have previously noted, in order to prevent them from competing with locally acting PTHrP molecules for PTHR1 receptors. However, one N-terminal PTH, hPTH-(1-37), but not hPTH-(1-34), is detectable in the blood of people suffering from chronic kidney failute (Hock, et al. 1997). This suggests that these hyperparathyroid people might have no circulating hPTH-(1-34)-like fragments to trigger osteosarcomas. An injected N-terminal PTH such as Forteo™ is in fact an unnatural addition to the circulating group of factors. Therefore we really still don't know whether Forteo™ or other osteogenic PTHs are oncogenic.

But there could be another danger—colon cancer! The proliferatively active colon cells, like osteoblasts, also have PTHR1 receptors (Gentili et al., 2001, 2002; Li et al., 1995). Signals from these receptors in rat duodenal cells rapidly (within just 5 minutes) cause a c-Src-kinase-mediated phosphorylation and activation of phosphatidylinositol-3 kinase which in turn triggers a mitogenic $ERK1 \Rightarrow ERK2$ cascade (see Pearson et al., 2001 for a review of these proliferation-stimulating enzymes) in rat intestinal cells (Gentili et al., 2001, 2002) just as the signals from PTHR1 receptors attached to NHERF-2 scaffolds do in PS120 fibroblasts (Mahon and Segre, 2002). Therefore, boluses of PTH could stimulate colon cell proliferation and at least promote colon carcinogenesis. Indeed there are scattered reports of primary hyperparathyroidism with its 2-10-fold higher level of circulating PTH-(1-84) being associated with increased colon cancer incidence in humans (Artru et al., 2001; Conteas et al., 1988; Farr, 1976; Feig and Gottesman, 1987; Kawamura et al., 1999; Strodel et al., 1988).

It seems that colon cancer may not be a problem in rats. The lifetime treatment of Lilly's Fisher 344 rats with huge doses of hPTH-(1-34) (Forteo™) seems to have only induced osteosarcomas which could spread to soft tissue (Vahle et al.,2002, 2004). But we still had to find out whether a realistic PTH treatment, i.e., one that was just long enough to stimulate bone growth but no longer could either by itself trigger colon carcinogenesis or enhance the triggering of carcinogenesis by a colon carcinogen. In collaboration with Dr R.P. Bird of the University of Manitoba we (Whitfield et al., 2003) have tested the ability of a strongly osteogenic 6-week treatment (i.e., once-daily injections of 20 nmoles/kg of body weight) of Sprague-Dawley rats with two PTH fragments, hPTH-(1-34) and $[Leu^{27}]cyclo(Glu^{22}\text{-}Lys^{26})hPTH\text{-}(1\text{-}31)NH_2$ (Ostabolin C™), starting 9 weeks after OVX, to *initiate* colon carcinogenesis or increase the initiatory activity of two pretreatment injections (15 mg/kg of body weight) of a potent colon carcinogen—AOM (azoxymethane). The initiation of colon carcinogenesis by AOM is indicated by the appearance within just 6 weeks of as many as 263-273 of strikingly visible and very easily counted ACFs (aberrant crypt foci) per colon, a few of which if we had waited long enough would have sprouted tumors (Bird and Good, 2000; Whitfield et al., 2003). (In other words one does not have to wait for very long to find out whether an agent can trigger colon carcinogenesis, but of course it would be necessary to wait for 2 years to know how many of the aberrant crypt foci could complete the process and generate tumors [Bird and Good, 2000]). While the daily PTH injections stimulated femoral bone formation, they did not cause the appearance of ACFs or affect the induction of ACFs by AOM or accelerate the progression of primary single ACFs to multiple ACFs (Whitfield et al., 2003). Nor did AOM affect the PTHs' abilities to stimulate bone formation (Whitfield et al., 2003).

While a PTH treatment that is no longer than needed to strongly stimulate bone formation does not by itself initiate colon carcinogenesis or enhance or accelerate the initiation of colon carcinogenesis by a potent carcinogen in rats, we can't lower our guard yet. There is a danger that a human (or primate)-type colon may respond differently to PTHs as indicated by the increased incidence of colon cancer in hyperparathyroid humans. Clearly we must find out whether PTHs stimulate ACFs and ultimately tumor formation at least in the colons of non-human primates.

ix. A Possible Threat

In Chapter 2 iv c, we learned that arterial plaques could be transformed into true bone by an active process involving migratory advential myofibroblasts, CVC- (calcifying vascular cell) and vascular smooth muscle cell (VSMC)-derived osteoblast-like plaque cells induced by endothelial cell-derived BMP-2 and the Msx2 expression it stimulates in vascular target cells (Doherty et al., 2002, 2003a, 2003b, 2004; Hruska et al., 2005; Shao et al., 2005; Whitfield, 2005a). In other words there are small "plaque bones" in atherosclerotic blood vessels which in some cases even have hemopoietic marrow. From the "working" model in Figure 23 the osteogenic PTHs and their PTHR1 receptors would be expected to stimulate "plaque bone" growth and with this aggravate coronary heart disease.

The expressions of PTHrP and signals from activated PTHR1 receptors are maturation program drivers for skeletal osteoblasts, with the P3 promoter of the the *pthr1* gene being one of the targets of osteoblastic maturation-driving Cbfa-1/Runx-2 (Fig. 9). PTHrP and the cyclic AMP signals from PTHrP-activated PTHR1 drive osteoblast maturation by upregulating the Jun B transcription factor and disconnecting the cyclin D1-CDK 4/CDK 6-mediated mechanism that starts the events leading to the initiation of chromosome replication and mitosis (Datta et al., 2005).

VSMCs also make PTHrP and decorate their surfaces with PTHR1 receptors in human and rat atherosclerotic lesions upon injury which is the site for atheroma initiation (Nakayama et al., 1994; Ozeki et al., 1996; Ishikawa et al., 2000; Martin-Ventura et al., 2003; Whitfield, 2005). The VSMCs in the "shoulder" region—the inflammation action center—of human carotid atherosclerotic plaques also overexpress PTHrP and the PTHR1 receptor (Martin-Ventura et al., 2003). The cyclic AMP signals from the PTHR1 receptors activated by *externally added* PTHrP stimulate the VSMCs to make MCP-1 chemokine to increase the migration of circulating monocytes through the vascular endothelium and swell the foam cell (FCELL) population in the subendothelial vascular intima (Charo and Traubman, 2004; Martin-Ventura et al., 2003) (Fig. 41).Therefore, the VSMCs' overexpressed PTHrPs load the plaque's intima with proinflammatory and osteoinductive factors from accumulating macrophages (Fig. 41). But endogenous PTHrP does a lot more.

You may recall from chapter 2 v that the full-length PTHrP (the holo-or H-PTHrP in Fig. 41), such as that made by the VSMCs in the plaque "shoulder", has something in its 86-106 region that no PTH has—a NLS (nuclear/nucleolar localizing signal sequence) that enables some emerging full-lenngth PTHrPs (H-PTHrP) to be shipped from their factory in the VSMC's endoplasmic reticulum to targets in the nucleus (Massfelder et al., 1997; De Miguel et al., 2001; Fiaschi-Taesch et al., 2004; Schordan et al., 2004). A full-length PTHrP translation product with a traditional upstream signal sequence is fed into the cell's secretion machinery, secreted and then with its N-terminal 1-36 region activates the producer cell's or a neighboring cell's PTHR1 receptors (Massfelder et al., 2004; De Miguel et al., 2001; Fiaschi-Taesch et al., 2004) (Fig. 41). But recall that there are other translation initiation sites on the transcript that disrupt the secretion signal sequence. This prevents the protein from being grabbed by the

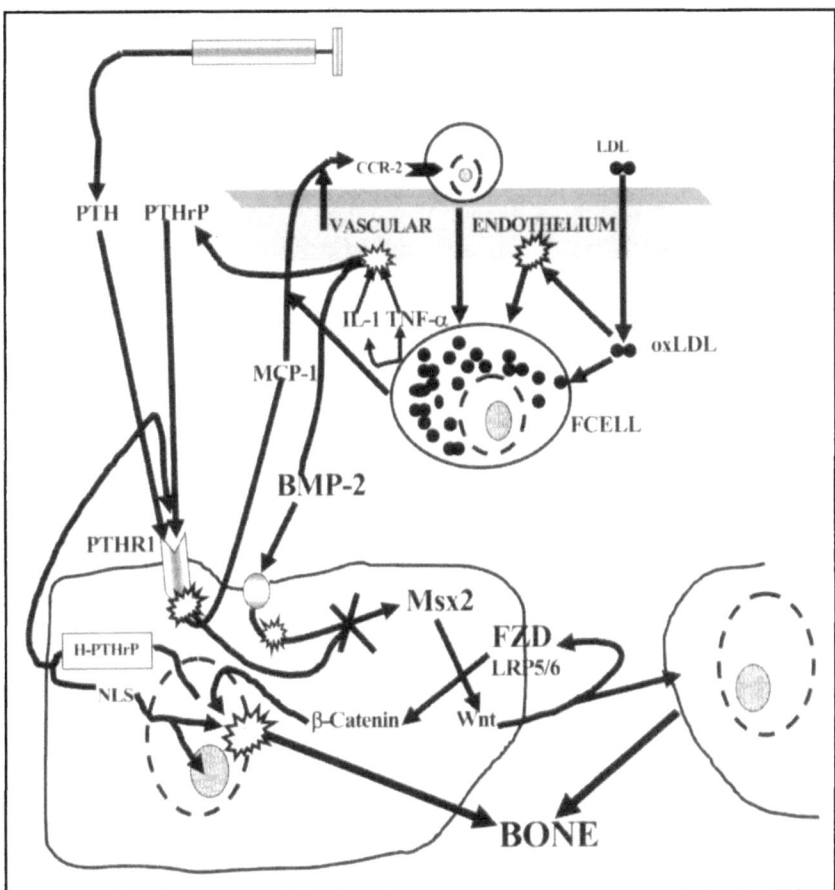

Figure 41. The involvement of PTH, PTHrP and their shared PTHR1 receptor in vascular ossification. Vascular bone formation starts in a hypercholestrolemic person with influx of LDL particles through the endothelium into the intima of a lesion-prone region of an arterial wall to start an inflammatory process. There the LDLs are modified by oxidation (oxLDL) and cause the endothelial cells to express surface molecules that grab passing mononuclear cells. The snared mononuclear cells are induced to move across the endothelium by diapedesis when the MCP-1 chemoattractant chemokine binds to their CCR-2 receptors. In the subendothelial intima the mononuclear cells cells transform into macrophages that feed on LDLs to become foam cells (FCELLs) which produce several cytokines, chemokines (including MCP-1) and growth factors (IL-1, TNF-α). One of the effects of the many factors from the endothelial cells and macrophages, particularly the osteogenic BMP-2, is to cause the endothelial cells to make PTHrP and the vascular smooth muscle cells to express both PTHrP and PTHR1 receptor. The full-length PTHrP (H-PTHrP; holo PTHrP) made by the cells gets into the nucleus via its NLS (nuclear localization signal sequence). Meanwhile BMP-2 stimulates the expression of Msx2 which in turn activates the Wnt mechanism which is promoted by the PTHrP and stimulates bone formation by β-catenin-activation of osteoblast-specific genes. However, an injected osteogenic PTH (which never has a NLS) that is being used to stimulate bone growth in an osteoporotic person, or the cell's own secreted PTHrP, or an injected N-terminal PTHrP fragment (e.g., hPTHrP-(1-34) that cannot get into the cell nucleus because it lacks the NLS only activates the cells PTHR1s, the signals from which prevent the smooth muscle cell from becoming an osteoblast by stopping BMP-2 from stimulating Msx2 and the Wnt mechanism. But the PTHR1's cyclic AMP signaling also stimulates the expression of MCP-1 which further drives the migration of mononuclear cells into the vascular wall and with this the atheromatous process.

secretory mechanism (Massfelder et al., 2004; De Miguel et al., 2001; Fiaschi-Taesch et al., 2004). The emerging *full-length* H-PTHrP's NLS is now grabbed and shipped into the nucleus where it stimulates proliferation by a process that requires its C-terminal 112-139 region and hyperphosphorylates and silences the inhibitory cell cycle G_1/S checkpoint regulator pRB (retinoblastoma protein) (Fiaschi-Taesch et al., 2004) (Fig. 41). VSMCs engineered to overexpress a NLS-less PTHrP,which cannot get into the nucleus, secrete the peptide which stimulates the PTHR1 receptors, the signals from which inhibit proliferation, carotid intimal thickening and vascular ossification (Massfelder et al., 1997, 1998; Massfelder and Helwig, 2003; Fiaschi-Taesch et al., 2004) (Fig. 41). In other words, preventing PTHrP from getting into the nucleus converts it from a proliferative and ossification stimulator to a PTHr1-activating inhibitor of VSMC proliferation.

It follows from this that a skeletally osteogenic PTH such as rhPTH-(1-34)OH (Lilly's Forteo™) or PTHrP-(1-34)OH should reduce vascular intimal thickening and ossification. Indeed Ishikawa et al. (2000) have shown that injected PTHrP-(1-34) prevented cuff placement from inducing intimal thickening in rat femoral artery. And Jono et al. (1997) reported that hPTH-(1-34)OH and hPTHrP-(1-34)OH inhibited calcification by bovine VSMCs. Then Shao et al.(2003) reported that intermittent injections of hPTH-(1-34)OH into LDLR (low-density lipoprotein receptor)-lacking mice prevented a diabetes-producing high-fat diet from ossifying aortas and heart valves as happens in untreated mice. But this flatly contradicts the fact that the potently osteogenic, PTHR1-activating N-terminal PTH and PTHrP fragments stimulate bone-lining cells to revert to bone making osteoblasts! Why are VSMCs so different?

You may recall from section iiie that PTH-(1-34)OH osteogenically activates lining cells by doing two things. First it stops osteocytes from making SOST that inhibits the Wnt/FZD/LRP5/6 mechanism that enables β-catenin to build-up in the cytoplasm and get into the nucleus and turn on a cluster of osteogenic genes (Fig. 27). And second, it stimulates the expression of FZD and LRPs and shuts down the expression of the FZD/LRP inhibiting Dikk-1 (Fig. 27). VSMCs are stimulated by BMP-2 from cytokine-stimulated endothelial cells to express Msx2 which in turn stimulates Wnt expression, secretion and with this the activation of the FDZ/LRPs in the Wnt producer cell and its neighbors (Fig. 41). Happily for osteoporotics, PTHR1's cyclic AMP signals inhibit BMP-2 stimulation of Msx2 expression and Wnt-driven ossification in the VSMCs (Shao et al., 2005) (Fig. 41).

x. PTHrP, PTHR1 Receptors and Stroke

Neurons in parts of the brain such as as the parietal cortex and the striatum make PTHrP (Funk et al., 2003; Weir et al., 1990). Neurons also express the PTHR1 receptor as do the smooth muscle cells of the brain microvasculature (Funk et al., 2003). When parts of the "stroked" brain are deprived of blood (i.e., when they have been rendered *ischemic*), both the glial cells and neurons promptly respond by making and secreting the pro-inflammatory cytokines IL-1β and TNF-α (Liu et al., 1993, 1994). These cytokines together stimulate both the endothelial cells of the regional blood vessels and the local glial cells to make PTHrP (Fig. 41; Funk et al., 2001, 2003). This cytokine-PTHrP response is an ingenious way to get more blood flowing into the ischemic regions (Funk et al., 2003; Whitfield et al., 1997a). The PTHrP from the vascular endothelial cells and associated glial cells binds to PTHR1 receptors on the vascular endothelial cells, stimulates adenylyl cyclase and triggers a cyclic AMP surge that causes the blood vessels to dilate and the blood flow to increase.

This mechanism has been illustrated by experiments reported by Funk et al. (2003). They found that blocking a rat's middle cerebral artery stimulated IL-1β and TNF-α production in the ischemic cerebral hemisphere. These cytokines increased PTHrP production by the cerebral vasculature which, by dilating the blood vessels, tripled the flow of blood into the ischemic hemisphere. They also showed that when boluses of PTHrP were delivered into the lateral cerebral ventricle at the rate of 5 μl/1-2 minutes for 30 minutes before and 90 minutes after arterial blockage, the infarct size was reduced by 34%. Thus PTHrP and its N-terminal receptor have the potential of being used to reduce damage in the "stroked" brain if ways can be found to get it to the ischemic tissue. However, intravenous injection should deliver the peptide to the walls of the blood vessels in the stroke penumbra and cause the vessels to dilate. The cyclic AMP surge caused in blood vessels by PTH and PTHrP might also relieve the severe local ischemia caused by the vasospasm triggered by the NO-binding hemoglobin in the blood from a burst cerebral blood vessel. But this might also make matters worse by dilating the ruptured blood vessel.

xi. Grand Summary

Now that you have bravely plowed through this mass of facts and necessarily a lot of fancy to reach this point, you might appreciate a working model to pull it all together. This model is in Figure 23. It is all there with Fis1/INI1, endothelin and its receptors, the several missionary factors, SOST, the appearance of PTHR1 receptors, etc. You can follow the cells and their drivers as they progress from the first progeny of a mesenchymal "stem cell" to transit amplifying progenitors, PTHR1-festooned osteoblasts and finally osteocytes and lining cells or apoptotic cadavers. You can even make a bone-lining cell revert to a functioning osteoblast by simply pushing its PTHR1 button.

CHAPTER 6

The Clinical Prospects of the Invincible PTHs

"Market was the key word here, for at big pharma, marketers, not scientists, decided which drugs the companies should pursue."—M. Shaynerson and M.J. Plotkin (2002)

The osteoporosis market is a rapidly growing "marketer's" dream. And several years ago there was no known bone-growing drug—only the ever-worrisome (for cancerophobics) estrogens and the other antiresorptives. So the marketeers at Eli Lilly decided that the old hPTH-(1-34) would be the ideal drug with which to break into this pharmaceutical gold mine. All they had to do was take it off the shelf, brush off the dust, spider webs and dead flies, and infect *Escherichia coli* with a suitable expression vector into which the gene had been stitched to make it, and then sell tons of it to the growing hordes of fracturing "greys" to toughen up their fragile bones and strengthen the anchorage of the artificial hips and knees they will also need. Thus was born Lilly's recombinant hPTH-(1-34) which was officially approved for clinical use under the trade name Forteo™ by the USFDA in November 2002.

It must clearly be understood that any PTH, such as hPTH-(1-34) (Forteo™), will never be used as a preventative like the much cheaper antiresorptives(anti-catabolics). This, of course does not please the marketers who love continuously given drugs, like the diabetics' insulin or some future drug that must be given regularly to drain the Alzheimer patient's brain of the continuously accumulating neuron-destroying β-amyloid protein, that produce a steady cash flow. Indeed, for example space travelers should not use an expensive PTH to maintain their bones during flight because weekly Fosamax™ would be a cheaper bone-maintaining agent. A PTH should used: (1) when a man or woman has a high risk of fracture; (2) when the person has had an osteoporotic fracture; (3) when the person's BMD is more than 3 standard deviations from the population mean; (4) when the person is of an advanced age; (5) when the person cannot, or will not, take bisphosphonates; (6) when the person has had a fracture while being treated with a bisphosphonate. Currently the only available hPTH-(1-34) (Forteo™) will be used for no more than a couple years to restore severly deteriorated microarchitecture and lost bone strength and then turn the job over to an antiresorptive to protect the new "PTH bone" in postmenopausal women and men. It is also likely that the PTHs will be used for short periods to accelerate fracture mending not just in osteoporotics and space travelers, but in people of all ages, especially children with their osteoblast-loaded, hence maximally PTH-responsive, bones (Koch, 2001; Walker, 1971), and they should be used to strengthen the depleted bones of astronauts returning from long space missions during which they had not been protected from bone loss in flight by taking an antiresorptive such as Fosamax™. But the PTH's may be less effective or ineffective *during* long space missions because of a progressive loss of the ability of the osteoprogenitor cells, genomically unprimed by strain, in unloaded bone to respond to PTH-induced IGF-I (Bikle et al., 1994; Kostenuik et al., 1999)!The PTH's should also be able to strengthen bone eroded by arthritis. And last, but by no means least for aging populations, the PTHs will soon be accelerating the osteointegration, and later reversing any loosening, of various orthopedic and dental implants (M..Allen et al., 2003; Kawane et al., 2002; Shirota et al., 2003; Skripitz and Aspenberg, 2001a, 2001b; Skripitz et al.,200a, 200b).

Growing Bone, Second Edition, by James F. Whitfield. ©2007 Landes Bioscience.

Beside Eli Lilly's "Forteo™" the first generation of PTHs include the recombinant hPTH-(1-84) (NPS Pharmaceuticals Inc.'s Preos™ in North America and Preotact™ in the European Union) which, like Forteo™, was also just picked off the shelf several years ago and put into bacteria by the now defunct Allelix Corporation (Toronto, Canada), has finished its phase III trial level, but in my experience is no better at building bone than the smaller peptides and is very unlikely to be modifiable for oral delivery (Whitfield et al., 1997c). However, the USFDA delayed its approval in March 2006 because the dose used in the trial caused excessive hypercalcemia (MorganStanley Equity Research North America, 2006) and then in June 2006 NPS stopped further expensive trialing and other development for the North American market. But, Preotact™ has been approved by the Committee for Medicinal Products for Human Use (CHMP) of the European Medicines Agency for marketing in the European Union.

These old PTHs will soon be followed onto the market by the new Ostabolin™ Family of designer PTH peptides such as Zelos Therapeutic's potently anabolic Ostabolin-C™([Leu^{27}]cyclo(Glu^{22}-Lys^{26})hPTH-(1-31)NH$_2$), high doses of which, as I said in Chapter 5 iva , actually seem to *decrease* bone resorption while strongly stimulating bone growth in cynomologous monkeys (Jolette et al., 2003, 2005). In this case there would not be the worrisome possibility of the PTH causing a stressful hypercalcemia in some patients and the huge advantage of being able to safely use higher doses. The first of this family, Ostabolin™ (hPTH-(1-31)NH$_2$, was "discovered" at the National Research Council of Canada during my group's project to locate the different signaling domains of hPTH-(1-34) using ROS 17/2 rat osteoblastic cells for signal testing (Jouishomme et al, 1992; Whitfield et al., 1996). The fragment, unlike hPTH-(1-34) (Forteo™) had a C-terminal NH$_2$ group instead of an OH group which gave it the ability to: associate with PTHR1 and stimulate adenylyl cyclase as strongly as hPTH-(1-34), but, because it lacked the C-terminal PKC2 region (Fig.20), not PKCs in ROS 17/2 cells; to resist serum carboxy peptidases; and to strongly stimulate bone growth in OVXed rats (Whitfield et al., 1996). Ostabolin™ was then made more effective by improving its receptor-binding ability in two ways (Barbier et al., 1997; Whitfield et al., 1998c). First the peptide's major, but unstable, receptor-binding α-helix between Ser^{17} and Gln^{29} was stabilized by linking Glu^{22} to Lys^{26} (a so-called lactam linkage) and second by increasing the hydrophobicity of the hydrophobic face of the amphiphilic α-helix by replacing the natural polar Lys at position 27 with non-polar Leu to produce ([Leu^{27}]cyclo(Glu^{22}-Lys^{26})hPTH-(1-31)NH$_2$) or Ostabolin-C™ (i.e., cyclized Ostabolin) (Barbier et al., 1997; Whitfield et al., 1998a, 1998b,1998c). The ability of such changes to increase the binding to PTHR1 receptors and adenylyl cyclase stimulation has been most dramatically demonstrated when hPTH-(1-28) with its very low receptor-binding and adenylyl cyclase-stimulating activities was changed into [Leu^{27}]cyclo(Glu^{22}-Lys^{26})hPTH-(1-28)NH$_2$ ("mini-C") with its receptor-binding and adenylyl cyclase-stimulating activities dramatically raised to those of hPTH-(1-34) (Whitfield et al., 2000c).

Then there will come newer oral, nasal/inhalant formulations of the smaller peptides (Morley, 2005). Those who are trying to develop non-injectable formulations of insulin and other peptide hormones face the very difficult problem of making a peptide that maintains a functional level of the hormone in the blood. BUT the goal for non-injectable PTH formulations is the exactly the opposite—a PTH formulation that peaks only briefly in the circulation to transiently flood the circulation with normally rare N-terminal fragments and mimic the transient PTH boluses from daily subcutaneous injections. In 2001, Mehta et al announced that a PTH pill may have been discovered. They found a way to make a recombinant peptide amide such as hPTH-(1-31)NH$_2$, my laboratory's Ostabolin™ (Whitfield et al.,1996; Mehta et al., 2001, 2002; Morley, 2005). Then they reported that this rhPTH-(1-31)NH$_2$ incorporated into an orally administered solid capsule formulation is far more able to get into the blood of dogs than Lilly's hPTH-(1-34) (Forteo™) in the same type of capsule (Mehta et al., 2001, 2002). And what is more it it had the *sine qua non* property for an osteogenic PTH of peaking only briefly in the blood between 120 and 150 minutes after giving the capsule!! In one example, the peak blood concentration of hPTH-(1-31)NH_2 was an impressive 2155 ± 456 (SEM) pg/ml compared to a much more modest 359 ± 152 (SEM) pg/ml obtained with an equimolar amount of hPTH-(1-34). This striking difference is probably due to hPTH-(1-31)NH_2's terminal NH_2 group making it more resistant to gut proteases than hPTH-(1-34). And I was told in September 2003 by Dr. N. Mehta (Unigene Laboratories, Fairfield NJ) that encapsulated hPTH-(1-34)NH_2 (might we call it *Forteoaminde?*) is indeed as good as Ostabolin™ in getting into the blood. And both we and Glaxo-SmithKline have confirmed that this recombinant preparation is at least as strongly osteogenic in the OVXed rat as hPTH-(1-34)NH_2. Then Glaxo-SmithKline reported results of an ongoing phase I human trial which show that the Unigene capsule or a variant of it can deliver intact and active recombinant Ostabolin™ into the bloodstream of humans as well as dogs (FDA News October 14, 2004). The desired brevity of the PTH peak was achieved by constructing the capsule so that it protected the enclosed PTH from the low gastric pH but released it upon contact with the high pH of the duodenum in a "puff" of acid which along with the terminal NH_2 group would hold off the high pH (i.e., alkaline)-requiring proteases long enough for some of the little PTH peptide to escape from the dangerous intestinal lumen either through (the transcellular route) or between (the paracellular route) the gut lining cells.

Matsumoto et al (2004) have reported that their nasal-spray formulation of hPTH-(1-34) safely and, when sniffed for 3 months, significantly increased bone formation in 92 osteoporotic persons. A nasally sprayed 1000-μg daily bolus of hPTH-(1-34) produced the same peak serum concentration of peptide as a subcutaneous injection of 20 μg of the peptide. During 3 months of a daily spray serum markers of bone formation procollagen fragments and osteocalcin and lumbar bone density increased with the 1000-μg boluses being the most effective. Only 5 of the 92 persons sniffing the 1000-μg boluses became hypercalcemic with blood calcium levels over 10.4 but lower than 11.0 mg /100 ml.

As I have said the most exciting thing about the new peptides, Ostabolin™ and Ostabolin-C™, is the reduced ability, or maybe inability, of daily boluses of even very large doses of them to stimulate osteoclasts. Thus, the marketers will likely say in their promotional brocures "*does not cause the major side effect of hypercalcemia*".

Other agents such as BMPs, FGF-2 and IGF-I are being considered as anabolic therpeutics. But as we have learned the PTHs actually operate through the same Cbfa1/Runx-2 protein as BMP-2 acts and stimulate osteoblastic cells to make FGF-2 and IGF-I to mediate their osteogenic action and stimulate osteoprogenitors that are too immature to make PTHR1 receptors. In other words the patient gets a triple whammy for the price of one in each PTH injection.

Another way to treat osteoporosis would be to try to block sclerostin, the osteocyte-manufactured homeostsic osteogenesis blocker. And indeed Warmington et al. (2005) have reported that injecting a monoclonal antibody to sclerostin into OVXed rats strongly increased BMD, trabecular number and thickness, and increased cortical thickness and with this reduced endosteal circumference. But the protein is no better than a PTH and moreover the osteogenic PTHs inhibit sclerostin expression (Bellido et al., 2005; Keller and Kneissel, 2005)! So why bother trying to develop an anti-sclerostin and then put it through expensive clinical trials to end up mimicking the already exhaustively trialed PTHs?

A possible drawback of the PTHs is that a lot of cells stretching from brain to bone to colon express PTHR1 receptors at critical stages of differentiation mainly for responding to locally produced PTHrP. However, no widespread effects have yet been seen. Obviously the best strategy in the long run would be to make a functional PTH analog perhaps by conjugating it with a bisphosphonate that localizes to bone. Such an analog would then have minmal sideffects.

It is also possible that a CaR-binding "calcilytic" with a short cirulating half-life will be invented that can trigger necessarily short, osteogenic spurts of endogenous PTH secretion by fooling the cells into believing that there is a dangerously low level of circulating Ca^{2+} (Gowan et al., 2000; Hebert, 2006; Nemeth and Fox, 2003). And.it appears that such "fast PTH rise-brief peak-sharp fall"-inducing calcilytics have been developed by Avery et al. (2005) at Bristol-Myers Squibb Company. But osteoclasts are normally turned off by Ca^{2+} stimulating their CaR, so a calcilytic would jam the "off switch". However, if an anticatabolic of some kind is used to prevent the calcilytic from stimulating osteoclasts, the calcilytic-induced pulses of endogenous PTH might be strongly osteogenic. However, why would we need such a drug when blocking CaRs might be too risky because a lot of other cells use these senors to control major body functions ranging from brain to gut and kidneys (Brown and MacLeod, 2001; Chattopadhyay and Brown, 2003)? The need for such a drug will vanish upon the arrival of the oral or inhaled non-hypercalcemia-causing PTHs such as Ostabolin™.

All of this means that the time has come when female and male osteoporotics or returning space travelers will be given (or, like diabetics, will give themselves) a daily shot of a PTH or swallow a daily Ostabolin™ or Ostabolin C™ pill or have a daily sniff of the peptide (a sort of osteo-snuff) to make new bone. Returning space travelers may also take a daily PTH shot, pill or sniff unless, of course, they were protected from bone loss by taking an antiresorptive drug such as alendronate (Fosamax™) during their flight. After, the PTH treatment he or she will take a weekly dose of Fosamax™ or maybe a subcutaneous shot of the very long-acting OPG to prevent the loss of the "PTH bone" (Bolon et al., 2002a, 2002b; Capparelli et al., 2003; Morony et al., 1999; Bekker et al., 2001; Black et al., 2004) for 1 or 2 years to extend the mineralization and thus the optimal hardening of the new bone. More recently, a better antiresorptive than Fosamax™ and a safer alternative to the now discontinued development of the RANKL-blocking OPG has appeared—AMG 162 human monoclonal anti-RANKL antibody ("denosumab" Bekker et al., 2004; Lacey, 2006; McClung et al., 2004). A single subcutaneous injection of 3 mg/kg of this antibody into postmenopausal women persists at a significant level in the blood for about 9 months and can reduce the serum level of N-terminal collagen breakdown products (NTX) from on-going bone turnover by as much as 84% for over 6 months (!) and significantly increase bone density at all sites (hip, spine, wrist and total body) without stimulating osteoblast activity as indicated by the the serum level of bone-specific alkaline phosphatase (Bekker et al., 2004; Lacey, 2006; McClung et al., 2004). Or maybe, the person may be able some day to take leptin (modified to not cross the blood-brain barrier and stimulate the central osteoblast-inhibiting adrenergic system) along with the PTH to collaborate with bone building and harden the new bone by promoting mineralization and restraining osteoclasts (Whitfield, 2002b). With respect to leptin's brain-centered association with β-adrenergic restraining of osteoblast activity, it is likely that the PTHs will be particularly effective in stimulating bone growth in fracture-prone elderly people taking β-blockers with or without diuretics such as the thiazides.

CHAPTER 7

OGP—The Osteogenic Growth Peptide

Scooping out the marrow or driving a nail into the marrow cavity of a bone such as the tibia, like a fracture, causes the marrow cavity to fill with a blood clot and releases a shower of osteogenic signalers such as β2-microglobulin, IGF-I, PDGF, TGF-βs, and VEGF from platelets, shocked bone-lining cells and injured vascular cells (Carano and Fivaroff, 2003; Bab and Einhorn, 1993; Gerstenfeld et al., 2003). They trigger and then drive a burst of blood vessel regeneration and bone formation (without an intermediate cartilaginous stage) that spreads from the cortical periphery to the center of the cavity and eventually fills the cavity with trabecular (cancellous) bone. At the peak of this osteogenic phase, the marrow cavity is filled with busy osteoblasts and both mineralized and unmineralized matrix. It is these that summon the osteoclasts and then the hematopoietic cells. The osteoclasts chew up the trabecular bone plug, which is replaced by normal marrow (Bab and Einhorn, 1993). In other words the primary bone plug and its builders establish the microenvironment for the new bone marrow with its stromal cells and their osteogenic progeny and hematopoietic stem cells and their osteoclastic and various other progeny.

Our hero, OGP, appears when the marrow is dug out or "nailed" and marrow regeneration starts from the hematopoietic stem cells lining the surviving trabeculae. It comes from the clot and regenerating marrow to locally stimulate osteoprogenitors and bone growth. But it also escapes into the general circulation and stimulates bone formation in such distant places as the other tibia or the lower jaw. It is actually OGP-(10-14) (reviewed most recently by Bab and Chorev, 2002).

OGP starts out as a 19-amino acid piece of histone H4—actually H4's 85-103 C-terminal tail. Histone H4 genes (there are 4 of them) are turned on by the repair signals because repair requires cell proliferation which in turn requires that all of the H1, H2A, H2B, H3 and H4 histone genes be switched on when the cells start replicating their chromosomes in preparation for later division. An ingenious, cell cycle-coupled, repair-amplifying by-product of the switch-on of the histone H4 genes is the translation of *some* of the genes' mRNA messages starting not at the normal first site to make full-length H4 for new chromosomal nucleosomes but at an alternative site—AUG-85—to produce the pre-OPG-(85-103) fragment which is then trimmed down to OGP-(90-103). The OGP-(90-103) is then dumped out of the cell along with OGP-(90-103)BP(binding protein) in an OGP•OGPBP complex. Outside the cell, OGP separates from the complex and is activated by being further trimmed down to OGP-(10-14). OPG-(10-14) can then bind to G-protein-coupled (specifically Gi) receptors on the producer cell as well as its neighbors (Bab and Chorev, 2002). Other OGPs make their way into the blood where they combine with α_2-macroglobulin to form a mobile OGP reservoir and travel to distant bones.

OGP stimulates trabecular bone formation in OVXed mice by stimulating the generation and activity of osteoblasts. But it seems to be not nearly as potent as the PTHs. In the presumably best example provided by Chen et al.(2000) it did not raise the trabecular volume in OVXed mice above the sham value in intact mice as do the PTHs in OVXed rats. It can *mildly* stimulate the proliferation of osteoblastic precursor cells and fibroblasts and it can stimulate the proliferation of hemopoietic stem cells (Bab and Chorev, 2002). And it can enhance fracture healing in rabbits under certain unstable conditions (immobilized with a plastic plate), but curiously *not* under "stable" conditions (a dynamic compression plate), but it does not increase callus size as do rhPTHs –(1-34) (Forteo™) and [Leu^{27}]cyclo[Glu^{22}-Lys^{26}]hPTH-(1-31)NH$_2$ (Ostabolin C™) (Andreassen et al., 1999, 2004; Sun and Ashhurst, 1998).

CHAPTER 8

The Statins

While the PTHs are by far the leading anabolic agents for treating osteoporosis and mending fractures, another family of drugs has been trying to challenge them but with very mixed results (Mundy, 2000; Whitfield, 2001, 2002a). These now old drugs were discovered by Endo and colleagues during a search for microbial weapons of mass destruction that microorganisms use to defend themselves by blocking key sterol syntheses in their enemies (Endo, 1992). They are produced by molds such as *Penicillium citrinum* (Endo 1972, et al., 1977; Kuroda et al., 1977; Veillard and Mach, 2002). This ability to inhibit sterol production turned out to be useful for reducing blood cholesterol and preventing heart attacks and stroke (Veillard and Mach, 2002).

There are at least eight statins and more on the big pharmas' drawing boards and in their test tubes. Four (lovastatin, mevaststin, pravastatin, simavstatin) are natural or fermentation-derived while the other four (atorvastatin, cerivastatin, fluvastatin, rosurvastatin) are synthetic (Veillard and Mach, 2002). The various statins differ in their lipophilicity which determines their ability to diffuse into the cell through its predominantly lipid plasma membrane. Thus, pravastatin has the lowest lipophilicity of the eight and thus cannot passively get through cell membranes as easily as, for example, cerivastatin or simvastatin.

Although the PTHs are very close to the ideal anabolic drugs, they must still be injected subcutaneously (though not for long) and they affect the many cells that share PTHR1 receptors with bone cells. But the injection drawback will soon disappear with advent of the Ostabolin™ family PTHs such as oral rhPTH-(1-31)NH$_2$ or inhalable [Leu^{27}]cyclo(Glu^{22}-Lys^{26})hPTH-(1-31)NH$_2$ (Ostabolin-C™) (Mehta et al., 2001, 2002; Morley, 2005).The ideal drug would be a small molecule that stimulates bone growth as strongly as a PTH by selectively stimulating bone cells when given orally or even better topically.

Such a drug should stimulate osteoblast progenitors to make BMP-2. At this point I must remind the reader of BMPs' main job—"*A fundamental function of the BMPs is to induce the differentiation of mesenchymal cells toward cells of the osteoblastic lineage to promote osteoblastic maturation and function*" (Canalis et al., 2003). The BMPs in turn, would set up an escalating positive feedback cycle by stimulating the cells to express the BMP-2/4-stimulating Cbfa1/Runx-2 and with this commit their great-grandcells to become mature osteoblasts (Canalis et al., 2003; Ji et al., 2000; Yamaguchi et al., 2000).

Therefore, Mundy and his colleagues tested more than 30,000 small 'natural' compounds for the ability to stimulate BMP-2 expression (Garrett et al., 2001b; Mundy, 2000; Mundy et al., 1999). They used 2T3-BMP-2-LUC mouse osteoblasts that had been immortalized by carrying the SV-40 tumor virus's large T-antigen attached to a BMP-2 gene promoter and a firefly luciferase gene which was also coupled to a BMP-2 gene promoter (Ghosh-Choudhury et al., 1996; Mundy et al., 1999). Their effort was rewarded when one of the compounds stimulated the luciferse gene's BMP-2 promoter at 0.1-5 µM (Garrett et al., 2001c; Mundy et al., 1999). This compound was lovastatin, now a member of a family of very widely used, serum cholesterol-lowering drugs. Later they found that other "lipophilic" statins—atorvastatin, mevastatin, simvastatin—were also effective. Currently the lipophilic synthetic cerivastatin is the most potent, but the old hydrophilic cardiovascular "work horse", pravastatin, does not stimulate the BMP-2 promoter (Garrett et al., 2001c).

Sugiyama et al. (2000), using human osteosarcoma cells with a BMP-2 or BMP-4 promoter-driven luciferase reporter gene, reported that simvastatin stimulated the BMP-2 promoter but not the BMP-4 promoter. Significantly, the statin did not affect BMP-2 expression in Chinese hamster ovary cells. Evidently osteoblasts, but not Chinese hamster ovary cells, had something statins needed to selectively stimulate a BMP-2 promoter. More recently, Maeda et al. (2001) have reported that simvastatin, at concentrations between 0.01 and 0.1 µM, stimulated MC3T3-E1 mouse preosteoblasts to express BMP-2 and active alkaline phosphatase and form mineralized nodules. .

This discovery was, and still is, very exciting because the statins were found at the end of the 1970s and have since then become very familiar drugs that have been given to many thousands of men and women throughout the world in an effort to reduce cholesterol levels and the risk of heart attacks and other cardiovascular problems (Herrington and Klein, 2001; Moghadasian and Frohlich, 2001; Whitfield, 2001,2002a). After being converted in the body from lactone pro-drugs to the β-hydroxy acid drugs, they stop cholestrol synthesis by inhibiting the synthesis of its mevalonic acid precursor from HMG-CoA by HMG-CoA reductase (Fisher et al., 1999; Garrett et al., 2001b; Horiuchi and Maeda, 2006; Shobab et al., 2005; Whitfield, 2001,2002a) (Fig. 14).

i. Animal Studies

Mundy and his colleagues (Garrett et al., 2001c) first confirmed that the statins, atorvastatin, cerivastatin, fluvaststain, mevaststatin, and simvastatin, but not pravastatin, stimulated bone cells to make BMP-2. Then they showed that 0.25 and 0.5 µM lovastatin and simvaststain stimulated bone formation in mouse calvarial organ cultures by 2.8- and 4.3-fold respectively and that the bone-building action could be inhibited by the BMP-2 blocker noggin which binds to BMP proteins and prevents them from binding to their receptors (reviewed by Garrett et al., 2001c; Mundy et al., 1999). Amazingly, only one 24-hour exposure to cerivastatin was enough to kick-start the osteogenic machine into working for *2 weeks!* They also gave simvastatin to ovary-intact and OVXed rats by oral gavage for 35 days. Giving the animals 5 or 10 mg of simvastatin/kg/day nearly doubled trabecular volume and increased the bone formation rate by about 50% (Garrett et al., 2001c; Mundy et al., 1999). Oxlund and Andreassen (2002) have since reported that giving OVXed Wistar rats simvastatin by gastric tube twice each day for 3 weeks significantly increased the cortical mineral apposition rate and the bending strength of the tibial diaphysis. And Skoglund et al. (2003) have reported that simvastatin, which as we shall see below does *not* stimulate the growth of femoral cortical and trabecular bone in OVXed Swiss-Webster mice (von Stechow et al., 2003), significantly increases callus size and the force required to refracture femurs of mature male BALB/c mice.

Oral delivery of statins is not the best way to stimulate bone growth in mice and rats because the drugs can be converted into inactive metabolites by the cytochrome P450 system during their first passage through the liver (Gutierrez et al., 2000; Shepherd, 2000). In fact, the pharmaceutical industry has designed these drugs to selectively target the liver—the main site of cholesterol synthesis (Garrett et al., 2001c; Horiuchi and Maeda, 2006). The ideal would be to initially avoid the liver as much as possible by delivering them through the skin instead of the gut. And this can be done. Gutierrez et al. (2000) have found that lovastatin was, on a per mg basis, a 50 times better bone maker when applied to rat skin than when given orally. And Whang et al. (2000) reported that lovastatin's potency as a mouse calvarial (skull bones) osteogen was increased as much as 80 times when delivered continuously from a poly[lactide-co-glycolide] scaffold implanted into the skin over the mouse skulls.

But things are by no means clear for statins and bone. Maritz et al (2001) have found that the effects of statins are dose-dependent. Thus, a low dose (1 mg/kg/day) of simvastatin for 12 weeks decreased bone formation and increased bone resorption with a net drop in BMD in young female Sprague-Dawley rats while a 20-fold higher dose of the statin stimulated both formation and resorption which cancelled out into no net change in BMD. Other groups couldn't get statins to stimulate bone growth in mice and rats or murine calvarial explants as did the unfailingly potent hPTH-(1-34) (Forteo™) or prevent OVX-induced bone loss in mice and rats (Crawford et al., 2001; Gasser, 2001; Sato et al., 2001; Staal et al., 2003; Yao et al., 2001)! For example, Von Stechow et al (2003) found that the mean trabecular bone volume in OVXed mice, after 8 weeks of daily injections of 80 μg of hPTH-(1-34)/kg, was 4 times higher than in vehicle-injected control mice but simvastatin did not increase the serum osteocalcin level, did not affect trabecular volume or stimulate bone growth. However, Staal et al. (2003) have shown that cerivastatin does inhibit osteoclastic bone resorption in vitro in proportion to its ability to inhibit HMG-CoA reductase, but not in the rat although it does hit the rat's bone cells. The lack of an effect of statins such as lovastatin and simvastatin in the rat could be due to the ability of the tissues to mount a compensatory upregulation of HMG-CoA reductase that overrides the statin action (Staal et al., 2003). But, as we shall see further on, it could also be due to the statins *inhibiting* osteoblast precursor proliferation (Reinholz et al., 2002). And Banu and Kalu (2002) have reported that cerivastatin, unlike hPTH-(1-34), does not prevent age-dependent bone loss in Sprague-Dawley rats.

ii. Osteogenic Mechanism

Statins start things going by inhibiting HMG-CoA reductase (i.e., hydroxymethyl glutaryl-coenzyme A) (Fisher et al., 1999; reviewed by Garrett et al., 2001c; Veillard and Mach, 2002; and Whitfield, 2001, 2002a) (Figs. 14 and 42). Besides preventing cholesterol, FPP (farnesyl pyrophosphate) and GPP (geranylgeranyl pyrophosphate) synthesis, the inhibition of HMG-CoA reductase by a statin such as lovastatin will inhibit osteoclast generation and trigger osteoclasts' apoptotic self-destruction exactly like the FPP synthase-inhibiting N-BPs (N-containing bisphosphonates) such as alendronate (Fosamax™) (Benford et al., 1999, 2001; Fisher et al., 1999; Rezka et al., 1999; van Beek, et al., 1999) (Figs. 14 and 15). However, in some cells the statins may do something else, because they also selectively stimulate the expression of BMP-2, which in turn mediates the osteogenic response probably by stimulating the expression of Cbfa1/Runx-2 and the shower of events it enables (Fig. 7). This would mean that *sometimes* in rodents the statins are 'double-barreled' bone anabolics that murder osteoclasts and stimulate osteoblasts. To get the same effect with a PTH it would have to be given along with anti-RANKL antibody, leptin, OPG or an anti-catabolic such as alendronate to prevent the PTH from stimulating osteoclasts.

At first sight it would seem that when they happen the stimulation of BMP-2 and osteogenesis should be independent from HMG-CoA-reductase inhibition simply because the N-BP alendronate that inhibits FPP synthase doesn't stimulate osteoblast-specific genes or bone formation in rats (Onyia et al., 2002). But it isn't—Garrett et al (2001b) have reported that mevalonate or geranylgeranyl pyrophoshate prevents lovastatin from stimulating BMP-2 expression and bone formation in rat bone cultures.

An exciting abstract by Garrett et al. (2000) suggested that the osteoblast-stimulating mechanism had been found. But this has turned out to be a false alarm. All that came from it was the discovery of different, but clinically useless, kinds of osteoblast stimulator!

The clue to the supposed new stimulators was the resemblance of the statin lactone ring to that of lactacystin, an inhibitor (from *Streptomyces*) of the cell's protein shredder/garburettor— the 26S proteasome (i.e., multicatalytic endopeptidase complex). They then found that lactacystin and other unrelated shredder inhibitors (the peptide aldehydes PS-1 and MG132;

the α,β-epoxyketone epoxomicin) *which* (according to them) *are unlikely to be HMG-CoA inhibitors*, were bone anabolics. They all stimulated BMP-2 expression, the growth of cultured neonatal mouse bone and trabecular bone in rats and mice. Moreover, pre-osteoblasts and osteoblasts could also be made to express alkaline phosphatase and osteocalcin and make bone by disabling the ubiquitin-proteasome system (UPS) either by having the cells over-express a dominant negative mutant of the E3 ubiquitin ligase gene or by inhibiting the E3 ligase with N-Leu-Ala. Significantly, the chymotrypsin of the osteoblasts' shredder was more responsive to the inhibitors than other cells' chymotrypsins. This disabling of osteoblast UPS would be expected to stop the cytoplasmic turnover of the β-catenin transcription enhancer which would then surge into the nucleus and stimulate genes for BMP-4, Cbfa1/Runx-2, cyclin D1, Myc and osteopontin (Hershko et al., 2000; Zhao et al., 2002). And Chen et al (2001) have shown that disabling the osteoblast UPS with epoxomicin or proteoasome inhibitor-1 also stops the Slimb ubiquitin ligase from inducing the cleavage of active big Gli3 into inactive little (the N-terminal short form) Gli3 which enables big Gli3 to accumulate and get into the nucleus to stimulate the *bmp*-2 gene and the production of autocrine/paracrine BMP-2. Operating through Smad second messengers, BMP-2, like the osteogenic PTHs operating through cyclic AMP (Fujita et al., 2000; Moore et al., 2000; Selvamurugan et al., 2000), would stimulate the cells to express Cbfa1/Runx-2 and drive the maturation of the cells into fully fledged bone-making osteoblasts. As expected, inhibiting BMP-2 gene expression with Noggin prevented both the UPS inhibitors and statins from stimulating bone formation (Garrett et al., 2001a).

This was all very elegant. But Garrett and his group had not addressed *THE KEY POINT—* do statins in fact inhibit the shredder and cause a Gli3 build up? If they do stimulate bone formation by inhibiting the UPS why are their anabolic actions, like the anti-catabolic actions of N-bisphosphonates such as alendronate, prevented by mevalonate or geranylgeranyl pyrophosphate (Garrett et al., 2001a)? It would be expected that alendronate, for example, would be a potent stimulator of BMP-2 expression and bone growth, which it is not (Onyia et al., 2002). In fact alendronate actually *inhibits* the expression of the genes for several bone formation markers in the proximal femoral metaphyses of OVXed rats (Onyia et al., 2002)! And now at least one lipophilic statin—lovastatin—has been shown to actually *stimulate* the shredder in MC3T3-E1 mouse calvarial preosteoblasts (Murray and Murray, 2001)! Alas, as so often happens, a beautiful idea has been murdered by an ugly fact. Obviously although we know that UPS shredder inhibitors can stimulate osteoblasts and maybe bone formation we must look elsewhere to find out how statins' stimulate bone growth. Maybe eNOS is the one to follow.

Since statins stimulate eNOS, the constitutive Ca^{2+}•calmodulin-activatable endothelial nitric oxide synthase, by extending the lifespan of its mRNA's by blocking the Rho GTPase's geranylgeranylation-dependent membrane-associated, eNOS-suppressing activity (e.g., Feron et al., 2001; Laufs and Liao, 1998); *since* knocking out the eNOS gene interferes with osteoblast differentiation and migration to microdamage sites and reduces bone formation in mice and the osteogenic response of calvarial preosteoblasts from these eNOS[-/-] mice (Afzal et al., 2000; Aguirre et al., 2001; Armour et al., 2001; Garrett et al., 2001b; van't Hof and Ralston, 2001; van't Hof et al., 2002); *since* inhibiting eNOS activity prevents statins from stimulating BMP-2 expression and bone formation in murine calvarial cultures (Garrett et al., 2001b), a better place to look for the statin targets are the eNOS-sequestering caveolae ("little caves"), which are small (60-100 μm in diameter) clathrin-free pouches in the membranes of most cells including human and murine osteoblastic cells (Lofthouse et al., 2001; Mineo and Anderson, 2001; Razani et al., 2002; Solomon et al., 2000) (Fig. 42).

Caveolar pouches are rich in sphingomyelin and glycosphinglolipids and their formation and function require cholesterol (Anderson and Jacobson, 2002; Mineo and Anderson, 2001; Razani et al., 2002). The key protein component of a caveola is one of 3 caveolins, particularly caveolin-1. Cholesterol has an extremely high affinity for caveolin-1 and palmitoylated

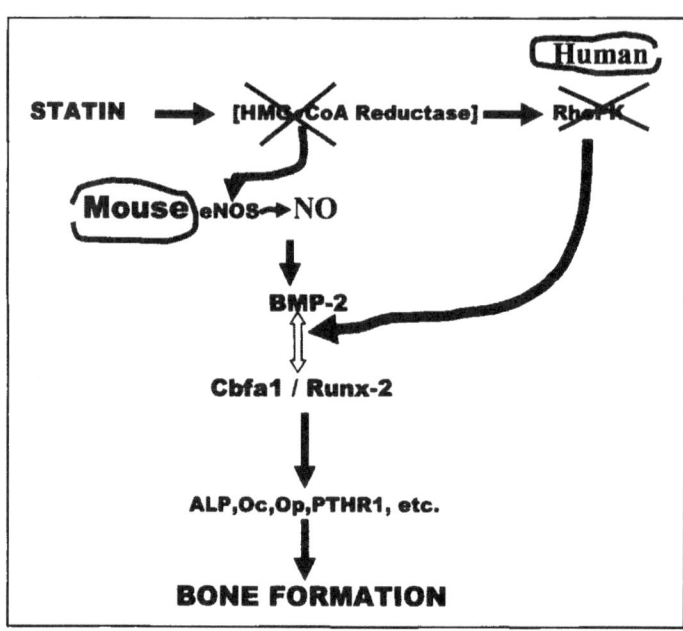

Figure 42. Lipophilic statins are frustratingly inconsistent stimulators of bone growth in humans, mice and rats. Actually it seems that they are most likely anti-catabolics like the N-bisphosphonate Fosamax™ (see text and Figs. 14 and 15). There is evidence that their ability to stimulate bone growth in mice (that is when they do) is due to activation of eNOS. Normally eNOS is isoprenylated and sequestered and inactivated in caveolar pouches in the cell membrane. Statins inhibit isoprenylation by inhibiting HMG•CoA reductase. This causes the release of eNOS (Ca^{2+}-stimulable nitric oxide synthase) which is subsequently activated and produces NO (nitric oxide) from the amino acid arginine that sets in motion the autocatalytic BMP-2/Cbfa1-Runx-2 cycle. However, in humans NO appears *not* to be involved in stimulating osteoblasts. Instead it seems that the mechanism involves the isoprenylation (geranylgeranylation) of the Rho GTPase that stimulates Rho-PK. The functioning Rho-PK in turn inactivates something that would otherwise turn on the BMP-2/Cbfa1-Runx-2 cycle. Statin works by preventing the isoprenylation and activation of Rho-GTPase which results in the inhibition of Rho-PK which in turn releases the Rho-PK-blocked mechanism that starts the BMP-2/Cbfa1-Runx-2 cycle and the switching on of the osteoblast gene set. Abbreviations: **ALP**, bone specific alkaline phosphatase; **Oc**, osteocalcin; **Op**, osteopontin; **PTHR1**, the receptor; **Rho-PK**, Rho-dependent protein kinase. A color version of this figure is available online at www.eurekah.com.

molecules of caveolin-1 bind to cholestrol in cholesterol-rich lipid rafts floating in the plasma membrane. The cholesterol-attached caveolin-1 molecules then bind to each other with their oligermerization domains and pull the raft down into a pouch. Caveolin-1 is one of a family of proteins known as MORFs (modifiers of raft function) that recruit and organize signal circuit boards or rafts into specifically functioning clusters. A large number of receptors, membrane proteins, non-receptor tyrosine kinases, GTP-ases, signal protein adapters, nuclear receptors (androgen and estrogen receptors), structural proteins and others (see Razani et al., 2002 for a recent list) are variously contained and arranged on the bottoms of these little pouches. eNOS is one of these and, like the others, is bound by its caveolin-binding motif consisting of Phe-Ser-Ala-Ala-Pro-Phe-Ser-Gly-Tryp amino acids to the CSDs (caveolin scaffolding domains) of caveolin-1 oligomers hanging down from the bottoms of the caveolae (Chambliss et al., 2000; Chen et al., 2001; Couet et al., 2001; Okamoto et al., 1998; Razandi et al., 1999, 2002; Razani et al., 1999, 2001, 2002; Solomon et al., 2000).

Binding to the CSDs *silences* various signaling proteins (Anderson, 1998; Couet et al., 2001; Feron et al., 2001; Fielding and Fielding, 2001; Razani et al., 2002; Xu et al., 2001). Thus, when palmitoylated eNOS is bound to the lipid raft domain, it is optimally active, but it shuts down when its CBD (caveolin-binding domain) in or near its catalytic region binds to caveolin-1's CSD. The caveolin-1•eNOS linkage must be broken if the eNOS is to function again. This linkage can be broken and a burst of eNOS activity triggered by Ca^{2+}•calmodulin generated by various signaling mechanisms (Razani et al., 2002). Prenylated (i.e., geranylgeranylated), caveola-bound Rho can also inhibit eNOS (van't Hof et al., 2002a). Since caveolin-1 expression is controlled by the free cholesterol level and the cholestreol level must be at a certain threshold level in order to form caveolae (Fielding and Fielding, 2001 Razani et al., 2002), and since eNOS is bound and repressed by prenylated Rho in the caveolae (Van't Hof et al., 2002a) reducing cholesterol production with a statin should, and indeed does, reduce caveolar formation as well as deprenylate and inactivate Rho and with this liberate eNOS which might still be clinging to the raft. This sidelining of Rho and reduction of caveolin-1 should raise the basal and Ca^{2+}-stimulable eNOS activity and NO production that stimulates BMP-2 expression and bone growth (Feron et al., 2001) (Fig. 42).

While the lipophilic statins and bisphosphates such as alendronate share the same ultimate osteoclast-killing action—a failure of FPP synthase to hold the deadly apoptosis-triggering caspase-3 in check (Fig. 14)—only the statins can stimulate eNOS and bone growth. And according to van't Hof et al (2002a) the anabolic action of mevastatin on mouse calvaria was reduced in eNOS-knock-out mice while the bisphosphonate-like antiresorptive action was unaffected by knocking the eNOS gene out. So while the anti-catabolic action of statins is mediated by the same lack of FPP synthase that is caused by alendronate and other N-bisphosphonates (Figs. 14 and 15), the statins' anabolic action, at least in mice, may be mediated by a stimulation of eNOS activity (Fig. 42).

A model for how statins might stimulate eNOS in bone cells has been found in vascular endothelial cells. In these cells, a statin such as atorvastatin starts by reducing cholesterol and with it caveolin-1. Released from the caveolin's inhibitory grip eNOS binds to hsp (heat-shock protein)90 (Brouet et al., 2001a, 2001b). The statin also stimulates PI-3 kinase (Dimmeler et al., 2001; Urbich et al., 2002) and induces the tyrosine phosphorylation of the hsp90 chaperone (Brouet et al., 2000). hsp90 then binds eNOS. PI-3 phosphorylates and thus activates Akt/PKB. The phospho-hsp90•eNOS complex then recruits the activated phospho-Akt/PKB by binding hsp90's 327-340 region to the phospho-Akt/PKBkinase's 229-309 region to form a phospho-hsp90•phospho-Akt/PKB•eNOS complex in which the phospho-Akt/PKB activates eNOS by phosphorylating eNOS's Ser 1177 (Brouet et al., 2001a, 2001b; S. Sato et al., 2000).

But there is a problem with eNOS and its NO product being responsible for statin-induced BMP-2 expression and osteogenesis (*when* it happens!). While inhibiting eNOs in mouse calvarial cells with L-NAME prevents the osteogenic response to statins, the same does not appear to apply to human cells (Fig. 42). Ohnaka et al (2001) have reported that inhibiting Rho-associated protein kinase—probably by preventing Rho-GTPase's isoprenylation and tethering to the caveolar scaffold—mediated pitavastatin's stimulation of BMP-2 and the turning on of the osteocalcin gene in both normal human osteoblasts from surgically discarded bone chips and MG-63 human osteosarcoma cells (Fig. 42). As would be expected if these statin-stimulated gene expressions were due to a failure of Rho-GTPase to be prenylated, plugged into the membrane and from there drive Rho-kinase, they were prevented from stimulating the osteoblasts by giving them geranylgeranyl pyrophosphate. But here's the whammy—the statin's actions were unaffected by inhibiting eNOS with L-NAME or, for that matter, by inhibitng various protein kinases such as PKCs, protein tyrosine kinases and MAP kinases. Mediation by Rho-kinase was supported by its specific inhibitor, hydroxyfasudil (Nakamura et al., 2001; Shimokawa et al., 19990, also increasing BMP-2 and osteocalcin gene expressions.

The likelihood of such a Rho-kinase-mediated restraint of bone-specific gene expression such as BMP-2 and Cbfa1/Runx2 has very recently been extended to the cells in fetal mouse calvarial cultures by Harmey et al. (2004). They have shown that stimulating Rho-kinase with *Pasteurella multocida* toxin blocks the progression of the calvarial cells into bone nodule formers while Rho-kinase inhibitors stimulated BMP-2 and Cbfa1/Runx2 expression and accelerated the differentiation of the calvarial cells into bone nodule-forming osteoblasts.

The identity of the geranylgeranylated, membrane-associated Rho-kinase is unknown. But it seems that geranylgeranylated Rho-protein kinase plugged into a lipid-raft/scaffold and activated by geranylgeranylated GTP•Rho phosphorylates and *inhibits* a protein that otherwise would stimulate BMP-2 and the all-important Cbfa1/Runx2 gene expressions and osteoblastic maturation. A model for this mechanism has been found in the rabbit basilar artery (Nakamura et al., 2001). Rho-kinase inhibits MLC(myosin light chain)phosphatase and thus promotes myosin light chain phosphorylation and the artery's contraction. But inhibiting the Rho-kinase with hydroxyfasudil releases MLC phosphatase which dephosphorylates myosin light chain and causes the artery to relax and dilate.

Recently another factor has been introduced into the statin story by Reinholz et al. (2002). They reported that as expected lipophilic statins (mevastatin and simvastatin) stimulated BMP-2 expression in human fetal osteoblasts. In addition, the statins stimulated the cells to express RANKL and OPG with the induction of RANKL being greater than OPG so that there was a drop in the OPG/RANKL ratio. But despite the BMP-2 stimulation there were *drops* in cell proliferation and matrix mineralization! Since anti-RANKL antibody partially prevented the reduction of hFOB cell proliferation but not the reduction of mineralization it appears that statins reduce hFOB cell proliferation via BMP-2-stimulated RANKL expression and matrix mineralization by a RANKL-independent mechanism. In other words these results, if confirmed, mean that RANKL can substantially affect osteoblasts as well as kill osteoclasts. This suggests that osteoblasts have RANK receptors! And this has been shown by Faccio et al. (2003) to be true for murine primary calvarial osteoblastic cells. *This BMP-2–RANKL interaction may be why statin anabolic action is so frustratingly uncertain.*

Finally we must not leave this section without mentioning that statins undoubtedly directly inhibit osteoclast generation as well as disable and murder mature osteoclasts just like N-bisphosphonates (Figs. 14 and 15). Also there is something else! you may remember that osteoclast generation is stimulated by signals from RANK receptors on osteoclast progenitors and that these signals activate the NF-κB (the "NK" in RANK) transcription factor. Statins have an anti-inflammatory action by preventing NF-κB activation in immune cells (Veillard and Mach, 2002). They may do this by inhibiting the prenylation (i.e., geranlygeranylation) of Rho-kinase, which prevents the kinase from being plugged into an appropriate membrane scaffold and triggering a kinase cascade that ultimately phosphorylates and switches off a nuclear receptor—PPARα (peroxisome proliferator-activated receptor-α). If not phosphorylated, PPARα combines with the nuclear retinoic acid X receptor (RXR) and the resulting complex activates genes with promoter-switch boxes having peroxisome proliferator response elements (Schoonjans et al., 1996; Veillard and Mach, 2002). But unphosphorylated PPARα also binds to NF-κB which prevents it from binding to and activating the promoters of its many target genes (Delerive et al., 1999; Veillard and Mach, 2002)! Therefore, without active, membrane-inserted isoprenylated Rho kinase, the unphosphorylated PPARα in statin-treated osteoclast progenitors can prevent RANK-activated NF-κB from driving their differentiation into mature osteoclasts.

iii. Human Studies

If statins stimulate bone formation in rodents there should be some indication of this in the thousands of people given these drugs. But, as we shall see, the anabolic effects of statins in postmenopausal women are just as uncertain as the effects in rodents.

iii a. Statins Reduce Fracturing

Bauer et al.(1999) reported the results of an analysis of the data from a study of osteoporotic fractures (8412 women 65 years and older) and a fracture intervention trial (6459 women aged 55 to 80 years). Most of the 599 statin users among the 14871 women in the two studies had been given lovastatin. The relative risk of hip fracture was only 0.30 (95% confidence interval [CI] 0.08-1.18) and the overall relative risk of non-spine fractures was 0.83 (95% CI 0.61-1.15).

Chan et al.(2000) reported that according to the records from six U.S. HMOs (health maintenance organizations) the relative risk of fracturing for patients who had had more than 13 dispensings of statins was 0.48 (95% CI 0.27-0.83). Meier et al. (2000) found the same from 300 practices in the UK-based General Practice Research Database. Their base population was 91611 patients who were at least 50 years old. Of these 28340 were taking lipid-lowering drugs. The adjusted odds ratio for fracturing in statin users relative to controls was 0.55 (95% CI 0.44-0.69). Other lipid-lowering drugs did not affect the odds ratio. Wang et al. (2000) found similarly lowered fractured risks in the records of 6110 New Jersey residents aged 65 years or more. Statin users had a relative hip fractures risk of 0.54 (95% CI 0.36-0.82). Chung et al. (2000) examined the records of 69 type 2 diabetics, 33 of whom had not been given statins and 36 of whom had been given lovastatin, pravastatin (which would not have affected bone cells) and simvastatin. The spinal BMD rose 2.9% in the statin-treated group during 14 months. During the same time the BMDs in the different regions of the femur and in the total hip increased by a significant 0.88-2.32% in the statin-treated men but by only 1.8% in the statin-treated women. Finally, M.H. Chan et al. (2001) have reported that giving simvastatin (20 mg/day for 4 weeks) to 17 hypercholesterolemic, non-osteoporotic patients increased one indicator of bone formation, the serum osteocalcin concentration. But it did not increase another formation indicator, the bone-specific alkaline phosphatase activity

Pasco et al. (2002) have reported the results of the small-scale "Geelong" (Geelong, southeastern Australia) osteoporosis study. In this study there were 16 statin users in the fracture group and 53 users in the non-fracture group with the fracture odds for the statin users (adjusted for BMD) at the femoral neck, spine and whole body being respectively 0.45(95% CI 0.25-0.80), 0.42 (95% CI, 0.24-0.75), 0.43 (95% CI, 0.24-0.78) of the values in the non-users. Adjusting for age, weight, medications, and "life-style" factors did not "substantially" change these facture odds. Surprisingly, in view of the statin users' large, 60% reduction in fracture risk there were at most only marginal (in fact non-significant) increases in BMD in the femoral neck, spine and whole body.

Montagnani et al. (2003) have treated 30 postmenopausal hypercholesterolemic women (of average age 61.2 ± 4.9 years) with simvastatin for one year. They found small increases in the femoral and lumbar BMDs by the end of the treatment, which was significantly higher than in the postmenopausal control women with normal blood cholesterol levels.

It seems from these reports that the statins are indeed safe, effective bone anabolics or pseudo-anabolics that can reduce the risk of hip and spine fractures by as much as 71% depending on the extent of use. However, there are other reports, which just as convincingly indicate that lipophilic statins do *not* stimulate bone growth or affect the fracture risk.

iii b. Statins Don't Reduce Fracturing

A problem with retrospectively assessed effects of giving statins on fracturing is that one statin, the popular hydrophilic pravastatin, does not affect cultured bone cells or stimulate bone growth in rodents (Garrett et al., 2001). This explains the findings of Reid et al (2001) on the effects of long-term pravastatin treatment on fracturing in 9014 patients with ischemic heart disease. There were 107 pravastatin-treated (40 mg/day) patients admitted to hospital for fractures compared with 101 in the placebo-treated group (fracture risk 1.05 [95% CI

0.80-1.16]). When fractures not requiring hospital admission were also included, the fracture risk was 0.94 (95% CI 0.77-1.16).

LaCroix et al. (2003) searched the huge database in the Women's Health Initiative Observational Study (WHI-OS) and found no reducton of fracturing in statin (atorvastatin, lovastatin, pravastatin, fluvastatin)-treated women. In a study funded by Proctor and Gamble, van Staa et al. (2000) also found that cholesterol-lowering doses of atorvastatin, pravastatin and simvastatin did not significantly lower the fracture risk. This was a much more extensive analysis of the UK General Practice Research database than that of Meier et al. which indicated that statins nearly halved the fracture risk. Van Staa et al.'s conclusion that were was no reduction was based on a population of 216,062 fracture patients from 686 general practices while Meier et al.'s conclsion was based on 91,611 fracture patients from "only" 300 practices. Sirola et al. (2001) and Solomon et al. (2001) could also find no evidence in the clinical databases that statins significantly lower fracture risk. Finally, Cauley et al. (2000) used the WHI-OS database to find out whether statins could enhance the BMDs of post-menopausal women. They found that although statin treatment did not affect fracture risk, treatment with a high-potency (based on cholesterol-lowering ability) lipophilic statin such as atorvastatin or simvastatin for more than 3 years only modestly protected hip and lumber vertebral BMD. Edwards et al. (2000) also found that treatment of 41 women with various statins (27 received the high potency simvastatin or atorvastatin) did not reduce fracturing, although there were 8.5 to 12% increases in the BMDs of hip and spine.

Cosman et al. (2001), like M. H. Chan et al. (2001), have carried out a small, short-term experiment on 14 postmenopausal women (mean age of 58 years) designed to find out what a 12-week treatment with 0.4 mg of cerivastatin/day might do to bone instead of serum cholesterol levels. Bone formation markers (type I procollagen propeptide and osteocalcin) did not change, but the resorption markers (urinary N- and C-terminal telopeptides) did drop slightly (<20%) within 6 weeks in the cerivastatin-treated group. They concluded that this statin did not detectably stimulate bone formation, but it might have had a modest bisphosphonate-like anti-catabolic action. Moreover, the very small increase Montagnani et al. (2003) found in BMD and bone alkaline phosphatase activity in 30 hypercholesterolemic postmenopausal women treated for 1 year with simvastatin could have been due to as much to osteoclast inhibition as to direct stimulation of bone growth.

Rejnmark et al. (2004) have reported the results of an experiment to determine whether simvaststin can stimulate bone formation. Eighty-two otherwise healthy, osteopenic or osteoporotic postmenopausal women were randomized for daily treatment for 1 year (52 weeks) with 40 mg of simvastatin/day or placebo. None of the women were hyperlipidemic, but simvastatin reduced the serum cholesterol by an average 27% but by 26 weeks after the end of the treatment the cholesterol level had returned to the baseline value. This obviously effective dose of the statin did not affect indicators of bone formation (bone-specific alkaline phosphatase or osteocalcin) or indicators of bone resorption. Nor did it significantly affect the BMD of lumbar spine or hip. There was, however, a small increase in the forearm BMD in the treated group. The authors conclude that *"our results do not support a general beneficial of simvastatin on bone"*.

However, statins are rather effective anti-osteoclast agents because they, like N-bisphosphonates such as alendronate (Fosamax™), can prevent the synthesis of the so-called isoprenoids farnesyl diphosphate and geranygeranyl diphosphate. *Indeed statins are better than N-BPs at inhibiting bone resorption by rabbit osteoclasts and by osteoclasts in mouse calvarial cultures* (Rogers, 2003). These lipids are tacked by appropriate transferases onto the C-terminal cysteine residue of each member of the small GTPases of the Ras, Rho and Rab families (Coxon and Rogers, 2003). This enables the "isoprenylated" GTPases to drive various osteoclast functions such as podosome formation, formation of the sealing ring, the transport to the apical ruffled border of key components, the fusion of the acid pumps to the ruffled border membrane, and the prevention of apoptosis (Coxon and Rogers, 2003;Stenback, 2002). Obviously, the prevention of the GTPases' isoprenylation when a statin inhibits an osteoclasts's HMG-CoA reductase will throw a very large "monkey wrench" into the digger's machinery with disastrous consequences including death for the digger but happily respite for the bone.

iv. Statins and Alzheimer's Disease

The swelling crowd of aging persons with deteriorating skeletons must also eventually face brains eroded by Alzheimer's disease. Therefore it is very important that another possible advantage of being given statins besides preventing the cardiovascular consequences of hypercholesterolemia and osteoporosis (*if its osteogenicity can be confirmed in future prospective trials)* is suggested by three things: an apparent relation between increased levels of cholesterol and Alzheimer's disease; the 70% lower incidence of late-onset Alzheimer's Disease among statin users particularaly those under 80 years of age ; and the ability of statins to significantly reduce the progression of the disease (Burns and Duff, 2002; Fassbender et al., 2001; Jick et al., 2000; Miller and Chacko, 2004; Panegyres, 2001; Rockwood et al., 2002; Shobab et al., 2006; Sjögren and Blennow, 2005; Sjögren et al., 2006; Sparks et al., 2005; Wolozin, 2004;Wolozin et al., 2000, 2004). Thus it appears that cholesterol and its metabolites contribute to late-onset Alzheimer's disease. Indeed the genes most closely associated with late-onset Alzheimer's disease, especially the e4 allele of the apolipoprotein E gene, are related to cholesterol metabolism and transport. The reason for this linkage is that cholesterol collaborates with sphingomyelin and the caveolin and flotillin proteins to form lipid rafts in internal membranes which can serve as platforms that enable neuron-killing raftophilic enzymes such as α-, β-, and γ-secretases to cluster and collaborate (Cordy et al., 2006).

The peptides that initiate Alzheimer's disease are cut out of APP, amyloid precursor protein, which is a transmembrane protein with FE65 protein and TIP60, a histone acetyltransferase clinging to its C-terminal tail (Cao and Shdhof, 2001; Nakata and Suzuki, 2006). The key initiators of this destructive process are the secreted $A\beta$-40 peptide with 40 aminoacids, but more importantly the self-aggregating, 42-aminoacid $A\beta$-42 peptide, that are cut out of the neuronal transmembrane-spanning region of the amyloid precursor protein (APP) by tandemly operating β- and γ-secretases which are either primarily or secondarily clustered with APP in the cholesterol-rich lipid rafts in membranes of the endoplasmic reticulum and/or Golgi network; $A\beta$-40 is generated only in the *trans*-Golgi network while $A\beta$-42 is generated in the endoplasmic reticulum as well as the Golgi apparatus (Cordy et al., 2006: Greenfield et al., 1999; Hartmann et al., 1997; Li, 2001; Shobab et al., 2006; Standridge, 2006; Urano et al., 2005; Wolfe, 2006). The first, β-secretase cut may be made by a cysteine protease or by BACE 1 (β-APP cleavage aspartic endopeptidase/ASP-2 [aspartic protease-2]) which is anchored to the raft by activity-determining glycosyl-phosphatidylinositol (Cordy et al., 2003). This releases the N-terminal part of APP. The γ-secretase complex with an aspartyl protease called presenilin at its core then operates inside the membrane to cut the $A\beta$ peptide out of the transmembrane part of the remaining APP stump (Cordy et al., 2006; Li, 2001; Wolfe, 2006). The $A\beta$s produced by the cysteine protease and γ-secretase are then secreted from the cell by regulated secretory mechanism while the BACE 1-γ-secretase-generated peptides are secreted by the constitutive secretory mechanism (Hook and Reisine, 2003).

Among very many other things, the accumulation of $A\beta$ in cortical and hippocampal neurons is responsible for deficits in synaptic transmission and LTP (long-term potentiation) (see the excellent review by Standridge, 2006). APP's liberated C-terminal fragment and FE65 and TIP60 are translocated into the nucleus (Nakaya and Suzuki, 2006). When they enter the nucleus the histone-acetylating TIP60 enables the gene-activating/suppressing FE65 to get at its target genes by decondensing the chromatin (Bruni et al., 2002). The C-terminal APP fragment also somehow stimulates GSK-3β (the very same glycogen synthase-3β that controls the WNT/β-catenin gene-signaling and control mechanism often discussed in Chapter 5) (Ryan and Pimplikar, 2005). This is a very nasty thing to do to a neuron because it abnormally phosphorylates the microtubule-associated tau protein which, instead of promoting microtubule assembly, self assembles into fibrillar tangles that wreck the cytoskeleton and prevent the assembly of tubulin into the microtubules that make up the all-important cellular railroad system (Li and Paudel, 2006 and, of course, Standridge, 2006).).

When not being tandemly carved up by the β- and γ-secretases, APP's normal job is to stimulate gene activities when it is phosphorylated in response to as yet undefined signalers. When phosphorylated it releases FE65 and TIP60 (Nakaya and Suzuki, 2006) which proceed to the nucleus to do their gene-modulating jobs. A hint of the importance of APP, FE65 and their relatives in murine brain development is indicated by the severe cortical dysplasia caused by impaired axonal pathfinding and neuronal migration by their deletion (GuJnette et al. 2006).

The initially soluble Aβs secreted by the neurons are kept from accumulating and aggregating into fibrillar amyloid plaques by being chewed up by proteases such as neprilysin and the insulin-degrading enzyme IDE, metabolized by astrocyters and especially by the microglial macrophages as well as being simply dumped out of the cell and into the circulation (Cuello, 2005 Selkoe, 2000 Shibata et al., 2000; Shiiki et al., 2004 Li. 2001 Tupo and Arias, 2005 Zlokovic, 2005a, 2005b). However, these Aβ-clearing mechanisms may be overwhelmed by excessive Aβ production, aging-dependent or genetically based clearance failures (Selkoe, 2001). This has been seen, for example, by Sparks et al. (2005) who found that circulating Aβ 40 rose progressively with the decline from high-normal brain function ⇒ low-normal function ⇒ to moderate cognitive impairment (MCI) but then dropped with fullscale Alzheimer's disease presumably as the soluble Aβ clearance mechanism failed and the fragments aggregated into insoluble fibrillar plaques. As the Aβ amyloid plaques appear they drive neurodegeneration when microglia hopelessly try to phagocytose ("eat") them and consequently suffer from severe cyto-indigestion which causes them to produce neuron-killing agents such as proinflammatory cytokines and NO. The situation worsens when these proinflammatory cytokines hit the astrocytes enveloping and servicing their "client" neurons. The astrocytes are then stimulated to make large amounts of NO which kill the juxtaposed their associated neurons (Chiarini et al., 2005).

Since cholesterol is a major raft ingredient, its level in the brain should and seemingly does affect the functioning of the secretases that product the deadly Aβs (Sidera et al., 2005; Shobab et al., 2006). (Circulating cholesterol is made in the liver but the brain makes its own large supply.) Indeed, the production of Aβs in APP-transfected cells is stimulated by increasing the level of intracellular cholesterol while the cholesterol-inhibitng lovastatin reduces it (Frears et al., 1999; Shobab et al., 2006). Raising the cholesterol level increases the formation of lipid rafts to which the raftophilic secretases attach, cluster, interact and produce Aβs (Sidera et al., 2005a, 2005b, 2005b; Shobab et al., 2006). As might be expected from the cholesterol-dependence of lipid raft formation and the isoprenylation–dependence of the anchorage of the γ-secretase complex to the lipid raft, the inhibition of cholesterol synthesis and its component-anchoring geranyl and farnesyl intermediates must disperse the lipid rafts and with this disrupt the functional secretase complexes (Sidera et al., 2005a; Urano et al., 2005; Wolozin et al., 2004). Thus, a progressively rising cholesterol production and handling can drive the accumulation of Aβ plaques in the brain by overwhelming the Aβ-clearing mechanisms especially as they weaken with age, but statins passing through the blood-brain barrier can indirectly affect this in two linked ways: by lowering cholesterol levels and wrecking the secretases' lipid rafts by preventing their isoprenylation-dependent anchorage. It is also likely that the build up of Aβ to the sporadic Alzheimer disease-triggering level in the aging brain is the result of the progressive reduction of the peptide's drainage from the brain because of the declining blood flow due to progressive cholesterol-associated microvascular deterioration (Whitfield, 2006c). Obviously long-term statin usage could slow or prevent the development of sporadic Alzheimer's disease by preventing this microvascular deterioration (Whitfield, 2006c).

Recently I have chanced upon a likely reason why statins might slow or even prevent the development of late-onset Alzheimer's disease. The accumulation of the deadly Ab peptides in the aging brain is the result in large part of fading abilities to destroy the peptides in the brain and to dump them out of the brain. A powerful Ab dumper in the vascular endothelial cells is LRP (lipoprotein receptor-related protein) which picks up Abs carries them through the blood-brain barrier and dumps them into the blood (Deane et al., 2004). And Deane et al. (2004) have shown that lovastatin and simvastatin strongly stimulate LRP expression in human cerebrovascular endothelial cells. Therefore, maintaining a high level of LRP in the brain's vascular endothelial cells by taking a statin could stop the aging brain's LRP from declining with age and consequently prevent Abs from building up to Alzheimer's disease-inducing level.

v. Statins' Clinical Prospects

As we learned above, some groups, but not others, have found statins to be potent bone anabolics for rats and mice and their anabolic potency for human bones is at best uncertain at best. There could be two reasons for this uncertainty. First, the statins are antiresorptives like the N-BP alendronate and owe their osteogenic effectiveness in young rodents to their potent osteoclast-disabling and killing abilities, which would leave the still-growing bones to the osteoblasts as happens in RANKL-knock-out mice. The spotty success in showing responses in rats could be due to compensatory surges of HMG-CoA reductase overriding the statin action, but humans don't have this (Staal et al., 2003). In mature humans with their tighly coupled osteoclast-osteoblast crack-repairing remodeling team, killing osteoclasts could counterproductively reduce BMU mobilization and thus impair microfracture repair. Alternatively, data such as those in the large WHI-OS and the UK's General Practice Research Database might be partly misleading because they come from patients treated for various lengths of time with different doses and statins such as pravastatin that were meant to reduce cholesterol rather than fracturing.

However there may be another explanation. As you learned above the anabolic as opposed to the anti-catabolic actions of statins may be mediated (at least in mice)through a burst of eNOS activity. If this be true, the anabolic effectiveness of statins would be affected by the lack of estrogen in both OVXed rodents and postmenopausal women which would drastically drop the eNOS level and increase the eNOS-inhibiting caveolin level (Farhat et al., 1996; Pavo et al., 2000; Stefano and Peter, 2001; Razandi et al., 2002; Tan et al., 1999; Whitfield et al., 2002; Zhu and Smart, 2003). In other words, in an estrogen-depleted OVXed rat or postmenopausal woman, there may not be enough eNOS for an optimally anabolic, fracture-lowering response to statins, but they could still kill osteoclasts and produce a N-BP-like antiresorptive response. An indication of a role for estrogen in controlling the responses to statins has been provided by Maritz et al. (2001) who have found that a high dose of simvastatin (20 mg/kg /day) only variably and slightly affected bone formation without affecting bone resorption in OVXed Sprague-Dawley rats, but it did significantly stimulate both formation and resorption in sham-operated control rats. However, we can't ignore Ohnaki et al. (2001)'s finding that inhibiting eNOS did not affect pitavastatin-induced stimulation of BMP-2 and osteocalcin expression in human osteoblasts.

Clearly there must be prospective placebo-controlled trials designed specifically to assess the osteogenicity rather than the lipid-lowering abilites of various doses of statins with reduced hepatoselectivity in postmenopausal patients. Moreover, the estrogen and eNOS statuses of the patients should be assessed along with BMD and fracturing. If topical delivery of a potent bone-anabolic statin turns out to be as feasible in humans as it has been in some rats we will have a double-barreled osteoclast-killing/osteoblast-stimulating drug, like leptin (Whitfield, 2002b), but with the immense bonus of protecting the patient from a wide variety of other things including Alzheimer's disease as well as cardiovascular diseases. However, if it should turn out that the statins require help from estrogens to be reliably anabolic for postmenopausal women they would be considerably less attractive than the potently anabolic PTHs which do not need help from estrogen or anything else (except for strain) to stimulate bone growth. However, the best of all possible Worlds for osteoporotics might some day include a short treatment with PTH to restore bone strength plus prolonged statin treatment to prevent cardiovascular and sporadic Alzheimer diseases.

CHAPTER 9

Surface Signaling Steroids—Real Anabolics or Pseudo-Anabolics?

i. Vitamin Ds

In 1998 Sicinski et al reported the synthesis of new 1α, 25-$(OH)_2$-19-nor-vitamin D_3 analogs such as 2MD (2-methylene-19-nor-(20S)-1α,$25(OH)_2D_3$) which are super-potent derivatives of the natural 1α, 25-$(OH)_2$-vitamin D_3. Four years later Shevde et al. (2002) reported that 2MD is a far more potent stimulator of osteoblast activities than 1α, 25-$(OH)_2$-vitamin D_3. Indeed, Shevde et al. (2002) have written "...*there is little evidence that vitamin D plays a direct role in new bone formation, apart from its action to maintain calcium and phosphorus levels in the blood*". But this osteoblast stimulation is compromised by 2MD being a 100-fold better stimulator of the RANKL gene (which has a vitamin D-responsive site in its promoter [Kitazawa et al., 2003]) and down-regulator of osteoprotegrin expression by presumably immature osteoblastic cells than the natural hormone. The result of this is the formation of more, bigger and better-bone-digging osteoclasts, which make 2 MD a 30-fold more effective mobilizer of bone Ca^{2+}.

As might be expected from this, orally giving 18.7 pmoles of 2 MD/kg of body weight to old OVXed retired breeder rats for 7 days a week for 23 weeks increased total bone mineral density to a mean value 7% higher than in sham-operated rats and 9% higher than in untreated OVXed rats (Shevde et al., 2002). The steroid actually increased femoral trabecular bone volume. And, most important, it didn't do this by decreasing bone resorption, but by actually increasing the rate of bone formation which 1α, 25-$(OH)_2$-vitamin D_3 can't do (Shevde et al., 2002). However, Shevde et al. did not say by how much 2MD increased the formation rate which would be a surprising omission if the stimulation had been significantly large.

How might 2MD stimulate osteoblast activity? 1α, 25-$(OH)_2$-vitamin D_3 activates two differently sited receptors (Boland et al., 2002; Farach-Carson, 2001; Farach-Carson and Xu, 2002; Nemere and Farach-Carson, 1998). One of these is the slowly acting, classical nVDR which operates in the nucleus and stimulates various genes and requires 1α, 25-$(OH)_2$-vitamin D_3's 1α OH group to be activated while the other, somehow related receptor—the almost instantly responding, fast-acting mVDR—is plugged into the osteoblast's cell membrane and does not need the 1α, 25-$(OH)_2$-vitamin D_3's 1α OH group to be activated . The rapidly triggered signals from the mVDR are similar, if not identical, to those from activated PTHR1— they include G-protein-coupled activations of c-Src protein-tyrosine kinase, adenylyl cyclase and phospholipases A,C and D which open Ca^{2+} channels and stimulate the release of Ca^{2+} from internal stores and activate PKCs, and set off MAP kinase cascades that stimulate various genes (Boland et al., 2002; Nemere and Farach-Carson, 1998). The fact that 2MD does *not* have a greater affinity for the nVDR (Shevde et al., 2002) suggests that the analog might have a greater affinitiy for the mVDR than 1α,25-$(OH)_2$-vitamin D_3. *This would mean that 2MD is actually a PTH mimic!* This also suggests that other analogs such as $25(OH)$-16ene-23yne-D_3

Growing Bone, Second Edition, by James F. Whitfield. ©2007 Landes Bioscience.

(Farach-Carson, 2001), which cannot activate nVDR because it lacks 1α (OH), should also be PTH-like bone anabolics.

2MD certainly looks like a good candidate for admission to the exclusive Bone Anabolics Club. It has the great marketing advantage of being an oral drug. But it is also unfortunately a super osteoclast generator in rats, which raises the worrisome specter of hypercalcemia in osteoporotic humans, which has prevented worldwide approval of other vitamin Ds for treating osteoporosis. Indeed as I said above the osteoclast-stimulating RANKL gene in osteoblasts has a vitamin D-responsive site in its promoter (Kitazawa et al., 2003). It is important now to find out whether 2MD is a direct stimulator of osteoblast generation and activity like PTH, or whether the anabolic response is just a reaction to strong osteoclast activity. Can 2MD stimulate bone formation when given along with osteoprotegerin to prevent osteoclastogenesis?

Then there is ED-71 (1α, 25-$(OH)_2$-2β-(3-hydroxypropoxy)vitamin D_3) (Kubodera et al., 2003). This is an excellent example of a pseudo-anabolic agent. In OVXed Wistar-Imamichi rats ED-71 lowered bone resorption and did not cause hypercalcemia without affecting ongoing bone formation and thus stopped and reversed the OVX-induced BMD drop (Kubodera et al., 2003). This differential effect resulted in a large increase (double the value in sham-operated rats) in the concentration of osteocalcin from the unaffected osteoblasts in the blood of OVXed rats given 0.2 μg/kg of body weight twice a week for 3 months. However, the authors of this report did not tell us what ED-71 did to bone in sham-operated rats.

ED-71 also reduces osteoclast activity and increases the BMD of the L2-L4 vertebrae of osteoporotic men and women without causing hypercalcemia (Kubodera et al., 2003).

ii. Unisex Steroids

It is possible to make politically correct unisex steroid analogs that can activate cell surface receptors in caveolar scaffolds but have no genomic (or genotropic) actions (Kousteni et al., 2001, 2002a, 2002b). One of these gender-nonspecific ligands is estren—4-estren-3α,17β-diol which does not stimulate the transcription of genes with promoters having EREs (estrogen-response elements) (Kousteni et al., 2002a, 2002b). Remarkably it can target the ligand-binding domains of androgen as well as estrogen receptors.

Starting continuous estren infusion (by implanting it in slow-release pellets) immediately after OVX into female Webster mice or after orchidectomy into male Webster mice significantly reduces or prevents any post-operative bone loss and prevents frequency of apoptotic osteoblasts from rising above the sham level (Kousteni et al., 2002a; Manolagas et al., 1999). As expected from the lack of genomic (genotropic) ability the analog cannot affect uterine or seminal vesicle weight or stimulate the proliferation MCF-7 mammary cells (Kousteni et al., 2002a; Manolagas et al., 1999). However, and this is a very important "however", the continuous pounding with estren reduces, but by no means stops as can estradiol, the loss of femoral trabeculae or raise the bone mass above the sham value as do the PTHs (Kousteni et al., 2002a; Manolagas et al., 1999). Furthermore, the analog does *not* at least hold the vertebral bone formation rate at the sham-operated value. In the experiment of Kousteni et al. (2002a), the bone formation rate/bone area (%/day) in the estren-treated OVXed mice was only 0.158 compared to 0.457 in the sham-OVXed mice. Nor does estren increase the number of osteoblasts per bone perimeter above the sham value. By contrast, an optimal injection of any of the anabolic PTHs starting immediately after OVX or several weeks after OVX can raise the trabecular thickness far above the sham level in rats (Whitfield et al., 1998c). Indeed it is important to know whether this unisex steroid analog also stimulates bone growth in OVXed rats as well as it apparently does in mice. And indeed Whitfield et al (2001a) have shown that estratriene-3-ol did not stimulate bone growth in OVXed rats while 6 weeks of daily injections of 0.8 nmole of Ostabolin-C™/100g of body weight raises the mean trabecular thickness in OVX 1.7 times above the normal thickness in vehicle-treated sham-operated animals.

A potent anabolic agent like a PTH stimulates bone growth in three ways—cause the reversion of lining cells to active osteoblasts, the secretion by PTHR1-expressing osteoblastic cells of missionary factors such as FGF-2, IGF-I and TGF-β1 that stimulate the proliferation of osteoprogenitor cells, and prevent osteoblast and osteocyte apoptosis. It appears that estren signaling only prevents osteoblast apoptosis, which is obviously by no means enough to cause supra-sham bone growth as do the PTHs. However, it is no more effective than dihydrotestosterone or estradiol in preventing osteoblast apoptosis. It seems that the anti-apoptotic action of estrogens on osteoblasts is due to the activation of the ERα, ERβ or androgen receptors' ligand-binding domain which triggers a Src/Shc/ERK1/2 signal cascade (Kousteni et al., 2002a, 2002b). Indeed an analog such as 1,2,5-tris(4-hydroxylphenyl)-4-propylpyrazole that only has a genomic (genotropic) action does not reduce osteoblastic apoptosis or increase osteoclast apoptosis as do the native sex steroids (Kousteni et al., 2002a).

All that can be said at the time of writing about this media-hyped unisex steroid estren is that it is only a weak pseudo-anabolic anti-resorptive (anti-catabolic) like the native sex steroids. But it could be a safe osteopenia/osteoporosis preventative or maintainer of new PTH-bone because it has none of the potential breast, seminal vesicle or uterine cancer-promoting activities of the native steroids (Kousteni et al., 2002a). However, it has clearly and necessarily shown how little an increase in bone growth can be produced by just blocking apoptosis.

CHAPTER 10

Strontium, Calcium's Big Brother

Strontium (Sr) has been very recently hyped as the latest "paradigm-changing" thing in the treatmen of osteoporosis as indicated by the title of a paper by Reginster et al (2003)—"*Strontium Ranelate: A New paradigm for the treatment of Osteoporosis*". When two atoms of strontium combine with ranelic acid, they form strontium ranelate, Sr-ranelate, that effectively stimulates the proliferation of fibroblasts and pre-osteoblasts from fetal Sprague-Dawley rat calvaria and less effectively the proliferation of mature osteoblasts from these tissues. (But of course this should have been expected because as we have learned mature osteoblasts can't proliferate—they are terminally differentiated.) Neither Ca-ranelate nor ranelic acid alone can mimic the effects of Sr-ranelate on these culured rat cells (Marie, 2006; Reginster et al., 2003). Sr-ranelate, also inhibits osteoclast activity on culture models without killing the cells or their ability to attach to bone (Reginster et al., 2003). Again neither Ca-ranelate nor ranelic acid alone mimicked Sr-ranelate. Thus, the stimulation of osteoblast precursor proliferation and the inhibition of osteoclast activity are specific for Sr-ranelate.

It might be expected that Sr ranelate could act via the CaR (the Ca-sensing receptor) (Brown, 2003). And it does activate the CaR (Brown, 2003; Reginster et al., 2003). As we learned in Chapter 2 activating the CaRs on immature osteoblastic cells stimulates their proliferation and their migration to osteoclast excavation sites and at the same time inhibits osteoclast activity (Brown, 2003; Huang et al., 2001). But this would seem not to be the case because Ca-ranelate does not mimic the stimulatory actions of Sr-ranelate on the rat calvarial cells. But wait! There is evidence from experiments on Chinese hamster ovary cells AtT20 cells that the responsiveness of the CaR to Sr-ranelate is enhanced by a high Ca^{2+} concentration such as 2 mM (Reginster et al., 2003). Therefore, since Sr-ranelate must operate in osteoclast excavation sites with extremely high Ca^{2+} concentrations (see Chapter 2), the Sr-ranelate action might operate at least in part through the CaRs on osteoblastic cells and osteoclasts. However, Pi and Quarles (2004, 2005) have produced evidence for bone cells having a novel Ca^{2+}-sensing receptor,Ob.CASR, that responds to strontium. Maybe this receptor belongs to the CaR-related mGluR (metabotropic glutamate receptor) family (Chattopadhyay and Brown, 2003; Hinoi et al., 2004; A.F. Taylor, 2002).

Of course these culture models suggested that Sr-ranelate might be a powerful, "double-barreled" osteoblast-stimulating/osteoclast-suppressing drug for stimulating bone growth in rats and humans (Marie, 2006). Indeed at a dose of 900 mg/kg/day Sr-ranelate increased the ultimate strength of midshaft femur and lumbar vertebrae of normal rats (Marie et al., 2001). At a dose range of 77-308 mg/kg/day, Sr-ranelate was able to substantially reduce the post ovariectomy loss of trabecular bone from an untreated 45% to 25% (Marie et al., 2001). However, unlike the powerfully anabolic PTHs, which can raise the post-ovariectomy trabecular bone volume in rats to twice the value in the sham-operated control rats, Sr-ranelate couldn't even get the trabecular volume back to the sham value. The Sr-ranelate also reduces the trabecular bone loss in rat long bones caused by immobilization (Marie et al., 2001).

Of course the big question is—what does Sr-ranelate do to osteoporotic women? After all young rats, unlike old postmenopausal women, have growing bones and the inhibition of osteoclasts alone can release the osteoclastic brakes on their bone growth. The answer has recently been published in the New England Journal of Medicine by Meunier et al. (2004). This paper reports the results of Sr-ranelate's Phase II trial. As expected the results are less dramatic than in rodents. There were 723 placebo-treated and 719 Sr-ranelate-treated post-menopausal women with an average age of 69 years with a relative compliance of 85 and 83% respectively. Two g of Sr-ranelate, dissolved in normal drinking water, were swallowed each day for 3 years. About 11% of the Sr in the 2190 g of ingested of ranelate would have been retained in their skeletons at the end of the treatment (Blake and Fogelman, 2005). At first sight, it appeared that the Sr-ranelate had increased the average spinal BMD as much as the approved daily dosage of hPTH-(1-34) (Forteo™). But this was not true because the bone mineral density should have been corrected because Sr is a heavier atom ($Z = 38$) than Ca ($Z = 20$). When this is done, Sr was only 15-50% as effective as the PTH (Blake and Fogelman, 2005). However, the Sr salt did reduce the new vertebral fracture rate fron a placebo value of 32.8 to 20.9 %. The data indicated that 9 patients would have to be treated in order to prevent 1 patient from having a vertebral fracture. The 41% reduction of the risk of new vertebral fractures by 3 years of treatment with Sr-ranalate, harmless though it is, can't match the much higher 65% reduction by 21 months of treatment with hPTH-(1-34) (Forteo™) (Meunier et al., 2004; Neer et al., 2001).

In conclusion Sr-ranelate can modestly increase bone formation in humans probably as in rats by stimulatng the proliferation of osteoprogenitor cells and somehow incapacitating osteoclasts without killing them. Swallowing Sr-ranelate may well be the the best thing to give astronauts to prevent microgravity-induced bone loss as they set out on long space voyages with initially strong bones. Sr-ranelate might also be a simple oral drug to swallow while self-injecting, swallowing or inhaling PTHs to enhance the PTHs' formidable osteogenicity by stopping the peptides from stimulating osteoclasts which the osteoclast-killing alendronate (Fosamax™) can't do because the bisphosphonate inhibits osteoblast activity (Onyia et al., 2002)..

CHAPTER 11

Afterword

Our aging and soon to be space-faring world needs drugs that can directly stimulate bone growth instead of merely stopping bone loss useful though this may be. As you have seen in these many pages, there are now some potent anabolic drugs. One of PTHs, the venerable hPTH-(1-34) (Forteo™), approved by the USFDA in October 2001, is being widely used and others such as Ostabolin-C™ are in the middle of their trials and destined to join Forteo™ in the important job of growing bone. They are very safe peptides when used for a couple of years which even use other candidate anabolic agents such as FGF-2 and IGF-I to partly drive bone growth. But at the moment they suffer from having to be injected subcutaneously for a couple of years to grow new bone before the injections can be stopped and the patient be switched over to an oral antiresorptive to keep their new "PTH bone". However, the experience from the several trials indicates that autoinjection by osteoporotics is not a significant deterrent, and there is one oral PTH already in phase II trial and inhalable PTHs are being developed by Lilly and Zelos Therapeutics. Close behind the PTHs may be the non-injectable, topically applicable, but so far frustratingly erratic, statins with some bone anabolic and antiresorptive promise from experiments on mice and rats, with very, beneficial cardiovascular actions and with the very important possibility of being anti-sporadic Alzheimer's disease in humans. What incredible drugs they might turn out to be if they could get their acts together! Then there is the "sleeper"—leptin—which to everyone's surprise may turn out to be another member of the exclusive bone builders' club. But as of late-2006 it is probably be too late for these contenders. There will be no competition for the new non-injectable, low- or maybe even non-hypercalcemogenic, potently anabolic PTHs which will be unbeatable when they arrive on the market.

Farewell dear reader, but do stay tuned for the next update so as not to miss the exciting things that will be happening at the cutting edge of research and development of bone-growing drugs. As you have seen this is really a work in progress loaded with rewards for researchers, big pharmas but most important for senior citizens with increasingly fracture-prone skeletons, loosening artificial knees and hips and dental implants.

References

Aarts MM, Davidson D, Corluka A et al. Parathyroid hormone-related protein promotes quiescence and survival of serum-deprived chondrocytes by inhibiting rRNA synthesis. J Biol Chem 2001; 276:37934-37943.

Aarts MM, Rix A, Guo J et al. The nucleolar targewting signal (NTS) of parathyroid hormone-related protein mediates endocytosis and nucleolar translocation. J Bone Miner Res 1999; 14:1493-14503.

Abedin M, Tintut Y, Demer LL. Vascular calcification. Arterioscler Vasc Biol 2004; 24:1161-1170.

Ackerman J. The downside of upright. National Geographic 2006; 210:126-145.

Adams C, Mansfield K, Perlot RL et al. Matrix regulation of skeletal cell apoptosis: Role of calcium and phosphate atoms. J Biol Chem 2001; 276:20316-20322.

Adams GB, Chabner KT, Alley IR et al. Stem cell engraftment at the endosteal niche is specifed by the calcium-sensing receptor. Nature 2006; 439:599-603.

Adams JM. Ways of dying: Multiple pathways to apoptosis. Genes Devel 2003; 17:2481-2495.

Ae K, Kobayashi N, Sakuma R et al. Chromatin remodeling factor encoded by ini1 induces G_1 arrest and apoptosis in ini1-deficient cells. Oncogene 2002; 21:3112-3120.

Afzal F, O'Shaughnessy M, Nohadani RM et al. Osteoblast growth and differentiation is retarded in endothelial nitric oxide synthase knockout mice (abstract). J Bone Miner Res 2000; 15:S217.

Aguirre J, Buttery L, O'Shaughnessy M et al. Endothelial nitric oxide synthase gene-deficient mice demonstrate marked retardation in postnatal bone formation, reduced bone volume, and defects in osteoblast maturation and activity. Am J Pathol 2001; 158:247-257.

Aguirre JL, Plotkin LI, Strotman B et al. The anti-apoptotic effects of mechanical stimulation in osteoblasts/osteocytes are transduced by the estrogen receptor (ER): A novel ligand-independent function of the ER. J Bone Miner Res 2003; 18:S71.

Aho S. Soluble form of Jagged1: Unique product of epithelial keratinocytes and a regulator of keratinocyte differentiation. J Cell Biochem 2004; 92:1271-1281.

Akhter MP, Kimmel DB, Recker RR. Effect of parathyroid hormone (HPTH[1-84]) treatment on bone mass and strength in ovariectomized rats. J Clin Densitom 2001; 4:13-23.

Al-Bhalal L, Akhbar M. Molecular basis of autosomal dominant polycystic kidney disease. Adv Anat Pathol 2005; 12:126-133.

Alieva IB, Gorgidze LA, Komarova YA et al. Experimental model for studying the primary cilia in tissue culture cells. Membr Cell Biol 1999; 12:895-905.

Ahima RS, Flier JS. Leptin. Annu Rev Physiol 2000; 62:413-437.

Ali AA, Plotkin LI, Foote IP et al. Bcl-2 is a pivotal mediator of the anti-apoptotic effect of PTH on osteoblasts: Evidence from RNA-silencing and Bcl-2-deficient mice (abstract). J Bone Miner Res 2003; 18:S73.

Alkhiary YM, Gerstenfeld LC, Cullimane DM et al. Parathyroid hormone (1-34; teriparitide) enhances experimental fracture healing (abstract). J Bone Miner Res 2003; 18:S24.

Allen CM, Chakravarthy B, Morley P et al. hPTH-(1-34) and hPTH-(1-31)NH$_2$ prevent phosphate from killing MC3T3-E1 preosteoblasts and HKRKB7 pig kidney cells (abstract). J Bone Miner Res 2002; 17:S390.

Allen MJ, Schoonmaker JE, Mann KA et al. PTH analogs enhance bone formation at a weight-bearing cement-bone interface. Trans Orthoped Res Soc 2004.

Allen MR, Iwata K, Phipps R et al. Alteration in canine vertebral bone turnover, mineralization, microdamage accumulation, and biomechanical propoerties following 1-year treatment with clinical treatment doses of risedronate or alendronate. Bone 2006; 38:S42.

Allen SC, Hebbes TR. Myb-induced myeloid protein 1 (Mim-1) is an acetyltransferase. FEBS Lett 2003; 534:119-124.

Alvarez J, Sohn P, Zeng X et al. TGFβ2 mediates the effects of hedgehog on the hypertrophic differentiation amd PTHrP expression. Development 2002; 129:1913-1924.

Alvarez M, Thunyakitpisal P, Morrison P et al. PTH-responsive osteoblast nuclear matrix architectural transcription factor binds to the rat type I collagen promoter. J Cell Biochem 1998; 69:336-352.

Alvarez-Buylla A, Garcia-Verdugo J, Tramontin AD. A unified hypothesis on the lineage of neural stem cells. Nature Rev Neurosci 2001; 2:287-293.

Amalric F, Baldin V, Bosc-Bierne I et al. Nuclear translocation of basic fibroblastic growth factor. Ann NY Acad Sci 1991; 638:127-138.

Amin D, Cornell SA, Gustafson SK et al. Bisphosphjonates used for the treatment of bone disorders inhibit squalene synthase and cholesterol biosynthesis. J Lipid Res 1992; 33:1657-1663.

Amizuka N, Karaplis AC, Henderson JE et al. Haploinsufficiency of parathyroid hormone related peptide (PTHrP) results in abnormal postnatal development. Dev Biol 1996; 175:166-176.

Amizuka N, Warshawsky H, Henderson JE et al. Parathyroid hormone-related peptide-depleted mice showabnormal epiphyseal cartilage development and altered endochondral one formation. J Cell Biol 1994; 126:1611-1623.

Amling M, Neff L, Tanaka S et al. Bcl-2 lies downstream of parathyroid hormone-related peptide in a signaling pathway that regulates chondrocyte maturation during skeletal development. J Cell Biol 1997; 136:205-213.

Anderson RGW. The caveolar membrane system. Annu Rev Biochem 1998; 67:199-225.

Anderson RGW, Jacobson K. A role for lipid shells in targeting proteins to caveolae, rafts, and other lipid domains. Science 2002; 296:1821-1825.

Anderson HC. Molecular biology of matrix vesicles. Clin Orthop 1995; 314:266-280.

Anderson HC. Matrix vesicles and calcification. Curr Rhematol Rep 2003; 5:222-226.

Andreassen TT, Ejersted C, Oxlund H. Intermittent parathyroid hormone (1-34) treatment increases callus formation and mechanical strength of healing rat fractures. J Bone Miner Res 1999; 14:960-968.

Andreassen TT, Oxlund H. The influence of combined parathyroid hormone and growth hormone treatment on cortical bone in aged ovariectomized rats. J Bone Miner Res 2000; 15:2266-2275.

Andeassen TT, Willick GE, Morley P et al. Treatment with parathyroid hormone hPTH(1-34),hPTH(1-31), and moncyclic hPTH-(1-31) enhances fracture strength and callus amount-after withdrawal fracture strength and callus mechanical quality continue to increase. Calcif Tissue Int 2004, (Epub Jan 23).

Anreatta RH, Hartmann A, Kamber JA et al. Synthese der sequenz 1-34 von menschlichem Parat-hormon. Helv Chim Acta 1973; 56:470-473.

Antonsson B, Martinou JC. The Bcl-2 protein family. Exp Cell Res 2000; 256:50-57.

Arai F, Hirano A, Omura M et al. Tie2/Angiopoietin-1 signaling regulates hematopoietic stem cell quiescence in the bone marrow niche. Cell 2004; 118:149-161.

Arai F, Hirano A, Omura M et al. Regulation of hemtopoietic stem cells by the niche. Trends Cardiovasc Med 2005; 15:75-79.

Archer CW, Francis-West P. The chondrocyte. Int J Biochem Cell Biol 2003; 35:401-404.

Arends RJ, Langerwerf PEJ, van de Klundert TMC et al. Responses of MC3T3-E1 osteoblast cells to parathyroid hormone are dependent on differentiaition stage and duration of treatment (abstract). J Bone Miner Res 2004; 19:S150.

Arey BJ, Seethala R, Ma Z et al. A novel calcium-sensing receptor antagonist transiently stimulates parathyroid hormone secretion in vivo. Endocrinology 2005; 146:2015-2022.

Arita S, Ikeda S, Ito et al. Intermittent hPTH administration increases bone strength in rat lumbar vertebral body after ovariectomey with cortical bone mass increase (abstract). J Bone Miner Res 2002; 17:S274.

Armour KE, Armour KJ, Gallagher ME et al. Defective bone formation and anabolic response to exogenous estrogen in mice with targeted disruption of endothelial nitric oxide synthase. Endocrinology 2001; 142:760-766.

Arnett TR, Gibbons DC, Utting JC et al. Hypoxia is a major stimulator of osteoclast formation and bone resorption. J Cell Physiol 2003; 196:2-8.

Artru P, Tournigand C, Mabro M et al. Hypoparathyroïdie primitive associée au cancer colique. Gastroenterol Clin Biol 2001; 25:208-209.

Ashby MC, Tepikin AV. ER calcium and the functions of intracellular organelles. Cell Devel Biol 2001; 12:11-17.

Atkins GT, Kostakis P, Pan B et al. RANKL expression is related to the differentiation state of human osteoblasts. J Bone Miner Res 2003; 18:1088-1098.

Aubin JE. Advances in the osteoblast lineage. Biochem Cell Biol 1998; 76:899-910.

Atkinson MJ, Hesch RD, Cade C et al. Parathyroid hormone stimulation of mitosis in rat thymic lymphocytes is independent of cyclic AMP. J Bone Miner Res 1987; 2:303-309.

Atkinson PJ, Hallsworth AS. The changing structure of aging human mandibular bone. Gerodontology 1983; 2:57-66.

Aubin JE. The role of osteoblasts. In: Henderson JE, Goltzman D, eds. The Osteoporosis Primer. Cambridge: Cambridge University Press, 2000:18-35.

Aubin JE. Regulation of osteoblast formation and function. Rev Endocrine Metab Dis 2001; 2:81-94.

Aubin JE, Bonnelye E. Osteoprotegerin and its ligand: A new paradigm for regulation of osteoclastogenesis and bone resorption. Medscape Women's Health 2000; 5, (http://www.medscape.com/).

Aubin JE, Triffitt JE. Mesenchymal stem cells and osteoblast differentiation. Principles of Bone Biology. 2nd ed. Vol. 1. San Diego: Academic Press, 2002:59-91.

Awata H, Huang C, Handlogten ME et al. Interactions of the calcium-sensing receptor and filamin, a potential scaffolding protein. J Biol Chem 2001; 276:34871-34879.

Azarani A, Goltzman D, Orlowski J. Parathyroid hormone and parathyroid hormone-related peptide inhibit the apical Na^+/H^+ exchanger NHE-3 isoform in renal cells (OK) via a dual signaling cascade involving protein kinase A and C. J Biol Chem 1995a; 270:20004-20010.

Azarani A, Orlowski J, Golzman D. Parathyroid hormone and parathyroid hormone-related peptide activate the Na^+/H^+ exchanger NHE-1 isoform in osteoblastic cells (UMR-106) via a cAMP-dependent pathway. J Biol Chem 1995b; 270:23166-23172.

Bab I, Chorev M. Osteogenic growth peptide: From concept to drug design. Biopolymers (Peptide Science) 2002; 66:33-48.

Bab I, Einhorn TA. Regulatory role of osteogenic growth polypeptides in bone formation and hemopoiesis. Crit Rev Eukaryot Gene Express 1993; 3:31-46.

Bakker AD, Klein-Nulend J, Burger EH. Mechanotransduction in bone cells proceeds via activation of COX-2, but not COX-1. Biochem Biophys Res Commun 2003; 305:677-683.

Bakker RT. The Dinosaur Heresies. New York: William Morrow and Compnay Inc., 1986.

Balcerzak M, Hamade E, Zhang L et al. The roles of annexins and alkaline phosphatase in mineralization process. Acta Biochim Pol 2003; 50:1019-1038.

Baldock PA, Sainsbury A, Couzeau M et al. Hypothalamic Y2 receptors regulate bone formation. J Clin Invest 2002; 109:915-921.

Bandorowicz-Pikula J, Buchet R, Pikula S. Annexins as nucleotide-binding proteins: Facts and speculations. BioEssays 2001; 23:170-178.

Banerjee C, Javed A, Choi JY et al. Differential regulation of the two principal Runx2/Cbfa1 n-terminal isoforms in response to protein morphogenic protein-2 during development of the osteoblast phenoptype. Endocrinology 2001; 142:4026-4039.

Banu J, Kalu DN. Effects of cerivastatin and parathyroid hormone in male Sprage-Dawley rats (abstract). J Bone Miner Res 2002; 17:S296.

Bhangu PS, Genever PG, Spencer GJ et al. Evidence for targeted vesicular exocytosis in osteoblasts. Bone 2001; 29:16-23.

Barber DL, Ganz MB. Guanine nucleotides regulate β-adrenergic activation of Na-H exchange independently of receptor coupling to Gs. J Biol Chem 1992; 267:20607-20612.

Barber DL, Ganz MB, Bongiorno PB et al. Mutant constructs of the β-adrenergic receptor that are uncoupled from adenylyl cyclase retain functional activation of Na-H exchange. Mol Pharmacol 1992; 41:1056-1060.

Barbier JR, Gardella TJ, Dean T et al. Backbone-methylated analogues of the principle receptor binding region of human parathyroid hormone. J Biol Chem 2006; 280:23771-23777.

Barbier JR, Neugebauer W, Morley P et al. Bioactivities and secondary structures of constrained analogues of human parathyroid hormone: Cyclic lactams of the receptor-binding region. J Med Chem 1997; 40:1373-1380.

Bauer DC, Mundy GR, Jamal SA et al. Statin use, bone mass and fracture: An analysis of two prospective studies (abstract). J Bone Miner Res 1999; 14:S179.

Barrett MG, Belinsky GS, Tashjian Jr AH. A new action of parathyroid hormone, receptor-mediated stimulation of extracellular acidification in human osteoblast-like SaOS-2 cells. J Biol Chem 1997; 272:26346-26353.

Barros SP, Silva MAD, Somerman MJ et al. Parathyroid hormone protects against periodontitis-associated bone loss. J Dent Res 2003; 82:791-795.

Barry JA, Tanner SJ, Peyton A et al. Mechanical loading and PTH stimulation of DNA synthesis (abstract). J Bone Miner Res 2002; 17:S329.

Bassilana F, Susa M, Keller HJ et al. Human mesenchymal cells undergoing osteogenic differentiation express leptin and functional leptin receptors (abstract). J Bone Miner Res 2000; 15:S378.

Bauer E, Aub JC, Albright F. Studies of calcium and phosphorus metabolism: V. Study of the bone trabeculae as a readily available reserve supply of calcium. J Exp Med 1929; 49:145-162.

Baylink DJ, Strong DD, Mohan S. The diagnosis and treatment of osteoporosis: Future prospects. Mol Med Today 1999; 5:133-140.

Becamel C, Alonso G, Galéotti N et al. Synaptic multiprotein complexes associated with 5-HT$_{2c}$ receptors: A proteomic approach. EMBO J 2002; 21:2332-2342.

Beck Jr GR. Inorganic phosphate as a signaling molecule in osteoblast differentiation. J Cell Biochem 2003; 90:234-243.

Beck Jr GR, Knecht N. Osteopontin regulation by inorganic phosphate is ERK 1 / 2, PKC and proteasome dependent. J Biol Chem 2003; 278:41921-41929.

Beck Jr GR, Moran E, Knecht N. Inorganic phosphate regulates multiple genes during osteoblast differentiation, including Nrf2. Exp Cell Res 2003; 288:288-300.

Beck Jr GR, Sullivan EC, Moran E et al. Relationship between alkaline phosphatase levels, osteopontin expression, and mineralization in differentiating MC3T3-E1 osteoblasts. J Cell Biochem 1998; 68:269-280.

Beck Jr GR, Zerler B, Moran E. Phosphate is a specific signal for induction of osteopontin gene expression. Proc Natl Acad Sci USA 2000; 97:8352-8357.

Becker AJ, McCulloch EA, Siminovitch L et al. The effect of differing demanda for blood cell production on DNA synthesis by hemopoietic colony-forming cells of mice. Blood 1965; 26:296-308.

Beir F, Ali Z, Mok D et al. TGFβ and PTHrP control chondrocyte proliferation by activating cyclin D1 expresssion. Mol Biol Cell 2001; 12:3852-3863.

Bekker PJ, Holloway D, Nakanishi A et al. The effect of a single dose of osteoprotegerin in postmenopausal women. J Bone Miner Res 2001; 16:348-360.

Bekker PJ, Holloway D, Rasmussen AS et al. A single-dose placebo-controlled study of AMG 162, a fully human monoclonal antibody to RANKL, in postmenopausal women. J Bone Miner Res 2004; 19:1059-1066.

Belinsky GS, Morley P, Whitfield JF et al. Ca^{2+} and extracellular acidification rate responses to parathyroid hormone fragments in rat ROS 17/2 and human SaOS-2 cells. Biochem. Biophys Res Commun 1999; 266:448-453.

Belinsky GS, Tashjian Jr AH. Direct measurement of hormone-induced acidification in intact bone. J Bone Miner Res 2000; 15:550-556.

Bellido T, Ali AA, Gubrij I et al. Chronic elevation of parathyroid hormone in mice reduces expression of sclerostin by osteocytes: A novel mechanism for hormonel control of osteoblastogenesis. Endocrinology 2005; 146:4577-4583.

Bellido T, Plotkin LI, Davis J et al. Protein kinase A-dependent phosphorylation and inactivation of the pro-apoptotic protein Bad mediates the anti-apoptotic effect of PTH on osteoblastic cells (abstract). J Bone Miner Res 2001; 16:S203.

Bellido T, Plotkin LI, O'Brien A et al. PTH-mediated control of proteasome-mediated degradation of Runx2/Cbfa1: A pivotal determinant of the longevity of PTH-initiated anti-apoptosis signaling in osteoblastic cells. J Bone Miner Res 2002; 17:S128.

Benford HL, Frith JC, Auriola S et al. Farnesol and geranylgeraniol prevent activation of caspases by aminobisphosphonates: Biochemical evidence for two distinct pharmacological classes of bisphosphonate drugs. Mol Pharmacol 2001; 56:131-140.

Benford HL, McGowan NW, Helfrich MH et al. Visualization of bisphosphonate-induced caspase-3 activity in apoptotic osteoclasts in vitro. Bone 2001; 28:465-473.

Bennett BD, Solar GP, Yuan JQ et al. A role for leptin and its cognate receptor in hematopoiesis. Curr Biol 1996; 6:1170-1180.

enten WP, Stephan C, Lieberherr M et al. Estradiol signaling via sequestrable surface receptors. Endocrinology 2001; 142:1669-1677.

Bentolila V, Boyce TM, Fyhrie DP et al. Intracortical remodeling in adult rat long bones after fatigue loading. Bone 1998; 23:275-281.

Bertiaume LG. Insider information: How palmitoylation of Ras makes it a signaling double agent. Science's STKE 2002, (http://www.stke.org/cgi/content/sigtrans;2002/152/pe41).

Betancourt M, Wirfel KL, Raymond AK et al. Osteosarcoma of bone in a patient with primary hyperparathyroidism: A case report. J Bone Miner Res 2003; 18:163-166.

Bianco P, Bradbeer JN, Riminucci M et al. Marrow stromal cells: Identification, morphometry, confocal imaging and changes in disease. Bone 1993; 14:315-320.

Bianco P, Riminucci M. The bone marrow stroma in vivo: Ontogeny, structure, cellular composition and changes in disease. In: Beresford JN, Owen ME, eds. Marrow Stromal Cell Culture. Cambridge: Cambridge University Press, 1998:10-25.

Bianco P, Riminucci M, Gronthos S et al. Bone marrow stromall stem cells: Nature, biology, and potential applications. Stem Cells 2001; 19:180-192.

Bianco P, Riminucci M, Kuznetsov S et al. Multipotential cells in the bone marrow stroma: Regulation in the context of organ physiology. Crit Revs Euk Gene Express 1999; 10:159-173.

Bidwell JP, Torrungruang K, Alvarez M et al. Involvement of the nuclear matrix in the control of skeletal genes: The NMP1 (YY1), NMP2 (Cbfa1), and NMP4 (Nmp4/CIZ) transcription factors. Crit Rev Eukaryotic Gene Express 2001; 11:279-297.

Bier E. The Coiled Spring. Cold Spring Harbor, New York: Cold Spring Harbor Laboratory Press, 2000.

Bijlsma MF, Spek CA, Peppelenbosch MP. Hedgehog: An unusual signal transducer. BioEssays 2004; 26:387-394.

Bikle DD, Harris J, Halloran BP et al. Skeletal unloading indices resistance to insulin-like growth factor. J Bone Miner Res 1994; 9:1789-1796.

Bilezikian JP. Sex steroids, mice, and men: When androgens and estrogens get very close to one another. J Bone Miner Res 2002; 17:563-566.

Bilezikian JP, Marcus MA, Levine MA, eds. The Parathyroid Hormone. 2nd ed. San Diego: Academic Press, 2001.

Bilton RL, Booker GW. The subtle side to hypoxia-inducible factor (HIFα) regulation. Eur J Biochem 2003; 270:791-798.

Bird RP, Good CK. The significance of aberrant crypt foci in understanding the pathogenesis of colon cancer. Toxicology Letters 2000; 112-113:395-402.

Bisello A, Sneddon WB, Friedman PA. PTHR1 endocytosis is not necessary for resensitization of cAMP signaling. J Bone Miner Res 2002; 17:S287.

Bitgood MJ, McMahon AP. Hedgehog and Bmp genes are coexpressed at many diverse sites of cell-cell interactions in the mouse embryo. Dev Biol 1995; 172:126-138.

Bjurholm A. Neuroendocrine peptides in bone. Int Orthop 1991; 15:325-329.

Black DM, Bilezikian JP, Ensrud KE et al. One year of alendronate after one year of parathyroid hormone (1-84) for osteoporosis. New Engl J Med 2005; 353:555-565.

Black DM, Greenspan SL, Ensrud KE et al. The effects of parathyroid hormone and alendronate alone or in combination in postmenopausal osteoporosis. N Engl J Med 2003; 349:1207-1215.

Black KM, Theriault BL, Anderson GI. Do bones have memory? Glutamate receptor expression in bone (abstract). J Bone Miner Res 2002; 17:S330.

Black TM, Theriault BL, Anderson GI. Mechanotransduction via glutamate receptors: Regulation of osteoblast activity in response to mechanical stimulation. J Bone Miner Res 2003; 18:S133.

Blain H, Vuillemin A, Guillemin F et al. Serum leptin level is a predictor of bone mineral density in postmenopausal women. J Clin Endocrinol Metab 2002; 87:1030-1035.

Blair HC. How the osteoclast degrades bone. BioEssays 1998; 20:837-846.

Blair HC, Zaidi M, Schlesinger PH. Mechanisms balancing skeletal matrix synthesis and degradation. Biochem J 2002; 364:329-341.

Blake GM, Fogelman I. Long-term effect of strontium ranelate treatment on BMD. J Bone Miner Res 2005; 20:1901-1904.

Blin-Wakkach C, Lezot F, Ghoul-Mazgar S et al. Endogenous Msx1 antisense transcript: In vivo and in vitro evidences, structures, and potential involvement in skeleton development in mammals. Proc Natl Acad Sci USA 2001; 98:7336-7341.

Bliziotes MM, Eshleman AJ, Zhang XW et al. Neurotransmitter action in osteoblasts: Expression of a functional system for serotonin receptor activation and reuptake. Bone 2001; 29:477-486.

Blomme EA, Sugimoto Y, Lin YC et al. Parathyroid hormone-related protein is a positive regulator of keratinocyte growth factor expression by normal dermal fibroblasts. Mol Cell Endocrinol 1999; 152:189-197.

Blum WF. Leptin: The voice of the adipose tissue. Hormone Res 1997; 48(Suppl. 4):2-8.

Bockaert J, Claeysen S, Bécamel C et al. G protein-coupled receptors: Dominant players in cell-cell communication. Int Rev Cytol 2002; 212:63-132.

Bockaert J, Marin P, Dumuis A et al. The 'magic tail' of G protein-coupled receptors: An anchorage for functional protein networks. FEBS Lett 2003; 546:65-72.

Boivin G, Meunier PJ. Effects of bisphosphonates on matrix mineralization. J Musculoskel Neuron Interact 2002; 2:538-543.

Boivin GY, Chavassieux PM, Santors AC et al. Alendronate increases bone strength by increasing the mean degree of mineralization of bone tissue in osteoporotic women. Bone 2000; 27:687-694.

Boland R, De Boland AR et al. Nongenomic stimulation of tyrposine phosphorylation cascades by 1,25(OH)$_2$D$_3$ by VDR-dependent and -independent mechanisms in muscle cells. Steroids 2002; 67:477-482.

Bolon B, Campagnuolo G, Feige U. Duration of bone protection by a single osteoprotegerin injection in rats with adjuvant-induced arthritis. Cell Mol Life Sci 2002a; 59:1569-1576.

Bolon B, Shaloub V, Kostenuik PJ et al. Osteoprotegerin, an endogenous antiosteoclast factor for protecting bone in rheumatoid arthritis. Arthritis Rheum 2002b; 46:3121-3135.

Bonadio J. Tissue engineering via local gene delivery: Update and future prospects for enhancing the technology. Adv Drug Del Rev 2000; 44:185-194.

Bonadio J, Smiley E, Path P et al. Localized, direct plasmid gene delivery in vivo: Prolonged therapy results in reproducible tissue regeneration. Nature Medicine 1999; 5:753-759.

Bone HG, Hosking D, Devogelaer JP et al. Ten years' experience with alendronate for osteoporosis in postmenopausal women. New Engld J Med 2004; 350:1189-1199.

Bonnet N, Brunet-Imbault B, Parnaud CJ et al. β2-Adrenergic agonists have negative effects on bone architecture and density in rat. J Bone Miner Res 2003; 18:S42.

Bonewald LF. Generation and function of osteocyte dendritic processes. J Musculoskelet Neuronal Interact 2005; 5:321-324.

Börcsök I, Schairer HU, Sommer U et al. Glucocorticoids regulate the expression of the human osteoblastic endothelin A receptor gene. J Exp Med 1998; 188:1563-1573.

Bosch P, Musgrave DS, Lee JY et al. Osteoprogenitor cells within skeletal muscle. J Orthop Res 2000; 18:933-944.

Boskey AL. Mineral-matrix interactions in bone and cartilage. Clin Orthop 1992; 281:244-274.

Boss JH, Misselevich I. Osteonecrosis of the femoral head of laboratory animals: The lessons learned from a comparative study of osteonecrosis in man and experimental animals. Vet Pathol 2003; 40:345-354.

Boström K, Demer LL. Regulatory mechanisms in vascular calcification. Critical Rev Eukaryot Gene Express 2000; 12:151-158.

Bouche G, Gas N, Prats H et al. Basic fibroblast growth factor enters the nucleolus and stimulates the transcription of ribosomal genes in ABAE cells undergoing G$_0$-G$_1$ transition. Proc Natl Acad Sci USA 1987; 84:6770-6774.

Bouletreau PJ, Warren SM, Spector JA et al. Hypoxia and VEGF up-regulate BMP-2 mRNA and protein expression in microvascular endothelial cells: Implications for fracture healing. Plast Reconstruct Surg 2002; 109:2384-2397.

Bouloumie A, Drexler HC, Lafontan M et al. Leptin, the product of Ob gene, promotes angiogenesis. Circ Res 1998; 83:1059-1066.

Boulter C, Mulroy S, Webb S et al. Comparison of the cardiovascular, skeletal, and renal defects in mice with a targeted disruption of the Pkd1 gene. Proc Natl Acad Sci USA 2001; 98:12174-12179.

Bouxsein ML. Etiology and biomechanics of hip and vertebral fractures. In: Marcus R, ed. Atlas of Clinical Endocrinology. Vol. 3. Osteoporosis, Philadelphia: Current Medicine, 1999:139-148.

Bowe EA, Notomi T, Horner A et al. Glutamate signalling and LTP-like mechanisms account for loading memory in osteoblasts. J Bone Miner Res 2004; 19:S10.

Bowler WH, Buckley KA, Gartland A et al. Extracellular nucleotide signaling: A mechanism for integrating local and systemic responses in the activation of bone remodeling. Bone 2001; 28:507-512.

Bowe AE, Finnegan R, Jan de Beur SM et al. FGF-23 inhibits renal tubular phosphate transport and is a PHEX substrate. Biochem Biophys Res Commun 2001; 284:977-981.

Boyce RW, Paddock CL, Franks AF et al. Effects of intermittent hPTH-(1-34) alone and in combination with 1,25(OH)$_2$D$_3$ or risedronate on endosteal bone remodeling in canine cancellous and cortical bone. J Bone Miner Res 1996; 11:600-613.

Brailov I, Bancila M, Brisorgeuil MJ et al. Localization of 5-HT$_6$ receptors at the plasma membrane of neuronal cilia in the rat brain. Brain Res 2000; 872:271-275.

Brandi ML, Collin-Osdoby P. Vascular biology and the skeleton. J Bone Miner Res 2006; 21:183-192.

Brandi ML, Crescioli C, Tanini A et al. Bone endothelial cells as estrogen targets. Calcif Tissue Int 1993; 53:312-317.

Brandon C, Eisenberg LM, Eisenberg CA. WNT signaling modulates the diversification of hematopoietic cells. Blood 2000; 96:4132-4141.

Briggs SD, Xiao T, Sun ZW et al. Gene silencing: Trans-histone regulatory pathway to chromatin. Nature 2002; 418:498.

Brighton CT, Hunt RM. Histochemical localization of calcium in growth plate mitochondria and matrix vesicles. Fed Proc 1976; 35:143-147.

Brighton CT, Hunt RM. The role of mitochondria in growth plate calcification as demonstrated in a rachitic model. J Bone Joint Surg 1978; 60A:630-639.

Brann DW, De Sevilla L, Zamorano PL et al. Regulation of leptin gene expression and secretion by steroid hormones. Steroids 1999; 64:659-663.

Bringhurst FR. PTH receptors and apoptosis in osteocytes. J Muscoskel Neuron Interact 2002; 2:245-251.

Bringhurst FR. Circulating forms of parathyroid hormone: Peeling back the onion. Clin Chem 2003; 49:1973-1975.

Bringhurst FR, Stern AM, Yotts M et al. Peripheral metabolism of PTH: Fate of biologically active amino terminus in vivo. Am J Physiol 1988; 255:E886-E893.

Brody T, Cravchick A. Drosophila melanogaster G protein-coupled receptors. J Cell Biol 2000; 150:F83-F88.

Brommage R, Hotchkiss CE, Lees CJ et al. Daily treatment with human recombinant parathyroid hormone-(1-34), LY33333, for 1 year increases bone mass in ovariectomized monkeys. J Clin Endocrinol Metab 1999; 84:3757-3763.

Brouet A, Sonveaux P, Dessy C et al. Hsp90 ensures the transition from the early Ca^{2+}-dependent to the late phosphorylation-dependent activation of the endothelial nitric-oxide synthase in vascular endothelial growth factor-exposed endothelial cells. J Biol Chem 2001a; 276:32663-32669.

Brouet A, Sonveaux P, Dessy C et al. Hsp90 and caveolin are key targets for the proangiogenic nitric oxide-mediated effects of statins. Circ Res 2001b; 89:866-873.

Brown EM. Is the calcium receptor a molecular target for the action of strontium on bone? Osteoporosis Int 2003; 14(Suppl. 3):S25-S34.

Brown EM, MacLeod RJ. Extracellular calcium sensing and extracellular calcium signaling. Physiol Revs 2001; 81:240-297.

Bruick RK. Oxygen sensing in the hypoxic response pathway: Regulation of the hypoxia-inducible transcription factor. Genes Dev 2003; 17:2614-2623.

Bucay N, Sarosi I, Dunstan CR et al. osteoportegerin-deficient mice develop early onset osteoporosis and arterial calcification. Genes Dev 1998; 12:1260-1268.

Bucay N, Sarosi I, Dunstan CR et al. Osteoprotegerin-deficient mice develop early onset osteoporosis and arterial calcification. Genes Dev 2006; 12:1260-1268.

Buckley KA, Dillon JP, Chen BYY et al. Low dose PTH, insufficient to induce cyclic AMP or c-fos, stimulates an anabolic gene response in human osteoblasts. Bone 2006; 38:S10-S11.

Bullough PG. L'Ostéonécrose. Ann Pathol 2001; 21:512-523.

Bunting CH. The formation of true bone with cellular (red) marrow in a sclerotic aorta. J Exp Med 1906; 8:365-376.

Burger E. Experiments on cell mechanosensitivity: Bone cells as mechanical engineers. In: Cowin SC, ed. Bone Mechanics Handbook. 2nd ed. Boca Raton: CRC Press, 2001:28-1-28-16.

Burger EH, Klein-Nulend J, Smit TH. Strain-derived canalicular fluid flow regulates osteoclast activity in a remodelling osteon—a proposal. J Biomechanics 2003; 36:1453-1459.

Burguera B, Hofbauer LC, Thomas T et al. Leptin reduces ovariectomy-induced bone loss in rats. Endocrinology 2001; 142:3546-3553.

Burns M, Duff K. Cholesterol in Alzheimer's disease and tauopathy. Ann NY Acad Sci 2002; 977:367-375.

Burr DB, Hirano T, Turner CH et al. Intermittently administered human parathyroid hormone (1-34) treatment increases intracortical bone turnover and porosity without reducing bone strength in the humerus of ovariectomize cynomolgus monkeys. J Bone Miner Res 2001; 16:157-165.

Burr DB, Martin RB, Schaffler MB et al. Bone remodeling in response to in vivo fatigue fatigue microdamage. J Biomech 1985; 18:189-200.

Burr DB, Robling AG, Turner CH. Effects of biomechanical stress on bone in animals. Bone 2002; 30:781-786.

Calvert JP. Cilia in PKD—letting it all hang out. J Am Soc Nephrol 2002; 13:2614-2616.

Byron JW. Cyclic nucleotides and the cell cycle of the hematopoietic stemcell. In: Abou-Sabé M, ed. Cyclic Nucleotides and the Regulation of Cell Growth. Stroudsburg, Dowden: Hutchinson and Ross Inc., 1977:81-92.

Calvi LM, Adams GB, Weibrecht KW et al. Osteoblastic cells regulate the haematopoietic stem cell niche. Nature 2003; 425:841-846.

Calvi LM, Sims NA, Hunzelman JL et al. Activated parathyroid hormone/parathyroid hormone-related protein receptor in osteoblastic cells differentially affects cortical and trabecular bone. J Clin Invest 2001; 107:277-286.

Cameron DA. The ultrastructure of bone. In: Bourne GH, ed. The Biochemistry and Physiology of Bone. 2nd ed. New York: Academic Press, 1972:191-236.

Canalis E, Economides AN, Gazzerro E. Bone morphogenic proteins, their antagonists, and the skeleton. Endocrine Rev 2003; 24:218-235.

Canatan H, Bakan I, Akbulut MA et al. Comparative analysis of plasma leptin levels in both genders of patients with essential hypertension and healthy subjects. Endocrine Res 2004; 30:95-105.

Candeliere GA, Liu F, Aubin JE. Individual osteoblasts in the developing calvaria express different gene repertoires. Bone 2001; 28:351-361.

Candeliere GA, Rao Y, Floh A et al. cDNA fingerprinting of osteoprogenitor cells to isolate differentiation stage-specific genes. Nucleic Acids Res 1999; 27:1079-1083.

Candeliere GA, Yoshiko Y, Aubin JE. CGI-135 is part of a new chromatin remodeling protein complex and it regulates osteoprogenitor proliferation-differentiation transition. J Bone Miner Res 2002; 17:S195.

Cann CE, Roe EB, Sanchez SD et al. PTH effects in the femur: Envelope-specific responses by 3DQCT in postmenopausal women (abstract). J Bone Miner Res 1999; 14:S137.

Cao R, Brakenhielm E, Wahlestedt C et al. Leptin induces vascular permeability and synergistically stimulates angiogenesis with FGF-2 and VEGF. Proc Natl Acad Sci USA 2001; 98:6390-6395.

Cao X, Südhof TC. A transcriptionally active complex of APP woth FE65 and histone acetyltransferase Tip60. Science 2001; 293:115-120.

Caporale LH. Darwin in the Genome. New York: McGraw-Hill, 2002.

Capparelli C, Morony S, Warmington K et al. Sustained antiresorptive effects after a single treatment with human recombinant osteoprotegerin (OPG): A pharmacodynamic and pharmacokinetic analysis in rats. J Bone Miner Res 2003; 18:852-858.

Carmignoto G. Reciprocal communication systems between astrocytes and neurons. Prog Neurobiol 2000; 62:561-581.

Carano RAD, Filvaroff EH. Angiogenesis and bone repair. Drug Discovery Today 2003; 8:980-989.

Carter WB, Uy K, Ward MD et al. Parathyroid-induced angiogenesis is VEGF-dependent. Surgery 2000; 128:458-464.

Carvalho RS, Scott JE, Suga DM et al. Stimulation of signal transduction pathways in osteoblasts by mechanical strain potentiated by parathyroid hormone. J Bone Miner Res 1994; 9:999-1011.

Caverzasio J, Bonjour JR. Characteristics and regulation of Pi transport in osteogenic cells for bone metabolism. Kidney Int 1996; 49:975-980.

Cenci S, Weitzmann MN, Roggia CR et al. Estrogen deficiency induces bone loss by enhancing T-cell production of TNF-α. J Clin Invest 2000; 106:1229-1237.

Center for Drug Evaluation and Research, Endocrinology snd Metabolic Drugs Advisory Committee Meeting, 2001, (Available from URL:http://www.fdavideo.com).

Chakravarthy BR, Durkin JP, Rixon RH et al. Parathyroid hormone fragment [3-34] stimulates protein kinase C (PKC) activity in rat osteosarcoma and murine T-lymphoma cells. Biochem Biophys Res Commun 1990; 171:1105-1110.

Chambliss KL, Yuhanna IS, Mineo C et al. Estrogen receptor alpha and endothelial nitric oxide synthase are organized into functional signaling module in caveolae. Circ Res 2000; 87:E44-E52.

Chan DC. Mitochondrial fusion and fission in mammals. Annu Rev Cell Dev Biol 2006; 22:79-99.

Chan K, Andrade SE, Boles M et al. Inhibitors of hydroxymethylglutaryl-coenzyme A reductase and risk of fracture among older women. Lancet 2000; 355:2185-2188.

Chan MH, Mak TW, Chiu RW et al. Simvastatin increases serum osteocalcin concentration in patients treated for hypercholesterolaemia. J Clin Endocrinol Metab 2001; 86:4556-4559.

Chan NX, Ryder KD, Pavalko FM et al. Ca^{2+} regulates fluid shear-induced cyto-skeletal reorganization and gene expression in osteoblasts. Am J Physiol Cell Physiol 2000; 278:C989-997.

Chang DJ, Ji C, Kim KK et al. Reduction in transforming growth factor beta receptor I expression and transcription factor CBFa1 on bone cells by glucocorticoid. J Biol Chem 1998; 273:4892-4896.

Chang W, Rodriguez L, Tu C et al. High extracellular Ca^{2+} enhances the differentiation of growth plate chondrocytes (abstract). J Bone Miner Res 2003; 18:S290.

Chang W, Tu C, Chen TH et al. Expression and signal transduction of calcium-sensing receptors in cartilage and bone. Endocrinology 1999; 140:5883-5893.

Chang W, Tu C, Pratt S et al. Extracellular Ca^{2+}-sensing receptors modulate matrix production and mineralization in chondrogenic RCJ3.1C5.18 cells. Endocrinology 2002; 143:1467-1474.

Chao DT, Korsmeyer SJ. BCL-2 family: Regulators of cell death. Annu Rev Immunol 1998; 16:395-419.

Charo IF, Taubman MB. Chemokines in the pathogenesis of vascular diseases. Circ Res 2004; 95:858-866.

Chatterjee O, Nakchbandi IA, Philbrick WM et al. Endogenous parathyroid hormone-related protein functions as a neuroprotective agent. Brain Res 2002; 930:58-66.

Chattopadhyay N, Brown EM, eds. Calcium-Sensing Receptor. Boston: Kluwer Academic Publishers, 2003.

Chattopadhyay N, Yano S, Tfelt-Hansen J et al. Mitogenic action of calcium-sensing receptor on rat calvarial osteoblasts. Endocrinology 2004; 145:3451-3462.

Chauvin S, Vilardaga J, Benecsik M et al. Parathyroid hormone receptor recycling: Regulation by specific structural features of the receptor (abstract). J Bone Miner Res 2001; 16:S228.

Chen C, Kalu DN. Modulation of intestinal estrogen receptor by ovariectomy, estrogen and growth hormone. J Pharmacol Exp Ther 1998; 286:328-333.

Chen D, Garrett IR, Qiao M et al. Proteasome inhibitors stimulate osteoblast differentiation and bone formation by inhibiting GLI3 degradation and enhancing BMP-2 expression (abstract). Bone 2001; 28:S74.

Chen D, Ji X, Harris MA et al. Differential roles for bone morphogenic protein (BMP) receptor type IB and IA in differentiation and specification of mesenchymal precursor cells to osteoblast and adipocyte lineages. J Cell Biol 1998; 142:295-305.

Chen D, Zangel AL, Zhao Q et al. Ovine caveolin-1: cDNA cloning, E-coli expression, and association with endothelial nitric oxide synthase. Mol Cell Endocrinol 2001; 175:41-56.

Chen H, Chan DC. Emerging functions of mammalian mitochondrial fusion and fission. Hum Mol Genet 2005; 14(Spec. No.2):R238-239.

Chen HL, Demiralp B, Schneider A et al. Parathyroid hormone and parathyroid hormone-related protein exert both pro-and anti-apoptotic effects in mesenchymal cells. J Biol Chem 2002; 277:19374-19381.

Chen Q. Mechanotransduction pathways in cartilage. In: Massaro EJ, Rogers JM, eds. The Skeleton. Totowa: Humana Press, 2004:89-98.

Chen Q, Johnson DM, Haudenschild DR et al. Progression and recapitulation the chondrocyte differentiation program: Cartilage matrix protein is a marker for cartilage maturation. Devel Biol 1995; 172:293-306.

Chen X, Dai J, Orellana SA et al. Is PKIγ responsible for termination of immediate-early gene expression after induction by PTH? J Bone Miner Res 2002; 17:S250.

Chen X, Dai J, Orellana SA et al. PKIγ knock down inhibits termination of immediate-early gene expression induced by PTH. J Bone Miner Res 2003; 18:S75.

Chen X, Macica CM, Liang G et al. Stretch-induced parathyroid hormone -Related peptide (PTHrP) gene expression in bone (abstract). J Bone Miner Res 2003; 18:S218.

Chen XD, Stewart SA, Manolagas SC et al. Dissection of PTH-induced signaling network in osteoblastic cells using a novel bioinformatics approach. J Bone Miner Res 2004; 19:S76.

Chenu C. Glutamatergic innervation in bone. Microscopy Res Tech 2002; 58:70-76.

Cherruau M, Morvan FO, Schirar A et al. Chemical sympathectomy-induced changes in TH-, VP-, and CGRP-immunoreactive fibers in the rat manible periosteum: Influence on bone resorption. J Cell Physiol 2003; 194:341-348.

Chiarini A, Dal Pra I, Menapace L et al. Soluble amyloid beta-peptide and myelin basic protein strongly stimulate, alone and in synergism with combined proinflammatory cytokines, the xpression of functional nitric oxide synthase-2 in normal adult human astrocytes. Int J Mol Med 2005; 16:801-807.

Chin KV, Yang WL, Ravatn R et al. Reinventing the wheel of cyclic AMP. Novel mechanisms of cAMP signaling. Ann NY Acad Sci 2002; 968:49-64.

Chinsamy A. The Microstructure of Dinosaur Bone. Baltimore: Johs Hopkins University Press, 2005.

Chinsamy A, Rich T, Vickers-Rich P. Polar dinosaur bone histology. J Vertebrate Paleontol 1998; 18:385-390.

Chorev M. Parathyroid hormone 1 receptor: Insights into structure and function. Receptors and Channels 2002; 8:219-242.

Chow JW, Fox S, Jagger CJ et al. Role of parathyroid hormone in mechanical responsiveness of bone. Am J Physiol 1998; 274:E146-E154.

Christopoulos A, Christopoulos G, Morfis M et al. Novel receptor partners and functions of receptor activity-modifying proteins. J Biol Chem 2003; 278:3293-3297.

Chu SC, Chou YC, Liu JY et al. Fluctuation of serum leptin level in rats after ovariectomy and the influence of estrogen supplement. Life Sci 1999; 64:2299-2306.

Chung CH, Baek SH. Deubiquiting enzymes: Their diversity and emerging roles. Biochem Biophys Res Commun 1999; 266:633-640.

Chung U, Schipani E, McMahon AP et al. Indian hedgehog couples chondrogenesis to osteogenesis in endochondral bone development. J Clin Invest 2001; 107:295-304.

Chung YS, Lee MD, Lee SK et al. HMG-CoA reductase inhibitors increase BMD in type 2 diabetes mellitus patients. J Clin Endocrinol Metab 2000; 85:1137-1142.

Cinamon U, Turcotte RE. Primary hyperparathyroidism and malignancy: "Studies by nature". Bone 2006.

Clack J. Getting a leg up on land. Sci Amer 2005; 293:100-107.

Coe MR, Summers TA, Parsons SJ et al. Matrix mineralization in hypertrophic chondrocyte cultures. β Glycerophosphate increases type X collagen messenger RNA and the specific activity of pp60-src kinase. Bone Miner 1992; 18:91-106.

Coen G. Leptin and bone metabolism. J Nephrol 2004; 17:187-189.

Cohen J, Stewart I. What does a Martian look like? Hoboken: John Wiley and Sons, 2003.

Cole JA. Parathyroid hormone activates mitogen-activated protein kinase in opossum kidney cells. Endocrinology 1999; 140:5771-5779.

Collett GDM, Canfield AE. Angiogenesis and pericytes in the initiation of extopic calcification. Cir Res 2005; 96:930-938.

Conner AC, Simms J, Hay DL et al. Heterodimers and family-B GPCRs: RAMPs, CGRP and adrenomedullin. Biochem Soc Trans 2004; 32(Pt5):843-846.

Conrads KA, Yi M, Simpson KA et al. A combined proteome and microarray investigation of inorganic phosphate-induced preosteoblast cells. Mol Cell Proteomics 2005; 4:1284-1296.

Consensus Development Conference: Prophylaxis and treatment of osteoporosis. Am J Med 1991; 90:107-110.

Consensus Development Conference: Diagnosis, Prophylaxis, and Treatment of Osteoporosis. Am J Med 1993; 94:646-650.

Conteas CN, Desai TK, Arlow FA. Relationship of hormones and growth factors to colon cancer. Gastroenterol Clin North America 1988; 17:761-772.

Cooper D. Regulation and organization of adenylyl cyclases and cAMP. Biochem J 2003; 375:517-529.

Coppock D, Kopman C, Gudas J et al. Regulation of the quiescence-induced genes: Quiescin Q6, decorin, and ribosomal protein S29. Biochem Biophys Res Commun 2000; 269:604-610.

Coppock D, Kopman C, Scandalis S et al. Preferential gene expression in quiescent human lung fibroblasts. Cell Growth Diff 1993; 4:483-493.

Corbit KC, Aanstad P, Singla V et al. Vertebrate Smoothened functions at the primary cilium. Nature 2005; 437:1018-1021.

Cordy JM, Hooper NM, Turner AJ. The involvement of lipid rafts in Alzheimer's disease. Mol Membrane Biol 2006; 23:111-122.

Cordy JM, Hussain I, Dingwall C et al. Exclusively targeting beta-secretase to lipid rafts by GPI-anchor addition up-regulates beta site processing of the amyloid precursor protein. Proc Natl Acad Sci USA 2003; 100:11735-11740.

Cornish J, Callon KE, Bava U et al. The direct actions of leptin on bone cells increase bone strength in vivo: An explanation of low fracture rates in obesity (abstract). Bone 2001; 28:S88.

Cornish J, Callon KE, Lin C et al. Stimulation of osteoblast proliferation by C-terminal fragments of parathyroid hormone-related protein. J Bone Miner Res 1999; 14:915-922.

Cornish J, Callon KE, Nicholson GC et al. Parathyroid hormone-related protein-(107-139) inhibits bone resorption in vivo. Endocrinology 1997; 138:1299-1304.

Corral DA, Amling M, Takeda S et al. Dissociation between bone resorption and bone formation in osteopenic transgenic mice. Proc Natl Acad Sci USA 1998; 95:13835-13840.

Corre J, Planat-Benard V, Coberand JX et al. Human bone marrow adipocytes support complete myeloid and lymphoid differentiation from human CD34+ cells. Brit J Haematol 2004; 127:344-347.

Corti R, Fuster V, Badimon JJ. Pathogenic concepts of acute coronary syndromes. J Am Coll Cardiol 2003; 41(Suppl S):7S-14S.

Cosman F, Lindsay R. Is parathyroid hormone a therapeutic option for osteoporosis? A review of the clinical evidence. Calcif Tissue Int 1998; 62:475-480.

Cosman F, Lindsay R. Parathyroid hormone as an anabolic treatment. In: Stevenson JC, Lindsay R, eds. Osteoporosis. London: Chapman and Hall, 1998:293-307.

Cosman F, Nieves J, Formica L et al. Parathyroid hormone in combination with estrogen dramatically reduces vertebral fracture risk (abstract). Osteoporosis Int 2000; 11(Suppl 2):S17.

Cosman F, Nieves J, Woelfert L et al. Parathyroid hormone added to established hormone therapy: Effects on vertebral fracture and maintenance of bone mass after parathyroid hormone withdrawal. J Bone Miner Res 2001a; 16:925-931.

Cosman F, Nieves J, Woelfert L et al. Alendronate does not block the anabolic effect of PTH in postmenopausal osteoporotic women. J Bone Miner Res 1998; 13:1051-1055.

Cosman F, Nieves J, Zion M et al. Effects of short-term cerivastatin on bone turnover (abstract). J Bone Miner Res 2001b; 16:S296.

Cosman F, Nieves JW, Zion M et al. Daily and cyclic parathyroid hormone in women receiving alendronate. New Engl J Med 2005; 353:566-575.

Couet J, Belanger MM, Roussel E et al. Cell biology of caveoli and caveolin. Adv Drug Deliv Rev 2001; 49:223-235.

Cowin SC. Bone poroelasticity. J Biomech 1999; 32:217-238.

Crawford DT, Qi H, Chisey-Frink KL et al. Statin increases cortical bone in young male rats by single, local administration but fails to restore bone to ovariectomized (OVX) rats by daily systemic administration (abstract). J Bone Miner Res 2001; 16:S295.

Crompton M. The mitochondrial permeability transition pore and its role in cell death. Biochem J 2000; 341:233-249.

Csordas G, Hajnoczky G. Plasticity of mitochondrial calcium signaling. J Biol Chem 2003; 278:42273-42282.

Cuello AC. Intracellular and extracellular Aβ, a tale of two neuropathologies. Brain Pathol 2005; 15:66-71.

Cui YF, Lord BI, Woolford LB et al. The relative spatial distribution of in vitro-CFCs in the bone marrow, responding to specific growth factors. Cell Prolif 1996; 29:243-257.

Cuthbertson RM, Kemp BE, Barden JA. Structure of osteostatin PTHrP[Thr 107](107-139). Biochim Biophys Acta 1999; 1432:64-72.

Daaka Y, Luttrell LM, Lefkowitz RJ. Switching of the coupling of the β_2-adrenergic receptor to different G proteins by protein kinase A. Nature 1997; 390:88-91.

Daifotis AG, Weir EC, Dreyer BE et al. Stretch-induced parathyroid hormone-related peptide gene expression in the rat uterus. J Biol Chem 1992; 267:23455-23458.

Dal Pra I, Chiarini A, Nemeth EF et al. The roles of Ca^{2+} and the Ca^{2+}-sensing receptor (CASR) in the expression of inducible NOS (nitric oxide synthase)-2 and its BH_4 (tetrahydrobiopterin)-dependent activation in cytokine-stimulated adult human astrocytes. J Cell Biochem 2005; 96:428-438.

Danks JA, Trivett MK, Power DM et al. Parathyroid hormone-related protein in lower vertebrates. Clin Exp Pharmacol Physiol 1998; 25:750-752.

Daub H, Wallasch C, Lankenau A et al. Signal characteristics of G protein-transactivated EGF receptor. EMBO J 1997; 16:7032-7044.

Datta NS, Chen C, Berry JE et al. PTHrP signaling targets cyclin D1 and induces osteoblastic cell growth arrest. J Bone Miner Res 2005; 20:1051-1064.

Davies JA. Mechanisms of Morphogenesis: The Creation of Biological Form. Burlington, MA: Elsevier Academic Press, 2005.

Davies JE. Bone Engineering. Toronto: em²inc publishers, 2000.

Davies JE. Understanding peri-implant endosseus healing. J Dental Ed 2003; 67:932-949.

David P, Nguyen H, Barbier A et al. The bisphosphonate tiludronate is a potent inhibitor of the osteoclast vacuolar H⁺-ATPase. J Bone Miner Res 1996; 11:1498-1507.

Dazzi F, Ramasamy R, Glennie S et al. The role of mesenchymal stem cells in haematopoiesis. Blood Reviews 2006; 20:161-171.

Deak F, Wagener R, Kiss I et al. The matrilins: A novel family of oligomeric matrix proteins. Matrix Biol 1999; 18:55-64.

Deane, R, Wu Z, Zlokovic BV. RAGE (Yin) versus LRP (Yang) balance regulates Alzheimer amyloid β-peptide clearance through transport across the blood-brain barrier. Stroke 2004; 35:2628-2631.

Debiais F, Lasmoles F, Lefevre G et al. Glycogen synthase kinase-3 (GSK-3) signaling is involved in the antiapoptotic effect of fibroblast growth factor-2 in human calvaria osteoblasts. J Bone Miner Res 2002; 17:S239.

de Cathelineau AM, Henson PM. The final step in programmed cell death: Phagocytes carry apoptotic cells to the grave. Essays in Biochemistry 2003; 39:105-117.

de Crombrugge B, Lefebvre V, Nakashima K. Regulatory mechanisms in the pathways of cartilage and bone formation. Curr Opin Cell Biol 2001; 13:721-727.

Deftos LJ, Burton D, Hastings RH et al. Comparative tissue distribution of the processing enzymes "prohormone thiol protease", and prohormone convertases 1 and 2, in human PTHrP-producing cell lines and mammalian neuroendocrine tissues. Endocrine 2001; 15:217-224.

de Gortazar AR, Alonso V, Esbrit P. Anabolic effects of intermittent parathyroid hormone-related protein (107-139) (Osteostatin) on human osteoblastic cells in vitro. J Bone Miner Res 2004; 19:S194.

Delerive P, De Bosscher K, Besnard S et al. Peroxisome proliferator-activated receptor α negatively regulates the vascular inflammatory gene response by negative cross-talk with transcription factors NF-κB and AP-1. J Biol Chem 1999; 274:32048-32054.

Delmas P. Polycystins: From mechanosensation to gene regulation. Cell 2004; 118:145-148.

Delmas PD, Vergnaud P, Arlot ME et al. The anabolic effect of human PTH (1-34) on bone formation is blunted when bone resorption is inhibited by the bisphosphonate tiludronate—Is activated resorptionn a prerequisite for the in vivo effect of PTH on formation in a remodeling system? Bone 1995; 16:603-610.

De Miguel F, Fiaschi-Taesch N, Lopez-Talavera JC et al. The C-terminal region of PTHrP, in addition to the nuclear localization signal, is essential for the intracrine stimulation of proliferation in vascular smooth muscle cells. Endocrinology 2001; 142:4096-4105.

Demer LL. Vascular calcification and osteoporosis: Inflammatory reponses to oxidized lipids. Int J Epidemiol 2002; 31:737-741.

Demer LL, Tintut Y. Mineral exploration: Search for the mechanism of vascular calcification and beyond. Arterioscler Thromb Vasc Biol 2003; 23:1739-1743.

Dempster DW. Bone remodeling. In: Riggs BL, Melton IIIrd LJ, eds. Osteoporosis: Etiology, Diagnosis, and Management. 2nd ed. Philadelphia: Lippincott-Raven Publishers, 1995:67-91.

Dempster DW. Exploiting and bypassing the bone remodeling cycle to optimize the treatment of osteoporosis. J Bone Miner Res 1997; 12:1152-1154.

Dempster DW. The contribution of trabecular architecture to cancellous bone quality. J Bone Miner Res 2000; 15:20-23.

Dempster DW, Cosman F, Parisien M et al. Anabolic action of parathyroid hormone. Endocrine Revs 1993; 14:690-709.

Dempster DW, Hughes-Begos CE, Plavetic-Chee K et al. Normal human osteoclasts formed from peripheral blood monocytes express PTH type 1 receptors and are stimulated by PTH in the absence of osteoblasts. J Cell Biochem 2005; 95:139-148.

Dempster DW, Parisien M, Silverberg SJ et al. On the mechanism of cancellous bone preservation in postmenopausal women with mild primary hyperparathyroidism. J Clin Endocrinol Metab 1999; 84:1562-1566.

Denhardt DT, Giachelli CM, Rittling SR. Role of osteopontin in cellular signaling and toxicant injury. Annu Rev Pharmacol Toxicol 2001; 41:723-749.

Denhardt DT, Noda M. Osteopontin expression and function: Role in bone remodeling. J Cell Biochem Suppl 1998; 30-31:92-102.

Dennett DC. Breaking the Spell. New York: Viking Penguin, 2006.

Dennis JE, Charbord P. Origin and differentiation of human and murine stroma. Stem Cells 2002; 20:205-214.

de Rooij J, Rehmann H, van Triest M et al. Mechanism of regulation of the Epac family of cAMP-dependent RapGEFs. J Biol Chem 2000; 275:20829-20836.

Dhillon H, Glatt V, Ferrrari SL et al. β-Adrenergic receptor KO mice have increased bone mass and strength but are not protected from ovary-induced bone loss (abstract). J Bone Miner Res 2004; 19:S32.

Dimmeler S, Aicher A, Vasa M et al. HMG-CoA reductase inhibitors (statins) increase endothelial progenitor cells via the PI 3-kinase/Akt pathway. J Clin Invest 2001; 108:391-397.

D'Ippolito G, Divieti P, Howard GA et al. Regulation of gap-junctional communication in bone-derived cells by PTH fragments with different C termini. J Bone Miner Res 2002; 17:S392.

Divieti P. PTH and osteocytes. J Musculoskelet Neuronal Interact 2005; 5:328-330.

Divieti P, Inomata N, Chapin K et al. Receptors for the carboxyl-terminal region of PTH(1-84) are highly expressed in osteocytic cells. Endocrinology 2001; 142:916-925.

Divieti P, John MR, Jüppner H et al. Human PTH-(7-84) inhibits bone resorption in vitro via actions independent of the type 1 PTH/PTHrP receptor. Endocrinology 2002a; 143:171-176.

Divieti P, Lotz O, Geller A et al. Inhibition of osteoclast formation by human PTH-(7-84) involves direct actions on hemopoietic cells (abstract). J Bone Miner Res 2002b; 17:S167.

Doan DN, Veal TM, Yan Z et al. Loss of the INI1 tumor suppressor does not impair the expression of multiple BRG1-depnedent genes of the assembly of SWI/SNF enzymes. Oncogene 2004; 23:3462-3473.

Dobnig H, Turner RT. Evidence that intermittent treatment with parathyroid hormone increases bone formation in adult rats by activation of bone lining cells. Endocrinology 1995; 136:3632-3638.

Dobnig H, Turner RT. The effects of programmed administration of human parathyroid hormone fragment (1-34) on bone histomorphometry and serum chemistry in rats. Endocrinology 1997; 138:4607-4612.

Dobson KB, Skerry TM. The NMDA-type glutamate receptor antagonist MK801 regulates the differentiation of rat bone marrow osteoprogenitors and influence adipogenesis (abstract). J Bone Miner Res 2000; 15:S272.

Dodd JS, Raleigh JA, Gross TS. Osteocyte hypoxia: A novel mechanotransduction pathway. Am J Physiol 1999; 277:C598-C602, (Cell Physiol 46).

Doherty MJ, Canfield AE. Gene expression during vascular pericyte differentiation. Crit Rev Eukaryot Gene Expr 1999; 9:1-17.

Doherty TM, Asotra K, Fitzpatrick LA et al. Calcification in atherosclerosis: Bone biology and chronic inflmammation at the arterial crossroads. Proc Natl Acad Sci USA 2003a; 100:11201-11206.

Doherty TM, Fitzpatrick LA, Inoue D et al. Molecular, endocrine, and genetic mechanisms of arterial calcification. Endocrine Rev 2004; 25:629-672.

Doherty TM, Shah PK, Rjavashisth TB. Cellular oirigins of atherosclerosis: Towards ontogenetic endgame? FASEB J 2003b; 17:592-597.

Doherty TM, Uzui H, Fitzpatrick LA et al. Rationale for the role of osteoclast-like cells in arterial calcification. FASEB J 2002; 16:577-582.

Donahue SW, Donahue HJ, Jacobs CR. Osteoblastic cells have refractory periods for fluid-flow-induced intracellular calcium oscillations for short bouts of flow and display multiple low-magnitude oscillations during long-term flow. J Biomech 2003; 36:35-43.

Donahue SW, Galley SA, Vaughan MR et al. Parathyroid hormone may maintain bone formation in hibernating black bears (Ursus americanus) to prevent disuse osteoporosis. J Exp Biol 2006; 209(Pt.9):1630-1638.

Donahue SW, Jacobs CR, Donahue HJ. Flow-induced calcium oscillations in rat osteoblasts are age, loading frequency, and shear stress dependent. Am J Physiol Cell Phsyiol 2001; 281:C1635-1641.

Doolan CM, Condliffe SB, Harvey BJ. Rapid nongenomic activation of cytosolic cyclic AMP-dependent protein kinase activity and $[Ca^{2+}]_I$ by 17β-oestradiol in female rat distal colon. Brit J Pharmacol 2000; 129:1375-1386.

Drescher W, Bunger MH, Weigert K et al. Methylprednisolone enhances contraction of procine femoral head arteries. Clin Orthop Relat Res 2004; 423:112-117.

Ducy P. Cbfa1: A molecular switch in osteoblast biology. Devel Dynamics 2000; 219:461-471.

Doxsey S. Reevaluating centrosome function. Nature Rev Mol Cell Biol 2001; 2:688-698.

Ducy P, Amling M, Takeda S et al. Leptin inhibits bone formation through a hypothalamic relay: A central control of bone mass. Cell 2000; 100:197-207.

Ducy P, Desbois C, Boyce B et al. Increased bone formation in osteocalcin-deficient mice. Nature 1996; 382:448-452.

Ducy P, Karsenty G. Two distinct osteoblast-specific cis-acting elements control expression of a mouse osteocalcin gene. Mol Cell Biol 1995; 15:1858-1869.

Ducy P, Schinke T, Karsenty G. The osteoblast: A sophisticated fibroblast under central surveillance. Science 2000; 289:1501-1504.

Dumollard R, Duchen M, Sardet C. Calcium signals and mitochondria at fertilization. Semin Cell Dev Biol 2006; 17:314-323.

Dumond H, Presle N, Terlain R et al. Evidence for a key role of leptin in osteoarthritis. Arthritis Rheum 2003; 48:3009-3012.

Duncan R, Misler S. Voltage-activated and stretch-activated Ba^{2+}-conducting channels in an osteoblast-like cell line (UMR 106). FEBS Lett 1989; 251:17-21.

Duncan AW, Rattis FM, Di Mascio LN et al. Intergration of Notch and Wnt signaling in hematopoietic stem cell maintenance. Nature Immunol 2005; 6:314-322.

Duncan RL, Hruska KA, Misler S. Parathyroid hormone activation of stretch-activated cation channels in osteosarcoma cells (UMR 106.01). FEBS Lett 1992; 307:219-223.

Dunford JE, Ebetino EH, Rogers MJ. The mechanism of inhibition of farnesyl diphosphate synthase by nitrogen-containing bisphosphonates. J Bone Miner Res 2002; 17:S255.

Dustin ML, Colman DR. Neural and immunological synaptic relations. Science 2002; 298:785-789.

Dvorak MM, Siddiqua A, Ward DT et al. Physiological changes in extracellular calcium concentration directly control osteoblast function in the absence of calciotropic hormones. Proc Natl Acad Sci USA 2004; 101:5140-5145.

Ebert BL, Furth JD, Ratcliffer PJ. Hypoxia and mitochondrial inhibitirs regulate expression of glucose transporter-1 via distinct cis-acting sequences. J Biol Chem 1995; 270:29083-29089.

Edlich M, Yellowley CE, Jacobs CR et al. Oscillating fluid flow regulates cytosolic calcium concentration in bovine articular chondrocytes. J Biomechanics 2001; 34:59-65.

Edwards CJ, Hart DJ, Spector TD. Oral status and increased bone mineral density in postmenopausal women. Lancet 2000; 355:2218-2219.

Ehehalt R, Keller P, Haass C et al. Amyloidogenic processing of the Alzheimer β-amyloid precursor protein depends on lipid rafts. J Cell Biol 2003; 160:113-123.

Einhorn TA. Biomechanics of bone. In: Bilzekian JP, Raisz LG, Rodan GA, eds. Principles of Bone Biology. San Diego: Academic Press, 1996:25-37.

Egan JJ, Gronowicz G, Rodan GA. Parathyroid hormone promotes the disassembly of cytoskeletal actin and myosin in cultured osteoblastic cells: Mediation by cyclic AMP. J Cell Biochem 1991; 45:101-111.

Eleftriou F, Ahn J, Takeda S et al. Leptin regulation of bone resorption by the sympathetic nervous system and CART. Nature 2005; 434:514-520.

Endo A. The discovery and development of HMG-CoA reductase inhibitors. J Lipid Res 1992; 33:1569-1582.

Endo A, Tsujita Y, Kuroda M et al. Inhibition of cholesterol synthesis in vitro and in vivo by ML-236A and ML-236B. Eur J Biochem 1997; 77:31-36.

Enjuanes A, Supervia A, Nogues X et al. Leptin receptor (OB-R) gene expression in human primary osteoblasts: Confirmation. J Bone Miner Res 2002; 17:1135.

Enlow DH, Brown SO. A comparative histological study of fossil and recent bone tissues. Part II. Texas J Sci 1957; 9:186-214.

Enlow DH, Brown SO. A comparative histological study of fossil and recent bone tissues. Part III. Texas J Sci 1958; 10:187-230.

Ehrlich PJ, Lanyon LE. Mechanical strain and bone cell function: A review. Osteoporosis Int 2002; 13:688-700.

Ekblad E, Edvinsson L, Wohlestedt C et al, Neurppeptide Y coexists and cooperates with noradrenaline in perivascular nerve fibers. Regul Pept 1984; 8:225-235.

Erdmann S, Burkhardt H, von der Mark K et al. Mapping of a carboxy-terminal active site of parathyroid hormone by calcium-imaging. Cell Calcium 1998; 23:413-421.

Erclik MS, Mitchell J. The role of protein kinase C-δ in PTH stimulation of IGF-binding protein-5 mRNA in UMR-106-01 cells. Am J Physiol Endocrinol Metab 2001; 282:E534-E541.

Erickson GM, Makovicky Y, Currie PJ et al. Gigantism and comparative life-history parameters of tyrannosaurid dinosaurs. Nature 2004; 430:772-775.

Eriksen EF, Axelrod DW, Melsen F. Bone Histomorphometry. New York: Raven Press, 1994.

Eriksen EF, Langdahl B, Klassen M. The cellular basis of osteoporosis. Spine: State Arts Revs 1994; 8:23-62.

Eriksen EF, Zeng QQ, Donley DW et al. Local IGF-II expression in human bone is greater in women treated with teriparatide than with placebo: A quantitative immunohistochemical study study of bone biopsies in the fracture prevention trial (abstract). J Bone Miner Res 2004; 19:S134.

Errazahi A, Bouizar Z, Lieberherr M et al. Functional type I PTH/PTHrP receptor in freshly isolated newborn rat keratinocytes: Identification by RT-PCR and immunohistochemistry. J Bone Miner Res 2003; 18:737-750.

Errazahi A, Bouizar Z, Rizk-Rabin M. RT-PCR identification of PTHrp PTH/PTHrp receptor mRNAs during the steps of differentiation pathway of rat newborn keratinocytes: A putative autocrine role of PTHrp. Bone 1998; 23:S248.

Esbrit P, Alvarez-Arroyo MV, De Miguel F et al. C-terminal parathyroid hormone-related protein increases vascular endothelial growth factor in himan osteoblastic cells. J Am Soc Nephrol 2000; 11:1085-1092.

Espinosa L, Itzstein C, Cheynel H et al. Active NMDA glutamate receptors are expressed by mammalian osteoclasts. J Physiol 1999; 518:47-53.

Ettinger B, San Martin J, Crans G et al. Differential effects of teriparatide on BMD after treatment with raloxifene or alendronate. J Bone Miner Res 2004; 19:745-751.

Evans BAJ, Elford C, Gregory JW. Leptin control of bone metabolism (abstract). Bone 2001; 28:S149.

Everts V, Delaissé JM, Korper W, Jansen D, Tigchelaar-Gutter W, Saftig P, Beertsen W. The bone lining cell: Its role in cleaning Howship's lacunae and initiating bone formation. J Bone Miner Res 2002; 17:77-90.

Faccio R, Lam J, Aya K et al. Oligomeric RANKL induces bone formation via the ERK pathway. J Bone Miner Res 2003; 18:S141.

Fadeel B, Zhivotovsky B, Orrenius S. All along the watchtower: On the regulation of apoptosis regulators. FASEB J 1999; 13:1647-1657.

Falany ML, Thames AM, McDonald JM et al. Osteoclasts secrete the cytokine MIM-1. Biochem Biophys Res Commun 2003; 281:180-185.

Falkenstein E, Tillmann HC, Christ M et al. Multiple actions of steroid hormones—a focus on rapid, nongenomic effects. Pharmacol Revs 2000; 52:513-555.

Fang J, Zhu YY, Smiley E et al. Stimulation of new bone formation by direct transfer of osteogenic plasmid genes. Proc Natl Acad Sci USA 1996; 93:5753-5758.

Fantuzzi G, Faggioni R. Leptin in the regulation of immunity, inflammation, and hematopoiesis. Leukoc Biol 2000; 68:437-446.

Farach-Carson MC, Xu Y. Microarray detection of gene expression changes induced by $1,25(OH)_2D_3$ and a Ca^{2+} influx-activating analog in osteoblastic ROS 17/2.8 cells. Steroids 2002; 67:467-470.

Farhat MY, Lavigne MC, Ramwell PW. The vascular protective effects of estrogen. FASEB J 1996; 10:615-624.

Farley JR, Stilt-Coffing B. Apoptosis may determine the release of skeletal alkaline phosphahtase activity from human osteoblast-line cells. Calcif Tissue Int 2001; 68:43-52.

Farr HW. Hyperparathyroidism and cancer. CA Cancer J Clin 1976; 26:66-74.

Farzaneh-Far A, Proudfoot D, Weisberg PL et al. Matrix gla protein is regulated by a mechanism functionally related to the calcium-sensing receptor. Biochem Biophys Res Commun 2000; 277:736-740.

Fassbender K, Simons M, Bergmann C et al. Simvaststin strongly reduces levels of Alzheimer's disease β amyloid peptides Aβ42 and Aβ40 in vivo and in vitro. Proc Natl Acad Sci USA 2001; 98:5856-5861.

Faucheux C, Horton MA, Price JS. Nuclear localization of type I parathyroid hormone/parathyroid hormone-related protein receptors in deer antler osteoclasts: Evidence for parathyroid hormone-related protein and receptor activator of NF-κB-dependent effects on osteoclast formation in regenerating mammalian bone. J Bone Miner Res 2002; 17:455-464.

Faulkner K. Bone matters: Are density increases necessary to reduce fracture risk? J Bone Miner Res 2000; 15:183-187.

Fei J, Viedt C, Soto U et al. Endothelin-1 and smooth muscle cells: Induction of Jun amino-terminal kinase through an oxygen radical-sensitive mechanism. Arterioscler Thromb Vasc Biol 2000; 20:1244-1249.

Feig DS, Gottesman IS. Familial hyperparathyroidism in association with colonic carcinoma. Cancer 1987; 60:429-432.

Feister HA, Torrungruang K, Thunyakitpisal P et al. NP/NMP4 transcription factors have distinct osteoblast nuclear matrix subdomains. J Cell Biochem 2000; 79:506-517.

Fenton AJ, Kemp BE, Hammonds RG et al. A potent inhibitor of osteoclastic bone resorption within a highly conserved pentapeptide region of parathyroid hormone-related protein; PTHrP[107-111]. Endocrinology 1991; 129:3424-3426.

Fermore B, Skerry TM. PTH/PTHrP receptor expression in osteoblasts and osteocytes but not resorbing bone surfaces in growing rats. J Bone Miner Res 1995; 10:1935-1943.

Feron O, Dessy C, Desager JP et al. Hydroxy-methylglutaryl-Coenzyme A reductase inhibition promotes endothelial nitric oxide synthase activation through a decrease in caveolin abundance. Circulation 2001; 103:113-118.

Fiaschi-Taesch N, Takane KK, Masters S et al. Parathyroid hormone-related protein as a regulator of pRB and the cell cycle of arterial smooth muscle. Circulation 2004; 110:177-185.

Fielding CJ, Fielding PE. Caveolae and intracellular trafficking of cholestero. Adv Drug Del Rev 2001; 49:251-264.

Fields RD, Stevens-Graham B. New insights into neuron-glia communication. Science 2002; 298:556-562.

Fietta P. Focus on leptin, a pleotropic hormone. Minerva Med 2005; 96:65-75.

Figenschau Y, Knutsen G, Shahazeydi S et al. Human articular chondrocytes express functional leptin receptors. Biochem Biophys Res Commun 2001; 287:190-197.

Finkel E. The mitochondrion: Is it central to apoptosis? Science 2001; 292:624-626.

Fiorelli G, Gori F, Frediani U et al. Membrane-binding sites and non genomic effects of estrogen in cultured human preosteoclastic cells. J Steroid Bio chem Molec Biol 1996; 59:233-240.

Finkelstein JS, Hayes A, Hunzelman JL et al. Effects of parathyroid hormone, alendronate, or both in men with osteoporosis. N Engl J Med 2003; 349:1216-1226.

Fischer DF, van Dijk R, Sluijs JA et al. Activation of the Notch pathway in Down syndrome: Cross-tals of Notch and APP. FASEB J 2005; 19:1451-1458.

Fisher JE, Rogers MJ, Halasay JM et al. Alendronate mechanism of action: Geranylgeraniol, an intermediate in the mevalonate pathway, prevents inhibition of osteoclast formation, bone resorption, and kinase activation in vitro. Proc Natl Acad Sci USA 1999; 96:133-138.

Fitzpatrick LA, Turner RT, Ritman ER. Endochondral bone formation in the heart: A possible mechanism of coronary calcification. Endocrinol 2003; 144:2214-2219.

Fiziola M, Weinstein H. Structural models for dimerization of G-protein coupled receptors: The opioid receptor homdimers. Biopolymers 2002; 66:317-325.

Flanagan JA, Power DM, Bendell LA et al. Cloning of the cDNA for sea bream (Sparus aurata) parathyroid hormone-related protein. Gen Comp Endocrinol 2000; 118:373-382.

Fleet JC. Leptin and bone: Does the brain control bone biology? Nutr Rev 2000; 58:209-211.

Fleisch H. Bisphosphonates in Bone disease. London: The Parthenon Publishing Group, 1997.

Fleisch H. Bisphosphonates: Mechanisms of action. Endocrine Revs 1998; 19:80-100.

Fletcher BS, Lim RW, Varnum BC et al. Structure and expression of TIS21, a primary response gene induced by growth factors and tumor promoters. J Biol Chem 1991; 266:14511-14518.

Fliedner TM, Graessle D, Paulsen C et al. Structure and function of bone marrow hemopoiesis; mechanisms of response to ionizing radiation exposure. Cancer Biother Raidopharm 2002; 17:405-426.

Fraher LJ, Avram R, Watson PH et al. A comparison of the biochemical responses to 1-31 hPTH and 1-34 hPTH given to healthy humans by slow infusion. J Clin Endocrinol Metab 1999; 84:2739-2743.

Franceschi RT. The developmental control of osteoblast-specific gene expression: Role of specific transcription factors and the extracellular matrix environment. Crit Rev Oral Biol Med 1999; 10:40-57.

Franceschi RT, Xiao G. Regulation of the osteoblast-specific transcrition factor, Runx2: Responsiveness to multiple signal transduction pathways. J Cell Biochem 2003; 88:446-454.

Fried H, Kutay U. Nucleocytoplasmic transport; taking an inventory. Cell Mol Life Sci 2003; 60:1659-1688.

Friedman J, Babu B, Clark RB. β_2-Adrenergic receptor lacking the cyclic AMP-dependent protein kinase consensus sites fully activates extracellular signal-regulated kinases 2 in human embryonic kidney 293 cells: Lack of evidence for G_s/G_i switching. Mol Pharmacol 2002; 62:1094-1102.

Friedman JM. Obesity in the new millennium. Nature 2000; 404:632-634.

Friedman PA, Gesek FA, Morley P et al. Cell-specific signaling and structure activity relations of parathyroid analogs in mouse kidney cells. Endocrinology 1999; 140:301-309.

Frisch SM, Screaton RA. Anoikis mechanisms. Curr Opin Cell Biol 2001; 13:555-562.

Frith JC, Monkkonen J, Auriola S et al. The molecular mechanism of action of the antiresorptive and anti-inflammatory drug clodronate: Evidence for the formation in vivo of a metabolitie that inhibits bone resorption and causes osteoclast and macrophage apoptosis. Arthritis Rheum 2001; 44:2201-2210.

Frith JC, Monkkonen J, Blackburn GM et al. Clodronate and liposome-encapsulated clodronate are metabolized to a toxic ATP analog, adenosine 5'-(β,γ-dichloromethylene) triphosphate by mammalian cells in vitro. J Bone Miner Res 1997; 12:1358-1367.

Frost HM. Presence of microscopic cracks in vivo in bone. Henry Ford Hospital Medical Bull 1960; 8:25-35.

Frost HM. Bone microdamage:factors that impair its repair. In: Uhthoff HK, ed. Current Concepts in Bone Fragility. Berlin: Springer Verlag, 1985:121-148.

Frost HM. Osteoporoses. In: Whitfield JF, Morley P, eds. Anabolic Treatments for Osteoporosis. Boca Raton: CRC Press, 1997:1-27.

Frühbeck G. Intracellular signalling pathways activated by leptin. Biochem J 2006; 393:7-20.

Fu Q, Jilka RL, Manolagas SC et al. Parathyroid hormone stimulates receptor activator of NFκB ligand and inhibits osteoprotegerin expression via protein kinase A activation of CREB. J Biol Chem 2002; 277:48868-48875.

Fuchs JL, Schwark HD. Neuronal primary cilia: A review. Cell Biol Int 2004; 28:111-118.

Fujimori A, Cheng SL, Avioli LV et al. Structurefunction relationship of parathyroid hormone: Activation of phospholipase-C, protein kinase-A and -C in osteosarcoma cells. Endocrinology 1992; 130:29-36.

Fujisawa H, Misaki K, Takabatake Y et al. Cyclin D1 is overexpressed in atypical teratopoid/rhabdoid tumor with hSNF5/INI1 gene inactivation. J Neurooncol 2005; 73:117-124.

Fujita T, Inoue T, Morii H et al. Effect of an intermittent weekly dose of human parathyroid hormone (1-34) on osteoporosis: A randomized double-masked prospective study using three dose levels. Osteoporosis Int 1999; 9:296-306.

Fujita T, Fukuyama R, Izumo N et al. Transactivation of of core binding factor α1 as a basic mechanism to trigger parathyroid hormone-induced osteogenesis. Jpn J Pharmacol 2001a; 86:405-416.

Fujita T, Izumo N, Fukuyama R et al. Phosphate provides an extracelluar signal that drives nuclear export of Runx2/Cbfa1 in bone cells. Biochem Biophys Res Commun 2001b; 280:348-352.

Fujita T, Meguro T, Fukuyama R et al. New signaling pathway for parathyroid hormone and cyclic AMP action on extracellular-regulated kinase and cell proliferation in bone cells. Checkpoint of modulation by cyclic AMP. J Biol Chem 2002; 277:22191-22200.

Fukata S, Hagino H, Kamcyama Y et al. Effects of intermittent administration of human parathyroid hormone on bone mineral density in rats with collagen-induced arthritis. J Bone Miner Res 2003; 18:S386.

Fukui N, Zhu Y, Maloney WJ et al. Stimulation of BMP-2 expression by pro-inflammatory cytokine IL-1 and TNF-α in normal and osteoarthritic chondrocytes. J Bone Joint Surg Am 2003; 85(A Suppl 3):59-66.

Fukumoto S. Localization and function of calcium-sensing receptors in bone cells. Nippon Rinsho 1998; 56:1419-1424.

Fuller K, Murphy C, Kirsten B et al. TNF-α potently activates osteoclasts, through a direct action independent of and strongly synergistic with RANKL. Endocrinology 2002; 143:1108-1118.

Funahashi H, Yada T, Suzuki R et al. Distribution, function, and properties of leptin receptors in the brain. Internat Rev Cytol 2003; 224:1-27.

Funk JL. A role for parathyroid hormone-related protein in the pathogenesis of inflammatory / autoimmune diseases. Internat Immunopharmacol 2001; 1:1101-1121.

Funk JL, Migliati E, Chen G et al. Parathyroid hormone-related protein induction in focal stroke: A neuroprotective vascular peptide. Am J Physiol Regul Integr Comp Physiol 2003; 284:R1019-1020.

Funk JL, Trout CR, Wei H et al. Parathyroid hormone-related protein (PTHrP) induction in reactive astrocytes following brain injury: A possible mediator of brain injury. Brain Res 2001; 915:195-209.

Funk JL, Wei H, Downey KJ et al. Expression of PTHrP and its cognate receptor in the rheumatoid synovial microcirculation. Biochem Biophys Res Commun 2002; 297:890-897.

Gabet Y, Kohavi D, Müller R et al. Human parathyroid hormone 1-34 reverses bone loss in orchidectomized adult rats mainly by vastly increasing trabecular thickness. J Bone Miner Res 2003; 18:S386-S387.

Gabet Y, Müller R, Levy J et al. Parathyroid hormone 1-34 enhances titanium implant anchorage in low-density trabecular bone: A correlative microcomputed tomographic and biochemical analysis. Bone 2006, (published online February, 2006).

Gainsford T, Alexander WS. A role for leptin in hemopoiesis? Mol Biotech 1999; 11:149-1548.

Galindo M, Stein JL, Stein et al. Bone-specific Runx2 levels are tightly regulated during the cell cycle to support a novel function in control of the switch from proliferation to quiescence in osteoblasts. J Bone Miner Res 2003; 18:S137.

Gallien-Lartigue O, Carrez D. Induction in vitro de la phase S dans les cellules souches multipotentes de la moelle osseuse par l'hormone parathroïdienne. C R Acad Sci Paris [D] 1974; 278:1765-1768.

Galvin RJS, Fuson TR, Yang X et al. hPTH (1-38) stimulation of bone resorption in murine calvariae is mediated in part by its ability to decrease OPG (abstract). J Bone Miner Res 2001; 16:S449.

Garrett IR, Chen D, Zhao M et al. Statins mediate bone formation by enhancing BMP-2 expression (abstract). Bone 2001a; 28:S75.

Garrett IR, Gutierrez G, Chen D et al. Specific inhibitors of the chymotryptic component of the proteasome are potent bone anabolic agents (abstract). J Bone Miner Res 2000; 15:S197.

Garrett IR, Gutierrez G, Chen D et al. Statins stimulate bone formation by enhancing eNOS expression (abstract). J Bone Miner Res 2001b; 16:S75.

Garrrett IR, Gutierrez G, Mundy GR. Statins and bone formation. Curr Pharma Design 2001c; 7:715-736.

Gasser JA. Quantitative assessment of bone mass and geometry by pQCT in rats in vivo and site specificity of changes at different skeletal sites. J Jpn Soc Bone Morphom 1997; 7:107-114.

Gasser JA. Fluvastatin and cerivastatin are not anabolic for bone after local or systemic administration of nontoxic doses in mice and rats (abstract). J Bone Miner Res 2001; 16:S295.

Gat-Yablonski G, Ben-Ari T, Shtaif B et al. Leptin reverses the inhibitory effect of caloric restriction on growth. Endocrinology 2004; 145:343-350.

Gat-Yablonski G, Shtaif B, Phillip SM. Leptin stimulates PTHrP expression in the endocondral growth plate. J Histochem Cytochem 2006, (in press).

Gazzero E, Ganji V, Canalis E. Bone morphogenic proteins induce the expression of noggin, which limits their activity in cultures rat osteoblasts. J Clin Invest 1998; 102:2106-2114.

Genant HK, Cann CE, Etting B et al. Quantitative computed tomography of vertebral spongiosa: A sensitive method for detecting early bone loss after oophorectomy. Ann Intern Med 1982; 97:699-705.

Gensure RC, Shimizu N, Tsang JC et al. Evidence that Bpa 19-modified parathyroid hormone-related peptide (PTHrP) agonist and antagonist analogs crosslink to distict sites in transmembrane domain 2 of the PTH/PTHrP receptor. J Bone Miner Res 2002; 17:S286.

Gentili C, Morelli S, Boland R et al. Parathyroid hormone activation of map kinase in rat duodenal cells is mediated by 3',5'-cyclic AMP and Ca^{2+}. Biochim Biophys Acta 2001; 1540:201-212.

Gentili C, Morelli S, de Boland AR. Involvement of PI3-kinase and its association with c-Src in PTH-stimulate rat enterocytes. J Cell Biochem 2002; 86:773-783.

Gerritsen ME, Wagner GF. Stanniocalcin: No longer just a fish tale. Vitamins and Hormones 2005; 70:105-135.

Ghosh-Choudhury N, Windle JJ, Koop BA et al. Immortalized murine osteoblasts derived from BMP2-T-antigen expressing transgenic mice. Endocrinology 1996; 137:331-339.

Gibbs WW. Untangling the roots of cancer. Sci Amer 2003; 289:57-65.

Gillespie BT. The influence of growth factors and cytokines on hypertrophy and primary cilia expression of articular chondrocytes. 1999, (http://www.realscience.breckschol.org/upper/research/research97-99pub/GF/gfa.html).

Gimble JM, Zvonic S, Floyd E et al. Playing with bone and fat. J Cell Biochem 2006; 98:251-256.

Glimcher MJ. Cell biology during repair of osteonecrosis: Implications for rational treatment. Acta Orthopod Belgica 1999; 65(Suppl.1):17-22.

Glimcher MJ, Kenzora JE. The biology of osteonecrosis of the human femoral head and its clinical implications. III. Discussion of the etiology and genesis of the pathological squelae; comments on treatment. Clin Orthop Rel Res 1979; 140:273-312.

Glimm H, Oh IH, Eaves CJ. Human hematopoietic stem cells stimulated to proliferate in vitro lose engraftment potential during their $S/G_2/M$ transit and do not reenter G_0. Blood 2000; 96:4185-4193.

Goetze S, Bungenstock A, Czupalla C et al. Leptin induces endothelial cell migration through Akt, which is inhibited by PPARγ-ligands. Hypertension 40:748-754.

Goodenough DA, Paul DL. Beyond the gap: Functions of unpaired connexon channels. Nature Rev Mol Cell Biol 2003; 4:1-10.

Gohel A, Gronowicz G. Glucocorticoids induce apoptosis in osteoblasts in mice by the regulation of BCL-2 and other cell cycle factors (abstract). J Bone Miner Res 1997; 12:S284.

Goldring SR. Bone and joint destruction in rheumatoid arthritis: What is really happening? J Rheumatol Suppl 2002; 65:44-48.

Goldring SR, Gravallelese EM. Pathogenesis of bone erosions in rheumatoid arthritis. Curr Opin Rheumatol 2000; 12:195-199.

Goldring SR, Gravallese EM. Pathogenesis of bone lesions in rheumatoid arthritis. Curr Rheumatol Reports 2002; 4:226-231.

Goldstein SA, Bonadio J. Potential role for direct gene transfer in the enhancement of fracture healing. Clin Orthop 1998; 355(Suppl):S154-162.

Goligorsky MS, Menton DN, Hruska KA. Parathyroid hormone-induced changes of the brush border topography and cytoskeleton in cultured renal proximal tubular cells. J Membr Biol 1986; 92:151-162.

Goltzman D, White JH. Developmental and tissue-specific regulation of parathyroid hormone (PTH)/PTH-related peptide receptor gene expression. Crit Rev Eukaryot Gene Expr 2000; 10:135-149.

Gomez S, Diaz-Curiel M, DeLa Piedra C et al. Effects of PTH on bone architecture and bone mass in orchidectomized rats (abstract). J Bone Miner Res 2004; 19, (in press).

Gooch KJ, Tennant CJ. Mechanical Forces: Their Effects on Cells and Tissues. New York: Springer-Verlag, 1997.

Gopalakrishnan R, Suttamanatwong S, Carlson AE et al. Role of matrix Gla protein in parathyroid hormone inhibition of osteoblast mineralization. Cells Tissues Organs 2005; 181:166-175.

Gordeladze JO, Drevon CA, Syversen U et al. Leptin stimulates human osteoblastic cell proliferation, de novo collagen synthesis, and mineralization: Impact on differentiation markers, apoptosis, and osteoclastic signaling. J Cell Biochem 2002a; 85:825-836.

Gordeladze JO, Reppe S, Gautvik KM et al. Direct leptin exposure and mechano-stimulation of human osteoblastic cells promote proliferation and differentiation, while inhibiting apoptosis. J Bone Miner Res 2002b; 17:S444.

Gordeladze JO, Reseland JE, Drevon CA. Pharmacological interference with transcriptional control of osteoblasts: A possible role for leptin and fatty acids in maintaining bone strength and body lean mass. Curr Pharma Design 2001; 7:275-290.

Gordon J, Bennett AR, Blackburn CC et al. Gcm2 and Foxn1 mark early parathyroid- and thymus-specific domains in the developing third pharyngeal pouch. Mech Devel 2001; 103:141-143.

Gori F, Hofbauer LC, Dunstan CR et al. The expression of osteoprotegerin and RANKL ligand and the support of osteoclast formation by stromal-osteoblast lineage cells is developmentally regulated. Endocrinology 2000; 141:4768-4776.

Gowen M, Stroup GB, Dodds RA et al. Antagonizing the parathyroid calcium receptor stimulates parathyroid hormone secretion and bone formation in osteopenic rats. J Clin Invest 2000; 105:1595-1604.

Grasso P, Leinung MC, Lee DW. Epitope mapping of sectreted mouse leptin utilizing peripherally administered synthetic peptides. Reg Pept 1999; 85:93-100.

Gravallese EM, Goldring SR. Cellular mechanisms and the role of cytokines in bone erosions in rheumatoid arthritis. Arthritis and Rheumatism 2000; 43:2143-2151.

Gray C, Marie, Arora M et al. Glutamate does not play a major role in controlling bone growth. J Bone Miner Res 2001; 16:742-749.

Greaves M. Cancer: The evolutionary legacy. Oxford: Oxford University Press, 2000.

Greenfield EM. ODF/OPGL is increased by IL-1 and IL-6 (abstract). J Bone Miner Res 1999; 14:S361.

Greenfield JP, Tasai J, Gouras GK et al. Endoplasmic reticulum and trans-Golgi network generate distinct populations of Alzheimer beta-amyloid peptides. Proc Natl Acad Sci USA 1999; 96:742-747.

Grey A, Reid IR. Emerging and potential therapies for osteoporosis. Expert Opin Investig Drugs 2005; 14:265-278.

Grimsley C, Ravichandran KS. Cues for apoptotic cell engulfment: Eat-me, don't eat me and come-get-me signals. Trends Cell Biol 2003; 13:648-656.

Gros TS, Rabaia NA, Moy NY et al. Osteopontin: A marker of osteocyte stress in response to loss of mechanical loading (abstract). J Bone Miner Res 2002; 17:S191.

Gross TS, Akeno N, Clemens TL et al. Selected contribution: Osteocytes upregulate HIF-α in response to acute disuse and oxygen deprivation. J Applied Physiol 2001; 90:2514-2519.

Gross TS, King KA, Rabaia NA et al. Upregulation of osteopontin by osteocytes deprived of mechanical loading or oxygen. J Bone Miner Res 2005; 20:250-256.

Gross TS, Poliachik SL, Ausk BJ et al. Why rest stimulates bone formation: A hypothesis based on complex adaptive phenomenon. Exerc Sport Sci Rev 2004; 32:9-13.

Gu Y, Publicover SJ. Expression of functional metabotropic glutamate receptors in primary cultured rat osteoblasts. J Biol Chem 2000; 275:34252-34259.

Guaradvaccaro D, Corrente G, Covone F et al. Arrest of G_1-S progression is Rb dependent and relies on the inhibition of of cyclin D1 transcription. Mol Cell Biol 2000; 20:1797-1815.

Gubrij I, Ali AA, Chambers TM et al. Decreased apoptosis and increased bone formation in implants of marrow-derived osteoblast progenitors overexpressing Bcl-2: In vivo evidence for a pivotal role of apoptosis in bone formation (abstract). J Bone Miner Res 2003; 18:S136.

Guénette S, Chang Y, Hiesberger T et al. Essential roles fore FE65 amyloid precursor protein-ineracting proteins in brain development. EMBOI J 2006; 25:420-431.

Guerreiro PM, Fuentes T, Power DM et al. Parathyroid hormone-related protein: A calcium regulatory factor in sea bream (Sparus aurata L) larvae. Am J Physiol Reg Int Comp Physiol 2001; 281:R855-R860.

Guise TA, Mohammad KS. Endothelins in bone cancer metastases. Cancer Treat Res 2004; 118:197-212.

Günther T, Chen ZF, Kim J et al. Genetic ablation of parathyroid glands reveals another source of parathyroid hormone. Nature 2000; 406:199-203.

Günther T, Karsenty G. Development of parathyroid glands. In: Naveh-Many T, ed. Molecular Biology of the Parathyrids. Austin: Landes Bioscience, 2005:1-7.

Gupta A, Miyauchi A, Fujimori A et al. Phosphate transport in osteoclasts: A functional and immunochemical characterization. Kidney Int 1996; 49:968-974.

Gurney JG, Severson RK, Davis S et al. Incidence of cancer in children in the United States. Sex-, race- and 1-year age-specific rates by histologic type. Cancer 1995; 75:2186-2195.

Gutierrez G, Garrett JR, Rossini G et al. Dermal application of lovastatin to rats causes greater increases in bone formation and plasma concentration than when administered by oral gavage (abstract). J Bone Miner Res 2000; 15:S427.

Hagino H, Okano T, Enokida M et al. The effect of parathyroid hormone on cortical bone response to in vivo external loading (abstract). J Bone Miner Res 2000; 15:806.

Hagiwara H, Ohwada N, Takata K. Cell biology of normal and abnormal ciliogenesis in the ciliated epithelium. Int Rev Cytol 2004; 234:101-141.

Hajnóczky G, Csordas G, Krishnamurthy R. Mitochondrial calcium signaling driven by the IP$_3$ receptor. J Bioenerg Biomembranes 2000; 32:15-25.

Hajnóczky G, Csordás G, Yi M. Old players in a new role: Mitochondria-associated membranes, VDAC, and ryanodine receptors as contributors to calcium signal propagation from endoplasmic reticulum to the mitochondria. Cell Calcium 2002; 32:363-377.

Hajnóczky G, Davies E, Madesh M. Calcium signaling and apoptosis. Biochem Biophys Res Commun 2003; 304:445-454.

Hakeda Y, Kobayashi Y, Yamaguchi K et al. Osteoclastogenesis inhibitory factor (OCIF) directly inhibits bone-resorbing activity of isolated mature osteoclasts. Biochem Biophys Res Commun 1998; 251:796-801.

Hall RA. β-adrenergic receptors and their interacting proteins. Semin Cell Devel Biol 2004; 15:281-288.

Hall RA, Ostedgaard LS, Premont RT et al. A C-terminal motif found in the β$_2$-adrenergic receptor, P2Y1 receptor and cyctic fibrosis transmembrane conductance regulator determines binding to the Na$^+$/H$^+$ exchanger regulatory factor family of PDZ proteins. Proc Natl Acad Sci USA 1998; 95:8496-8501.

Halladay DL, Miles RR, Gornik SA et al. Identification of signal transduction pathways and promoter sequences that mediate parathyroid hormone 1-38 inhibition of osteoprotegerin gene expression. J Cell Biochem 2002; 84:1-11.

Halleen JM, Raisanen S, Alatalo SL et al. Potential function for the ROS-generating activity of TRACP. J Bone Miner Res 2003; 18:1908-1911.

Halleen JM, Raisanen S, Salo JJ et al. Intracellular fragmentation of bone resorption products by reactive oxygen species generatewd by osteoclastic tartrate-resistant acid phosphatases. J Biol Chem 1999; 274:22907-22910.

Hamon M, Doucet E, LeFevre K et al. Antibodies and antisense oligonucleotides for probing the distribution and putative functions of 5-HT$_6$ receptors. Neuropsychopharmacol 1999; 21:68S-76S.

Hamrick MW, Pennington C, Newton D et al. Leptin deficiency produces contrsting phenotypes in bones of the limb and spine. Bone 2004; 34:369-371.

Hanes DR, Crotti TN, Loric M et al. Osteoprotegrin and receptoactivator of nuclear factor kappB ligand (RANKL) regulate osteoclast formation by cells in the human rheumatoid arthritic joint. Rheumatology (Oxford) 2001; 40, 623-630.

Hanafin NM, Chen TC, Heinrich G et al. Cultured human fibroblasts and not cultured human keratinocytes express a PTH/PTHrP receptor mRNA. J Invest Dermatol 1995; 105:133-137.

Harbour JW, Dean DC. The Rb/E2F pathway: Expanding roles in emerging paradigms. Genes Devel 2000; 14:2393-2409.

Harmey D, Stenbeck G, Nobes CD et al. Regulation of osteoblast differentiation by Pasteurella multocida toxin (PMT): A role for the Rho GTPase in bone formation. J Bone Miner Res 2004; 19:661-670.

Harris BZ, Lim WA. Mechanism and role of PDZ domains in signaling complex assembly. J Cell Sci 2001; 114:3219-3231.

Hartmann T, Bieger SC, Bruhl B et al. Disitnct sites of ijntracellular production for Alzheimer's disease Abeta 40/42 amyloid peptides. Nat Med 1997; 3:1016-1020.

Hauge EM, Qvesel D, Eriksen EF et al. Cancellous bone remodeling occurs in specialized compartments lined by cells expressing osteoblast markers. J Bone Miner Res 2001; 16:1575-1582.

Havers C. Osteologia nova, or some new observations of the bones, and the parts belonging to them, with the manner of their accretion and nutrition. Samuel Smith, London: Communicated to the Royal Society in Several Discourses, 1691.

Haylock DN, Nilsson SK. Stem cell regulation by the stem cell niche. Cell Cycle 2005; 4:1353-1355.

Hazenberg JG, Lee TC, Taylor D. The role of osteocytes in functional bone adaptation. BoneKEy-Osteovision 2006; 3:10-16.

Heaney RP. Is the paradigm shifting? Bone 2003; 33:457-465.

Hebert SC. Therapeutic use of calcimimetics. Annu Rev Med 2006; 57:349-364.

He X, Semenov M, Tamai K et al. LDL receptor-related proteins 5 and 6 in Wnt/β-catenin signaling: Arrows point the way. Development 2004; 131:1663-1677.

Hefferan TE, Rehman O, Lane NE et al. Monitoring the response to PTH by analysis of peripheral leukocyte gene expression (abstract). J Bone Miner Res 2002; 17:S214.

Heino TJ. Osteocytes as regulators of bone resorption and bone formation. Anatomy. Turku, Finland: University of Turku, 2005.

Hellio de Graverand M, Rattner J, Eggerer J et al. Primary cilia are a hallmark of rabbit meniscal cells. Proc. 47th Annual Meeting of the Orthopaedic Research Society, 2001:43.

Henriksen K, Karsdal M, Delaissé JM et al. RANKL and vascular endothelial growth factor (VEGF) induce osteoclast chemotaxis through an ERK1/2-dependent mechanism. J Biol Chem 2003; 278:48745-48753.

Herrington DM, Klein KP. Statins, hormones, and women: Benefits and drawbacks for athersclerosis and osteoporosis. Curr Atherosclerosis Reports 2001; 3:35-42.

Hershko A, Ciechanover A, Varshavsky A. The ubiquitin system. Nature Medicine 2000; 6:1073-1081.

Herzog H. Hypothalamic Y2 receptors: Central coordinator of energy homeostasis and bone mass regulation. Drug News Perspect 2002; 15:506-510.

Hesch RD, Busch U, Prokop M et al. Increase of vertebral density by combination therapy with pulsatile 1-38 PTH and sequential addition of calcitonin nasal spray in osteoporotic patients. Calcif Tissue Int 1989; 44:176-180.

Hesp R, Hulme P, Williams D et al. The relationship between changes in femoral bone density and calcium balance in patients with involutional osteoporosis treated with human PTH fragment 1-34. Metab Bone Dis Rel Res 1981; 2:331-334.

Hewitt SC, Deroo BJ, Korach KS. A new mediator for an old hormone? Science 2005; 307:1572-1573.

Hill PA, Tumber A, Meikle MC. Multiple extracellular signals promote osteoblast survival and apoptosis. Endocrinology 1997; 138:3849-3858.

Himms-Hagen J. Physiological roles of the leptin endocrine system: Differences between mice and humans. Crit Rev Clin Lab Sci 1999; 36:575-655.

Hino S, Tanji C, Nakayama KI et al. Phosphorylation of β-catenin by cyclic AMP-dependent protein kinase stabilizes β-catenin through inhibition of its ubiquitination. Mol Cell Biol 2005; 25:9063-9072.

Hinoi E, Fujimori S, Nakamura Y et al. Group III metabotropic glutamate receptors in rat cultured calvarial osteoblasts. Biochem Biophys Res Commun 2001; 281:341-346.

Hinoi E, Fujimori S, Takarada T et al. Facilitation of glutamate release by ionotropic glutamate receptors in osteoblasts. Biochem Biophys Res Commun 2002; 297:452-458.

Hinoi E, Fujimori S, Yoneda Y. Modulation of cellular differentiation by N-methyl-D-aspartate receptors in osteoblasts. FASEB J 2003; 17:1532-1534.

Hinoi E, Takarada T, Yoneda Y. Glutamate signaling system in bone. J Pharmacol Sci 2004; 94:215-220.

Hinson TK, Damodaran TV, Chen J et al. Identification of putative trans-membrane receptor sequences homologous to the calcium-sensing G-protein-coupled receptor. Genomics 1997; 45:279-289.

Hirano T, Burr DB, Cain RL et al. Changes in geometry and cortical porosity in adult, ovary-intact rabbits after 5 months treatment with LY333334 (hPTH 1-34). Calcif Tissue Int 2000; 66:456-460.

Hjälm G, MacLeod J, Kifor O et al. Filamin-A binds to the carboxyl-terminal tail of the calcium-sensing receptor, an interaction that participates in CaR-mediated activation of mitogen-activated protein kinase. J Biol Chem 2001; 276:34880-34887.

Hoare RJ, Gardella TJ, Usdin TB. Evaluating the signal transduction mechanism of the parathyroid hormone 1 receptor: Effect of receptor G-protein interaction on the ligand binding mechanism and receptor conformation. J Biol Chem 2001; 276:7741-7753.

Hoare SRJ, Usdin TB. Molecular mechanisms of ligand recognition by parathyroid hormone 1 (PTH1) and PTH2 receptors. Curr Pharma Design 2001; 7:689-713.

Hock D, Mägerlein M, Heine G et al. Isolation and characterization of the bioactive circulating human parathyroid hormone, hPTH-1-37. FEBS Lett 1997; 400:221-225.

Hock JM. Stemming bone loss by suppressing apoptosis. J Clin Invest 1999; 104:371-373.

Hodsman AB, Drost D, Fraher LJ et al. The addition of raloxifene analog (LY117018) allows for reduced PTH(1-34) dosing during reversal of osteopenia in ovariectomized rats. J Bone Miner Res 1999; 14:675-679.

Hodsman AB, Fraher LJ, Watson PH. Parathyroid hormone: The clinical Experience and prospects. In: Whitfield JF, Morley P, eds. Anabolic Treatments for Osteoporosis. Boca Raton: CRC Press, 1997:83-108.

Hodsman AB, Fraher LJ, Ostbye T et al. An evaluation of several biochemical markers for bone formation and resorption in a protocol utilizing cyclical cyclical parathyroid hormone and calcitonin therapy for osteoporosis. J Clin Invest 1993; 91:1138-1148.

Hodsman AB, Fraher LJ, Watson PH et al. A randomized controlled trial to compare the efficacy of cyclical parathyroid hormone versus cyclical parathyroid hormone and sequential calcitonin to improve bone mass in post-menopausal women with osteoporsis. J Clin Endocrinol Metab 1997; 82:620-628.

Hodsman AB, Hanley DA, Ettinger MP et al. Efficacy and safety of human parathyroid hormone-(1-84) in increasing bone mineral density in postmenopausal osteoporosis. J Clin Endocrinol Metab 2003; 88:5152-5220.

Hodsman AB, Steer BM. Early histomorphometric changes in response to parathyroid hormone therapy in osteoporosis: Evidence for de novo bone formation on quiescent cancellous surfaces. Bone 1993; 14:523-527.

Hodsman AB, Watson PH, Drost D et al. Assessment of maintenance therapy with reduced doses of PTH(1-34) in combination with a raloxifene analogue (LY117018) following anabolic therapy in the ovariectomized rat. Bone 1999; 24:451-455.

Hodsman AB, Watson PH, Fraher LJ et al. Increased bone turnover without change in cortical thickness or porosity after 2 years of cyclical PTH therapy for postmenopausal osteoporosis (abstract). J Bone Miner Res 2000; 15:799.

Hofbauer LC, Khosla S, Dunstan CR et al. Estrogen stimulates gene expression and protein production of osteoprotegerin in human osteoblastic cells. Endocrinology 1999; 140:4367-4370.

Hofbauer LC, Khoslab S, Dunstan CR et al. The roles of osteoprotegerin and osteoprotegerin ligand in the paracrine regulation of bone resorption. J Bone Miner Res 2000; 15:2-12.

Hofstaetter JG, Wang J, Glimcher MJ. The use of alendronate in the treatment of osteonecrosis of the femoral head to reduce degenerative osteoarthritis of the hip. J Bone Miner Res 2004; 19:S324.

Holden C. What's in a tooth? Science 2005; 310:1900.

Holick MF, Chimeh FN, Ray S. Topical PTH (1-34) is a novel, safe and effective treatment for psoriasis: A randomized self-controlled trial and an open trial. Brit J Dermatol 2003; 149:370-376.

Holick MF, Ray S, Chen TC et al. A parathyroid hormone antagonist stimulates epidermal proliferation and hair growth in mice. Proc Natl Acad Sci USA 1994; 91:8014-8016.

Hollinger S, Hepler JR. Cellular regulation of RGS proteins: Modulators and integrators of G protein signaling. Pharmacol Rev 2002; 54:527-559.

Hook VY, Burton D, Yasothornsrikul S et al. Proteolysis of PTHrP(1-141) by "prohormone thiol protease" at multibasic residues generates PTHrP-related peptides: Implications for PTHrP peptide production in lung cancer cells. Biochem Biophys Res Commun 2001; 285:932-938.

Hook VY, Reisine TD. Cysteine proteases are the major beta-secretase in the regulated secretory pathway that provides most of the beta-amylod in Alzheimer's disease: Role of BACE 1 in the constitutive secretory pathway. J Neurosci Res 2003; 74:393-405.

Hopyan S, Gokgoz N, Poon R et al. A mutant PTH/PTHrP type 1 receptor in endochondromatosis. Nat Genet 2002; 30:306-310.

Horiuchi N, Maeda T. Statins and bone metabolism. Oral Diseases 2006; 12:85-101.

Hoshi K, Ozawa H. Matrix vesicle calcification in bones of adult rats. Calcif Tissue Int 2000; 66:430-434.

Hruska KA, Mathew S, Saab G. Bone morphogenic proteins in vascular calcification. Circ Res 2005; 97:105-114.

Huang LE, Bunn HF. Hypoxia-inducible factor and its biomedical relevance. J Biol Chem 2001; 278:19575-19578.

Huang Y, Baker RT, Fischer-Vize JA. Control of cell fate by a deubiquitinating enzyme encoded by the fat facets gene. Science 1995; 270:1828-1831.

Huang Z, Cheng SL, Slatopolsky E. Sustained activation of the extracellular signal-regulated kinase pathway is required for extracellular calcium stimulation of human osteoblast proliferation. J Biol Chem 2001; 276:21351-21358.

Hunagfu D, Anderson K. Cilia and Hedgehog responsiveness in the mouse. Proc Natl Acad Sci USA 2005; 102:11325-11330.

Hunt JL, Fairman R, Mitchell ME et al. Bone formation in carotid plaques. Stroke 2002; 33:1214-1219.

Hurley MM, Marcello K, Abreu C et al. Transcriptional regulation of the collagenase gene by basic fibroblast growth factor in osteoblastic MC3T3-E1 cells. Biochem Biophys Res Commun 1995; 214:331-339.

Hurley MM, Okada Y, Sobue T et al. The anabolic effect of parathyroid hormone is impaired in bones of Fgf2 null mice (abstract). J Bone Miner Res 2002; 17:S140.

Hurley MM, Tetradis S, Huang YF et al. Parathyroid hormone regulates the expression of fibroblast growth factor-2 mRNA and fibroblast growth factor receptor mRNA in osteoblastic cells. J Bone Miner Res 1999; 14:776-783.

Hurley MM, Yao W, Arnaud CD et al. hPTH(1-34) treatment increased serum FGF-2 levels in glucocorticoid induced osteoporosis patients. J Bone Miner Res 2004; 19:S459.

Iacopetti P, Barsacchi G, Tirone F et al. Developmental expression of PC3 gene is correlated with neuronal cell birthday. Mech Devel 1994; 47:127-137.

Iacopetti P, Michelini M, Stuckmann I et al. Expression of the antiproliferative gene TIS21 at the onset of neurogenesis identifies single neuroepithelial cells that switch from proliferative to neuron-generating division. Proc Natl Acad Sci USA 1999; 96:4639-4644.

Iannotti JP, Naidu S, Noguchi Y et al. Growth plate matrix vesicle biogenesis. Clin Orthop Rel Res 1994; 306:222-229.

Iba K, Takada J, Yamashita T. The serum level of bone-specific alkaline phosphatase activity is associated with aortic calcification in osteoporosis patients. J Bone Miner Res 2004; 22:594-596.

Iguchi T, Ziff M. Electron microscopic study of rheumatoid synovial vasculature. J Clin Invest 1986; 77:355-361.

Iida-Klein A, Hughes C, Lu SS. Effects of cyclic versus daily hPTH-(1-34) regimens on bone strength in association with BMD, biochemical markers and bone structure in mice. J Bone Miner Res 2006; 21:274-282.

Ilmbalzano AN, Jones SN. Snf5 tumor suppressor couples chromatin remodeling, checkpoint control, and chromosoma stability. Cancer Cell 2005; 7:294-295.

Ingelton PM. Parathyroid hormone-related protein in lower vertebrates. Comp Biochem Physiol B Biochem Mol Biol 2002; 132:87-95.

Ingleton PM, Bendell LA, Flanagan JA et al. Calcium-sensing receptors and parathyroid hormone-related protein in the caudal neurosecretory system of the flounder (Platichthys flesus). J Anat 2002; 200:487-497.

Isales CM, Ding KH, Divieti P et al. A specific high-affinity carboxyterminal parathyroid hormone receptor is present in vascular endothelial cells and utilizes a calcium-dependent signaling pathway (abstract). J Bone Miner Res 2004; 19:S338.

Ishikawa M, Akishita M, Kozaki K et al. Expression of parathyroid hormone-related protein in human and experimental atherosclerotic lesions: Functional role in in arterial intimal thickening. Atherosclerosis 2000; 152:97-105.

Ito Y. Molecular basis of tissue-specific gene expression mediated by the Runt domain transcription factor PEBP2/CBF. Genes to Cells 1999; 4:685-696.

Itzstein C, Cheynel H, Burt-Pichat B et al. Molecular identification of NMDA glutamate receptors expressed in bone cells. J Cell Biochem 2001; 82:134-144.

Itzstein C, Espinosa L, Delmas PD et al. Specific agonists of NMDA receptors prevent osteoclasts sealing zone formation required for bone resorption. Biochem Biophys Res Commun 2000; 268:201-209.

Iwamoto R, Mekada E. Heparin-binding EGF-like growth factor: A juxtacrine growth factor. Cytokine Growth Factor Rev 2000; 11:335-344.

Iwaniec UT, Mosekilde L, Mitova-Caneva NG et al. Sequential treatmert with basic fibroblast growth factor and PTH is more efficacious than treatment with PTH alone for increasing vertebral bone mass and strength in osteopenic ovariectomizrd rats. Endocrinology 2002; 143:2515-2526.

Iwaniec UT, Shearon CC, Heaney RP et al. Leptin increases the number of mineralized bone nodules in vitro. J Bone Miner Res 1998; 13:S212.

Iwasa S, Fan J, Miyauchi T et al. Blockade of endothelin receptors reduces diet-induced hypercholesterolemia and atherosclerosis in apolipoprotein E-deficient mice. Pathobiology 2001; 69:1-10.

Iwata K, Allen MR, Phipps R et al. Microcrack initiation occurs more easily in vertebrae from beagles treated with alendronate than with risedronate. Bone 2006; 38:S42.

Jakoby MG, Semenkovich CF. The role of osteoprogenitors in vascular calcification. Curr Opin Nephrol Hypertens 2000; 9:11-15.

Jans DA, Hassan G. Nuclear targeting by growth factors, cytokines, and their receptor: A role in signaling? BioEssays 1998; 29:400-411.

Jans DA, Thomas RJ, Gillespie MT. Parathyroid hormone-related protein (PTHrP): A nucleoplasmic shuttling protein with distinct paracrine and intracrine roles. Vitamins and Hormones 2003; 66:345-384.

Janus M, Tembe V, Favus MJ. Role of protein kinase C in parathyroid hormone stimulation of renal 1,25-dihydroxyvitamin D_3 secretion. J Clin Invest 1992; 90:2278-2283.

Janulis M, Wong MS, Favus MJ. Structure-function requirements of parathyroid hormone for stimulation of 1,25-dihydroxyvitamin D_3 production by rat renal proximal tubules. Endocrinology 1993; 133:713-719.

Jee W. Integrated bone tissue physiology: Anatomy and physiology. In: Cowin SC, ed. Bone Mechanics Handbook. 2nd ed. Boca Raton: CRC Press, 2000:1-1-1-68.

Jensen AA, Greenwood JR, Brauner-Osborne H. The dance of the the clams: Twists and turns in the family C GPCR homodimer. Trends Pharmacol Sci 2002; 23:491-493.

Jensen CG, Poole CA, McGlashan SR et al. Ultrastructural tomographic and confocal imaging of the chondrocyte primary cilium in situ. Cell Biol Int 2004; 28:101-110.

Jerome CP, Burr DB, Van Bibber T et al. Treatment with human parathyroid hormone (1-34) for 18 months increases cancellous bone volume and improves trabecular architecture in ovariectomized cynomolgus monkeys (Macaca fascicularis). Bone 2001; 28:150-159.

Jerome CP, Carlson CS, Register TC et al. Bone functional changes in intact, ovariectomized, and ovariectomized, hormone-supplemented adult cynomolgus monkeys (Macaca fascicularis) evaluated by serum markers and dynamic histomorphometry. J Bone Miner Res 1994; 9:527-540.

Jerome CP, Johnson CS, Vafai HT et al. Effect of treatment for 6 months with human parathyroid hormone (1-34) peptide in ovariectomized cynomolgus monkeys (Macaca fascicularis). Bone 1999; 25:301-309.

Ji X, Chen D, Xu C et al. Patterns of gene expression associated with BMP-2-induced osteoblasts and adipocyte differentiation of mesenchymal progenitor cell 3T3-F442A. J Bone Miner Res 2000; 15:132-139.

Jick H, Zornberg GL, Jick SS et al. Statins and the risk of dementia. Lancet 2000; 356:1627-1631.

Jilka RL. Cytokines, bone remodeling, and estrogen deficiency. Bone 1998; 23:75-78.

Jilka RL, Weinstein RS, Bellido T et al. Increased bone formation by prevention of osteoblast apoptosis with parathyroid hormone. J Clin Invest 1999; 104:439-446.

Johansson A, Rosen CJ. The insulin-like growth factors: Potential anabolic agents for the skeleton. In: Whitfield JF, Morley P, eds. Anabolic Treatments for Osteoporosis. Boca Raton: CRC Press, 1997:185-205.

Jofuku A, Ishihara N, Mihara K. Analysis of functional domains of rat mitochondrial Fis 1, the mitochondrial fission-stimulating protein. Biochem Biophys Res Commun 2005; 333:650-659.

Johnson LC. The kinetics of skeletal remodeling. Birth defects Original Article Series 1966; 2:66-142.

Jolette J, Smith SY, Mayer J et al. Ostabolin-C™: A novel parathyroid hormone analogue uncouples bone turnover in monkeys (abstract). J Bone Miner Res 2003; 18:S386.

Jolette J, Smith SY, Moreau IA et al. Ostabolin-C™: A novel parathyroid hormone analogue uncouples bone turnover in monkeys treated for 1-year. J Bone Miner Res 2005; 20:S410.

Jono S, McKee MD, Murry CE et al. Phosphate regulation of vascular smooth muscle calcification (abstract). Circ Res 2000; 87:E10-E17.

Jono S, Nishizawa Y, Shioi A et al. Parathyroid hormone-related peptide as a local regulator of vascular calcification. Arterioscler. Thromb Vasc Biol 1997; 17:1135-1142.

Jordan BA, Cvejic S, Devi LA. Opioids and their complicated receptor complexes. Neuropsychopharmacol 2000; 23:S5-S18.

Jordan BA, Trapaidze N, Gomes I et al. Oligomerization of opioid receptors with β2-adrenergic receptors: A role in trafficking and mitogen-activated protein kinase activation. Proc Natl Acad Sci USA 2001; 98:343-348.

Jouishomme H, Whitfield JF, Chakravarthy B et al. The protein kinase-C activation domain of the parathyroid hormone. Endocrinology 1992; 130:53-60.

Jouishomme H, Whitfield JF, Gagnon L et al. Further definition of the protein kinase C activation domain of the parathyroid hormone. J Bone Miner Res 1994; 9:943-949.

Ju W, Hoffman A, Verschueren K et al. The bone morphogenic protein 2 signaling mediator Smad 1 participates predominantly in osteogenic and not in chondrogenic differentiation in mesenchymal progenitors C3H10T1/2. J Bone Miner Res 2000; 15:1889-1899.

Juhlin L, Hagforsen E, Juhlin C. Parathyroid hormone related protein is localized in the granular layer of normal skin and in the dermal infiltrates of mycosis fungoides but is absent in psoriatic lesions. Acta Derm Venereol 1992; 72:81-83.

Juul A. The effects of oestrogens linear bone growth. Hum Reprod Update 2001; 7:303-313.

Kalpana GV, Marmon S, Wang W et al. Binding and stimulation of HIV-1 integrase by a human homolog of yeast transcription factor SNF5. Science 1994; 266:2002-2006.

Kalu DN, Doyle FH, Pennock J et al. Parathyroid hormone and experimental osteoslerosis. Lancet 1970; 1:1363-1366.

Kameda T, Mano H, Yameda Y et al. Calcium-sensing receptors in mature osteoclasts which are bone-resorbing cells. Biochem Biophys Res Commun 1998; 245:419-422.

Kamioka H, Honjo T, Takano-Yamamoto T. A three-dimensional distribution of osteocyte processes revealed by the combination of confocal laser scanning microscopy and differential interference contrast microscopy. Bone 2001; 28:145-149.

Kamycheva E, Sundsfjord J, Jorde R. Serum parathyroid hormone levels predict coronary heart disease: The Tromsø study. Eur J Cardiovasc Prev Rehabil 2004; 11:69-74.

Kanazaki M, Zhang YQ, Mashima H et al. Translocation of a calcium-permeable cation channel induced by insulin-like growth factor-I. Nature Cell Biol 1:165-170.

Kanatami M, Sugimoto T, Takahashi Y et al. Estrogen via the estrogen receptor blocks cAMP-mediated parathyroid hormone (PTH)-stimulated osteoclast function. J Bone Miner Res 1998; 13:854-862.

Kanazawa M, Sugimoto T, Kanatami M et al. Involvement of osteoprotegerin/osteoclastogenesis inhibitory factor in the stimulation of osteoclast formation by parathyroid hormone in mouse bone cells. Eur J Endocrinol 2000; 142:661-664.

Karaplis AC. PTHrP: Novel roles in skeletal biology. Curr Pharma Design 2001; 7:655-670.

Kapalan FS, Smith RM. Fibrodysplasia ossificans progressive (FOP). J Bone Miner Res 1997; 12:855.

Karaplis AC, Luz A, Glowacki J et al. Lethal skeletal dysplasia from targeted disruption of the parathyroid hormone-related peptide gene. Genes Dev 1994; 8:277-289.

Karnik SK, Gogonea C, Patil S et al. Activation of G-protein coupled receptors: A common molecular mechanism. Trends Endocrinol Metab 2003; 14:431-437.

Karperien M, Farih-Sips H, Papapoulos SE et al. Involvement of Cbfa1 in transcritional regulation of the type I PTH/PTHrP-receptor in KS483 osteoblasts (abstract). J Bone Miner Res 2000; 15:S175.

Karsdal MA, Larsen L, Engsig MT et al. Matrix metalloproteinase-dependent activation of latent transforming growth factor-β controls the conversion of osteoblasts into osteocytes by blocking osteoblast apoptosis. J Biol Chem 2002; 277:44061-44067.

Karsenty G. The genetic transformation of bone biology. Genes Dev 1999; 13:3037-3051.

Karsenty G. Role of Cbfa1 in osteoblast differentiation and function. Seminars Cell Dev Biol 2000a; 11:343-346.

Karsenty G. The central regulation of bone remodeling. Trends Endocrinol Metab 2000b; 11:437-439.

Karsenty G. Chondrogenesis just ain't what it used to be. J Clin Invest 2001; 107:405-407.

Kartsogiannis V, Moseley J, McKelvie B et al. Temporal expression of PTHrP during endochondral bone formation in mouse and intramembranous bone formation in an in vivo rabbit model. Bone 1997; 21:385-392.

Kasperk CH, Börcsök I, Schairer HU et al. Endothelin-1 is a potent regulator of human bone cell metabolism in vitro. Calcif Tissue Int 1997; 60:368-374.

Katagiri T, Takahashi N. Regulatory mechanisms of osteoblast and osteoclast differentiation. Oral Dis 2002; 8:147-159.

Kawakami A, Eguchi K, Matsuoka N et al. Fas and Fas ligand interaction is necessary for human osteoblast apoptosis. J Bone Miner Res 1997; 12:1637-1646.

Kawakami A, Nakashima T, Tsuboi M et al. Insulin-like growth factor I stimulates proliferation and Fas-mediated apoptosis of human osteoblasts. Biochem Biophys Res Commun 1998; 247:46-51.

Kawamura YJ, Kazama S, Miyahara T et al. Sigmoid colon cancer associated with primary hyperparathyroidism: Report of a case. Surg Today 1999; 29:789-790.

Kawane T, Takahashi S, Saitoh H et al. Anabolic effects of recombinant human parathyroid hormone (1-84) on the mandibles of osteopenic ovariectomized rats with maxillary molar extraction. Horm Metab Res 2002; 34:293-302.

Ke HZ, Steppan CM, Swick AG. Treatment of skeletal disorders. United States Patent 2001; 6:352,970, (B1).

Ke HZ, Xu G, Brault AL et al. Deletion of the β2 adrenergic receptor prevents bone loss induced by isoproterenol in mice (abstract). J Bone Miner Res 2004; 19:S3.

Ke HZ, Steppan CM, Swick AG. Treatment of skeletal disorders. United States Patent Application Publication, 2002, (0019351 A1).

Keenan SM, Baldassare JJ. Molecular scaffold protein and cellular respon ses. Trends Endocrinol Metab 2001; 12:184-186.

Kellenberger S, Muller K, Richener H et al. Formoterol and isoproterenol induce c-fos gene expression in osteoblast-like cells by activating β2-adrenergic receptors. Bone 1998; 22:471-478.

Keller H, Kneissel M. SOST is a target for PTH in bone. Bone 2005; 37:148-158.

Kelly MJ, Levin ER. Rapid actions of plasma membrane estrogen receptors. Trends Endocrino Metab 2001; 12:152-156.

Kempf H, Chen RE, Kerley ER. The microscopic determination of age in human bone. Am J Phys Anthropol 1965; 23:149-164.

Kern B, Shen J, Starbuck M et al. Cbfa1 contributes to the osteoblast-specific expression of type I collagen genes. J Biol Chem 2001; 276:7101-7107.

Kerr JB. Atlas of Functional Histology. London: Mosby International Limited, 1999.

Khosla S. Oestrogen, bones and men: When testosterone just isn't enough. Clin Endocrinol (Oxford) 2002; 56:291-293.

Kifor O, Diaz R, Butters R et al. The calcium-sensing receptor is localized in caveolin-rich plasma membrane domains of bovine parathyroid cells. J Biol Chem 1998; 273:21708-21713.

Kifor O, Kifor I, Brown EM. Signal transduction in the parathyroid. Curr Opin Nephrol Hypertens 2002; 11:397-402.

Kifor O, Kifor I, Moore FD et al. m-Calpain colocalizes with the calcium-sensing receptor (CaR) in caveolae in parathyroid cells and participates in degradation of the CaR. J Biol Chem 2003; 278:31167-31176.

Kim DW, Kempf H, Chen RE et al. Characterization of Nkx3.2 DNA binding specificity and its requirement for somatic chondrogenesis. J Biol Chem 2003; 278:27532-27539.

Kim HW, Jahng JS. Effect of intermittent administration of parathyroid hormone on fracture healing in ovariectomized rats. Iowa Orthop J 1999; 19:71-77.

Kim, HKW, Randall TS, Bian H et al. Ibandronate decreases femoral head deformity following ischemic osteonecrosis of the femoral head in iummature pigs. J Bone Miner Res 2004; 19:S328.

Kim HP, Lee JY, Jeong JK et al. Nongenomic stimulation of nitric oxide release by estrogen is mediated by estrogen receptor alpha localized in caveolae. Biochem Biophys Res Commun 1999; 263:257-262.

Kim J, Jones BW, Zock C et al. Isolation and characterization of mammalian homologs of the Drosophila gene glial cells missing. Proc Natl Acad Sci USA 1998; 95:12364-12369.

Kinoshita T, Kobayashi S, Ebara S et al. Phosphodiesterase inhibitors, Pentoxyfylline and Rolipram,
increase bone mass mainly by promoting bone formation in normal mice. Bone 2000; 27:811-817.

Kirsch T, Harrison G, Golub EE et al. The roles of annexins and types II and X collagens in matrix
vesicle-mediated mineralization of growth plate cartilage. J Biol Chem 2000; 275:35577-35583.

Kirsch T, Nah HD, Shapiro IM et al. Regulated production of mineralization-competent matrix vesicles
in hypertrophic chondrocytes. J Cell Biol 1997; 137:1149-1160.

Kishida Y, Hirao M, Tamai N et al. Leptin regulates chondrocyte differentiation and matrix maturation
during endochondral ossification. Bone 2005; 37:607-621.

Kitazawa R, Kitazawa S, Maeda S et al. Promoter structure of mouse RANKL/TRANCE/OPGL/ODF
gene. Biochim Biophys Acta 1999; 1445:134-141.

Kitazawa S, Kajimoto K, Kondo T et al. Vitamin D_3 supports osteoclastogenesis via functional vitamin
D response element of human RANKL gene promoter. J Cell Biochem 2003; 89:771-777.

Kizer N, Guo KL, Hruska K. Reconstitution of stretch-activated cation channels by expression of the
epithelial sodium channel cloned from osteoblasts. Proc Natl Acad Sci USA 1997; 94:1013-1018.

Klein RF. Osteoporosis in men. In: Marcus R, ed. Atlas of Clincal Endocrinology, Vol.3, Osteoporosis.
Philadelphia: Current Medicine, 1999:85-99.

Klein-Nulent J, Nijweide PJ. Osteocyte and bone structure. Curr Osteoporosis Rep 2003; 1:5-10.

Kneissel M, Boyde A, Gasser JA. Bone tissue and its mineralization in aged estrogen-depleted rats after
long-term intermittent treatment with parathyroid hormone (PTH) analog SDZ PTS 893 or human PTH-(1-34). Bone 2001; 28:237-250.

Knothe-Tate M. Interstitial fluid flow. In: Cowin SC, ed. Bone Mechanics Handbook. 2nd ed. Boca
Raton: CRC Press, 2001:22-1-22-29.

Knothe-Tate ML. Whither flows the fluid in bone? An osteocyte's perspective. J Biomechanics 2003;
36:1409-1424.

Knothe-Tate ML, Adamson JR, Tami AE et al. The osteocyte. Int J Biochem Cell Biol 2004; 36:1-8.

Knothe Tate ML, Knothe U, Niederer P et al. Experimental elucidation of mechanical load-induced
fluid flow and its potential role in bone metabolism and functional adaptation. Am J Med Sci
1998; 316:189-195.

Knothe Tate ML, Steck R, Forwood MR et al. In vivo demonstration of load-induced fluid flow in the
rat tibia and its potential implications for processes associated with functional adaptation. J Exp
Biol 2000; 203:2737-2745.

Kobayashi K, Takahashi N, Jimi E et al. Tumor necrosis factor alpha stimulates osteoclast differentiation by a mechanism independent of the ODF/RANKL-RANK interaction. J Exp Med 2000;
191:275-286.

Kobayashi T, Chung U, Schipani E et al. PTHrP and Indian hedgehog control differentiation of growth
plate chondrocytes at multiple steps. Development 2002; 129:2977-2986.

Koch A, Yoon Y, Bonekamp NA et al. A role for Fis 1 in oth mitochondrial and perisomal fission in
mammalian cells. Mol Biol Cell 2005; 16:5077-5086.

Koch CA. Rapid increases in bone mineral density in a child with osteoporsis and autoimmune hypo-
parathyroidism treated with PTH 1-34. Exp Clin Endocrinol Diabetes 2001; 109:350-354.

Komori T. A fundamental transcription factor for bone and cartilage. Biochem Biophys Res Commun
2000; 276:813-816.

Kong YY, Felge U, Sarosi I et al. Activated T cells regulate bone loss and joint destruction in adjuvant
arthritis through osteoprotegerin ligand. Nature 1999; 402:304-309.

Kopp HG, Avecilla ST, Hooper AT et al. The bone marrow vascular niche: Home of HSC differentiation and mobilization. Physiology 2005; 20:349-356.

Kostenuik PJ, Capparelli C, Morony S et al. OPG and PTH-(1-34) have additive effects on bone density and mechanical strength in osteopenic ovariectomized rats. Endocrinology 2001; 142:4295-4304.

Kostenuik PJ, Harris J, Halloran BP et al. Skeletal unloading causes resistance of osteoprogenitor cells to
parathyroid hormone and to insulin-like growth factor-I. J Bone Miner Res 1999; 14:21-31.

Kostenuik PJ, Shalhoub V. Osteoprotegerin: A physiological and pharmacological inhibitor of bone resorption. Curr Pharma Design 2001; 7:613-635.

Kousteni S, Bellido T, Plotkin LI et al. Nongenotropic, sex-nonspecific signaling through the estrogen or androgen receptors: Dissociation from transcriptional activity. Cell 2001; 104:719-730.

Kousteni S, Chen JR, Bellido T et al. Reversal of bone loss in mice by nongenotropic signaling of sex steroids. Science 2002a; 298:843-846.

Kousteni S, Han L, Plotkin LJ et al. Nongenotropic regulation of CREB-, C/EBPβ-, as well as ELK-1 and AP-1-mediated transcription by estrogens: Downstream effects of ERK and JNK kinase modulation required for anti-apoptosis. J Bone Miner Res 2002b; 17:S169.

Koyama Y, Kitahara K, Ritting SR et al. Osteopontin plays a role in regulating apoptosis and mineralization under the control of inorganic phosphate. J Bone Miner Res 2003; 18:S137.

Krainock R, Murphy S. Nitric oxide. In: Conn PM, Means AR, eds. Principles of Molecular Regulation. Totowa: Humana Press, 2000:219-227.

Kreienkamp HJ. Organization of G-protein receptor signaling complexes by scaffolding proteins. Curr Opin Pharmacol 2002; 2:581-586.

Krempien B, Friedrich E, Ritz E. Effect of PTH on osteocyte ultrastructure. Adv Exp Med Biol 1978; 103:437-450.

Krishnan V, Heat H, Bryant HU. Mechanism of action of estrogens and selective estrogen receptor modulators. Vitamins and Hormones 2001; 60:123-147.

Krishnan V, Moore TL, Ma YL et al. Parathyroid hormone bone anabolic action requires cbfa1/runx-2 -dependent signaling. Mol Endocrinol 2003; 17:423-435.

Kronenberg HM, Bringhurst FR, Segré GV et al. Parathyroid hormone—Biosynthesis and Metabolism. In: Bilezikian JP, ed. The Parathyroids. 2nd ed. San Diego: Academic Press, 2001:17-30.

Kubota K, Sakikawa C, Katsumata M et al. Platelet-derived growth factor BB secreted from osteoclasts acts as an osteoblastogenesis inhibitory factor. J Bone Miner Res 2002; 17:257-265.

Kubodera N, Tsuji N, Uchiyama Y et al. A new active vitamin D analog, ED-71, causes increase in bone mass with preferential effects on bone in osteoporotic patients. J Cell Biochem 2003; 88:286-289.

Kufahl RH, Saha S. A theoretical model for stress-generated fluid flow in the canaliculi-lacunae network in bone tissue. J Biomech 1990; 23:171-180.

Kukreja SC, D'Anza JJ, Wimbiscus SA et al. Inactivation by plasma may be responsible for lack of efficacy of parathyroid hormone antagonists in hypercalcemia of malignancy. Endocrinology 1994; 134:2184-2188.

Kulkarni NH, Halladay DL, Miles RR et al. Effects of parathyroid hormone on Wnt signaling pathway in bone. J Cell Biochem 2005; 95:1178-1190.

Kurland EH, Cosman F, McMahon DJ et al. Parathyroid hormone as a therapy for idiopathic osteoporosis in men: Effects on bone mineral density and bone markers. J Clin Endocrinol Metab 2000; 85:3069-3076.

Kuroda M, Hazama-Shimada Y, Endo A. Inhibition of sterol synthesis by citrinin in a cell-free system from rat liver and yeast. Biochim Biophys Acta 1977; 486:254-259.

Kume K, Satomura K, Nishisho S et al. Potential role of leptin in endochondral ossification. J Histochem Cytochem 2002; 50:159-169.

Kuznetsov SA, Riminucci M, Ziran N et al. The interplay of osteogenesis and hematopoiesis: Expression of a constitutively active PTH/PTHrP receptor in osteogenic cells perturbs the establishment of hematopoiesis in bone amd of skeletal stem cells in the bone marrow. J Cell Biol 2004; 167:1113-1122.

Lacey DL. Antibody to Rank ligand—Overview of Rank ligand. J Bone Miner Res 2006; 20(Suppl 2):P23-P24.

LaCroix AZ, Cauley JA, Pettinger M et al. Statin use, clinical fracture, and bone density in postmenopausal women: Results from Womwn's Health Initiative Observational Study. Ann Intern Med 2003; 139:97-104.

Laharrague P, Larrouy D, Fontanilles AM et al. High expression of leptin by human bone marrow adipocytes in primary culture. FASEB J 1998; 12:747-752.

Laharrague P, Oppert JM, Brousset P et al. High concentration of leptin stimulates myeloid differentiaition from human bonemarrow CD34+ progenitors: Potential involvement in leukocytosis of obese subjects. Int J Obesity 2000; 24:1212-1216.

Laketic-Ljubojevic I, Suva LJ et al. Functional characterization of the N-methyl-D-aspartic acid-gated channels in bone. Bone 1999; 25:631-637.

Lam MH, House CM, Tiganis T et al. Phosphorylation at the cyclin-dependent kinase site (Thr 85) of parathyroid hormone-related protein negatively regulates its nuclear localization. J Biol Chem 1999; 274:18559-18566.

Lam MH, Thomas RJ, Loveland KL et al. Nuclear transport of parathyroid hormone (PTH)-related protein is dependedent on microtubules. Mol Endocrinol 2002; 16:390-401.

Lam MHC, Thomas RJ, Martin TJ et al. Nuclear and nucleolar localization of parathyroid hormone-related protein. Immunol Cell Biol 2000; 78:395-402.

Lamghari M, Tavares L, Camboa N et al. Leptin effect on RANKL and OPG expression in MC3T3-E1 osteoblasts. J Cell Biochem 2006, (published online Feb.14).

Lanchoney TF, Cohen RB, Rocke DM et al. Permanent heterotopic ossification at the injection site after diphtheria-tetanus-pertussis immunizations in children who have fibrodysplasia ossificans progressiva. J Pediatr 1995; 126:762-764.

Lando D, Gorman JJ, Whitelaw ML et al. Oxygen-dependent regulation of hypoxia-inducible factors by prolyl and aspariginyl hydroxylation. Eur J Biochem 2003; 270:781-790.

Lane N. Power, Sex, Suicide: Mitochondria and the Meaning of Life. Oxford: The University Press, 2005.

Lane NE, Sanchez S, Modin GW et al. Bone mass continues to increase at the hip after parathyroid hormone treatment is discontinued in glucocorticoid-induced osteoporosis: Results of a random-ized controlled clinical trial. J Bone Miner Res 2000; 15:944-951.

Lane NE, Kimmel DB, Nilsson MH et al. Bone-selective analogs of human PTH-(1-34) increase bone formation in an ovariectomized rat model. J Bone Miner Res 1996; 11:614-625.

Lane NE, Thompson JM, Strewler GJ et al. Intermittent treatment with parathyroid hormone (hPTH 1-34) increased trabecular bone volume but not connectivity in osteopenic rats. J Bone Miner Res 1995; 10:1470-1477.

Lane NE, Yao W, Arnaud CD. Changes in serum RANKL, OPG ans IL-6 during PTH (1-34) admin-istration in patients with glucocorticoid induced osteoporosis. J Bone Miner Res 2003; 18:S95.

Lane NE, Yao W, Balooch M et al. Both hPTH (1-34) and bFGF increase bone mass in osteopenic animals with different effects on trabecular bone architecture (abstract). J Bone Miner Res 2003; 18:S95.

Langub MC, Monier-Faugere MC, Qi Q et al. Parathyroid hormone/parathyroid hormone-related pep-tide type1 receptor in human bone. J Bone Miner Res 2001; 16:448-456.

Lanske B, Amling M, Neff L et al. Ablation of the PTHrP gene or the PTH/PTHrP receptor gene leads to distinct abnormalities in bone development. J Clin Invest 1999; 104:399-407.

Lanske B, Karaplis AC, Lee K et al. PTH/PTHrP receptor in early development and Indian hedgehog-regulated bone growth. Science 1996; 273:663-666.

Laroche M. Intraosseous circulation from physiology to disease. Joint Bone Spine 2002; 69:262-269.

Latchman DS. Eukaryotic Transcription Factors. 4th ed. San Diego: Elsevier Academic Press, 2004.

Laufs U, Liao JK. Post-transcriptional regulation of endothelial nitric oxide synthase mRNA stability by Rho GTP ase. J Biol Chem 1998; 273:24266-24271.

Laval-Jeantet AM, Bergot C, Carrol R et al. Cortical bone senescence and mineral density of the hu-merus. Calcif Tissue Int 1983; 35:268-272.

Lavasseur R, Sabatier JP, Potrel-Burgot C et al. Sympathetic nervous system as trasmitter of mechanical loading in bone. Joint Bone Spine 2003; 70:515-519.

Lawler OA, Miggin SM, Kinsella BT. Protein kinase A mediated phos-phorylation of serin 357 of the mouse prostacyclin receptor regulates its coupling to G_s-, to G_i- and to G_q-coupled effector signal-ing. J Biol Chem 2001; 276:33596-33607.

Leaffer D, Sweeny M, Kellerman LA et al. Modulation of osteogenic cell ultrastructure by RS-23581, an analog of human parathyroid hormone (PTH)-related peptide-(1-34), and ovine PTH-(1-34). En-docrinology 1995; 136:3624-3631.

Lebwohl M. Psoriasis. The lancet 2003; 361:1197-1205.

Lee CW, Nam JS, Park YK et al. Lysophosaphatidic acid stimulates CREB through mitogen- and stress-activated protein kinase-1. Biochem Biophys Res Commun 2003; 305:455-461.

Lee DK, George SR, O'Dowd BF. Unravelling the roles of the apelin system: Prospective therapeutic applications in heart failure and obesity. Trends Pharmacol Sci 2006; 27:190-194.

Lee E, HU N, Yuan SSF et al. Dual roles of the retinoblastoma protein in cell cycle regulation and neuron differentiation . Genes Devel 1994; 8:2008-2021.

Lee K, Deeds JD, Chiba S et al. Parathyroid hormone induces sequential c-fos expression in vivo: In situ localization of its receptor and c-fos messenger ribonucleic acids. Endocrinology 1994; 134:441-450.

Lee K, Deeds JD, Segre GV. Expression of parathyroid hormone-related peptide and its receptor messenger ribonucleic acid during fetal development of rats. Endocrinology 1995; 136:453-463.

Lee K, Lanske B, Karaplis AC et al. Parathyroid hormone-related peptide delays terminal differentiation of chondrocytes during endochondral bone development. Endocrinology 1996; 137:5109-5118.

Lee SP, Xie Z, Varghese G et al. Oligomerization of dopamine and serotonin receptors. Neuropsychopharmacol 2000; 23:S32-S40.

Lee SW, Kwak HB, Lee HC et al. The anti-proliferative gene TS21 is involved in osteoclast differentiation. J Biochem Mol Biol 2002; 35:609-614.

Lee YJ, Park JH, Ju SK et al. Leptin receptor isoform expression in rat osteoblasts and their functional analysis. FEBS Lett 2002; 528:43-47.

Lee ZH, Kim HH. Signal transduction by receptor activator of nuclear factor kappa B in osteoclasts. Biochem Biophys Res Commun 2003; 305:211-214.

Lefort K, Dotto GP. Notch signaling in the integrated control of keratinocyte growth /differentiation and tumor suppression. Semin Cancer Biol 2004; 14:374-386.

Lennon DP, Edmison JM, Caplan AI. Cultivation of rat marrow-derived mesechymal stem cells in reduced oxygen tension: Effects on in vitro and in vivo osteochondrogenesis. J Cell Physiol 2001; 187:345-355.

Levin ER. Nuclear receptor versus plasma membrane oestrogen receptor. Novartis Found Symp 2000; 230:41-55.

Li F. Survivin study: What is the next wave? J Cell Physiol 2003; 197:8-29.

Li T, Paudel HK. Glycogen synthase kinase 3 βphosphorylates Alheimer's disease-specific Ser 396 of microtubule-associated protein tau by a sequential mechanism. Biochemistry 2006; 45:3125-3133.

Li X, Zhang Y, Kang H et al. Sclerostin binds to LRP5/6 and antagonizes canonical Wnt signaling. J Biol Chem 2005; 280:19883-19887.

Lian JB, Javed A, Zaidi SK et al. Regulatory controls for osteoblast growth and differentiation: Role of Runx/Cbfa/AML factors. Crit Rev Eukaryotic Gene Express 2004; 14:1-41.

Li YM. γ-Secretase, a catalyst of Alzheimer's disease and signal transduction. Mol Interventions 2001; 1:198-207.

Lian JB, Stein GS. Runx2/Cbfa1: A multifunctional regulator of bone formation. Curr Pharma Design 2003; 9:2677-2685.

Lian JB, Stein GS, Stein JL et al. Transcriptional control of osteoblast differentiation. Biochem Soc Trans 1998; 26:14-21.

Liang H, Pun S, Wronski TJ. Bone anabolic effects of basic fibroblast growth factor in ovariectomized rats. Endocrinology 1999; 140:5780-5788.

Liang JD, Hock JM, Sandusky GE et al. Immunohistochemical localization of selected early response genes expressed in trabecular bone of young rats given PTH 1-34. Calcif Tissue Int 1999; 65:369-373.

Liebmann C. G protein-coupled receptors and their signaling pathways: Classical therapeutical targets susceptible to novel concepts. Curr Pharm Des 2004; 10:1937-1958.

Libby P. Inflammation in atherosclerosis. Nature 2002; 420:868-874.

LiChong KP, Li DY, Orlowski J et al. Selective domains of PTH and PTHrP activate the protein kinase A and protein kinase C pathways and differentially influence the Na^+/K^+ exchanger (NH3) and the type II Na/P_i cotransporter in renal OK cells (abstract). Bone 1998; 23:S356.

Lieberherr M, Grosse B, Kachkache M et al. Cell signaling and estrogens in female rat osteoblasts: A possible involvement of unconventional nonnuclear receptors. J Bone Miner Res 1993; 8:1365-1376.

Lim IK. TIS21 (/BTG2/PC3) as a link between ageing and cancer: Cell cycle regulator and endogenous cell death molecule. J Cancer Res Clin Oncol 2006; 132:417-426.

Lin SY, Makino K, Xia W et al. Nuclear localization of EGF receptor and its potential new role as a transcription factor. Nat Cell Biol 2001; 3:802-808.

Lincoln J, Hoyle CHV, Burnstock G. Nitric Oxide in Health and Disease. Cambridge: Cambridge University Press, 1997.

Lindsay R, Hodsman A, Genant H et al. A randomized controlled multi-center study of 1-84hPTH for treatment of postmenopausal osteoporosis (abstract). Bone 1998; 23:S175.

Lindsay R, Nieves J, Formica C et al. Randomised controlled study of effect of parathyroid hormone on vertebral bone mass and fracture incidence among postmenopausal women on oestrogen with osteoporosis. Lancet 1997; 550:555.

Lindsay R, Nieves J, Henneman E et al. Subcutaneous administration of the amino-terminal fragment of human parathyroid hormone (1-34): Kinetics and biochemical response in estrogenized osteoporotic patients. J Clin Endocrinol Metab 1993; 77:1535-1539.

Liu T, Clark RK, McDonnell PC et al. TNF-α expression in ischemic neurons. Stroke 1994; 25:1481-1488.

Liu T, McDonnell PC, Young PR et al. IL-1β mRNA expression in ischemic rat cortex. Stroke 1993; 24:1746-1751.

Llinás RR. I of the Vortex: From Neurons to Self. Cambridge, MA: The MIT Press, 2001.

Locklin RM, Khosla S, Turner RT et al. Mediators of the biphasic responses of bone to intermittent and continuously administered parathyroid hormone. J Cell Biochem 2003; 89:180-190.

Loeser RF. Systemic and local regulation of articular cartilage metabolism: Where does leptin fit in the puzzle. Arthritis Rheum 2003; 48:3009-3012.

Lofthouse RA, Davis JR, Frondoza CG et al. Identification of caveoli in normal human osteoblasts. J Bone Joint Surg Br 2001; 83:124-129.

Lomri A, Marie PJ. Changes in cytoskeletal proteins in response to parathyroid hormone and 1,25-dihydroxyvitamin D in human osteoblastic cells. Bone Miner 1990; 10:1-12.

Long JA. The Rise of Fishes. Baltimore: Johns Hopkins University Press, 1995.

Lord BI. The architecture of bone marrow cell populations. Int J Cell Cloning 1990; 8:317-331.

Lord BI, Testa NG, Hendry JH. The relative spatial distributions of CFU$_S$ and CFU$_c$ in the normal mouse femur. Blood 1975; 46:65-72.

Lorget F, Kamel S, Mentaverri R et al. High extracellular calcium concentrations directly stimulate osteoclast apoptosis. Biochem Biophys Res Commun 2000; 268:899-903.

Lowell, Le Roux I, Dunne J et al. Stimulation of human epidermal differentiation by Delta-Notch signalling at the boundaries of stem cell clusters. Curr Biol 2000; 10:491-500.

Lu SS, Ducayen-Knowles M, Dempster DW et al. Effects of parathyroid hormone on gene expression of RANK ligand (RANKL), osteoprotegerin (OPG) and the cognate receptor for PTH in mice. J Bone Miner Res 2001; 16:S426.

Lu W, Shen X, Pavlova X et al. Comparison of pkd1-targeted mutants reveals that loss of polycystin-1 causes cystogenesis and bone defects. Human Mol Genet 2001; 10:2385-2396.

Luckman SP, Hughes DE, Coxon FP et al. Nitorgen-containing bisphosphonates inhibit the mevalonate pathway and prevent post-translational prenylation of GTP-binding proteins, including Ras. J Bone Miner Res 1998; 13:581-589.

Lum L, Beachy PA. The Hedgehog response network: Sensors, switches, and routers. Science 2004; 304:1755-1759.

Luttrell LM, Roudabush FL, Choy EW et al. Activation and targeting of extracellular signal-regulated kinases by beta-arrestin scaffolds. Proc Natl Acad Sci USA 2001; 98:2449-2454.

LY333334 (Teriparatide Injection) Briefing Document. Indianapolis, Indiana: Lilly Reserach Laboratories, Eli Lilly and Companay, Lilly Corporate Center, 46285.

Ma YL, Bryant HU, Zeng Q et al. New bone formation with teriparatide [human parathyroid hormone-(1-34)] is not retarded by long-term pretreatment with alendronate, estrogen, or raloxifene in ovariectomized rats. Endocrinology 2003; 144:2008-2015.

Ma Y, Jee W, Yuan Z et al. Parathyroid hormone and mechanical usage have a synergistic effect in rat tibial diaphyseal cortical bone. J Bone Miner Res 1999; 14:439-448.

Ma YL, Zeng QQ, Cain RL et al. PTH induces similar anabolic effects in lumbar vertebrae of adult ovary-intact and osteopenic ovariectomized rats (abstract). J Bone Miner Res 2000; 15:813.

Ma YL, Cain RL, Halladay DL et al. Catabolic effects of continuous human PTH (1-38) in vivo is associated with sustained stimulation of of RANKL and inhibition of osteoprotegerin in gene-associated bone formation. Endocrinology 2001; 142:4047-4054.

Machwate M, Rodan SB, Rodan GA et al. Sphingosine kinase mediates cyclic AMP suppression of apoptosis in rat periosteal cells. Mol Pharmacol 1998; 54:70-77.

Macica CM, Broadus AE. PTHrP regulates cerebral blood flow and is neuroprotective. Am J Physiol Regulat Integr Comp Physiol 2003; 284:R1019-R1020.

MacLean HE, Guo J, Knight MC et al. The cyclin-dependent kinase inhibitor p57^{Kip2} mediated proliferative actions of PTHrP in chondrocytes. J Clin Invest 2004; 113:1334-1343.

MacLeod RJ, Chattopadhyay N, Brown EM. PTHrP strimulated by the calcium-sensing receptor requires MAP kinase activation. Am J Physiol Endocrinol Metab 2003; 284:E435-E442.

Maeda T, Matsunuma A, Kawane T et al. Simvastatin promotes osteoblast differentiation and mineralization in MC3T3 cells. Biochem Biophys Res Commun 2001; 280:874-877.

Maeder T. A few hundred people turned to bone. The Atlantic Monthly, 1998.

Magne D, Bluteau G, Faucheux C et al. Phosphate is a specific signal for ATDC5 chondrocyte maturation and apopotosis-associated mineralization: Possible implications of apoptosis in the regulation of endochondral ossification (abstract). J Bone Miner Res 2003; 18:1430-1442.

Mahon MJ, Bonacci TM, Divietti P et al. A docking site for G protein βγ subunits on the parathyroid hormone 1 receptor supports signaling through multiple pathways. Mol Endocrinol 2006; 20:36-46.

Mahon MJ, Donowitz M, Yun CC et al. Na$^+$/H$^+$ exchanger regulatory factor 2 directs parathyroid hormone 1 receptor signaling. Nature 2002; 417:858-861.

Mahon MJ, Segre GV. Parathyroid hormone-mediated activation of MAPK is dependent on NHERF2 in PS120 cells (abstract). J Bone Miner Res 2002; 17:S215.

Mahon MJ, Segre GV. NHERF-1 is required for phospholipase-C-dependent calcium influx induced by parathyroid hormone in opossum kidney cells. J Bone Miner Res 2003; 18:S29.

Majeska RJ, Minkowitz B, Bastian W et al. Effects of β-adrenergic blockade in an osteoblast-like cell line. J Orthop Res 1992; 10:379-384.

Makhluf HA, Mueller SM, Mizuno S et al. Age-related decline in osteoprotegerin expression expression by human bone marrow cells cultured in three-dimensional collagen sponges. Biochem Biophys Res Commun 2000; 268:669-672.

Malone ADT, Anderson CT, Temiyasathit S et al. Primary cilia: mechanosensory organelles in bone cells. J Bone Miner Res 2006; 21:S39.

Mannella CA. Our changing views of mitochondria. J Bioener Biomembranes 2000; 32:1-4.

Manolagas SC. Advances in the treatment of osteoporosis. Medscape Endocrinology Journal 1999; 1, (http://www.medscape.com).

Manolagas SC. Birth and death of bone cells: Basic regulatory mechanisms and implications for the pathogenesis and treatment of osteoporosis. Endocrine Revs 2000; 21:115-137.

Manolagas SC, Kousteni S, Jilka RL. Sex steroids and bone. Recent Prog Horm Res 2002; 57:385-409.

Manolagas SC, Weinstein RS, Bellido T et al. Activators of nongenomic estrogen-like signaling (ANGELS): A novel class of small molecules with bone anabolic properties (abstract). J Bone Miner Res 1999; 14:S180.

Mansfield K, Teixeira CC, Adams CS et al. Phosphate ions mediate chondrocyte apoptosis through a plasma membrane transporter mechanism. Bone 2001; 28:1-8.

Mansfield K, Rajpurohit R, Shapiro IM. Extracellular phosphate ions cause apoptosis of terminally differentiated epiphyseal chondrocytes. J Cell Physiol 1999; 179:276-286.

Maor GA, Rochwerger M, Segev Y et al. Leptin acts as a skeletal growth factor on chondrocytes of skeletal growth centers. J Bone Miner Res 2002; 17:1034-1043.

Marie PJ. Strontium ranelate: A physiological approach for optimizing bone formationand resorption. Bone 2006; 38:S10-S14.

Marie PJ, Ammann P, Boivin G et al. Mechanism of action and the therpeutic potential of strontium in bone. Calcif Tissue Int 2001; 69:121-129.

Marie PJ, Jones D, Vico L et al. Osteobiology, strain, and microgravity: Part 1. Studies at the cellular level. Calcif Tissue Int 2000; 67:2-9.

Marenholz I, Zirra M, Fischer DF et al. Identification of human epidermal differentiation complex (EDC)-encoded genes by subtractive hybridization of entire YACs to a gridded keratinocyte cDNA library. Genome Res 2001; 11:341-355.

Maritz FJ, Conradie MM, Hulley PA et al. Effect of statins on bone mineral density and bone histomorphometry in rodents. Arterioscler Thromb Vasc Biol 2001; 21:1636-1641.

Mark AL, Shaffer RA, Correia ML et al. Contrasting blood pressure effects of obesity in leptin-deficient ob/ob mice and agouti yellow obese mice. J Hypertens 1999; 17:1949-1953.

Marotti G. The structure of bone tissues and the cellular control of their deposition. Ital J Anat Embryol 1996; 101:25-79.

Marotti G. The osteocyte as a wiring transmission system. J Musculoskelet Neuron Interact 2000; 1:133-136.

Marijanovic I, Kronenberg MS, Erceg I et al. Dlx3,5 and 6 may have redundant effects on one development. J Bone Miner Res 2002; 17:S443.

Martin RB. Toward a unifying theory of bone remodeling. Bone 2000a; 26:1-6.

Martin RB. Does osteocyte formation cause the nonlinear refilling of osteons? Bone 2000b; 26:71-78.

Martin RB. Is all cortical bone remodeling initiated by microdamage? Bone 2002; 30:8-13.

Martin RB. Fatigue microdamage as an essential element of bone mechanics and biology. Calcif Tissue Int 2003; 73:101-107.

Martin RB, Burr DB, Sharkey NA. Skeletal Tissue Mechanics. New York: Springer Verlag, 1998.

Martin TJ. Does bone resorption inhibition affect the anabolic response to parathyroid hormone? Trends Endocrinol Metab 2004; 15:49-50.

Martin TJ. Osteoblast-derived PTHrP is a physiological regulator of bone formation. J Clin Invest 2005; 115:2322-2324.

Martin TJ, Romas E, Gillespie MT. Interleukins in the control of osteoclast differentiation. Crit Rev Eukaryot Gene Expr 1998; 8:107-123.

Martin-Ventura JL, Ortego M, Esbrit P et al. Possible role of parathyroid hormone-related protein as a proinflammatory cytokine in atherosclerosis. Stroke 2003; 34:1783-1789.

Marx J. How cells endure low oxygen. Science 2004; 303:1454-1456.

Mashiba T, Hirano T, Turner CH et al. Suppressed bone turnover by bisphosphonates increases microdamage accumulation and reduces some biomechanical properties in dog rib. J Bone Miner Res 2001; 15:613-620.

Mason DJ. Glutamate signalling and its potential applications to tissue engineering of bone. Eur Cells Materials 2004; 7:12-26.

Mason DJ, Suva LJ, Genever PG et al. Mechanically regulated expression of a neural glutamate transporter in bone: A role for excitatory amino acids as osteotropic agents? Bone 1997; 20:199-205.

Mason TM, Lord BI, Hendry JH. The relative spatial distributions of CFU-S and in vitro CFC in femora of mice of different ages. Brit J Haematol 1989; 73:455-461.

Massfelder T, Dann P, Wu TL et al. Opposing mitogenic and anti-mitogenic actions of parathyroid hormone-related protein in vascular smooth muscle cells: A critical role for nuclear targeting. Proc Natl Acad Sci USA 1997; 94:13630-13635.

Massfelder T, Fiaschi-Taesch N, Stewart AF et al. Parathyroid hormone-related peptide—A smooth muscle tone and proliferation regulatory protein. Curr Opin Nephrol Hypertens 1998; 7:27-32.

Massfelder T, Helwig JJ. The parathyroid hormone-related protein system: More data but more insolved questions. Curr Opin Nephrol Hypertens 2003; 12:35-42.

Matsuda S, Rouault JP, Magaud JP et al. In search of a function for the TIS21/PC3/BTG1/TOB family. Febs Lett 2001; 497:67-72.

Matsumoto T, Shiraki M, Nakamura T et al. Daily nasal spray of hPTH(1-34) for 3 months increases bone mass in osteoporotic subjects. J Bone Miner Res 2004; 19:S44.

Matthews JL, Martin JH. Intracellular transport of calcium and its relationship to homeostasis and mineralization. Am J Med 1971; 50:589-597.

Maudsley S, Pierce KL, Zamah AM et al. The beta(2)-adrenergic receptor mediates extracellular signal-regulated kinase activation via assembly of a multi-receptor complex with the epidermal growth factor receptor. J Biol Chem 2000; 275:9572-9580.

Mayr-Wohlfart U, Waltenberger J, Hausser H et al. Vascular endothelial growth factor stimulates chemotactic migration of primary human osteoblasts. Bone 2002; 30:472-477.

Mazzali M, Kipari T, Ophascharoensuk V et al. Osteopontin—A molecule for all seasons. Q J Med 2002; 95:3-13.

McCauley LK, Koh AJ, Beecher CA et al. PTH/PTHrP receptor is temporally regulated during osteoblasts differentiation and is associated with collagen synthesis. J Cell Biochem 1996; 61:638-647.

McCauley LK, Koh AJ, Beecher CA et al. Proto-oncogene c-fos is transcriptionally regulated by parathyroid hormone (PTH) and PTH-related protein in a cyclic adenosine monophosphate-dependent manner in osteoblastic cells. Endocrinology 1997; 138:5427-5433.

McCauley LK, Koh-Paige H, Chen CC et al. Parathyroid hormone stimulates fra-2 expression in osteoblastic cells in vitro and in vivo. Endocrinology 2001; 142:1975-1981.

McClung MR, Lewiecki EM, Bolognese MA et al. AMG 162 increases bone mineral density (BMD) within 1 month in postmenopausal women with low BMD. J Bone Miner Res 2004; 19:S20.

McKee MD, Murray TM. Binding of intact parathyroid hormone to chicken renal plasma membranes evidence for a second binding site with carboxyl terminal specificity. Endocrinology 1985; 117:1930-1939.

McKee MD, Nanci A. Osteopontin: An interfacial extracellular matrix protein in mineralized tissues. Connective Tissue Res 1996; 35:197-205.

Medjkane S, Novikov E, Versteege I et al. The tumor suppressor hSNF5/INI1 modulates cell growth and actin cytoskeleton organization. Cancer Res 2004; 64:3406-3413.

Mehta NM, Gilligan JP, Stern B et al. Biological activity of recombinant PTH analog UGL7841(abstract). J Bone Miner Res 2002; 17:S273.

Mehta N, Stern W, Sturmer A et al. Oral delivery of PTH analogs by a solid dosage formulation (abstract). J Bone Miner Res 2001; 16:S540.

Meir CR, Schlienger RC, Kraenzlin ME et al. HMG-CoA reductase inhibitors and the risk of fractures. JAMA 2000; 283:3205-3210.

Meir-Vismara E, Walker N, Vogel A. Single cilia in the articular cartilage of the cat. Expl Cell Biol 1979; 47:161-171.

Meleti Z, Shapiro IM, Adams CS. Inorganic phosphate induces apoptosis of osteoblast-like cells in culture. Bone 2000; 27:359-366.

Menon GK, Elias PM. Ultrastructural localization of calcium in psoriatic and normal human epidermis. Arch Dermatol 1991; 127:57-63.

Mentaverri R, Kamel S, Wattel A et al. Regulaion of bone resorption and osteoclast survival by nitric oxide: Possible involvement of NMDA receptor. J Cell Biochem 2003; 88:1145-1156.

Menton DN, Simmons DJ, Chang SL et al. From bone lining cells to osteocyte—An SEM study. Anat Rec 1984; 209:29-39.

Merle B, Itzstein C, Delmas PD et al. NMDA glutamate receptors are expressed by osteoclast precursors and involved in the regulation of osteoclastogenesis. J Cell Biochem 2003; 90:424-436.

Metz LN, Martin RB, Turner AS. Histomorphometric analysis of the effects of osteocyte density on osteonal morphology and remodeling. Bone 2003; 33:753-759.

Meunier PJ. Evidence-based medicine and osteoporosis: A comparsion of fracture risk reduction data from osteoporosis randomized clinical trials. Int J Clin Pract 1999; 53:122-129.

Meunier PJ, Roux C, Seeman E et al. The effects of strontium ranelate on the risk of vertebral fracture in women with postmenopausal osteoporosis. New Engl J Med 2004; 350:459-468.

Miao D, He B, Tong XK et al. Conditional knockout of PTHrP in osteoblasts leads to premature osteoporosis (abstract). J Bone Miner Res 2002; 17:S138.

Midura RJ, Su X, Morcuende JA et al. Parathyroid hormone rapidly stimulates hyaluronan synthesis by periosteal osteoblasts in the tibial diaphysis of the growing rat. J Biol Chem 2003; 278:51462-51468.

Midy V, Plouet J. Vasculotropin/vascular endothelial growth factor induces differentiaition in cultured osteoblasts. Biochem Biophys Res Commun 1994; 199:380-386.

Miki T, Nakatsuka K, Naka H et al. Effect and safety of intermittent weekly administration of human parathyroid hormone 1-34 in patients with primary osteoporosis evaluated by histomorphometry and microstructural analysis of the trabecular bone and after 1 year of treatment. J Bone Miner Metab 2004; 22:569-576.

Miles RR, Sluka JP, Halladay DL et al. Parathyroid hormone (hPTH-(1-38) stimulates the expression of UBP41, an ubiquitispecific protease, in bone. J Cell Biochem 2002; 85:229-242.

Miles RR, Sluka JP, Santerre RF et al. Dynamic regulation of RGS-2 in bone: Potential new insights into parathyroid hormone signaling mechanisms. Endocrinology 2000; 141:28-36.

Miller LJ, Chacko R. The role of cholesterol and statins in Alzheimer's disease. Ann Pharmacother 2004; 38:91-98.

Miller MD, Bushman FD. Ini1 for integration? Curr Biol 1995; 5:368-370.

Miller WE, McDonald PH, Cai SF et al. Identification of a motif in the carboxyl terminus of β-arrestin 2 responsible for activation of JNK3. J Biol Chem 2001; 276:27770-2777.

Minamino T, Kurihara H, Takahashi M et al. Endothelin-converting enzyme expression in the rat vascular injury model and human coronary atherosclerosis. Circulation 1997; 95:221-230.

Mineo C, Anderson RGW. Potocytosis. Histochem Cell Biol 2001; 116:109-118.

Minina E, Wenzel HM, Kreschel C et al. BMP and Ihh/PTHrP signaling interact to coordinate chondrocyte proliferation and differentiation. Development 2001; 128:4523-4534.

Miranti CK, Brugge JS. Sensing the environment: A historical perspective on intergrin signal transduction. Nature Cell Biol 2002; 4:83-90.

Misof BM, Roschger P, Cosman F et al. Effects of intermittent parathyroid hormone administration on bone mineralization density in iliac crest biopsies from patients with osteoporosis: A paired study before and after treatment. J Clin Endocrinol Metab 2003; 88:1150-1156.

Mitchell DR. The evolution of eukaryotic cilia and flagella as motile and sensory organelles. Origins and Evolution of Eukaryotic Endomembranes and Cytoskeleton. Austin: Landes Bioscience, 2006:1-11.

Mithal EM, Brown EM. An overview of extracellular calcium homeostasis and the roles of the CaR in parathyroid and C-cells. In: Chattopadhyay N, Brown EM, eds. Calcium-Sensing Receptors. Boston: Kluwer Academic Publishers, 2003.

Miyakoshi N, Kasukawa Y, Linkhart TA et al. Evidence that anabolic effects of PTH on bone require IGF-I in growing mice. Endocrinology 2001; 142:49-56.

Myakoshi N, Quin X, Kasukawa Y et al. Systemic administration of insulin-like growth factor (IGF)-binding protein-4 (IGFBP-4) increases bone formation parameters in mice by increasing IGF bioavailability via an IGFBP-4 protease-dependent mechanism. Endocrinology 2001; 142:2641-2648.

Miyamoto S, Teramoto H, Gutkind JS et al. Integrins can collaborate with growth factors for phosphorylation of receptor tyrosine kinases and MAP kinase activation; roles of integrin aggregation and occupancy of receptors. J Cell Biol 1996; 135:1633-1642.

Moerman EJ, Teng K, Lipschitz DA et al. Aging activates adipogenic and suppresses osteogenic programs in mesenchymal marrow stroma/stem cells: The role of PPAR-γ2 transcription factor and TGF-β / BMP signaling pathways. Aging Cell 2004; 3:379-389.

Moghadasian MH, Frohlich JJ. Statins and bones. Can Med Assoc J 2001; 164:803-805.

Mohammad KS, Guise TA. Mechanisms of osteoblastic metastases: Role of endothelin-1. Clin Orthop Relat Res 2003; 415(Suppl):S67-S74.

Mohammad KS, Guise TA, Chirgwin JM. PTHrP stimulates new bone formation by molecular mimicry of endothelin-1 (abstract). J Bone Miner Res 2003; 18:S26.

Mohan S, Baylink DJ. IGF system components and their role in bone metabolism. In: Rosenfeld RG, Roberts Jr CT, eds. The IGF System: Molecular Biology, Physiology, and Clincal Applications. Totowa, New Jersey: Humana Press, 1999:457-496.

Mohan S, Kutilek S, Zhang C et al. Comparison of bone formation responses to parathyroid hormone(1-34), (1-31), and 2-34 in mice. Bone 2000; 27:471-478.

Montagnani A, Gonnelli S, Cepollaro C et al. Effect of simvastatin on bone mineral density and bone turnover in and bone density in hypercholesterolemic postmenopausal women: A 1-year longitudinal study. Bone 2003; 32:427-433.

Montero A, Okada Y, Tomita M et al. Disruption of the fibroblast growth factor-2 gene results in decreased bone mass and bone formation. J Clin Invest 2000; 105:1085-1093.

Moonga BS, Dempster DW. Effects of peptide fragments of protein kinase C on isolated rat osteoclasts. Exp Physiol 1998; 83:717-725.

Moonga BS, Sun L, Corisdeo S et al. Novel mechanistic insights into the regulation of mature osteoclasts by osteoprotegerin (OPG) and osteoprotegerin-ligand (OPGL). Evidence for resorption stimulation through inhibition of extracellular Ca^{2+} sensing (abstract). J Bone Miner Res 1998; 14:S363.

Moore KA, Lemischka IR. "Tie-ing" down the hematopoietic niche. Cell 2004; 118:139-143.

Moore KA, Lemischka IR. Stem cells and their niches. Science 2006; 311:1880-1885.

Moore RE, Smith CK, Bailey CS et al. Characterization of β-adrenergic receptors on rat and human osteoblast-like cells and demonstration that β-receptor agonists can stimulate bone resorption in organ culture. Bone Miner 1993; 23:301-315.

Moore TL, Krishnan VG, Onyia JE et al. A mechanism for bone anabolic activity of PTH through Cbfa1/Osf-2 (abstract). J Bone Miner Res 2000; 15:S158.

Morfis M, Christopoulos A, Sexton PM. RAMPs: 5 years on, where to now? Trends Pharmacol Sci 2003; 24:596-601.

Morgan Stanley. NPS Pharmaceutical PREOS approvable letter worse than expected. Equity Research North America, 2006.

Mori H, Kitazawa R, Mizuki S et al. RANK ligand, RANK, and OPG expression in type II collagen-induced arthritis mouse. Histochem Cell Biol 2002; 117:283-292.

Mori S, Burr DB. Increased intrcortical remodeling following fatgue damage. Bone 1993; 14:103-109.

Morley P. Delivery of parathyroid hormone for the treatment of osteoporosis. Exp Opin Drug Deliv 2005; 2:993-1002.

Morley P, Whitfield JF, Willick G. Design and application of parathyroid hormone analogues. Curr Med Chem 1999; 11:1095-1106.

Morley P, Whitfield JF, Willick GE et al. The effect of monocyclic and bicyclic analogs of human parathyroid hormone hPTH-(1-31)NH$_2$ on bone formation and mechanical strength in ovariectomized rats. Calcif Tissue Int 2001; 68:95-101.

Morley P, Whitfield JF, Vanderhyden BC et al. A new, nongenomic estrogen action: The rapid release of intracellular calcium. Endocrinology 1992; 131:1305-1312.

Morony S, Capparelli C, Lee R et al. A chimeric form of osteoprotegerin inhibits hypercalcemia and bone resoption induced by IL-1β, TNF-α, PTH, PTHrP, and 1,25(OH)$_2$D3. J Bone Miner Res 2002; 14:1478-1485.

Morroni M, DeMatteis R, Palumbo C et al. In vivoleptin expression in cartilage and bone cells of growing rats and adult humans. J Anat 2004; 205:291-296.

Mosekilde L. Osteoporosis: Mechanisms and models. In: Whitfield JF, Morley P, eds. Anabolic Treatments for Osteoporosis. Boca Raton: CRC Press, 1997:31-58.

Mosekilde L, Thomsen JS, McOsker JE. No loss of biomechanical effects after withdrawal of short-term PTH treatment in an aged, osteopenic, ovariectomized rat model. Bone 1997; 20:429-437.

Moss ML, Cowin SC. Mechanosensory mechanisms in bone. In: Lanza R, Chick W, eds. Principles of Tissue Engineering. Austin: R.G. Landes Co./Landes Bioscience, 1997:645-659.

Muchardt C, Yaniv M. The mammalian SWI/SNF complex and the control of cell growth. Semin Cell Devel Biol 1999; 10:189-195.

Muchardt C, Yaniv M. When the SWI/SNF complex remodels the cell cycle. Oncogene 2001; 20:3067-3075.

Mundy G. Pathogenesis of osteoporosis and challenges for drug delivery. Adv Drug Del Res 2000; 42:165-173.

Mundy G, Garrett R, Harris S et al. Stimulation of bone formation in vitro and in rodents by statins. Science 1999; 286:1946-1949.

Murakami T, Yamamoto M, Ono K et al. Transforming growth factor-β1 increases mRNA levels of osteoclastogenesis inhibitory factor in osteoblastic /stromal cells and inhibits the survival of of murine osteoblast-like cells. Biochem Biophys Res Commun 1998; 252:747-752.

Murray EJB, Bentley GV, Grisanti MS et al. The ubiquitin-proteasome system and cellular proliferation and regulation in osteoblastic cells. Exp Cell Res 1998; 242:460-469.

Murray EJB, Murray SS. Lovastatin stimulates rather than inhibits proteasome in vitro and in osteoblasts (abstract). J Bone Miner Res 2001; 16:S206.

Murrills RJ, Bhat BM, Bodine PVN et al. Dissociation of functional response from cyclic AMP stimulation in a substituted and C-terminally truncated PTH peptide. J Bone Miner Res 2002; 17:SA432.

Nadal A, Ropero AB, Laribi O et al. Nongenomic actions of estrogens and xenoestrogens by binding at a plasma membrane recepor unrelated to estrogen receptor α and estrogen receptor β. Proc Natl Acad Sci USA 2000; 97:11603-11608.

Nadal A, Rovira JM, Laribi O et al. Rapid insulinotropic effect of 17β-estradiol via a plasma membrane receptor. FASEB J 1998; 12:1341-1348.

Nakade O, Takahashi K, Takuma T et al. Effect of extracellular calcium on the gene expression of bone morphogenic protein-2 and -4 of normal human bone cells. J Bone Miner Metab 2001; 19:13-19.

Nakajima A, Shimoji N, Shiomi K et al. Mechanisms for the enhancement of fracture healing in rats treated with intermittent low-dose human parathyroid hormone (1-34). J Bone Miner Res 2002; 17:2038-2047.

Nakajima M, Ejiri S, Tanaka M et al. Effects of intermittent administration of human parathyroid hormone (1-34) on mandibular condyle of ovariectomized rats. J Bone Miner Metab 2000; 18:9-17.

Nakajima R, Inada H, Koike T et al. Effects of leptin to cultured growth plate chondrocytes. Horm Res 2003; 60:91-98.

Nakamura K, Nishimura J, Hirano K et al. Hydroxyfasudil, an active metabolite of fasudil hydrochloride, relaxes the rabbit basilar artery by disinhibition of myosin light chain phosphatase. J Cereb Blood Flow Metab 2001; 21:876-885.

Nakashima K, Zhou X, Kunkel G et al. The novel zinc finger-containing transcription factor osterix is required for osteoblast differentiation and bone formation. Cell 2002; 108:17-29.

Nakayama T, Ohtsuru A, Enomoto H et al. Coronary atherosclerotic smooth muscle cells over-express human parathyroid hormone related peptides. Biochem Biophys Res Commun 1994; 200:1028-1035.

Nakorn TN, Traver D, Weissman IL et al. Myoerythroid-restricted progenitors are sufficient to confer radioprotection and provide the majority of day 8 CFU-S. J Clin Invest 2002; 109:1579-1585.

Nauli SM, Alenghat FJ, Luo Y et al. Polycystins 1 and 2 mediate mechnosensation in the primary cilium of kidney cells. Nature Genet 2003; 33:129-137.

Nauli SM, Zhou J. Polycystins and mechanosensation in renal and nodal cilia. Bio Essays 2004; 26:844-856.

Naveiras O, Daley GQ. Stem cells and their niche: A matter of fate. Cell Mol Life Sci 2006; 63:760-766.

Neer RM, Arnaud CD, Zanchetta JR et al. Effect of parathyroid hormone (1-34) on fractures and bone mineral density in postmenopausal women with osteoporosis. New Engl J Med 2001; 344:1434-1441.

Neer R, Hayes A, Rao A et al. Effects of parathyroid hormone, alendronate, or both on bone density in osteoporotic postmenopausal women (abstract). J Bone Miner Res 2002; 17:S135.

Neer R, Slovik D, Daly M et al. Treatment of postmenopausal osteoporosis with daily parathyroid hormone plus calcitriol. In: Christiansen C, Overgaard K, eds. Osteoporosis. Copenhagen: Osteo press, 1991:1314-1317.

Nemere I, Farach-Carson MC. Membrane receptors for steroid hormones: A case for specific cell surface binding sites for vitamin D metabolites and estrogens. Biochem Biophys Res Commun 1998; 248:443-449.

Nemeth EF, Delmar EG, Heaton WL et al. Calcilytic compounds: Potent and selective Ca^{2+} receptor antagonists that stimulate secretion of parathyroid hormone. J Pharmacol Exp Ther 2001; 299:323-331.

Nemeth EF, Fox J. Compounds acting on the parathyroid calcium receptor as novel therapies for hyperparathyroidism or osteoporosis. In: Chattopadhyay N, Brown EM, eds. Calcium-Sensing Receptor. Boston: Kluwer Academic Publishers, 2003:173-202.

Netelenbos C. Osteoporosis: Intervention options. Maturitas 1998; 30:235-239.

Netter FH. Atlas of Human Anatomy. 2nd ed. East Hanover, NJ: Novartis, 1997.

Neugebauer W, Barbier JR, Sung WL et al. Solution structure and adenylyl cyclase stimulating activities of C-terminal truncated human parathyroid hormone analogues. Biochemistry 1995; 34:8835-8842.

Nguyen T, Sherratt PJ, Pickett CB. Regulatory mechanisms controlling gene expression mediated by the antioxidant response element. Annu Rev Pharmacol Toxicol 2003; 43:233-260.

Ni CY, Murphy MP, Golde TE et al. γ-Secretase cleavage and nuclear localization of ErbB-4 receptor tyrosine kinase. Science 2002; 294:2179-2181.

Nickoloff BJ, Qin JZ, Chuturvedi V et al. Jagged-1 mediated activation of notch signaling induces complete maturation of human keratinocytes through NF-κB and PPARγ. Cell Death Differ 2002; 9:842-855.

Nickoloff BJ, Schröder JM, von den Driesch P et al. Is psoriasis a T-cell disease? Exp Dermatol 2000; 9:359-375.

Nicolas M, Wolfer A, Raj K et al. Notch 1 functions as a tumor suppressor in mouse skin. Nat Gen 2003; 33:416-421.

Nicolella DP, Lankford J. Microstructural strain near osteocyte lacuna in cortical bone in vitro. J Musculoskelet Neuronal Interact 2002; 2:261-263.

Nielsen LB, Pedersen FS, Pedersen L. Expression of type III sodium-dependent phosphate transporters/ retroviral receptors mRNAs during osteoblast differentiation. Bone 2001; 28:160-166.

Nishida S, Endo N, Yamagiwa H et al. Number of osteoprogenitor cells in human bone marrow markedly decreases after skeletal maturation. J Bone Miner Metab 1999; 17:171-177.

Nishida S, Yamaguchi A, Tanizawa T et al. Increased bone formation by intermittenet parathyroid hormone administration is due to the stimulation of proliferation and differentiation of osteoprogenitor cells in bone marrow. Bone 1994; 15:717-723.

Nishimura G, Kim OK, Sato S et al. Ischiospinal dysostosis with cystic kidney disease: Report of two cases. Clin Dysmorphol 2003; 12:101-104.

Nissenson RA. Parathyroid hormone-related protein. Rev Endo Metab Dis 2000; 1:343-352.

Noble BS. Osteocyte death: Its biological significance (abstract). J Bone Miner Res 2000; 15:823.

Noble BS, Reeve J. Osteocyte function, osteocyte death and bone fracture resistance. Mol Cell Endocrinol 2000; 159:7-13.

Noda M, Takuwa Y, Katoh T et al. Mechanical force regulation of vascular parathyroid hormone-related peptide expression. Kidney Int Suppl 1996; 55:S154-S155.

Nomura S, Takano-Yamamoto T. Molecular events caused by mechanical stress in bone. Matrix Biol 2000; 19:91-96.

Nusse R. Making hear or tail of Dickkopf. Nature 2001; 411:255-256.

Nusse R. The Wnt page. 2005, ⟨http://www.stanford.edu/_rnusse/wntwindow⟩.

O'Brien CA, Kern B, Gubrij I et al. Cbfa1 does not regulate RANKL gene activity in stromal/osteoblastic cells. Bone 2002; 30:453-462.

Ohishi K, Katayama N, Shiku H et al. Notch signaling in hematopoiesis. Semin Cell Devel Biol 2003; 14:143-150.

Ohnaka K, Shimoda S, Nawata H et al. Pitavastatin enhanced BMP-2 and osteocalcin expression by inhibition of rho-associated kinase in human osteoblasts. Biochem Biophys Res Commun 2001; 287:337-342.

Okabe M, Graham A. The origin of the parathyroid gland. Proc Natl Acad Sci USA 2004; 101:17716-17719.

Okada-Ban M, Thiery JP, Jouanneau J. Fibroblast growth factor-2. Int J Biochem Cell Biol 2000; 32:263-267.

Okamoto T, Schlegel A, Scherer PE et al. Caveolins, a family of scaffolding proteins for organizing "preassembled signaling complexes" at the plasma membrane. J Biol Chem 1998; 273:5419-5422.

Okazaki R, Inoue D, Shibata M et al. Estrogen promotes ealy osteoblast differentiation and inhibits adipocyte differentiation in mouse bone marrow stromal cell lines that express estrogen receptor (ER) α or β. Endocrinology 2002; 143:2349-2356.

Okuyama R, Nguyen BC, Talora C et al. High commitment of embryon-ic keratinocytes to terminal differentiation through a Notch1-caspase 3 regulatory me-anism. Dev Cell 2004; 6:551-562.

Olsnes S, Klingenberg O, Wiedlocha A. Transport of exogenous growth factors and cytokines to the cytosol and to the nucleus. Physiol Rev 2003; 83:163-182.

Olszak IT, Poznansky MC, Evans RH et al. Extracellular calcium elicits a chenmokinetic response from monocytes in vitro and in vivo. J Clin Invest 2000; 105:1299-1305.

Onyia JE, Dow E, Adams C et al. Gene array analysis of the bone effects of raloxifene and alendronate show that alendronate strongly inhibits the expression of bone formation marker genes (abstract). J Bone Miner Res 2002; 17:S157.

Onyia JE, Gelbert I, Zhang M et al. Analysis of gene expression by DNA microarray reveals novel clues to the mechanism of the catabolic and anabolic actions of PTH in bone (abstract). J Bone Miner Res 2001a; 16:S227.

Onyia JE, Ma YL, Galbreath E et al. ADAMTS-1: A cellular disinetgrin and metalloprotease with thrombospondin motifs is essential for normal bone growth and PTH regulated bone metabolism (abstract). J Bone Miner Res 2001b; 16:S158.

Oniya JE, Miles RR, Halladay DL et al. In vivo demonstration that parathyroid hormone (hPTH 1-38) inhibits the expression of osteoprotegerin (OPG) in bone with the kinetics of an immediate early gene. J Bone Miner Res 2000; 15:863-871.

Ontiveros CS, McCauley LK, McCabe LR. Gravitational force alters intracellular signaling and gene transactivation. J Bone Miner Res 2002; 17:S308.

Opas EE, Gentile MA, Rossert JA et al. Parathyroid hormone and prostaglandin E_2 preferentially increase luciferase levels in bone of mice harboring a luciferase transgene transgene controlled by the elements of the pro-α1(I) collagen promoter. Bone 2000; 27:27-32.

Orloff JJ, Kats Y, Urena P et al. Further evidence for a novel receptor for amino-terminal parathyroid hormone-related protein on keratinocytes and squamous carcinoma cell lines. Endocrinology 1995; 136:3016-3023.

Ornitz DM, Itoh N. Fibroblast growth factors. Genome Biol 2001; 2, (http://genomebiology.com/2001/2/3/reviews/3995.1).

O'Rourke MF, Staessen JA, Vlachopoulos C et al. Clinical applications of arterialstiffness. Am J Hypertens 2002; 15:426-444.

Orwoll E, Scheele WH, Paul S et al. The effect of parathyroid hormone (1-34) therapy on bone density in men with osteoporosis. J Bone Miner Res 2003; 18:9-17.

Otero M, Lago R, Lago F et al. Leptin from fat to inflammation: Old questions and new insights. FEBS Lett 2005; 579:295-301.

Ott SM. Sclerostin and Wnt signaling—The pathway to bone strength. J Clin Endocrinol Metab 2005; 90:6741-6743.

Otto F, Lübbert M, Stock M. Upstream and downstream targets of RUNX proteins. J Cell Biochem 2003, (Online 11 Mar 2003).

Oyajobi BO, Anderson DM, Traianedes K et al. Therapeutic efficacy of a soluble receptor activator of nuclear factor κB-IgG Fc fusion protein in suppressing bone resorption and hypercalcemia in a model of humoral hypercalcemia of malignancy. Cancer Res 2001; 61:2572-2578.

Ozeki S, Ohtsuru A, Seto S et al. Evidence that implicates the parathyroid hormone related peptide in vascular stenosis. Increased gene expression in the intima of injured carotid arteries and human restenosis coronoar lesions. Arterioscler Thromb Vasc Biol 1996; 16:565-575.

Palumbo C, Palazini S, Zaffe D et al. Osteocyte differentiation in the tibia of newborn rabbit: An ultrastructural study of the formation of cytoplasmic processes. Acta Anat 1990; 137:350-358.

Pan X, Song Z, Zhai L et al. Chromatin-remodeling factor INI1/hSNF5/BAF47 is involved in activation of the colony stimulating factor promoter. Mol Cells 2005; 20:183-188.

Panegyres PK. The functions of the amyloid precursor gene. Revs Neurosci 2001; 12:1-39.

Parfitt AM. Osteoclast precursors as leukocytes: Importance of the area code. Bone 1998; 23:491-494.

Parfitt AM. BMU origination and progression: Relationship to targeted and nontargeted remodeling (abstract). J Bone Miner Res 2000a; 15:823.

Parfitt AM. The mechanism of coupling: A role for the vasculature. Bone 2000b; 26:319-323.

Parfitt AM. The bone remodeling compartment: A circulatory function for bone lining cells. J Bone Miner Res 2001; 16:1583-1585.

Parfitt AM. Misconceptions (2): Turnover is always higher in cancellous than cortical bone. Bone 2002; 30:807-809.

Parfitt AM. What is the normal rat of bone remodeling? Bone 2004; 35:1-3.

Parhami F, Morrow AD, Balucan J et al. Lipid oxidation products have opposite effects on calcifying vascular cell and bone cell differentiation. Apossible explanation for the paradox of arterial calcification in osteoporotic patients. Arterioscler Thromb Vasc Biol 1997; 17:680-687.

Park HY, Kwon HM, Lim HJ et al. Potential role of leptin in angiogenesis: Leptin induces endothelial cell proliferation and expression of matrix metalloproteinases in vivo and in vitro. Exp Mol Med 2001; 33:95-102.

Paschalis EP, Glass EV, Donley DW et al. Bone mineral and collagen quality in iliac crest biopsies of patients given teriparatide: New results from fracture prevention trial. J Clin Endocrinol Metab 2005; 90:4644-4649.

Pasco JA, Hemnry MJ, Sanders KM et al. β-Adrenergic blockers reduce the risk of fracture partly by increasing bonemineral density: Geelong osteoporosis study. J Bone Miner Res 2004; 19:19-34.

Pasco JA, Kotowicz MA, Henry MJ et al. Statin use, bone mineral density, and fracture risk. Geelong osteoporosis study. Arch Intern Med 2002; 162:537-540.

Patton AJ, Genever PG, Birch MA et al. Expression of an N-methyl-D-aspartate receptor by human and rat osteoblasts and osteoclasts suggests a novel glutamate signaling pathway in bone. Bone 1998; 22:645-649.

Pavalko FM, Chen NX, Turner CH et al. Fluid shear-induced mechaniciical signaling in MC3T3-E1 osteoblasts requires cytoskeleton-integrin interactions. Am J Physiol 1998; 275:C1591-1601.

Pavalko FM, Norvell SM, Burr DB et al. A model for mechanotransduction in bone cells: The load-bearing mechanosomes. J Cell Biochem 2003; 88:104-112.

Pavo I, Laszlo F, Morschl E et al. Raloxifene, an estrogen-receptor modulator, prevents decreased constitutive nitric oxide and vasoconstriction in ovariectomized rats. Eur J Pharmacol 2000; 410:101-104.

Pazour GJ, Witman GB. The vertebrate primary cilium is a sensory organelle. Curr Opin Cell Biol 2003; 15:1-6.

Pearson D, Miller CG, eds. Clinical Trials in Osteoporosis. London: Springer-Verlag, 2002.

Pearson G, Robinson F, Beers T et al. Mitogen-activated protein (MAP)kinase pathways: Regulation and physiological functions. Endocrine Rev 2001; 22:153-183.

Pedrazzini T, Prolong F, Grouzmann E. Neuropeptide Y: The universal soldier. Cell Mol Life Sci 2003; 60:350-377.

Peet NM, Grabowski PS, Laketic-Ljubojevic I et al. The glutamate receptor antagonist MK801 modulates bone resorption in vitro by a mechanism predominantly involving osteoclast differentiation. FASEB J 1999; 13:2179-2185.

Péhu M, Policard A, Dufort A. L'Ostopétrose ou maladie des os marmoréens. La Presse Médicale 1931; 53:999-1003.

Pereira R, Econonmides AN, Canalis E. Bone morphogenic proteins induce gremlin, a protein that limits their activity in osteoblasts. Endocrinology 2000a; 141:4558-4563.

Pereira R, Rydziel S, Canalis E. Bone morhpogenic protein-4 regulates its own expression in cultured osteoblasts. J Cell Physiol 2000b; 182:239-246.

Perris AD, MacManus JP, Whitfield JF et al. Parathyroid glands and mitotic stimulation in rat bone marrow after hemorrhage. Am J Physiol 1971; 220:773-778.

Perris AD, Whitfield JF. Calcium and the control of mitosis in the mammal. Nature 1967; 216:1350-1351.

Perris AD, Whitfield JF. Calcium homeostasis and erythropoietic control in the rat. Can J Physiol Pharmacol 1971; 49:22-35.

Perris AD, Whitfield JF, Rixon RH. Stimulation of mitosis in bone marrow and thymus of normal and irradiated rats by divalent cations and parathyroid extract. Radiation Res 1967; 32:550-563.

Peterson CL. Chromatin remodelling enzmes: Taming the machines. EMBO Reports 2002; 3:319-322.

Petsko GA, Ringe D. Protein Structure and Function. London: New Science Press Ltd, 2004.

Pfeilschrifter J, Eberhardt W, Huwiler A. Nitric oxide and mechanisms of redox signaling: Matrix and matrix-metabolizing enzymes as prime oxide targets. Eur J Pharmacol 2001; 429:279-286.

Pfeilschrifter J, Laukhuf F, Müller-Beckmann B et al. Parathyroid hormone increases the concentration of insulin-like growth factor-1 and transforming growth factor beta 1 in rat bone. J Clin Invest 1995; 96:767-774.

Pfister MF, Lederer E, Forgo J et al. Parathyroid hormone-dependent degradation of type II Na^+/Pi cotransporters. J Biol Chem 1997; 272:20125-20130.

Pi M, Quarles LD. A novel cation-sensing mechanism in osteoblasts is a molecular target for strontium. J Bone Miner Res 2004; 19:862-869.

Pi M, Quarles LD. Osteoblast calcium-sensing receptor has characteristics of ANF/2005; 7TM receptors. J Cell Biochem 95:1081-1092.

Pi M, Garner SC, Flannery P et al. Sensing of extracellular cations in CasR-deficient osteoblasts. Evidence for a novel cation-sensing mechanism. J Biol Chem 2000; 275:3256-3263.

Pi M, Hinson TK, Quarles LD. Failure to detect the extracellular calcium-sensing receptor (CasR) in human osteoblast cell lines. J Bone Miner Res 1999; 14:1310-1319.

Picheret C, Horcajada MN, Mathey J et al. Isoflavone consumption does not increase the bone mass in osteopenic obese female Zuker rats. Ann Nutr Metab 2003; 47:70-77.

Pickard BW, Hodsman AB, Fraher LJ et al. Type 1 parathyroid hormone receptor nuclear trafficking: Association of PTHR1 with importin alpha₁and beta. Endocrinology 2006; 147:3326-3332.

Picotto G, Vazquez G, Boland R. 17β-Oestradiol increases intracellular Ca^{2+} concentration in rat enterocytes. Potential role of phospholipase C-dependent storeoperated Ca^{2+} influx. BioChem J 1999; 339:71-77.

Piekarski K, Munro M. Transport mechanism operating between blood supply and osteocytes in long bones. Nature 1977; 269:80-82.

Pierroz DD, Glatt V, Rizzoli R et al. Sustained cAMP signaling in primary osteoblasts from βarrestin2 KO mice leads to altered skeletal response to intermittent PTH (abstract). J Bone Miner Res 2003; 18:S175.

Pierroz DD, Bouxsein ML, Rizzoli R et al. Combined treatment with a β-blocker and intermittent PTH improves bone mass and microarchitecture in ovariectomized mice. Bone 2006, (published on line March 9).

Pilbeam C, Rao Y, Alander C et al. Downregulation of mRNA expression for the 'decoy' interleukin-1 receptor 2 by ovariectomy (abstract). J Bone Miner Res 1997; 12(Suppl 1):S433.

Pirola CJ, Wang HM, Strgacich MI et al. Mechanical stimuli induce vascular parathyroid hormone-related protein gene expression in vivo and in vitro. Endocrinology 1994; 134:2230-2236.

Plate U, Tkotz T, Wiesmann HP et al. Early mineraliztion of matrix vesicles in the epiphyseal growth plate. J Microsc 1996; 183(Pt. 1):102-107.

Plotkin LI, Bellido T, Ali AA et al. Runx2/Cbfa1 is essential for the anti-aopotosis effect of PTH on osteoblasts (abstract). J Bone Miner Res 2002; 17:S166.

Ponomareva LV, Wang W, Koszewski NJ et al. Mim-1, an osteoclast secreted chemokine, stimulates differentiation, matrix mineralization and increased Vitamin D receptor binding at the VDRE of osteoblastic precursor cells. J Bone Miner Res 2002; 17:S155.

Ponting CP, Phillips C, Davies KE et al. PDZ domains: Targeting signalling molecules to sub-membranous sites. BioEssays 1997; 19:469-477.

Poole CA, Flint MH, Beaumont BW. Analysis of the morphology and function of primary cilia in connective tissues: A cellular cybernetic probe. Cell Motil 1985; 5:175-193.

Poole CA, Jensen CG, Snyder JA et al. Confocal analysis of primary cilia structure and colocalization with the Golgi apparatus in chondrocytes and aortic smooth muscle cells. Cell Biol Int 1997; 21:483-494.

Poole CA, Zhang ZJ, Ross J. The differential distribution of acetylated and detyrosinated alpha-tubulin in the microtubular cytoskeleton and primary cilia of hyaline cartilage chondrocytes. J Anat 2001; 199:393-405.

Poole KES, Reeve J. Parathyroid hormone - A bone anabolic and catabolic agent. Curr Opin Pharmacol 2005; 5:612-617.

Poole KES, van Bezooijen RL, Loveridge N et al. Sclerostin is a delayed secreted product of osteocytes that inhibits bone formation. FASEB J 2005; 19:1842-1844.

Potten C, Wilson J. Apoptosis. Cambridge: Cambridge University Press, 2004.

Potts Jr JT, Tregear GW, Keutmann HT et al. Synthesis of a biologically active N-terminal tetratriacontapeptide of parathyroid hormone. Proc Natl Acad Sci USA 1971; 68:63-67.

Pouysségur J. Signal transduction: An arresting start for MAPK. Science 2000; 290:1574-1577.

Power DM, Ingleton PM, Flanagan J et al. Genomic structure and expression of parathyroid hormone-related protein gene (PTHrP) in a teleost., Fugu rubripes. Gene 2000; 250:67-76.

Power RA, Iwaniec UT, Wronski TJ. Changes in gene expression associted with the bone anabolic effects of basic fibroblast growth factor in aged ovariectomized rats. Bone 2002; 31:143-148.

Pozzan T, Magalh>es P, Rizzuto R. The comeback of mitochondria to calcium signaling. Cell Calcium 2002; 28:279-283.

Praetorius HA, Spring KR. A physiological view of the primary cilium. Annu Rev Physiol 2005; 67:515-529.

Praetorius HA, Spring KR. Bending the MDCK cell primary cilium increases intracellular calcium. J Membrane Biol 2001; 184:71-79.

Praetorius HA, Spring KR. Removal of the MDCK cell primary cilium abolishes flow sensing. J Membrane Biol 2003a; 191:69-76.

Praetorius HA, Spring KR. The renal cell primary cilium functions as a flow sensor. Curr Opin Nephrol Hypertens 2003b; 12:517-520.

Prenzel N, Zwick E, Daub H et al. EGF receptor transactivation by G-protein-coupled receptors requires metalloproteinase cleavage of pro-HB-EGF. Nature 1999; 402:884-888.

Ptashne M, Gann A. Genes and Signals. Cold Spring Harbor, NY: Cold Spring Harbor Laboratory Press, 2002.

Pugazhenthi S, Miller E, Sable C et al. Insulin-like growth factor-I induces bcl-2 promoter through the transcription factor cAMP-response element-binding protein. J Biol Chem 1999; 274:27529-27535.

Pugsley LT, Selye H. The histological changes in the bone responsible for the action of parathyroid hormone on the calcium metabolism of the rat. J Physiol 1933; 79:113-117.

Pun S, Dearden RL, Ratkus AM et al. Decreased bone anabolic effect of basic fibroblast growth factor at fatty marrow sites in ovariectomized rats. Bone 2001; 28:220-226.

Qin L, Raggatt L, Li X et al. Amphiregulin: A possible mediator of parathyroid hormone's anabolic action in bone (abstract). J Bone Miner Res 2003; 18:S100.

Qiu P, Qin L, Sorrentino RP et al. Comparative promoter analysis and its application in analysis of PTH-regulated gene expression. J Mol Biol 2003; 326:1327-1336.

Qiu S, Rao DS, Fyhrie DP et al. The morphological association between microcracks and osteocyte lacunae in human cortical bone. Bone 2005; 37:10-15.

Quinn PG. Mechanisms of basal and kinase-inducible transcription activator by CREB. Progr Nucleic Acid Res Mol Biol 2002; 72:269-305.

Radeff JM, Singh ATK, Dossing DA et al. Evidence that activation of phospholipase D is an early event in downstream signaling by PTH in UMR-106 osteoblasts. J Bone Miner Res 2002; 17:S342.

Radman DP, McCudden C, James K et al. Evidence for calcium-sensing receptormediated stanniocalcin secretion in fish. Mol Cell Endocrinol 2002; 186:111-119.

Raggatt LJ, Om L, Partrideg NC. Parathyroid hormone stimulation of IL-18 in osteoblastic cells. J Bone Miner Res 2003; 18:S398.

Rahmouni K, Haynes WG. Leptin and the central neural mechanisms of obesity hypertension. Drugs of Today 2002; 38:807-817.

Rahmouni K, Haynes WG. Leptin and the cardiovascular system. Recent Progr Horm Res 2004; 59:225-244.

Rangarajan A, Talora C, Okuyama R et al. Notch signaling is a direct determinant of keratinocyte growth arrest and entry into differentiation. EMBO J 2001; 20:3427-3436.

Rao LG, Murray TM. Binding of intact parathyroid hormone to rat osteosarcoma cells: Major contribution of binding sites for the carboxyterminal region of the hormone. Endocrinology 1985; 117:1632-1638.

Raouf A, Seth A. Discovery of osteoblast-associated genes using cDNA microarrays. Bone 2002; 30:463-441.

Razandi M, Pedram A, Greene GL et al. Cell membrane and nuclear estrogen receptors (ERs) originate from a single transcript: Studies of ERα and ERβ expressed in Chinese hamster ovary cells. Mol Endocrinol 1999; 13:307-319.

Razandi M, Oh P, Pedram A et al. ERs associate with and regulate the production of calveolin: Implications for signaling and cellular actions. Mol Endocrinol 2002; 16:100-115.

Razani B, Rubin B, Lisanti MP. Regulation of cAMP-mediated signal transduction via interaction of caveolins with the catalytic subunit of protein kinase A. J Biol Chem 1999; 274:26353-23360.

Razani B, Lisanti MP. Caveolin-deficient mice: Insights into caveolar function and human disease. J Clin Invest 2001; 108:1553-1561.

Razani B, Woodman SE, Lisanti MP. Caveolae: From cell biology to animal physiology. Pharmacol Rev 2002; 54:431-467.

Razani B, Zhang XL, Bitzer M et al. Caveolin-1 regulates transforming growth factor (TGF)-β/SMAD signaling through an interaction with the TGF-β type I receptor. J Biol Chem 2001; 276:6727-6738.

Re RN. The origins of intracrine hormone action. Am J Med Sci 2002a; 323:43-48.

Re RN. Toward a theory of intracrine hormone action. Regul Pept 2002b; 106:1-6.

Re RN. The intracrine hypothesis and intracellular peptide hormone action. BioEssays 2003; 25:401-409.

Re RN, Cook JL. The intracrine hypothesis: An update. Regul Pept 2005; 133:1-9.

Redruello B, Estevao MD, Rotllant J et al. Isolation and characterization of piscine osteonectin and downregulation of its expression by PTH-related protein. J Bone Miner Res 2005; 20:682-692.

Reeve J, Arlot M, Wooton R et al. Skeletal blood flow, iliac histomorphometry, and strontium kinetics in osteoporosis: A relationship between blood flow and corrected apposition rate. J Clin Endocrinol Metab 1988; 66:1124-1131.

Reeve J, Hesp R, Williams D et al. Anabolic effect of low doses of a fragment of human parathyroid hormone fragment on the skeleton in postmenopausal osteoporosis. Lancet 1976; I:1035-1036.

Reeve J, Meunier PJ, Parsons JA et al. Anabolic effect of human parathyroid hormone fragment on trabecular bone in involutional osteoporosis: A multicentre trial. Brit Med J 1980; 280:1340-1344.

Reeve J, Mitchell A, Tellez M et al. Treatment with parathyroid peptides and estrogen replacement for severe postmenopausal vertebral osteoporosis; prediction of long-term responses in spine and femur. J Bone Miner Metab 2001; 19:102-114.

Reeve J, Tregear GW, Parsons JA. Preliminary trial of low doses of human parathyroid hormone fragment 1-34 in treatment of osteoporosis. Clin Endocrinol 1976; 21:469-477.

Reginster JY, Burlet N. Osteoporosis: A still increasing prevalence. Bone 2006; 38:S4-S9.

Reginster JY, Deroisy R, Jupsin I. Strontium ranelate: A new paradigm in the treatment of osteoporosis. Drugs of Today 2003; 39:89-101.

Rehman Q, Lang TF, Arnaud CD et al. Daily treatment with parathyroid hormone is associated with an increase in vertebral cross-sectional area in postmenopausal women with glucocorticoid-induced osteoporosis. Osteoporosis Int 2003; 14:77-81.

Reid I, Comish J. Direct action of leptin on bone remodeling. Calcif Tissue Int 2004; 74:313-316.

Reid I, Cornish J, Baldock PA. Nutrition-related peptides and bone homeostasis. J Bone Miner Res 2006; 21:495-500.

Reid IR, Gamble GD, Grey AB et al. β-Blocker use, BMD, and fractures in the study of osteoporotic fractuRes. J Bone Miner Res 2005; 20:613-618.

Reid IR, Hague W, Emberson J et al. Effect of pravastatin on frequency of fracture in the LIPID study: Secondary analysis of a randomized controlled trial.Long-term intervention with pravastatin in ischaemic disease. Lancet 2001; 357:509-512.

Reinholz GG, Reinholz MM, Getz B et al. Lipophilic statins increase RANKL and OPG mRNA levels in human osteoblast cells (abstract). J Bone Miner Res 2002; 17:S445.

Rejnmark L, Buus NH, Vestergaard P et al. Effects of simvastatin treatment on bone turnover and BMD: A 1-year randomized controlled trial in postmenopausal osteopenic women. J Bone Miner Res 2003; 19:737-744.

Rensberger JM, Watanabe M. Fine structure of bone in dinosaurs, birds and mammals. Nature 2000; 406:619-622.

Reseland JE, Gordeladze JO. Role of leptin in bone growth: Central player or peripheral supporter? FEBS Lett 2002; 528:40-42.

Reseland JE, Gordeladze JO, Drevon CA. Leptin receptor (OB-R) gene expression in human primary osteoblasts: Reaffirmation. J Bone Miner Res 2002; 17:1136.

Reseland J, Syversen U, Bakke I et al. Leptin is expressed in and secreted from primary cultures of human osteoblasts and promotes bone mineralization. J Bone Miner Res 2001; 16:1426-1433.

Reszka AA, Halasy-Nagy JM, Masarachia PJ et al. Bisphosphoinates act directly on the osteoclast to induce caspase cleavage of Mst1 kinase during apoptosis. J Biol Chem 1999; 274:34967-34973.

Revankar CM, Cimino DF, Sklar LA et al. A transmembrane intracellular estrogen receptor mediates rapid cell signaling. Science 2005; 307:1625-1630.

Reya T, Duncan AW, Allies L et al. A role for Wnt signalling in self-renewal of haematopoietic stem cells. Nature 2003; 423:409-414.

Richardson ML. Osteonecrosis.www.rad.washington.edu/mskbook/osteonecrosis.html.

Riggs BL. The mechanism of estrogen regulation of bone resorption. J Clin Invest 2000; 106:1203-1204.

Riggs BL. Drugs used to treat osteoporosis: The critical need for a uniform nomenclature based on their action on bone remodeling. J Bone Miner Res 2005; 20:177-184.

Rihani-Bisharat S, Maor G, Lewinson D. In vivo anabolic effects of parathyroid hormone (PTH)28-48 and N-terminal fragments of PTH and PTH-related protein on neonatal mouse bones. Endocrinology 1998; 139:974-981.

Riordan M. The Hunting of the Quark. New York: Simon and Schuster, 1987:31.

Rittmaster RS, Bolognese M, Ettinger MP et al. Enhancement of bone mass in osteoporotic women with parathyroid hormone followed by alendronate. J Clin Endocrinol Metab 2000; 85:2129-2134.

Rixon RH, Whitfield JF. The radioprotective action of parathyroid extract. Int J Radiat Biol 1961; 3:361-367.

Rixon RH, Whitfield JF, Gagnon L et al. Parathyroid hormone fragments may stimulate bone growth in ovariectomized rats by activating adenylyl cyclase. J Bone Miner Res 1994; 9:1179-1189.

Rixon RH, Whitfield JF, MacManus JP. Stimulation of mitotic activity in rat bone marrow and thymus by exogenous adenosine 3',5'-monophosphate (cyclic AMP). Exp Cell Res 1970; 63:110-116.

Rixon RH, Whitfield JF, Youdale T. Increased survival of rats irradiated with X-rays and treated with parathyroid extract. Nature 1958; 182:1374.

Robinson JA, Susulic V, Zhao W et al. PTH anabolic induced gene in bone (PAIGB), a novel gene which may play an important rone in bone formation. J Bone Miner Res 2003; 18:S174.

Robling AG, Castillo AB, Turner CH. Biomechanical and molecular regulation of bone remodeling. Annu Rev Biomed Engin 2006; 8:455-498.

Rocheville M, Lange DC, Kumar U et al. Receptors for dopamine and somatostatin: Formation of heterooligomers with enhanced functional activity. Science 2000a; 288:154-157.

Rocheville M, Lange DC, Kumar U et al. Subtypes of the somatostatin receptor assemble as fuctional hom- and heterodimers. J Biol Chem 2000; 275:7862-7869.

Rockwood K, Kirkland S, Hogen DB et al. Use of lipid-lowering agents, indication bioas, and the risk of dementia in community-dwelling elderly people. Arch Neurol 2002; 59:223-227.

Rodan GA, Reszka AA. Bisphosphonate mechanism of action. Curr Mol Med 2002; 2:71-577.

Rodland KD. Calcium receptor-mediated signaling. In: Chattopadhyay N, Brown EM, eds. Calcium-Sensing Receptor. Boston: Kluwer Academic Publishers, 2003:53-67.

Roe EB, Chiu KM, Arnaud CD. Selective estrogen receptor modulators and postmenopausal health. Adv Intern Med 2000; 45:259-278.

Roodman GD. Advances in bone biology: The osteoclast. Endocrine Revs 1996; 17:308-332.

Rogers MJ. New insights into the molecular mechanisms of action of bisphosphonates. Curr Pharma Design 2003; 9:2643-2658.

Rogers MJ, Gordon S, Benford HL et al. Cellular and molecular mechanisms of action of bisphosphonates. Cancer 2000; 88:2961-2978.

Rose EB, Sanchez SD, Del Puerto GA et al. Parathyroid hormone 1-34 (hPTH 1-34) and estrogen produce dramatic bone density increases in postmenopausal osteoporosis (abstract). J Bone Miner Res 1999; 14:S137.

Rosen CJ, Bilezikian JP. Anabolic therapy for osteoporosis. J Clin Endocrinol Metab 2001; 86:957-964.

Roschger P, Rinnerthaler S, Yates J et al. Alendronate increases degree and uniformity of mineralization in cancellous bone and decreases the porosity in cortical bone of osteoporotic women. Bone 2001; 29:185-191.

Rotllant J, Redruello B, Guerreiro PM et al. Calcium mobilization from fish scales is mediated by parathyroid hormone related protein via the parathyroid type 1 receptor. Reg Peptides 2005; 132:33-40.

Rouault JP, Prévôt D, Berthet C et al. Interaction of BTG1 and p53-regulated BTG2 gene products with mCaf1, the murine homolog of a component of the yeast CCR4 transcription regulatory complex. J Biol Chem 1998; 273:22563-22569.

Roufflet J, Coxam V, Gaumet N et al. Preserved bone mass in ovarect-omized rats treated with parathyroid-hormone-related peptide (1-34) and (107-111) fragments. Reprod Nutr Dev 1994; 34:473-481.

Roy B, Li WW, Lee AS. Calcium-sensitive transcriptional activation of the proximal CCAAT regulatory element of the grp78/BiP promoter by the human nuclear factor CBF/NF-Y. J Biol Chem 1996; 271:28995-29002.

Rubin DA, Jüppner H. Zebrafish express the common parathyroid hormone/parathyroid hormone-related peptide receptor (PTH1R) and a novel receptor (PTH3R) that is preferentially activated by mammalian and fugufish parathyroid hormone-related peptide. J Biol Chem 1999; 274:28185-28190.

Rubin J, Murphy TC, Zhu L et al. Mechanical strain differentially regulates eNOS and RANKL expression via ERK 1 / 2 in primary bone stromal cells (abstract). J Bone Miner Res 2003; 18:S10-S11.

Rubin MR, Silverberg SJ. Vascular calcification and osteoporosis. Atherosclerosis 2004; 147:213-225.

Ruoslahti E, Rajotte D. The address system in the vasculature of normal tissues and tumors. Annu Rev Immunol 2000; 18:813-827.

Ryan KA, Pimplikar SW. Activation of GSK-3 and phosphorylation of CRMP2 in transgenic mice expressing APP intracellular domain. J Cell Biol 2005; 171:327-335.

Ryder KD, Duncan RL. Parathyroid hormone enhances fluid shear-induced $[Ca^{2+}]_i$ signaling in osteoblastic cells through activation of mechanosensitive and voltage-sensitive Ca^{2+} channels. J Bone Miner Res 2001; 16:240-248.

Ryoo HM, Hoffmann HM, Beumer T et al. Stage-specific expression of Dlx-5 during osteoblasts differentiation:involvement in regulation of osteocalcin gene expression. Mol Endocrinol 1997; 11:1681-1694.

Sabbagh Y, Carpenter TO, Demay MB. Hypophosphatemia leads to rickets by impairing caspase-mediated apoptosis of hypertrophic chondrocytes. Proc Natl Acad Sci USA 2005; 102:9637-9642.

Sakai A, Sakata T, Ikeda S et al. Intermittent administration of human parathyroid hormone(1-34) prevents immobilization-related bone loss by regulating bone marrow capacity for bone cells in ddY mice. J Bone Miner Res 1999; 14:1691-1699.

Sakai H, Kobayashi Y, Sakai E et al. Cell adhesion is a prerequisite for osteoclast survival. BioChem Biophys Res Commun 2000; 270:550-556.

Sakata T, Wang Y, Halloran BP et al. Skeletal unloading induces resistance to insulin-like growth factor-I (IGF-I) by inhibiting activation of IGF-I signal pathways. J Bone Miner Res 2004; 19:436-446.

Samnegard E, Iwaniec UT, Cullen DM et al. Maintenance of cortical bone in human parathyroid hormone(1-84)-treated ovariectomized rats. Bone 2001; 28:251-260.

Sanders JL, Chattopadhyay N, Kifor O et al. Ca^{2+}-sensing receptor expression and PTHrP secretion in PC-3 human prostate cancer cells. Am J Physiol Endocrinol Metab 2001; 281:E1267-E1274.

Satir P, Christensen ST. Overview of structure and function of mammalian cilia. Annu Rev Physiol 2007; 69:377-400.

Sato M, Schmidt A, Cole H et al. The skeletal efficacy of statins do not compare with low-dose parathyroid hormone (abstract). Bone 2001; 28:S80.

Sato M, Vahle J, Schmidt A et al. Abnormal bone architecture and biomechanical properties with near-lifetime treatment of rats with PTH. Endocrinology 2002; 143:3230-3242.

Sato M, Westmore M, Clendenon J et al. Three-dimensional modeling of the effects of parathyroid hormone on bone distribution in lumbar vertebrae of ovariectomized cynomolgus macaques. Osteoporos Int 2001; 11:871-880.

Sato M, Westmore M, Ma YL et al. Teriparatide [hPTH(1-34)] strengthens the proximal femur of ovariectomized nonhuman primates despite increasing porosity. J Bone Miner Res 2004; 19:623-629.

Sato S, Fujita N, Tsuruo T. Modulation of Akt kinase activity by binding to Hsp 90. Proc Natl Acad Sci USA 2000; 97:10832-10837.

Schaffler MB. The role of osteocytes in targeting microdamage-induced remodeling (abstract). J Bone Miner Res 2000; 15:823.

Schaffler MB, Choi K, Milgrom C. Aging and matrix microdamage accumulation in human compact bone. Bone 1995; 17:521-525.

Scherft JP, Daems W. Single cilia in chondrocytes. J Ultrastructural Res 1967; 19:546-555.

Schiller PC, D'Ippolito G, Roos BA et al. Anabolic or catabolic responses of MC3T3-E1 osteoblastic cells to parathyroid hormone depends on time and duration of treatment. J Bone Miner Res 1999; 14:1504-1512.

Schipani E, Chiusaroli R, Maier A et al. In vivo impairment of collagenase activity does not affect bone formation in transgenic mice expressing constitutively active PTH/PTHrP receptors in osteoblasts (abstract). J Bone Miner Res 2002; 17:S308.

Schipani E, Kruse K, Jüppner H. A constitutively active mutant PTH-PTHrP receptor in Jansen-type metaphyseal chondrodysplasia. Science 1995; 268:98-100.

Schipani EE, Lanske B, Hunzelman J et al. Targeted expression of constitutively active receptors for parathyroid hormone and parathyroid hormone-related peptide delays endochondral bone formation and rescues mice that lack parathyroid hormone-related peptide. Proc Natl Acad Sci USA 1997; 94:13689-13694.

Schipani E, Ryan HE, Didrickson S et al. Hypoxia in cartilage: HIF-1α is essential for chondrocyte growth arrest and survival. Genes Dev 2001; 15:2865-2876.

Schlienger RG, Kraenzlin ME, Jick SS et al. Use of beta-blockers and risk of fractures. J Amer Med Assoc (JAMA) 2004; 292:1326-1332.

Schneider A, Taboas JM, McCauley LK et al. Skeletal homeostasis in tissue-engineered bone. J Orthop Res 2003; 21:859-864.

Schneider L, Clement CA, Teilmann SC et al. PDGFRαα signaling is regulated through the primary cilium in fibroblasts. Curr Biol 2005; 15:1861-1866.

Schofield R. The relationship between the spleen colony-forming cell and haematopoietic stem cell. Blood Cells 1978; 4:7-25.

Schofield R. The pluripotent stem cell. Clin Haematol 1979; 8:221-237.

Schoneberg T, Schulz A, Gudermann T. The structural basis of G-protein-coupled receptor function and dysfunction in human diseases. Rev Physiol BioChem Pharmacol 2002; 144:145-227.

Schoojans K, Staels B, Auwerx J. The peroxisome proliferator activated receptors (PPARs) and their effects on lipid metabolism and adipocyte differentiation. Biochim Biophys Acta 1996; 1302:93-109.

Schordan E, Welsch S, Rothhut S et al. Role of parathyroid hormone-related protein in the regulation of stretch-induced renal vascular smooth muscle cell proliferation. J Amer Soc Nephrol 2004; 15:3016-3025.

Schultz E, Arfai K, Liu X et al. Aortic calcification and the risk of osteoporosis and fractures. J Clin Endocrinol Metab 2004; 89:4246-4253.

Schwartz B, Smirnoff P, Shany S et al. Estrogen controls expression and bioresponse of 1,25-dihydroxyvitamin D receptors in the rat colon. Mol Cell BioChem 2000; 203:87-93.

Schwartz EA, Leonard ML, Bizios R et al. Analysis and modeling of the primary cilium bending response to fluid shear. Am J Physiol 1997; 272:F132-F138.

Schwartz TW, Frimurer TM, Holst B et al. Molecular mechanism of 7TM receptor activation — A global toggle switch model. Annu Rev Pharmacol Toxicol 2006; 46:481-519.

Schweisguth F. Regulation of Notch signaling activity. Curr Biol 2004; 14:R129-R138.

Schweitert HR, Groen EWJ, Sollie FAE et al. Single dose subcutaneous administration of recombinant human parathyroid hormone [rhPTH(1-84)] in healthy postmenopausal volunteers. Clin Pharmacol Ther 1997; 61:360-376.

Schweitzer MH, Wittmeyer JL, Horner JR et al. Soft-tissue vessels and cellular preservation in Tyrannosaurus rex. Science 2005; 307:1952-1955.

Seebach C, Skripitz R, Andreassen TT et al. Intermittent parathyroid hormone (1-34) enhances mechanical strength and density of new bone after distraction osteogenesis in rats. J Orthop Res 2004; 22:472-478.

Seeman E. How do antiresorptive agents reduce fracture? IBMS BoneKEy 2001, (Jan 23 10.1138/ibmske; 20011012).

Segars JH, Driggers PH. Estrogen action and cytoplasmic signaling cascades. Part I: Membrane-associated signaling complexes. Trends Endocrinol Metab 2002; 13:349-354.

Selkoe DJ. Clearing the brains's cobwebs. Neuron 2001; 32:177-180.

Selvamurugan N, Pulumati MR, Tyson DR et al. Parathyroid hormone regulation of the rat collagenase-3 promoter by protein linase A-dependent transactivation of core binding factor α1. J Biol Chem 2000; 275:5037-5042.

Selye H. On the stimulation of new bone formation with parathyroid extract and irradiated ergosterol. Endocrinology 1932; 16:547-558.

Selz T, Caverzasio J, Bonjour JP. Regulation of Na-dependent P_i transport by parathyroid hormone in osteoblast-like cells. Am J Physiol 1989; 256:E93-E100.

Semenov M, Tamai K, He X. SOST is a ligand for LRP5/LRP6 and a Wnt signaling inhibitor. J Biol Chem 2005; 280:26770-26775.

Serre CM, Farlay D, Delmas PD et al. Evidence for a dense and intimate innervation of the bone tissue, including glutamate-containing fibers. Bone 1999; 25:623-629.

Sevetson B, Taylor S, Pan Y. Cbfa1/RUNX2 directs specific expression of the sclerostenosis gene (SOST). J Biol Chem 2004; 279:13849-13858.

Shafritz AB, Shore EM, Gannon FH et al. Overexpression of an osteogenic morphogen in fibrodysplasia ossificans progressiva. N Engl J Med 1996; 335:555-561.

Shanahan CM. Vascular calcification. Curr Opin Nephrol Hypertens 2005; 14:361-367.

Shao JS, Cheng SL, Charlton-Kachigian N et al. Teriparatide (human parathyroid hormone (1-34)) inhibits osteogenic vascular calcification in diabetic low density lipoprotein receptor-deficient mice. J Biol Chem 2003; 278:50195-50202.

Shao JS, Cheng SL, Pingsterhaus JM et al. Msx2 promotes cardiovascular calcification by activating paracrine Wnt signals. J Clin Invest 2005; 115:1210-1220.

Sharpe GR, Dillon JP, Durham B et al. Human keratinocytes express transcripts for three isoforms of parathyroid hormone-related protein (PTHrP), but not for the parathyroid hormone/PTHrP receptor: Effects of 1,25(OH)$_2$ vitamin D3. Br J Dermatol 1998; 138:944-951.

Shaynerson M, Plotkin MJ. The Killers Within—The Deadly Rise of Drug-Resistant Bacteria. Boston: Little Brown and Company, 2002:94.

Shenolikar S, Voltz JW, Minkoff CM et al. Targeted disruption of the mouse NHERF-1 gene promotes internalization of proximal tubule sodium-phosphate cotransporter type IIa and renal phosphate wasting. Proc Natl Acad Sci USA 2002; 99:11470-11475.

Shepherd J. Lipid lowering: Statins and the future. Heart 2000; 84:46-47.

Shevde NK, Bendixen AC, Dienger KM et al. Estrogens suppress RANK ligand-induced osteoclast differentiation via stromal cell independent mechqanism invoving c-Jun expression. Proc Natl Acad Sci USA 2000; 97:7829-7834.

Shevde NK, Plum LA, Clagett-Dame M et al. A potent analog of 1α,25-dihydroxyvitamin D$_3$ selectively induces bone formation. Proc Natl Acad Sci USA 2002; 99:13487-13491.

Shiba D, Takamatsu T, Yokoyama T. Primary cilia of inv/inv mouse renal epithelial cells sense physiological fluid flow: Bending of primary cilia and Ca^{2+} influx. Cell Struct Function 2005; 30:93-100.

Shibata M, Yamada S, Kumar SR et al. Clearancew of Alzheimer's amyloid-β$_{(1-40)}$ peptide from the brain by LDL receptor-related protein at the blood-brain barrier. J Clin Invest 2000; 106:1489-1499.

Shih NR, Jo OD, Yanagawa N. Effects of PHEX antisense in human osteoblast cells. J Am Soc Nephrol 2002; 13:394-399.

Shiiki T, Ohtsuki S, Kurihara A et al. Brain insulin impairs amyloid-β (1-40) clearance from the brain. J Neurosci 2004; 24:9632-96-37.

Shimokawa H, Seto M, Katsumata N et al. Rho-kinase-mediated pathway induces enhanced myosin light chain phosphorylation in a swine model of coronary artery spasm. Cardiovasc Res 1999; 43:1029-1039.

Shirota T, Tashiro M, Ohno K et al. Effect of intermittent parathyroid hormone (1-34) treatment on the bone response after placement of titanium implants into the tibia of ovariectomized rats. J Oral Maxillofac Surg 2003; 61:471-480.

Shobab LA, Hsiung GYR, Feldman HH. Cholesterol in Alzheimer's disease. Lancet Neurol 2006; 4:841-852.

Shum L, Nuckolls G. The life cycle of chondrocytes in the develponing skeleton. Arthritis Res 2001; 4:94-106.

Sierra-Honigman MR, Nath AK, Murakami C et al. Biological action of leptin as an angiogenic factor. Science 1998; 281:1683-1686.

Sidera S, Parsons R, Austen B. The regulation of β-secretase by cholesterol and statins in Alzheimer's disease. J Neurol Sci 2005a; 229-230:269-273.

Sidera S, Parsons R, Austen B. Post-translational processing of beta-secretase in Alzheimer's disease. Proteomics 2005b; 5:1533-1543.

Sietsema WK. Animal models of cortical porosity. Bone 1995; 17:297S-305S.

Simonet WS, Lacey DL, Dunstan CR et al. Osteoprotegerin: A novel secreted protein involved in the regulation of bone density. Cell 1997; 89:309-319.

Singh ATK, Bhattacharyya RS, Radeff JM et al. Regulation of parathyroid hormone-stimulated phospholipase D in UMR-106 cells by calcium, MAP kinase and small G proteins. J Bone Miner Res 2003; 18:1453-1460.

Singh ATK, Gilchrist A, Voyno-Yasentskaya T et al. Parathyroid hormone activates a Galpha12/Galpha13-Rho A-phospholipase D signaling pathway in osteoblastic cells. J Bone Miner Res 2004; 19:S478.

Singh AT, Kunnel JG, Strieleman PJ et al. Parathyroid hormone (PTH)-(1-34), [Nle(8,18),Tyr34] PTH-(3-34) amide, PTH-(1-31)amide, and PTH-related peptide-(1-34) stimulate phosphatidylcholine hydrolysis in UMR-106 osteoblastic cells: Comparison with effects of phorbol 12,13-dibutyrate. Endocrinology 1999; 140:131-137.

Singla V, Reiter JF. The primary cilium as the cell's antenna: Signaling at a sensory organelle. Science 2006; 313:629-633.

Sire JY, Huysseume A. Formation of dermal skeletal and and dental tissues in fish: A comparative and evolutionary approach. Biol Rev Camb Philos Soc 2003; 78:219-243.

Sirola J, Honkanen R, Kröger H et al. Effects of HMG-CoA reductase inhibitors, statins, on bone loss: A prospective population-based cohort study in early postmenopausal women (abstract). Bone 2001; 28:S220.

Sjögren M, Blennow K. The link between cholesterol and Alzheimer's disease. World J Biol Psychiatry 2005; 6:85-97.

Sjögren M, Mielke M, Gustafson D et al. Chloesterol and Alzheimer's disease—Is there a relation? Mech Ageing Dev 2006; 127:138-147.

Skerry TM. Identification of novel signaling pathways during functional adaptation of the skeleton to mechanical loading: The role of glutamate as a paracrine signaling agent in the skeleton. J Bone Miner Metab 1999; 17:66-70.

Skerry TM, Genever PG. Glutamate signaling in nonneuronal tissues. Trends Pharmacol Sci 2001; 22:174-181.

Skerry TM, Genever P, Taylor A et al. Absence of evidence is not evidence of absence. The shortcomings of the GLAST knockout mouse. J Bone Miner Res 2001; 16:742-749.

Skerry TM, Suva LJ. Investigation of the regulation of bone mass by mechanical loading; from quantitative cytochemistry to gene array. Cell BioChem Funct 2003; 21:223-229.

Skerry TM, Taylor AF. Glutamate signaling in bone. Curr Pharma Des 2001; 7:737-750.

Skoglund B, Forslund C, Aspenberg P. Simvastatin improves fracture healing in mice. J Bone Miner Res 2002; 17:2004-2008.

Skriptiz R, Andreassen TT, Aspenberg P. Strong effect of PTH(1-34) on regenerating bone. Acta Orthop Scand 2000a; 71:619-624.

Skriptiz R, Andreassen TT, Aspenberg P. Parathyroid hormone (1-34) increases the density of rat cancellous bone in a bone chamber. J Bone Joint Surg [Br] 2000b; 82B:138-141.

Skripitz R, Aspenberg P. Early effect of parathyroid hormone (1-34) on implant fixation. Clin Orthop Rel Res 2000a; 392:427-432.

Skripitz R, Aspenberg P. Implant fixation enhanced by intermittent treatment with parathyroid hormone. J Bone Joint Surg [Br] 2000b; 83B:437-440.

Skripitz R, Aspenberg P. Parathyroid hormone (1-34) increases attachment of PMM cement to bone. J Orthop Sci 2001; 6:540-544.

Skripitz R, Böhling S, Rüther W et al. Stimulation of implant fixation by parathyroid hormone (1-34)—A histomorhometric comparison of PMMA cement and stainless steel. J Orthopaed Res 2005; 23:1266-1270.

Smit TH, Burger EH, Huyghe JM. A case for strain-induced fluid flow as a regulator of BMU-coupling and osteonal alignment. J Bone Miner Res 2002; 17:2021-2029.

Smith BB, Cosenza ME, Mancini A et al. A toxicity profile of osteoprotegerin in the cynomologous monkey. Int J Toxicol 2003; 22:403-412.

Smith MM, Hall BK. Development and evolutionary origin of vertebrate skeletogenic and odontogenic tissues. Biol Revs 1990; 65:277-373.

Smith J, Huvos AG, Chapman M et al. Hyperparathyroidism associated with sarcoma of bone. Skeletal Radiol 1997; 26:107-112.

Smythies J. The Dynmaic Neuron. Cambridge, MA: The MIT Press, 2002.

Soerensen L, Eriksen EF. Endothelial cells mediate osteoblastic responses to sex steroids and selective estrogen receptor modulators (SERMS) via NO independent pathways (abstract). Bone 1998; 23:S369.

Sogaard CH, Mosekilde L, Thomsen JS et al. A comparison of the effects of two anabolic agents (fluoride and PTH) on ash density and bone strength assessed in an osteopenic rat model. Bone 1997; 20:439-449.

Solomon DH, Finkelstein JS, Wang PS et al. Statin lipid-lowering drugs and bone density (abstract). J Bone Miner Res 2001; 16:S293.

Solomon KR, Danciu TE, Adolphson LD et al. Caveolin-enriched membrane signaling complexes in human and murine osteoblasts. J Bone Miner Res 2000; 15:2380-2390.

Sompayrac L. How the Immune System Works. Oxford: Blackwell Science Ltd, 2003.

Sorkin AM, Dee KC, Knothe-Tate ML. "Culture shock" from the bone cell's perspective: emulating physiological conditions for mechanobiological investigations. Am J Physiol Cell Physiol 2004; 287:C1527-C1536.

Sowa H, Kaji H, Fway M et al. Parathyroid hormone-Smad3 axis exerts anti-apoptotic action and augments anabolic action of transforming growth factor β in osteoblasts. J Biol Chem 2003; 278:52240-52252.

Sowa H, Kaji H, Yamaguchi T et al. Activation of ERK1/2 and JNK by transforming growth factor β negatively regulate Smad3-induced alkaline phosphatase activity and mineralization in mouse osteoblastic cells. J Biol Chem 2002; 277:36024-36031.

Sparks LD, Petanceska S, Sabbagh M et al. Cholesterol, copper and Aβ in controls, MCI, AD and AD cholesterol-lowering treatment trial (ADCLT). Curr Alzheimer Res 2:527-539.

Spencer GJ, Genever PG. Long-term potentiation in bone—A role for glutamate in strain-induced cellular memory. BMC Cell Biol 2003; 4(1):9.

Spencer GJ, Hitchcock IS, Genever PG. Emerging neuroskeletal signalling pathways: A review. FEBS Lett 2004, (published online January 2004).

Spinella-Jaegle S, Rawadi G, Kawai S et al. Sonic hedgehog increases the commitment of pluripotent mesenchymal cells into the osteoblastic lineage and abolishes adipocyte differentiation. J Cell Sci 2001; 114:2085-2094.

Sprague SM, Popovtzer MM, Dranitzki-Elhalel M et al. Parathyroid hormone-induced calcium efflux from cultured bone is mediated by protein kinase C translocation. Am J Physiol 1995; 271:F1139-F1146.

Springer TA. Traffic signals for lymphocyte recirculation and leukocyte emigration: The multistep paradigm. Cell 1994; 76:301-314.

Srinivasan S, Gross TS. Canalicular fluid flow by bending of a long bone. Medical Engineering and Physics 2000; 22:127-133.

Srinivasan S, Agans SC, King KA et al. Enabling bone formation in the aged skeleton via rest-inserted mechanical loading. Bone 2003; 33:946-955.

Sriussadaporn S, Wong MS, Whitfield JF et al. Structurefunction relationship of human parathyroid hormone in the regulation of vitamin D receptor expression in osteoblast-like cells (ROS17/2.8). Endocrinology 1995; 136:3735-3742.

Staal A, Frith JC, French MH et al. The ability of statins to inhibit bone resorption is directly related to their inhibitory effect on HMG-CoA reductase activity. J Bone Miner Res 2003; 18:88-96.

Staal FJT, Clevers HC. Wnt signalling and haematopoiesis: a Wnt-Wnt situation. Nature Revs Immunol 2005; 5:21-30.

Stachowiak MK, Fang X, Myers JM et al. Integrative nuclear FGFR1 signaling (INFS) as part of a unigversal "feed-forward-and-gate" signaling module that controls cell growth and differentiation. J Cell Biochem 2003; 90:662-691.

Standridge JB. Vicious cycles within the neuropathophysiologic mechanisms of Alzheimer's disease. Curr Alzheimer Res 2006; 3:95-107.

Stanislaus D, Yang X, Liang D et al. In vivo regulation of apoptosis in metaphyseal trabecular bone of young rats by synthetic human parathyroid hormone (1-34) fragment. Bone 2000; 27:209-218.

Stayner C, Zhou J. Polycystin channels and kidney disease. Trends Pharmacol Sci 2001; 22:543-546.

Steers WD, Broder SR, Persson K et al. Mechanical stretch increases secretion of parathyroid hormone-related protein by cultured bladder smooth muscle cells. J Urol 1998; 160:908-912.

Stefano GB, Peter D. Cell surface estrogen receptors coupled to eNOS mediate immune and vascular tissue regulation: Therapeutic implications. Med Sci Monit 2001; 7:1066-1074.

Steggerda SM, Paschal BM. Regulation of nuclear import and export by the GTPase Ran. Int Rev Cytol 2002; 217:41-91.

Stein GS, Lian JB, Stein JL et al. Transcriptional control of osteoblast growth and differentiation. Physiol Rev 1996; 76:593-629.

Steinberg D. Atherogenesis in perspective: Hypercholesterolemia and inflammation as partners in crime. Nature Med 2002; 8:1211-1217.

Steitz SA, Speer MY, Curinga G et al. Smooth muscle cell phenotypic transition associated with calcification: Upregulation of Cbfa1 and downregulation of smooth muscle lineage markers. Circ Res 2001; 89:1147-1154.

Stem cells. Scientific progress and future research directions. Chapter 5: Hematopoietic stem cells. Dept. of Health and Human Services, 2001, (http://stemcells.nih.gov/info/scireport).

Stenbeck G. Formation and function of the ruffled border in osteoclasts. Sem Cell Devel Biol 2002; 13:285-292.

Steppan CM, Crawford DT, Chidsey-Frink KL et al. Leptin is a potent stimulator of bone growth in ob/ob mice. Regul Peptides 2000; 92:73-78.

Steppan CM, Ke HZ, Swick AG et al. Leptin administration cause an increase in brain size and bone growth in ob/ob mice (abstract). J Obes Rel Metab Disord 1999; 22(Suppl. 1):O131.

Stevenson JC, Lindsay R. Osteoporosis. London: Chapman and Hall Medical, Philadelphia: Current Medicine, 1999.

Stier S, Ko Y, Forkert R et al. Osteopontin is a hematooietic stem cell niche component that negatively regulates stem cell pool size. J Exp Med 2005; 201:1781-1791.

Stilgren LS, Reppe S, Abrahamsen B et al. Differential effects of PTH peptides on OPG and RANK-L mRNA expression in himan OHS osteosarcoma cells: A possible pathway of osteoblast dependent bone resorption (abstract). J Bone Miner Res 2001; 16:S545.

St-Jacques B, Hammerschmidt M, McMahon AP. Indian hedgehog signaling regulates proliferation and differentiation of chondrocytes and is essential for bone formation. Genes Dev 1999; 13:2072-2086.

Stocker R, Keaney JF. The role of oxidative modification in atherosclerosis. Physiol Rev 2004; 84:1381-1478.

Stokstad E. Dinosaurs under the knife. Science 2004; 306:962-965.

Stork PJS, Schmitt JM. Crosstalk between cAMP and MAP kinase signaling in the regulation of cell proliferation. Trends Cell Biol 2002; 12:258-266.

Streeten EA, Brandi ML. Biology of bone endothelial cells. Bone and Mineral 1990; 10:85-94.

Strein K. Are animal studies with bisphosphonates and PTH fragments predictive for the clinical studies? In: Russell RG, ed. Bone Diseases and Osteoporosis. article 12. London: IBC Technical Services Limited, 1994.

Strodel WE, Thompson NW, Eckhauser FE et al. Malignancy and concomitant primary hyperparathyroidism. J Surg Oncol 1988; 37:10-12.

Stupack DG, Cheresh DA. Get a ligand, get a life: Integrins, signaling and cell survival. J Cell Sci 2002; 115:3729-3738.

Su D, Ellis S, Napier A et al. Hoxa3 and Pax1 regulate epithelial cell death and proliferation during thymus and parathyroid organogenesis. Dev Biol 2001; 236:316-329.

Suda T, Takahashi N, Udagawa N et al. Modulation of osteoclast differentiation and function by the new members of the tumor necrosis factor receptor and ligand families. Endocrine Revs 1999; 20:345-357.

Sugiyama M, Kodama T, Konishi K et al. Compactin and simvastatin, but not pravastatin, induce bone morphogenic protein-2 in human osteosarcoma cells. Biochem Biophys Res Commun 2000; 271:688-692.

Suh PG, Hwang JI, Ryu SH et al. The roles of PDZ-containing proteins in PLC-β-mediated signaling. Biochem Biophys Res Commun 2001; 288:1-7.

Sun YQ, Ashhurst DE. Osteogenic growth pwptide enhances the rate of fracture healing in rabbits. Cell Biol Int 1998; 22:313-319.

Sun ZW, Allis CD. Ubiquitination of histone H2B regulates H3 methylation and gene silencing in yeast. Nature 2002; 418:104-108.

Sung WL, Chan BS, Luk CK et al. High-yield expression of fully bioactive N-terminal parathyroid hormone analogue in Escherichia coli. IUBMB Life 2000; 49:131-135.

Sunyer T, Lewis J, Collin-Osdoby P et al. Estrogen's bone-protective effects may involve differential IL-1 receptor regulation in human osteoclast-like cells. J Clin Invest 1999; 103:1409-1418.

Suzuki T, Chiba S. Notch signaling in hematopoietic stem cells. Int J Hematol 2005; 82:285-294.

Swarthout JT, Doggett TA, Lemken JL et al. Stimulation of extracellular signal-regulated kinases and proliferation in rat osteoblastic cells by parathyroid hormone is protein kinase C dependent. J Biol Chem 2001; 276:7586-7592.

Sweeny G. Leptin signaling. Cellular Signalling 2002; 14:655-663.

Swierenga SHH, MacManus JP, Braceland BM et al. Regulation of the primary immune response in vivo by parathyroid hormone. J Immunol 1976; 117:1608-1611.

Szczesniak AM, Gilbert RW, Mukhida M et al. Mechanical loading modulates glutamate receptor subunit expression in bone. Bone 2005; 37:63-73.

Takai H, Kanematsu M, Yano K et al. Transforming growth factor-β stimulates the production of osteoprotegerin/osteoclastogenesis inhibitory factor by bone marrow stromal cells. J Biol Chem 1998; 273:27091-27096.

Takaoki M, Murakami N, Gyotoku J. C-fos expression of osteoblast-like MC3T3-E1 cells induced either by cooling or by fluid flow. Biol Sci Space 2005; 18:181-182.

Takasu H, Gardella TJ, Luck MD et al. Amino-terminal modifications of human parathyroid hormone (PTH) selectively alter phospholipase C signaling via the Type 1 PTH receptor: Implications for design of signal-specific PTH ligands. Biochemistry 1999; 38:13453-13460.

Takeda E, Taketani Y, Morita K et al. Sodium-dependent phosphate co transporters. Int J Biochem Cell Biol 1999; 31:377-381.

Takeda S, Elefteriou F, Karsenty G. Common endocrine control of body weight, reproduction , anbone mass. Annu Rev Nutr 2003; 23:403-411.

Takeda S, Elefteriou F, Levasseur R et al. Leptin regulates bone formation via the sympathetic nervous system. Cell 2002; 111:305-317.

Taketani Y, Nashik K, Sawada N et al. Subcellular localization and PTH-dependent transclation of NaPi-IIa in renal proximal tubular cells. J Bone Miner Res 2003; 18:S170, (abstract).

Taketani Y, Takeichi T, Nashiki K et al. PTH-stimulated signaling molecules that are involved in the endocytosis of NaPi-IIa phosphate transporter are compartmentalized and activated in caveolae-like microdomains. J Bone Miner Res 2004; 19:S331.

Takeuchi T, Tsuboi T, Arai M et al. Adrenergic stimulation of osteoclastogenesis mediated by expression of osteoclast differentiation factor in MC3T3-E' osteoblast-like cells. Biochem Pharmacol 2000; 61:579-586.

Tam CS, Heersche JN, Murray TM et al. Parathyroid hormone stimulates the bone apposition rate independently of its resorptive action: Differential effects of intermittent and continuous administration. Endocrinology 1982; 110:506-512.

Tamasi JA, Arey BJ, Bertolini DR et al. Characterization of bone structure in leptin receptor-deficient Zuker (fa /fa) rats. J Bone Miner Res 2003; 18:1605-1611.

Tami AE, Nasser P, Verborgt MB et al. The role of interstitial fluid flow in the remodeling response to fatigue loading. J Bone Miner Res 2002; 17:2030-2037.

Tamura T, Udgawa N, Takahashi N et al. Soluble interleukin-6 receptor triggers osteoclast formation by interleukin-6. Proc Natl Acad Sci USA 1993; 90:11924-11928.

Tan E, Gurjar MV, Sharma RV et al. Estrogen receptor-α transfer into bovine aortic endothelial cells induces eNOS gene expression and inhibits cell migration. Cardiovasc Res 1999; 43:788-797.

Tanega C, Radman DP, Flowers B et al. Evidence for stanniocalcin and a related receptor in annelids. Peptides 2004; 25:1671-1679.

Takaoki M, Murakami N, Gyotoku J. C-fos expression of osteoblast-like MC3T3-E1 cells induced either by cooling or by fluid flow. Biol Sci Space 2005; 18:181-182.

Taurin S, Sandbo N, Qin Y et al. Phosphorylation of β-catenin by cyclic AMP-dependent protein kinase. J Biol Chem 2006, (published online February 13).

Tawfeek HA. PTH stimulation of mitogen-activated protein kinase does not require PTH/PTHrP receptor phosphorylation and internalization or phospholipse C activation (abstract). J Bone Miner Res 2002; 17:S392.

Taylor AF. Functional osteoblastic ionotropic glutamate receptors are a prerequisite for bone formation. J Musculoskel Neuron Interact 2002; 2:415-422.

Taylor AF, Brabbs AC, Peet NM et al. Bone formation /resorption and osteoblast/adipocytes plasticity mediated by AMPA/kainate glutamate receptors in vitro and in vivo (abstract). J Bone Miner Res 2000; 15:S275.

Taylor D, O'Brien F, Prima-Mello A et al. Compression data on bovine bone confirm that a "stressed volume" principle explains the variability of strength results. J Biomech 1999; 32:1199-1203.

Teichtahl AJ, Wluka AE, Proietto J et al. Obesity and the female sex, risk factors for knee osteoarthritis that may be attributable to systemic or local leptin biosynthesis and its cellular effects. Med Hypotheses 2005; 65:312-315.

Terlain B, Dumond H, Presle N et al. La leptine est-elle le maillonmanquant entre l'arthrose et obésité? Ann Phar Fr 2005; 63:186-193.

Tfelt-Hansen J, Brown EM. The calcium-sensing receptor in normal physiology and pathophysiology. Crit Rev Clin Lab Sci 2005; 42:35-70.

Tfelt-Hansen J, MacLeod RJ, Chattopadhyay N et al. Calcium-sensing receptor stimulates PTHrP secretion by PKC-dependent p38 pathway. J Bone Miner Res 2002; 17:S494.

Thélu J, Rossio P, Favier B. Notch signalling is linked to epidermal cell differentiation level in basal cell carcinoma, psoriasis and wound healing. BMC Dermatol 2002; 2(7):(1-12).

Thirunavukkarnasu K, Halladay DL, Miles RR et al. The osteoblast-specific transcription factor Cbfa1 contributes to the expession of osteoprotegerin, a potent inhibitor of osteoclast differentiation and function. J Biol Chem 2000; 275:25163-25172.

Thirunavukkarnasu K, Halladay DL, Miles RR et al. Analysis of regulator of G-protein signaling-2 (RGS-2) expression and function in osteoblastic cells. J Cell Biochem 2002; 85:837-850.

Thomas D, Cart SA, Piscopo DM et al. The retinoblastoma protein acts as a transcriptional coactivator required for osteogenic differentiation. Mol Cell 2001; 8:303-316.

Thomas GP, Baker SUK, Eisman JA et al. Changing RANKL/OPG mRNA expression in differentiating murine osteoclasts. J Endocrinol 2001; 170:451-460.

Thomas T, De Vittoris R, David VN et al. Leptin prevents disuse-induced bone loss in tail-suspended female rats (abstract). J Bone Miner Res 2001; 16:S143.

Thomas T, Gori F, Khosla S et al. Leptin acts on human marrow stromal cells to enhance differentiation of osteoblasts and to inhibit differentiation of adipocytes. Endocrinology 1999; 140:630-1638.

Till JE, McCulloch EA. A direct measurement of the radiation sensitivity of normal mouse bone marrow. Radiat Res 1961; 14:1419-1430.

Tintutt Y, Parhami F, Bostrom K et al. cAMP stimulates osteoblast-like differentiation of calcifying vascular cells. J Biol Chem 1998; 273:7547-7553.

Tintutt Y, Parhami F, Le V et al. Inhibition of osteoblast-specific transcription factor Cbfa1 by the cAMP pathway inotseoblastic cells. Ubiquitin/ proteasome-dependent regulation. J Biol Chem 1999; 274:28875-28879.

Tirone F. The gene PC3$^{TIS21/BTG2}$, prototype member of the PC3/BTG/TOB family: Regulator in control of cell growth, differentiation, and DNA repair? J Cell Physiol 2001; 187:155-165.

Tiyyagura SR, Kazerounian S, Schulz S et al. Receptrocal regulation of signaling by intracellular calcium and cyclic GMP. Vitamins and Hormones 2004; 69:69-94.

Tobimatsu T, Kaji H, Sowa H et al. Parathyroid hormone increases β-catenin levels through Smad3 in mouse osteoblastic cells. Endocrinology 2006, (published online February 16, 2006).

Togari A, Arai M, Mizutani S et al. Expression of mRNAs for neuropeptide receptors and β-adrenergic receptors in human osteoblasts and human osteogenic sarcoma cells. Neurosci Lett 1997; 233:125-128.

Tomkinson A, Gevers EF, Wit JM et al. The role of estrogen in the control of osteocyte apoptosis. J Bone Miner Res 1998; 13:12143-12150.

Toran-Allerand CD. Novel sites and mechanisms of oestrogen action in the brain. Novartis Found Symp 2000; 230:56-69.

Tonna EA, Lampen NM. Electron microscopy of aging skeletal cells. I. Centrioles and solitary cilia. J Gerontol 1972; 27:316-324.

Torday JS, Sun H, Wang L et al. Leptin mediates the parathyroid hormone-related protein paracrine stimulateion of fetal lung maturation. Am J Physiol Lung Cell Mol Pysiol 2002; 282:L405-L410.

Toromanoff A, Ammann P, Mosekilde L et al. Parathyroid hormone increases bone formation and improves mineral balance in vitamin D-deficient female rats. Endocrinology 1997; 138:2449-2457.

Tovar-Sepulveda VA, Shen X, Falzou M. Intracrine PTHrP protects against serum starvation-induced apoptosis and regulates the cell cycle in MCF-7 breast cancer cells. Endocrinology 2002; 143:596-606.

Tregear GW, van Rietschoten J, Greene E et al. Solid-phase synthesis of the biologically active N-terminal 1-34 peptide of human parathyroid hormone. Hoppe Seylers Z Physiol Chem 1974; 355:415-421.

Trivett MK, Walker TI, Macmillan DL et al. Parathyroid hormone-related protein (PTHrP) production sites in elasmobranchs. J Anat 2002; 201:41-52.

Troen BR. Molecular mechanisms underlying osteoclast formation amd activation. Exp Gerontol 2003; 38:605-614.

Tsingotjidou A, Nervina JM, Pham L et al. Parathyroid hormone induces RG-2 expression by a cyclic adenosine 3',5'-monophosphate-mediated pathway in primary neonatal murine osteoblasts. Bone 2002; 30:677-684.

Tumber A, Meikle MC, Hill PA. Autocrine signals promote osteoblast survival in culture. J Endocrinol 2000; 167:383-390.

Tucker M, Valencia-Sanchez MA, Staples RR et al. The transcription factor associated Ccr4 annd Caf1 proteins are components of the major cytoplasmic mRNA deadenylase in Saccharomyces cerevisiae. Cell 2001; 104:377-386.

Turco AE, Padovani EM, Chiaffoni GP et al. Molecular genetic diagnosis of autosomal dominant polycystic kidney disease in a newborn with bilateral cyctic kidneys detected prenatalyy and multiple skeletal malformations. J Med Genet 1993; 30:973.

Turco AE, Peissel B, Rossetti S et al. Skeletal malformations and polycystic kidney disease. J Med Genet 1994; 31:741-742.

Turner BM. Chromatin and Gene Regulation. Oxford: Blackwell Science, 2001.

Turner CH, Burr DB, Hock JM et al. The effects of PTH-(1-34) on bone structure and strength in ovariectomized monkeys. Adv Exp Med Biol 2001; 496:165-179.

Turner CH, Robling AG. Exercise as an anabolic stimulus for bone. Curr Pharma Des 2004; 10:2629-2641.

Turner CH, Robling AG, Duncan RL et al. Do bone cells behave like a neuronal network? Calcif Tissue Int 2002; 70:435-442.

Turner HE, Harris AL, Melmed S et al. Angiogenesis in endocrine tumors. Endocr Rev 2003; 24:600-632.

Tyson DR, Swarthout JT, Partridge NC. Increased osteoblastic c-fos expression by parathyroid hormone requires protein kinase A phosphorylation of the cyclic adenosine 3', 5'-monophosphate response element-binding protein at serine 133. Endocrinology 1999; 140:1255-1261.

Udawela M, Hay DL, Sexton PM. The receptor activity modifying protein family of G protein coupled receptor accessory proteins. Semin Cell Devel Biol 2004; 15:299-308.

Urano Y, Hayashi I, Isoo N et al. Association of active γ-secretase complex with lipid rafts. J Lipid Res 2005; 46:904-912.

Urbich C, Dernbach E, Zeiber AM et al. Double-edged role of statins in angiogenesis signaling. Circ Res 2002; 90:737-744.

Usdin TB, Bonner TI, Hoare SRJ. The parathyroid hormone 2 (PTH2) receptor. Receptors and Channels 2002; 8:211-218.

Vagner S, Gensac MC, Maret A et al. Alternative translation of human fibroblast growth factor 2 mRNA occurs by internal entry of ribosomes. Mol Cell Biol 1995; 15:35-44.

Utting JC, Robins SP, Brandao-Burch A et al. Hypoxia inhibits the growth, differentiation and bone-forming capacity of rat osteblasts. Exp Cell Res 2006; 312:1693-1702.

Uzawa K, Grzesik WJ, Nishiura T et al. Differential expression of human lysyl hydroxylase genes, lysine hydroxylation, and cross-linking of type I collagen during osteoblastic differentiation in vitro. J Bone Miner Res 1999; 14:1272-1280

Vahle JL, Long GG, Sandusky G et al. Bone neoplasms in F344 rats given teriparatide [rhPTH(1-34)] are dependent on duration of treatment and dose. Toxicol Pathol 2004; 32:426-438.

Vahle JL, Sato M, Long CC et al. Skeletal changes in rats given daily subcutaneous injections of recombinant human parathyroid hormone (1-34) for 2 years and relevance to human safety. Toxicol Pathol 2002; 30:312-321.

Valenta A, Roschger P, Fratzl-Zelman N et al. Combined treatment with PTH(1-34) and OPG increases bone volume and uniformity of mineralization in aged ovariectomized rats. Bone 2005; 37:87-95.

Van Beek E, Pieterman E, Cohen L et al. Farnesyl pyrophosphate synthase is the molecular target of nitrogen-containing bisphosphonates. Biochem Biophys Res Commun 1999; 264:108-111.

Van Bezooijen RL, Papapoulos SE, Lowik CW. Bone morphogenetic proteins and their antagonists: The sclerostin paradigm. J Endocrinol Invest 2005a; 28(8 Suppl.):15-17.

Van Bezooijen RL, Papapoulos SE, Hamdy NA et al. Control of bone formation by osteocytes? Lessons from rare skeletal disorders sclerostosis and van Buchem disease. BoneKEy-Osteovision 2005b; 2:33-38.

Van Bezooijen RL, Roelen BAJ, Visser A et al. Sclerostin is an osteocyte-expressed negative regulator of bone formation, but not a classical BMP antagonist. J Exp Med 2004; 199:805-814.

Van Den Berg DJ, Sharma AK, Bruno E et al. Role of members of the Wnt gene family in human hematopoiesis. Blood 1998; 92:3189-3202.

Vanderschueren D, Boonen S, Ederveen GH et al. Skeletal effects of estrogen deficiency as induced by an aromatase inhibitor in an aged rat model. Bone 2000; 27:611-617.

Van Es JH, van Gijn ME, Riccio O et al. Notch/γ-secretase inhibition turns proliferative cells in intestinal crypts and adenomas into goblet cells. Nature 2005; 435:959-963.

Van Staa TP, Wegman SLJ, DeVries F et al. The use of statins and risk of fractures (abstract). J Bone Miner Res 2000; 15:S155.

Van't Hof RJ, Ralston SH. Nitric oxide and bone. Immunology 2001; 103:255-261.

Van't Hof RJ, Armour KJ, Helfrich MH et al. The anabolic, but not anti-resorptive, effect of mevaststin in vivo is depenedent on endothelial nitric oxide synthase (abstract). J Bone Miner Res 2002a; 17:1191.

van't Hof RJ, Armour KJ, Penman S et al. Mice deficient in nNOS have increased bone mass; possible indications for a central mechanism of regulation for bone turnover(abstract). J Bone Miner Res 2002b; 17:S442.

Vashishth D. Age-dependent biomechanical modifications in bone. Crit Rev Eukaryotic Gene Express 2005; 15:343-357.

Vattikuti R, Towler DA. Osteogenic regulation of vascular calcification: An early perspective. Am J Physiol Endocrinol Metab 2004; 286:E686-E696.

Veillard NR, Mach E. Statins: The new aspirins? Cell Mol Life Sci 2002; 59:1771-1786.

Vezquez-Prado J, Casas-Gonzalez P, Garcia-S<inz JA. G protein-coupled receptor cross-talk: Pivotal roles of protein phosphorylation and protein-protein interactions. Cell Signalling 2003; 15:549-557.

Veillette CJ, von Schroeder HP. Endothelin-1 down-regulates the expression of vascular endothelial growth factor-A associated with osteoprogenitor proliferation and differentiation. Bone 2004; 34:288-296.

Verborgt O, Gibson GJ, Schaffler MB. Loss of osteocyte integrity in association with microdamage and bone remodeling after fatigue in vivo. J Bone Miner Res 2000; 15:60-67.

Verborgt O, Tatton NA, Majeska RJ et al. Spatial distribution of Bax and Bcl-2 in osteocytes after bone fatigue: Complementary roles in bone remodeling regulation? J Bone Miner Res 2002; 17:907-914.

Vermes C, Glant TT, Hallab NJ et al. The potential role of the osteoblast in the development of periprosthetic osteolysis. J Arthroplasty 2001; 16(Suppl. 1):95-100.

Versteege I, Medjkane S, Rouillard D et al. A key role of the hSNF5/INI1 tumour suppressor in the control of the G1-S of the cell cycle. Oncogene 2002; 21:6403-6412.

Vico L, Hinsenkamp M, Jones D et al. Osteobiology, strain, and microgravity. Part II: Studies at the tissue level. Calcif Tissue Int 2001; 68:1-10.

Vilagra A, Gutiérrez J, Paredes R et al. Reduced CpG methylation is associated with transcriptional activation of the bone-specific rat osteocalcin gene in osteoblasts. J Cell Biochem 2002; 85:112-122.

Virchow R. Cellular Pathology: As Based Upon Physiological and Pathological Histology. New York: Dover, 1863/1971.

Virdee K, Parone PA, Tolkovsky AM. Phosphorylation of the pro-apoptosis protein BAD on serine 155, a novel site, contributes to cell survival. Curr Biol 2000; 10:1151-1154.

Visnjic D, Kalajzic Z, Rowe DW et al. Hemopoiesis is severely altered in mice with an induced osteoblast deficiency. Blood 2004; 103:3258-3264.

Voltz JW, Weinman EJ, Shenolikar S. Expanding the role of NHERF, a PDZ-domain containing protein adapter, to growth regulation. Oncogene 2001; 20:6309-6314.

Von Schroeder HP, Veillette CJ, Payandeh J et al. Endothelin-1 promotes osteoprogenitor proliferation and differentiation in fetal rat calvarial cell cultures. Bone 2003; 33:673-684.

Von Stechow D, Fish S, Yahalom D et al. Does simvastatin stimulate bone formation in vivo? BMC Musculoskelet Disord 2003; 4:8.

Von Stechow D, Libermann T, Hattersley G et al. Transcriptionally profiling of estrogen- and PTH-regulated gene during bone formation in OVX mice. J Bone Miner Res 2003; 18:S138.

Vortkamp A, Lee K, Lanske B et al. regulation of rate of cartilage differentiation by Indian hedgehog and PTH-related protein. Science 1996; 273:613-622.

Vortkamp A, Pathi S, Peretti GM et al. Recapitulation of signals regulating embryonic bone formation during postnatal growth and in fracture repair. Mech Dev 1998; 71:65-76.

Vries RGL, Bezrookov V, Zuijderduijn LMP et al. Cancer-associated mutations in chromatin remodeler hSNF5 promote chromosomal instability by compromising the mitotic checkpoint. Genes Dev 2005; 19:665-670.

Walker D. The induction of osteopetrotic changes in hyphophysectomized, thyroparathyroidectomized, and intact rats of various ages. Endocrinology 1971; 89:1389-1406.

Walker G. Snowball Earth. New York: Crown Publishers, 2003.

Wallin R, Wajih N, Greenwood GT et al. Arterial calcification: A review of mechanisms, animal models, and the prospects for therapy. Med Res Rev 2001; 21:274-301.

Walsh CA, Bowler WB, Bilbe G et al. Effects of PTH on PTHrP gene expression in human osteoblasts: Upregulation with the kinetics of an immediate early gene. Biochem Biophys Res Commun 1997; 239:155-159.

Walsh K, Takahashi A. Transcriptional regulation of vascular smooth muscle phenotype. Z Kardiol 2001; 90(Suppl. III):12-16.

Wang D, Canaff L, Davidson D et al. Alteration in the sensing and transport of phosphate and calcium by differentiating chondrocytes. J Biol Chem 2001; 276:33995-34005.

Wang DS, Yamazaki K, Nohtomi K et al. Increase of vascular endothelial growth factor mRNA expression of 1,25-dihydroxyvitamin D3 in human osteoblast-like cells. J Bone Miner Res 1996; 11:472-479.

Wang L, Cowin SC, Weinbaum S et al. Modeling tracer transport in an osteon under cyclic loading. Ann Biomed Eng 2000; 29:810-816.

Wang L, Fritton SP, Cowin SC et al. Fluid pressure relaxation depends upon osteonal microstructure modeling an oscillatory bending experiment. J Biomech 1999; 32:663-672.

Wang PS, Solomon DH, Mogun H et al. HMG-CoA reductase inhibitors and the risk of hip fractures in elderly patients. JAMA 2000; 283:3211-3216.

Wang Y, Liu Y, Lee S et al. PTH promotes osteoblastic differentiation/commitment but not proliferation in GFP-marked primary osteoblast cultures (abstract). J Bone Miner Res 2003; 18:S77.

Wang Y, Nishida S, Burghardt A et al. The IGF-I receptor is required for the anabolic actions of parathyroid hormone on bone. J Bone Miner Res 2004; 19:S106.

Warmington K, Ominsky M, Bolon B et al. Sclerostin monoclonal antibody treatment of osteoporotic rats completely reverses one year of ovariectomy-induced systemic bone loss. J Bone Miner Res 2005; 20:S22.

Watson CS, ed. The Identities of Membrane Steroid Receptors. Boston: Kluwer Academic Publishers, 2003.

Watson PH, Fraher LJ, Hendy GN et al. Nuclear localization of the type 1 PTH/PTHrP receptor in rat tissues. J Bone Miner Res 2000a; 15:1033-1044.

Watson PH, Fraher LJ, Kisiel M et al. Enhanced osteoblast development after continuous infusion of hPTH-(1-84) in the rat. Bone 1999; 24:89-94.

Watson PH, Fraher LJ, Natale BV et al. Nuclear localization of the type 1 parathyroid/parathyroid hormone-related peptide receptor in MC3T3-E1 cells: Assocoation with serum-induced cell proliferation. Bone 2000b; 26:221-225.

Watson PH, Kisiel M, Patterson EK et al. Expression of type I PTH/PTHrP receptors in the rat osteoclast and osteoclast-like RAW 264.7 cells. J Bone Miner Res 2002; 17:S287.

Watson P, Lazowski D, Han V et al. Parathyroid hormone restores bone mass and enhances osteoblast insulin-like growth factor I gene expression in ovariectomized rats. Bone 1995; 16:375-365.

Weber JM, Forsythe SR, Christianson CA et al. Parathyroid hormone stimulates the expression of the Notch ligand Jagged1 in osteoblastic cells. Bone 2006, (published online, April 27).

Weiss LA, Barrett-Connor E, von Mühlen D et al. Leptin predicts bone density and bone resorption in older women but not older men: The Rancho Bernardo study. J Bone Miner Res 2006; 21, (published online).

Weisser J, Riemer S, Schmidt M et al. Four distinct chondrocyte populations in the fetal bovine growth plate: Highest expressionlevels of PTH/PTHrP receptor, Indian hedgehog, and MMP-13 in hypertrophic chondrocytes and their suppression by PTH (1-34) and PTHrP (1-40). Exp Cell Res 2002; 279:1-13.

Wen W, Meinkoth JL, Tsien R et al. Identification of a signal for rapid export of proteins from the nucleus. Cell 1995a; 82:463-473.

Wen W, Taylor SS, Meinkoth JL. The expression and intracellular distribution of the heat-stable protein kinase inhibitor is cell cycle regulated. J Biol Chem 1995b; 270:2041-2046.

Wenger RH. Cellular adaptation to hypoxia: O_2-sensing protein hydroxylases, hypoxia-inducible transcription factors, and O_2-regulated gene expression. FASEB J 2002; 16:1151-1162.

Weinstein RS, Jilka RL, Parfitt M et al. Inhibition of osteobastogenesis and promotion of apoptosis of osteoblasts and osteocytes by glucocorticoid. J Clin Invest 1998; 102:274-282.

Weir EC, Philbrick WM, Amling M et al. Targeted overexpression of parathyroid hormone-related peptide in chondrocytes causes chondrodysplasia and delayed endochondral bone formation. Proc Natl Acad Sci USA 1996; 93:10240-10245.

Westbroek I, van der Plas A, de Rooij KE et al. Expression of serotonin receptors in bone. J Biol Chem 2001; 276:2896-28968.

Whang K, Zhao M, Qiao M et al. Administration of lovastatin locally in low doses in a novel delivery ystem induces prolonged bone formation (abstract). J Bone Miner Res 2000; 15:S225.

Wheatley DN. Landmarks in the first hundred years of primary (9+0)cilium research. Cell Biol International 2005; 29:333-339.

Wheatley DN, Wang AM, Strugnell GE. Expression of primary cilia in mammalian cells. Cell Biol Int 1996; 20:73-81.

Whitfield JF. Statins; new drugs for treating osteoporosis? Exp Opin Invest Drugs 2001a; 10:409-415.

Whitfield JF. Leptin, brains and bones. Exp Opin Invest Drugs 2001b; 10:1617-1622.

Whitfield JF. Statins and the stimulation of bone growth—Do they or don't they? Geriatric Times 2002a; 3:23-28.

Whitfield JF. Leptin—A new member of the bone builder's club? Medscape Women's Health eJournal 2002b; 7(4), (http://www.medscape.com/viewarticle/439243).

Whitfield JF. The primary cilium—Is it an osteocyte's strain-sensing flowmetere? J Cell Biochem 2003a; 89:233-237.

Whitfield JF. How to grow bone to treat osteoporosis and mend fractures. Curr Rheumatol Rep 2003b; 5:45-56, (Curr Osteoporosis Rep 1:32-40).

Whitfield JF. The neuronal primary cilium — An extrasynaptic signaling device. Cell Signal 2004; 16:763-767.

Whitfield JF. Taming psoriatic keratinocytes—PTHs' uses go up another. notch. J Cell Biochem 2004a; 91:251-256.

Whitfield JF. Osteogenic PTHs and vascular ossification—Is there a danger for osteoporotics? J Cell Biochem 2005a; 95:437-444.

Whitfield JF. Parathyroid hormone (PTH) and hematopoiesis: New support for some old observations. J Cell Biochem 2005b; 96:278-284.

Whitfield JF. Parathyroid hormone: A novel tool for treating bone marrow depletion in cancer patients caused by chemotherapeutic drugs and ionizing radiation. Cancer Lett 2006a; 244:8-15.

Whitfield JF. Osteoporosis-treating parathyroid hormone peptides: What are they? what do theydo? How might they do it? Curr Opin Invest Drugs 2006b; 7:349-359.

Whitfield JF. Can statins put the brakes on Alzheimer's disease? Expert Opin Investig Drugs 2006c; 15:1479-1485.

Whitfield JF, Bird RP, Morley P et al. The effects of parathyroid hormone (PTH) fragments on bone formation and thei lack of effects on the initiation of colon carcinogenesis in rats as indicated by preneoplastic aberrant crypt formation. Cancer Lett 2003; 200:107-113.

Whitfield JF, Chakravarthy B. Calcium: The Grand-Master Cell Signaler. Ottawa: NRC Research Press, 2001.

Whitfield JF, Chakravarthy BR, Durkin JP et al. Parathyroid hormone stimulates protein kinase C but not adenylate cyclase in mouse epidermal keratinocytes. J Cell Physiol 1992; 150:299-303.

Whitfield JF, Isaacs RJ, Chakravarthy B et al. Stimulation of protein kinase-C activity in cells expressing human parathyroid hormone (PTH) receptors by C- and N-terminally truncated fragments of PTH 1-34. J Bone Miner Res 2001; 16:441-447.

Whitfield JF, Isaacs RJ, Jouishomme H et al. C-terminal fragment of parathyroid hormone-related protein, PTHrP-(107-111), stimulates membrane-associated protein kinase C activity and modulates the proliferation of human and murine skin keratinocytes. J Cell Physiol 1996a; 166:1-11.

Whitfield JF, Isaacs RJ, MacLean S et al. Stimulation of membrane-associated protein kinase-C activity in spleen lymphocytes by hPTH-(1-31)NH$_2$, its lactam derivative, hPTH-(1-31)NH$_2$ hPTH-(1-31)NH$_2$, and hPTH-(1-30)NH$_2$. Cell Signal 1999a; 11:159-164.

Whitfield JF, MacManus JP, Youdale T et al. The roles of calcium and cyclic AMP in the stimulatory action of parathyroid hormone on thymic lymphocyte proliferation. J Cell Physiol 1971; 78:355-368.

Whitfield JF, Morley P, Fraher L et al. The stimulation of vertebral and tibial bone growth by the parathyroid hormone fragments, hPTH-(1-31)NH$_2$, [Leu^{27}]cyclo(Glu^{22}-Lys^{26})hPTH-(1-31)NH$_2$,and hPTH0(1-30)NH$_2$. Calcif Tissue Int 2000a; 66:307-312.

Whitfield JF, Morley P, Langille RM et al. Adenylyl cyclase-activating agents: Parathyroid hormone and prostaglandins E. In: Whitfield JF, Morley P, eds. Anabolic Treatments for Osteoporosis. Boca Raton: CRC Press, 1998a:108-149.

Whitfield JF, Morley P, Ross V et al. Restoration of severely depleted femoral trabecular bone in ovariectomized rats by parathyroid hormone-(1-34). Calcif Tissue Int 1995; 56:227-231.

Whitfield JF, Morley P, Ross V et al. The hypotensive actions of opsteogenic and nonosteogenic parathyroid hormone fragments. Calcif Tissue Int 1997a; 60:302-308.

Whitfield JF, Morley P, Willick G et al. Stimulation of the growth of femoral trabecular bone in ovariectomized rats by the novel parathyroid hormone fragment hPTH-(1-31)NH$_2$ (ostabolin). Calcif Tissue Int 1996; 58:81-87.

Whitfield JF, Morley P, Willick GE. The Parathyroid Hormone: An Unexpected Bone builder for Treating Osteoporosis. Austin: Landes Bioscience, 1998b.

Whitfield JF, Morley P, Willick G et al. Cyclization by a specific lactam Increases the ability of human parathyroid hormone hPTH-(1-31)NH$_2$ to stimulate bone growth in ovariectomized rats. J Bone Miner Res 1997b; 12:1246-1252.

Whitfield JF, Morley P, Willick G et al. Comparison of the ability of recombinant humanparathyroid hormone, rhPTH-(1-84), and hPTH-(1-31)NH$_2$ to stimulate femoral trabecular bone growth on ovariectomized rats. Calcif Tissue Int 1997c; 60:26-29.

Whitfield JF, Morley P, Willick G et al. Comparison of the abilities of human parathyroid hormone (1-31)NH$_2$ and human parathyroid hormone-related protein (1-31)NH$_2$ to stimulate femoral trabecular bone growth in ovariectomized rats. Calcif Tissue Int 1997d; 61:322-326.

Whitfield JF, Morley P, Willick G et al. Comparison of the abilities of human parathyroid hormone (hPTH)-(1-34) and [Leu^{27}]cyclo(Glu^{22}-Lys^{26})hPTH-(1-31)NH$_2$ to stimulate femoral trabecular bone growth in ovariectomized rats. Calcif Tissue Int 1998c; 63:423-428.

Whitfield JF, Morley P, Willick G. The bone-building actions of the parathyroid hormone: Implications for the treatment of osteoporosis. Drugs Aging 1999b; 15:117-129.

Whitfield JF, Morley P, Willick G. Stimulation of femoral tabecular bone growth in ovariectomized rats by human parathyroid hormone (hPTH)-(1-30)NH$_2$. Calcif Tissue Int 1999c; 65:143-147.

Whifield JF, Morley P, Willick GE. The parathyroid hormones: Bone-forming agents for treatment of osteoporosis. Medscape Women's Health 2000b; 5, (http://womenshealth.medscape.com).

Whitfield JF, Morley P, Willick G et al. Lactam formation increases receptor binding, adenylyl cyclase stimulation and bone growth stimulation by human parathyroid hormone (hPTH)-(1-28)NH$_2$. J Bone Miner Res 2000c; 15:964-970.

Whitfield JF, Morley P, Willick G. The parathyroid hormone, its fragments and analogs—Potent bone-builders for treating osteoporosis. Exp Opin Invest Drugs 2000d; 9:1293-1315.

Whitfield JF, Morley P, Willick G et al. Estratriene-3-ol, an activator of nongenomic estrogen-like signaling (ANGELS), does not stimulate bone growth in ovariectomized rats. J Bone Miner Res 2001a; 16:S309.

Whitfield JF, Morley P, Willick G. The control of bone growth by parathyroid hormone (PTH), leptin and statins. Crit Rev Eukaryotic Gene Express 2002a; 12:23-51.

Whitfield JF, Morley P, Willick GE. Parathyroid hormone, its fragments and their analogs for the treatment of osteoporosis. Treat Endocrinol 2002b; 1:175-190.

Whitfield JF, Morley P, Willick G. Bone growth stimulators: New tools for treating bone loss and mending fractures. Vitamins and Hormones 2002c; 65:1-80.

Whitfield JF, Rixon RH, MacManus JP et al. Calcium, cyclic adenosine 3',5'-monophosphate, and the control of cell proliferation. In Vitro 1973; 8:257-278.

Whitfield JF, Youdale T, Perris AD. Early postirradiation changes leading to the loss of nuclear structure in rat thymocytes. Exp Cell Res 1967; 48:461-472.

Wiig JN, Sand Bakker T. Hyperparathyroidism with multiple malignant tumors of bone with giant cells. Acta Chir Scand 1971; 137:391-393.

Wilding JP, Gilbery SG, Bailey CJ et al. Increased neuropeptide-Y messenger ribonucleic acid (mRNA) and decreased neurotensin mRNA in the hypothalamus of the obese (ob/ob) mouse. Endocrinology 1993; 132:1939-1944.

Wiley JC, Wailes LA, Idzerda RL et al. Role of regulatory subunits and protein kinase inhibitor (PKI) in determining nuclear localization and activity of the catalytic subunit of protein kinase A. J Biol Chem 1999; 274:6381-6387.

Willert K, Brown JB, Danenberg E et al. Wnt proteins are lipid-modified and can act as stem cell growth factors. Nature 2003; 423:448-452.

Wilsman NJ. Cilia of adult canine articular chondrocytes. J Ultrastruct Res 1978; 64:270-281.

Wilsman NJ, Fletcher TF. Cilia of neonatal articular chondrocytes: Incidence and morphoilogy. Anat Rec 1978; 190:871-890.

Winkler DG, Sutherland MK, Geoghegan JC et al. Osteocyte control of bone formation via sclerostin, a novel BMP antagonist. EMBO J 2003; 22:6267-6276.

Wolfe MS. Shutting down Alzheimer's. Sci American 2006; 73-79.

Wolozin B. Cholesterol, statins and dementia. Curr Opin Lipidol 15:667-672.

Wolozin B, Brown IIIrd J, Theisler C et al. The cellular biochemistry of cholesterol and statins: Insights into pathophysiology and therapy of Alzheimer's disease. CNS Drug Rev 2004; 10:126-146.

Wolozin B, Kellman W, Ruosseau P et al. Decreased prevalence of Alzheimer disease associated with 3-hydroxy-3-methyglutaryl coenzyme A reductase inhibitors. Arch Neurol 2000; 57:1439-1443.

Wong SKF. G protein selectivity is regulated by multiple intracellura regions of GPCRs. Neurosignals 2003; 12:1-12.

Woolf V. A Room of One's Own. Harmonsworth: Penguin Books Ltd., 1975:28.

World Health Organization. Assessment of Fracture Risk and its Application to Screening fpr Postmenopausal Osteoporosis. Report of a WHO study Group. Geneva: World Health Organization, 1994.

Wronski TJ, Li M. PTH: Skeletal effects in the ovariectomized rat model for Osteoporosis. In: Whitfield JF, Morley P, eds. Anabolic Treatments for Osteoporosis. Boca Raton: CRC Press, 1997:59-81.

Wronski TJ, Ratkus AM, Thomsen JS et al. Sequential treatment with basic fibroblast growth factor and parathyroid hormone restores lost cancellous bone mass and strength in the proximal tibia of aged ovariectomized rats. J Bone Miner Res 2001; 16:1399-1407.

Wu LNY, Ishikawa Y, Sauer GR et al. Morphological and biochemical characterization of mineralizing primary culturers of avian growth plate chondrocytes: Evidence for cellular processing of Ca^{2+} and Pi prior to matrix mineralization. J Cell Biochem 1995; 57:218-237.

Wu S, Yoshiko Y, De Luca F. Stanniocalcin 1 acts as a paracrine regulator of growth plate chondrogenesis. J Biol Chem 2006; 281:5120-5127.

Wu Y, Kumar R. Parathyroid hormone regulates transforming growth factor β1 and β2, synthesis in osteoblasts via divergent signaling pathways. J Bone Miner Res 2000; 15:879-884.

Xian-Guang H, Aldridge RJ, Bergström J et al. The Cambrian Fossils of Chengjiang, China. Oxford: Blackwell Science Ltd, 2004.

Xiao Z, Zhang S, Mahlios J et al. Cilia-like structures and polycystin-1 in osteoblasts/osteocytes and associated abnormalities in skeletogenesis and Runx2 expression. J Biol Chem 2006; 281:30884-30895.

Xie H, Tang S, Cui R et al. Apelin and its receptor are expressed in human osteoblasts. Regulatory Peptides 2006; 134:118-125.

Xie Y, Gibbs TC, Meier KE. Lysophosphatidic acid as an autocrine and paracrine mediator. Biochim Biophys Acta 2002; 1582:270-281.

Xing L, Carlson L, Story B et al. Expression of either NF-κB p50 or p52 in osteoclast precursors is required for IL-1-induced bone resorption. J Bone Miner Res 2003; 18:260-269.

Xu HL, Galea E, Santizo RA et al. The key role of caveolin-1 in estrogen-mediatedregulation of endothelial nitric oxide function in cerebral arterioles in vivo. J Cereb Blood Flow Metab 2001; 21:907-913.

Yagi K, Tsuji K, Nifuji A et al. Bone morphogenic protein-2 enhances osterix gene expression in chondrocytes. J Cell Biochem 2003; 88:1077-1083.

Yamagishi T, Otsuka E, Hagiwara H. Reciprocal control of expression of mRNAs for osteoclast differentiation factor and OPG in osteogenic stromal cells by genistein: Evidence for the involvement of topoisomase II in osteoclastogenesis. Endocrinology 2001; 142:3632-3637.

Yamaguchi T. Calcium-sensing receptor in bone. In: Chattopadhyay N, Brown EM, eds. Calcium-Sensing Receptor. Boston: Kluwer Academic Publishers, 2003:103-124.

Yamaguchi A, Komori T, Suda T. Regulation of of osteoblast differentiation mediated by bone morphogenetic proteins, hedgehog and Cbfa 1. Endocrinology 2000; 21:393-411.

Yamaguchi T, Chattopadhyay N, Kifor O et al. Mouse osteoblastic cell line (MC3T3-E1) expresses extracellular calcium (Ca^{2+}o)-sensing receptor and its agonists stimulate chemotaxis and proliferation of MC3T3-E1 cells. J Bone Miner Res 1998a; 13:1530-1538.

Yamaguchi T, Chattopadhyay N, Kifor O et al. Activation of p42/44 and p38 mitogen-activated protein kinases by extracellular calcium-sensing receptor agonists induces mitogenic responses in the mouse osteoblastic MC3T3-E1 cell line. Biochem Biophys Res Commun 2000; 279:363-368.

Yamaguchi T, Chattopadhyay N, Kifor O et al. Expression of extracellular calcium-sensing receptors in human osteoblastic MG-63 cell line. Am J Physiol Cell Physiol 2001; 280:C382-C393.

Yamaguchi T, Kifor O, Chattopadhyay N et al. Expression of extracellular calcium (Ca^{2+}o)-sensing receptor in the clonal osteoblast-like cellines, UMP-106 and SAOS-2. Biochem Biophys Res Commun 1998b; 243:753-757.

Yamakawa M, Liu LX, Date T et al. Hypoxia-inducible factor-1 mediates activation of cultured vascular endothelial cells by inducing multiple angiogenic factors. Circ Res 2003; 93:664-673.

Yamamoto S, Morimoto I, Zeki K et al. Centrally administered parathyroid hormone (PTH)-related protein (1-34) but not PTH(1-34) stimulates arginine-vasopressin secretion and its messenger ribonucleic acid expression in supraoptic nucleus of the conscious rats. Endocrinology 1998; 139:383-3888.

Yamasaki Y, Yamaguchi T, Watahiki J et al. The role of craniofactial growth in leptin-deficient (ob / ob) mice. Orthod Craniofacial Res 2003; 6:233-241.

Yamashiro T, Fukunaga T, Yamashita K et al. Gene and protein expression of brain-derived neurotrophic factor and TrkB in bone and cartilage. Bone 2001; 28:404-409.

Yamashita H, Gao P, Cantor T et al. Large carboxy-terminal parathyroid hormone (PTH) fragment with a relatively longer half-life than 1-84 PTH is secreted directly from the parathyroid gland in humans. Eur J Endocrinol 2003; 149:301-306.

Yang D, Guo J, Divieti P et al. Parathyroid hormone activates PKC-δ and regulates osteoblastic differentiaition via a PLC-independent pathway. Bone 2005; 38:485-496.

Yao W, Hadi T, Zhou J et al. Basic FGF increases trabecular bone connectivity and improves bone strength in the lumbar vertegral body of osteopenic rats (abstract). J Bone Miner Res 2004; 19:S461.

Yao W, Li CY, Farmer RW et al. Simvastatin did not prevent bone loss in ovariectomized rat (abstract). J Bone Miner Res 2001; 16:S294.

Yellowley CE, Zhongyong LI, Zhou Z et al. Functional gap junctions between osteocyctic and osteoblastic cells. J Bone Miner Res 2000; 15:209-217.

Yin H, Morioka H, Towle CA et al. Evidence that HAX-1 is an interleukin-1α N-terminal binding protein. Cytokine 2001; 15:122-137.

Yin T, Li L. The stem cell niches in bone. J Clin Invest 2006; 116:1195-1201.

Yin XM, Ding WX, Zhao Y. Bcl-2 family proteins. Master regulators of apoptosis. In: Yin XM, Dong Z, eds. Essentials of Apoptosis. Totowa, NJ: Humana Press, 2003:13-27.

Yoder BK, Hou X, Guay-Woodford LM. The polycystic disease proteins, polycystin-1, polycystin-2, and cyctin, are colocalized in renal cilia. J Am Soc Nephrol 2002; 13:2508-2516.

Yoon Y, McNiven MA. Mitochondrial division: New partners in membrane pinching. Curr Biol 2001; 11:R67-R70.

You L, Cowin SC, Schaffler MB et al. A model for strain amplification in the actin cytoskeleton of osteocytes due to fluid drag on pericellular matrix. J Biomech 2001; 34:1375-1386.

Yoshiko Y, Maeda N, Aubin JE. Stanniocalcin 1 stimulates osteoblkast differentiaition in rat calvaria cultures. Endocrinology 2003; 144:4134-4143.

Young DW, Pratrap J, Javed A et al. SWI/SNF chromatin remodeling complex is obligatory for BMP-induced Runx2-dependent skeletal gene expression that controls osteoblast differentiation. J Cell Biochem 2005; 94:720-730.

Zabeau L, Lavens D, Peelman F et al. The ins and outs of leptin receptor activation. FEBS Lett 2003; 546:45-50.

Zabel U, Kleinschnitz C, Oh P et al. Calcium-dependent membrane association sensitizes soluble guanylyl cyclase to nitric oxide. Nature Cell Biol 2002; 4:307-311.

Zaid A, Adebanjo OA, Moonga BS et al. Emerging insights into the role of calcium ions in osteoclast regulation. J Bone Miner Res 1999; 14:669-674.

Zaidi SK, Javed A, Choi JY et al. A specific targeting signal directs Runx2/Cbfa1 to subnuclear domains and contributes to transactivation of the osteocalcin gene. J Cell Sci 2001; 114:3093-3102.

Zamah AM, Delahunty M, Luttrell LM et al. Protein kinase A-mediated phosphorylaion of the β2-adrenergic receptor regulates its coupling to Gs and Gi. Demonstration in a reconstituted system. J Biol Chem 2002; 277:31249-31256.

Zanchetta JR, Bogado CE, Ferretti JL et al. Effects of teriparatide [recombinant human parathyroid hormone (1-34)] on cortical bone in postmenopausal women with osteoporosis. J Bone Miner Res 2003; 18:539-543.

Zao M, Zhang J, Feng JQ et al. β-Catenin directly activates bone-specific genes in osteoblasts and chondrocytes. J Bone Miner Res 2002; 17:S182.

Zaragoza Z, Soria E, Lopez E et al. Activation of the mitogen activated protein extracellular signal-regulated kinase 1 and 2 by the nitric oxide-cGMP-cGMP-dependent protein kinase axis regulates the expression of matrix metalloproteinase 13 in vascular endothelial cells. Mol Pharmacol 2002; 62:927-935.

Zecchi-Orlandini S, Formigli L, Tani A et al. 17β-estradiol induces apoptosis in the preosteoclastic FLG29.1 cell line. Biochem Biophys Res Commun 1999; 255:680-685.

Zerega B, Ceremelli S, Bianco P et al. Parathyroid hormone [hPTH-(1-34)] and parathyroid hormone-related protein [hPTHrP-(1-34)] promote reversion of hypertrophic chondrocytes to a prehypertrophic proliferating phenotype and prevent terminal differentiation of osteoblast-like cells. J Bone Miner Res 1999; 14:1281-1289.

Zhang F, Chen Y, Heiman M et al. Leptin: Structure, function and biology. Vitamins Horm 2005; 71:345-372.

Zhang J, Niu C, Yo L et al. Identification of the haematopoietic stem cell niche and control of niche size. Nature 2003; 425:836-841.

Zhang RW, Supowit SC, Xu X et al. Expression of selected osteogenic markers in the fibroblast-like cells of rat marrow stroma. Calcif Tissue Int 1995; 56:283-291.

Zhang X, Schwarz EM, Young DA et al. Cyclooxygenase-2 regulates mesenchymal cell differentiation into the osteoblast lineage and is critically involved in bone repair. J Clin Invest 2002a; 109:1405-1415.

Zhang X, Sobue T, Hurley MM. FGF-2 increases colony formation, PTHreceptor, and IGF-1 mRNA in mouse marrow stromal cells. Biochem Biophys Res Commun 2002b; 290:526-531.

Zhang ZK, Davies KP, Allen J et al. Cell cycle arrest and respression of cyclin D1 transcription by INI1/hSNF5. Mol Cell Biol 2002; 22:5975-5988.

Zheng B, Clemmons DR. Blocking ligand occupancy of the αVβ3 integrin inhibits insulin-like growth factor I signaling in vascular smooth muscle cells. Proc Natl Acad Sci USA 1998; 95:11217-11222.

Zheng MH, McCaughan HB, Papadimitriou JM et al. Tartrate resistant acid phosphatase activity in rat cultured osteoclasts is inhibited by a carboxyl terminal peptide (osteostatin) from parathyroid hormone-related protein. J Cell Biochem 1994; 54:145-153.

Zhou H, Dempster DW, Lindsay R et al. Apoptosis of osteoblasts is increased in patients with mild primaryhyperparathyroidism and preserved cancellous bone (abstract). J Bone Miner Res 2004; 19:S186.

Zhou YX, Xu X, Chen L et al. A Pro250Arg substitution in mouse Fgfr 1 causes increased expression of Cbfa1 and premature fusion of calvarial sutures. Hum Mol Genet 2000; 9:2001-2008.

Zhu J, Emerson SG. A new bone to pick: Osteoblasts and the haematopoietic stem-cell niche. BioEssays 2004; 26:595-599.

Ziff M. Role of endothelium in the pathogenesis of rheumatoid synovitis. Int J Tissue React 1993; 15:135-137.

Zimmer C. At the Water's Edge. New York: The Free Press, 1998.

Zimmermann KW. Beiträge zur Kenntnis einiger Drüsen und Epithelien. Arch Mikr Entwicklungsmech 1898; 52:552-706.

Zlokovic BV. Neurovascular mechanisms of Alzheimer's neurodegeneration. Trends Neurosci 2005a; 28:202-208.

Zlokovic BV, Deane R, Sallstrom J et al. Neurovascular pathways and Alzheimer amyloid beta-peptide. Brain Pathol 2005b; 15:78-83.

Zuscik MJ, D'Souza MD, Ionescu AM et al. Growth plate chondrocyte maturation is regulated by basal intracellular calcium. Exp Cell Res 2002; 276:310-319.

Zwartkruis FJT, Bos JF. Ras and Rap 1: Two highly related small GTPases with distinct function. Exp Cell Res 1999; 253:157-165.

Index

A

Abberant crypt foci (ACFs) 163, 164

Adenylyl cyclase 16, 45, 50, 61, 87-94, 96-103, 105, 114, 117, 121, 122, 126, 127, 130-133, 137-144, 147, 148, 157, 158, 166, 169, 187

Adipocyte 23, 39, 40, 81, 106, 107, 129, 137, 139, 145

Alendronate 55, 57-59, 76, 77, 79, 83, 84, 171, 176, 177, 179, 182, 185, 191

Alkaline phosphatase (ALP) 31, 33, 37, 48, 55, 57, 65, 75, 76, 83, 101, 106, 109, 112, 115, 116, 121, 122, 125, 131, 132, 150, 151, 154, 171, 175, 177, 178, 181, 182

Alzheimer's disease 2, 100, 168, 183-186, 193

AMPA (DL-α-amino3-hydroxy-5-methylisoxasole-4-propionic acid) 3, 22, 23, 31

AMPA/kainate receptor/channel 31

Amphiregulin (AR) 104, 118, 128

Amyloid β peptides (Aβs) 183-185

Amyloid precursor protein (APP) 183, 184

Annexin I 25

Annexin V 132, 150, 151, 154, 156

Annexin VIII 156

Antioxidant response element (ARE) 131

Aortic calcification 37

AP-1 29, 52, 53, 99, 122, 144, 145, 156

APAF-1 33

Apelin (APL) 123, 127, 129, 132, 133, 137

Apoptosis 11-13, 18-25, 27, 33, 50, 52, 53, 55, 57, 58, 60, 65, 81, 106, 112, 116, 118, 119, 123, 124, 126, 132, 136-138, 140, 141, 149-151, 153-155, 157, 179, 183, 189

Apoptosis-inducing factor (AIF) 33

APRO1 111-113, 116, 122, *see also* TIS21

Aromatase 1, 2

Atherosclerosis 36

Atorvastatin 174, 175, 179, 182

ATP 9, 17, 19, 20, 24, 27, 57, 87, 97, 108, 133, 149, 154, 155

Azoxymethane (AOM) 163, 164

B

β-adrenergic nerve 43, 45, 48

β1-adrenoreceptors 45

β2-adrenoreceptors 45

β-arrestin 89, 91, 93, 99, 122, 143

β-blocker 45, 46, 50, 171

β-catenin 18, 36, 112-114, 117, 123, 124, 130, 131, 137-139, 165, 166, 177, 184

β-glycerophosphate 131

BACE1 (β-secretase) 100, 138, 157, 183, 184

BAD 118, 123, 124, 151

Basic Multicellular Unit (BMU) 3, 5, 6, 10-12, 18, 19, 21, 24-27, 30, 32, 33, 35, 39-41, 51, 54, 55, 59, 60, 64, 68-71, 83, 84, 103, 106, 117, 129, 130, 162, 185

Bcl-2 11, 13, 20, 21, 33, 34, 48, 118, 119, 123-125, 143, 151, 154

Bcl-X$_L$ 118, 123

Betacellulin 128

Bisphosphonate 55-60, 69, 77, 79, 83, 84, 160, 168, 171, 176-180, 182, 191

Blister 25, 26, 29, 32, 120

BMP-2 24, 27, 29, 36, 38, 53, 71, 82, 107, 120, 129, 150- 152, 164-166, 170, 174-180, 185

BMP-4 36, 38, 39, 71, 175, 177

Bone marrow niches 136

Bone mineral density (BMD) 39, 46, 49, 57, 66, 70-72, 74-84, 148, 160, 168, 170, 176, 181, 182, 185, 187, 188, 191

Bone morphogenic protein (BMP) 24, 26-29, 36, 38, 39, 53, 71, 82, 107, 109, 111, 120, 129, 150-152, 156, 164-166, 170, 174-180, 185

Bone porosity 6, 37, 79

Bone-remodeling compartment 25, 29

Bone sialoprotein (BSP) 27, 115, 121, 143

Brain-derived neurotrophic factor (BDNF) 32